T0191520

Communications
in Computer and Information Science   **873**

Commenced Publication in 2007
Founding and Former Series Editors:
Phoebe Chen, Alfredo Cuzzocrea, Xiaoyong Du, Orhun Kara, Ting Liu,
Dominik Ślęzak, and Xiaokang Yang

Editorial Board

Simone Diniz Junqueira Barbosa
  Pontifical Catholic University of Rio de Janeiro (PUC-Rio),
  Rio de Janeiro, Brazil
Joaquim Filipe
  Polytechnic Institute of Setúbal, Setúbal, Portugal
Igor Kotenko
  St. Petersburg Institute for Informatics and Automation of the Russian
  Academy of Sciences, St. Petersburg, Russia
Krishna M. Sivalingam
  Indian Institute of Technology Madras, Chennai, India
Takashi Washio
  Osaka University, Osaka, Japan
Junsong Yuan
  University at Buffalo, The State University of New York, Buffalo, USA
Lizhu Zhou
  Tsinghua University, Beijing, China

More information about this series at http://www.springer.com/series/7899

Kangshun Li · Wei Li
Zhangxing Chen · Yong Liu (Eds.)

Computational Intelligence and Intelligent Systems

9th International Symposium, ISICA 2017
Guangzhou, China, November 18–19, 2017
Revised Selected Papers, Part I

Springer

Editors
Kangshun Li
College of Mathematics and Informatics
South China Agricultural University
Guangzhou
China

Wei Li
Jiangxi University of Science
and Technology
Ganzhou, Jiangxi
China

Zhangxing Chen
Chemical and Petroleum Engineering
University of Calgary
Calgary, AB
Canada

Yong Liu
School of Computer Science
and Engineering
The University of Aizu
Aizu-Wakamatsu, Fukushima
Japan

ISSN 1865-0929   ISSN 1865-0937 (electronic)
Communications in Computer and Information Science
ISBN 978-981-13-1647-0   ISBN 978-981-13-1648-7 (eBook)
https://doi.org/10.1007/978-981-13-1648-7

Library of Congress Control Number: 2018948807

© Springer Nature Singapore Pte Ltd. 2018
This work is subject to copyright. All rights are reserved by the Publisher, whether the whole or part of the material is concerned, specifically the rights of translation, reprinting, reuse of illustrations, recitation, broadcasting, reproduction on microfilms or in any other physical way, and transmission or information storage and retrieval, electronic adaptation, computer software, or by similar or dissimilar methodology now known or hereafter developed.
The use of general descriptive names, registered names, trademarks, service marks, etc. in this publication does not imply, even in the absence of a specific statement, that such names are exempt from the relevant protective laws and regulations and therefore free for general use.
The publisher, the authors, and the editors are safe to assume that the advice and information in this book are believed to be true and accurate at the date of publication. Neither the publisher nor the authors or the editors give a warranty, express or implied, with respect to the material contained herein or for any errors or omissions that may have been made. The publisher remains neutral with regard to jurisdictional claims in published maps and institutional affiliations.

This Springer imprint is published by the registered company Springer Nature Singapore Pte Ltd.
The registered company address is: 152 Beach Road, #21-01/04 Gateway East, Singapore 189721, Singapore

Preface

Volumes CCIS 873 and CCIS 874 comprise proceedings of the 9th International Symposium on Intelligence Computation and Applications (ISICA 2017) held in Guangzhou, China, during November 18–19, 2017. ISICA 2017 successfully attracted over 180 submissions. After rigorous reviews and plagiarism checking, 51 high-quality papers are included in CCIS 873, while another 50 papers are collected in CCIS 874. ISICA conferences are one of the first series of international conferences on computational intelligence that combines elements of learning, adaptation, evolution, and fuzzy logic to create programs as alternative solutions to artificial intelligence.

ISICA 2017 featured the most up-to-date research in analysis and theory of evolutionary computation, neural network architectures and learning, neuro-dynamics and neuro-engineering, fuzzy logic and control, collective intelligence and hybrid systems, deep learning, knowledge discovery, learning, and reasoning. ISICA 2017 provided a venue to foster technical exchanges, renew everlasting friendships, and establish new connections. Prof. Yuanxiang Li, one of the pioneers in parallel and evolution computing at Wuhan University, wrote a beautiful poem in Chinese for the ISICA 2017 event. It is our pleasure to translate his poem with the title of "Computational Intelligence Debate on the Pearl River":

Wear a smile on a bright face;
Under the night light on the Pearl River;
You are like star and moon shining on the Tower Small Slim Waist;
Ride waves on the cruise ship;
Leave bridges behind in a boundless moment.

You are from far away;
A journey of thousand miles;
Meet in the Guangzhou City;
Brighten up the field of intelligent evolution;
Explore the endless road to intelligence.

Prof. Li's poem points out one of ISICA's missions of pursuing the truth that a complex system inherits the simple mechanism of evolution, while simple models may lead to the evolution of complex morphologies. Following the success of the past eight ISICA events, ISICA 2017 continued to explore the new problems emerging in the fields of computational intelligence.

On behalf of the Organizing Committee, we would like to thank warmly the sponsors, South China Agricultural University, who helped in one way or another to achieve our goals for the conference. We wish to express our appreciation to Springer for publishing the proceedings of ISICA 2017. We also wish to acknowledge the dedication and commitment of both the staff at the Springer Beijing Office and the CCIS editorial staff. We would like to thank the authors for submitting their work, as well as the Program Committee members and reviewers for their enthusiasm, time, and expertise. The invaluable help of active members from the Organizing Committee, including Wei Li, Hui Wang, Lei Yang, Yan Chen, Lixia Zhang, Weiguang Chen, Zhuozhi Liang, Junlin Jin, Ying Feng, and Yunru Lu, in setting up and maintaining the online submission systems by EasyChair, assigning the papers to the reviewers, and preparing the camera-ready version of the proceedings is highly appreciated. We would like to thank them personally for helping to make ISICA 2017 a success.

March 2018

Kangshun Li
Yong Liu
Wei Li
Zhangxing Chen

Organization

Honorary Chairs

Hisao Ishibuchi   Osaka Prefecture University, Japan
Qingfu Zhang   City University of Hong Kong, SAR China
Yang Xiang   Deakin University, Australia

General Chairs

Kangshun Li   South China Agricultural University, China
Zhangxing Chen   University of Calgary, Canada
Yong Liu   University of Aizu, Japan

Program Chairs

Aniello Castiglione   University of Salerno, Italy
Jing Liu   Xidian University, China
Han Huang   South China University of Technology, China
Hailin Liu   Guangdong University of Technology, China

Local Arrangements Chairs

Wei Li   South China Agricultural University, China
Yan Chen   South China Agricultural University, China

Publicity Chairs

Lei Yang   South China Agricultural University, China
Lixia Zhang   South China Agricultural University, China

Program Committee

Aimin Zhou   East China Normal University, China
Allan Rocha   University of Calgary, Canada
Dazhi Jiang   Shantou University, China
Dongbo Zhang   Guangdong University of Science and Technology, China
Ehsan Aliabadian   University of Calgary, Canada
Ehsan Amirian   University of Calgary, Canada
Feng Wang   Wuhan University, China
Guangming Lin   Southern University of Science and Technology, China
Guoliang He   Wuhan University, China

Hailin Liu	Guangdong University of Technology, China
Hu Peng	Jiujiang University, China
Hui Wang	Nanchang Institute of Technology, China
Iyogun Christopher	University of Calgary, Canada
Jiahai Wang	Sun Yet-Sen University, China
Jing Wang	Jiangxi University of Finance and Economics, China
Jun He	Aberystwyth University, UK
Jun Zou	The Chinese University of Hong Kong, SAR China
Kangshun Li	South China Agricultural University, China
Ke Tang	Southern University of Science and Technology, China
Kejun Zhang	Zhejiang University, China
Lingling Wang	Wuhan University, China
Lixin Ding	Wuhan University, China
Lu Xiong	South China Agricultural University, China
Maoguo Gong	Xidian University, China
Mohammad Zeidani	University of Calgary, Canada
Rafael Almeida	University of Calgary, Canada
Sanyou Zeng	China University of Geosciences, China
Shenwen Wang	Shijiazhuang University of Economics, China
Wayne Li	University of Calgary, Canada
Wei Li	South China Agricultural University, China
Wensheng Zhang	Chinese Academy of Sciences, China
Xiangjing Lai	University of Angers, France
Xin Du	Fujian Normal University, China
Xinyu Zhou	Jiangxi Normal University, China
Xuesong Yan	China University of Geosciences, China
Xuewen Xia	East China Jiaotong University, China
Ying Huang	Gannan Normal University, China
Yong Liu	The University of Aizu, Japan
Zahra Sahaf	University of Calgary, Canada
Zhangxing Chen	University of Calgary, Canada
Zhun Fan	Shantou University, China

Contents – Part I

Evolutionary Multi-objective and Dynamic Optimization – Optimal Control and Design

Evolutionary Multi-objective and Dynamic Optimization – Hybrid Methods

Data Mining – Association Rule Learning

Data Mining – Data Management Platforms

Everywhere Connectivity – Wireless Sensor Networks

Contents – Part II

Complex Systems Modeling – Multimedia Simulation

Intelligent Information Systems – Information Retrieval

Intelligent Information Systems – E-commerce Platforms

Artificial Intelligence and Robotics – Query Optimization

Artificial Intelligence and Robotics – Intelligent Engineering

Virtualization – Motion-Based Tracking

Virtualization – Image Recognition

Neural Networks and Statistical Learning – Neural Architecture Search

A New Recurrent Neural Network with Fewer Neurons for Quadratic Programming Problems

Sanfeng Chen[1], Xin Han[1,2], Fei Tang[1], and Guangming Lin[1(✉)]

[1] Shenzhen Institute of Information Technology,
Shenzhen 518029, Guangdong, China
Linggm@sziit.edu.cn
[2] School of Mechanical and Electrical Engineering,
Jiangxi University of Science and Technology, Ganzhou 341000, Jiangxi, China

Abstract. A new recurrent neural network is presented to solve a general quadratic programming problem in real time. In contrast with the available neural networks, the new neural network is with fewer neurons for solving quadratic programming problems. The global convergence of the model is proven with contraction analysis. The discrete time model and an alternative model for solving the problem under irredundant equality constraints are also studied. Simulation results demonstrate that the proposed recurrent neural networks are effective.

Keywords: Recurrent neural network · Quadratic programming
Contraction analysis · Global convergence

1 Introduction

There are a lot of studies of recurrent neural networks focusing on the filed of signal processing [1, 2], pattern classification [3, 4], robotics [5, 6], optimization [7], and so on. Especially, with the invention of the Hopfield [8], it was specially invented for solving online optimization. Recurrent neural networks are becoming a popular research branch in the field of online optimization. They are with powerful parallelism and online solving capability. Recurrent neural networks have made huge advances for online optimization in both theory and application. A recurrent neural network [9] is developed for nonlinear programming problems, where a penalty term is introduced as equality and inequality constraints, and it converges to an approximate optimal solution. A switched-capacitor neural network is proposed [10] for solving nonlinear convex programming problems. However, the model will be unstable in the case that the optimal solution is outside the feasible region. A neural network is proposed for solving linear quadratic programming problems [11]. The optimal solution is proven globally converged. Some slack variables is introduced to the problem, which leads the dimension of model is too large. A dual neural network is proposed for reducing the dimension. It is composed of a single layer of neurons, and the dimension of the dual network is equal to its neurons. The model and its modifications [13, 14] are introduced to kinematic control of

© Springer Nature Singapore Pte Ltd. 2018
K. Li et al. (Eds.): ISICA 2017, CCIS 873, pp. 3–16, 2018.
https://doi.org/10.1007/978-981-13-1648-7_1

robot [12, 15]. A simplified dual neural network is proposed [16]. It much reduces complexity while the convergence property is sound. The model is applied to the KWTA problem in real time [17], which is just a single neuron. However, it just deals with quadratic programming problem with a square quadratic term in box constraints and cost function. A recurrent neural network for solving general quadratic programming problems is proposed. It is with fewer neurons, and the dimension of the model is greatly reduced while keeping sound accuracy and efficiency.

The remainder of this paper is organized as follows. In Sect. 2, A neural network model is presented for solving quadratic programming problems. In Sect. 3, the convergence of the neural network is analyzed and it is proven to be globally convergent to the optimal solution of the quadratic programming problems. A discrete-time model in Sect. 4 for solving the same problem and an alternative neural network model for solving the quadratic programming problem under irredundant equality constraints are studied. In Sect. 5, numerical examples are given to demonstrate the effectiveness of our method. Section 6 is the conclusion.

In this paper, $\mathbb{R}$ denotes the real number field, A^T represents the transport matrix of A, I denotes a unitary matrix.

2 Mathematical Model

The general quadratic programming problem as following is studied:

$$
\min(\frac{1}{2}x^T Wx + c^T x)
$$
$$
\text{s.t.} \quad \begin{aligned} Ax &= b \\ Ex &\le e \end{aligned}
\tag{1}
$$

Where $W \in \mathbb{R}^{n \times n}$, $x \in \mathbb{R}^n$ denotes a positive definite matrix, $c \in \mathbb{R}^n$, $A \in \mathbb{R}^{m \times n}$, $b \in \mathbb{R}^m$, $e \in \mathbb{R}^q$ and $E \in \mathbb{R}^{q \times n}$. The equality constraint $Ax = b$ is transformed to two inequalities equivalently: $-Ax \le -b$ and $Ax \le b$. The Eq. (1) is transferred as a quadratic programming problem subject to inequality constraints. It is as following:

$$
\min(\frac{1}{2}x^T Wx + c^T x)
$$
$$
\text{s.t.} \quad \frac{1}{2}Bx \le d
\tag{2}
$$

$$
B = \begin{bmatrix} A \\ -A \\ E \end{bmatrix}, \, d = \begin{bmatrix} b \\ -b \\ e \end{bmatrix}
\tag{3}
$$

Where $B \in \mathbb{R}^{p \times n}$, $d \in \mathbb{R}^p$ and $p = 2 \times m + q$. The inequality constraint of (2) is transformed as $\max(Bx - d) \leq 0$. The $\max(x)$ denotes the largest element of vector x. The problem (1) is equivalent to the Eq. (4):

$$\min(\frac{1}{2}x^T Wx + c^T x)$$
$$\text{s.t. } B_\sigma^T x - d_\sigma \leq 0 \tag{4}$$

In the Eq. (4), σ denotes the row No. of the biggest element of $Bx - d$, and B_σ^T represents the σth row of B, d_σ denotes the σth element of d. According to the KKT terms, the solution of problem (4) meets the requirements:

$$Wx + c - \mu B_\sigma = 0$$

$$\begin{cases} B_\sigma^T x - d_\sigma = 0 \text{ if } \mu \leq 0 \\ B_\sigma^T x - d_\sigma \leq 0 \text{ if } \mu = 0 \end{cases} \tag{5}$$

The dual variable of inequality constraint in the Eq. (4) is represented with $\mu \in \mathbb{R}$. The Eq. (5) is simplified with an upper saturation function as following:

$$Wx + c - \mu B_\sigma = 0$$
$$B_\sigma^T x - d_\sigma = g(B_\sigma^T x - d_\sigma - \mu) \tag{6}$$

Where the upper saturation function $g(.)$ is as following:

$$g(.) = \begin{cases} 0 & x > 0 \\ x & x \leq 0 \end{cases} \tag{7}$$

The W is positive definite, x could be explicitly solved with μ and the first equality in Eq. (6) as following:

$$x = \mu W^{-1} B_\sigma - W^{-1} c \tag{8}$$

A dynamic neuron is used to solve μ in Eq. (6) as following:

$$\in \dot{\mu} = g(B_\sigma^T x - d_\sigma - \mu) - B_\sigma^T x + d_\sigma \tag{9}$$

Where $\in > 0$ is a scaling parameter. Substituting (8) into (9) generates the neural network dynamics with the following state equation and output equation,
State equation:

$$\in \dot{\mu} = g(-\mu + B_\sigma^T W^{-1} B_\sigma \mu - B_\sigma^T W^{-1} c - d_\sigma) - B_\sigma^T W^{-1} B_\sigma \mu + B_\sigma^T W^{-1} c + d_\sigma \tag{10}$$

Output equation:

$$x = \mu W^{-1} B_\sigma - W^{-1} c \tag{11}$$

Where σ is the row No. of $\max(Bx - d)$, d and B and are as Eq. (3).

Remark 2.1: Only one dynamic neuron is required in the neural network (10), which is nothing to do with the conditions of Eq. (1). There are at least q dynamic neurons in recurrent neural networks for solving a general quadratic programming problem in [13–16, 18]. It is the No. of inequalities of problem (1). However, the proposed model is just a single dynamic neuron, which greatly reduces the number of neurons and computational complexity.

Remark 2.2: The neural network dynamic modeled by (10) is a switched dynamic system. It switches in a family of dynamic systems under the endogenous switching signal σ (signal flow in the neural network is plotted in Fig. 1)

$$\begin{aligned} &\in \dot{\mu} = f_\sigma(\mu), \sigma \in S = \{1, 2 \dots, p\} \\ &\sigma : \mu \to S \end{aligned} \tag{12}$$

Where $f_\sigma(\mu) = g(B_\sigma^T x - d_\sigma - \mu) - B_\sigma^T x + d_\sigma$

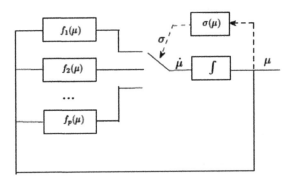

Fig. 1. Signal flow of the simplified neural network.

Remark 2.3: The proposed model is liable to be implemented with hardware. In comparison with the research [16], the new part with hardware implementation is a switching of the proposed network (10). The σ is a switching signal, which is equal to the row No. of $\max(Bx - d)$. It is the winner of $Bx - d$. σ is gotten with a WTA circuit [19]. It is worked as the switching signal for B_σ and d_σ in (10). There are a lot of studies on WTA circuits, and it is easily implemented with CMOS [20]. It is also easily implemented by a recurrent neural network just a single dynamic neuron [19]. There is just two dynamic neurons in the neural network (10), which is much less than the solutions of [13–18].

3 Convergence Analysis

It is a feasible way to prove the global convergence of the proposed system by constructing a common negative definite Lyapunov function. However, choosing such a common Lyapunov function is a difficult problem. Some researches on the contraction theory [21, 22] greatly simplify the proof process with virtual dynamics of the system. In this paper, the contraction analysis is made to prove the proposed model (10) convergent. The proof process is based on one definition and two lemmas:

Definition 3.1 ([22]): Given the system equations $\dot{x} = f(x, t)$, a region of the state space is called a contraction region with respect to a uniformly positive definite metric $M(x, t) = \Theta^T \Theta$ if $\frac{\partial f^T}{\partial x} M + M \frac{\partial f}{\partial x} + \dot{M} \leq - \beta M$ (with $\beta > 0$) in that region.

Lemma 3.1 [22]: Given the system equations
  $\dot{x} = f(x, t)$, any trajectory, which starts in a ball of constant radius with respect to the metric $M(x, t)$, center at a given trajectory and contained at all times in a contraction region with respect o $M(x, t)$, remains in that ball and converges exponentially to this trajectory. furthermore, global exponential convergence to the given trajectory is guaranteed if the whole state space is a contraction region with respect to the metric $M(x, t)$.

Remark 3.1: As pointed out in [21], Lemma 3.1 holds for all switched systems in case that all subsystems in the switched family are within a same contraction region $M(x, t)$.

Lemma 3.2: The solution to problem (6) (represented by μ^*, x^*) is the optimal solution to problem (1). At the same time, μ^* is an equilibrium point of the state equation of the proposed model, which is as following:

$$g(-\mu + B_\sigma^T W^{-1} B_\sigma \mu - B_\sigma^T W^{-1} c - d_\sigma) - B_\sigma^T W^{-1} B_\sigma \mu + B_\sigma^T W^{-1} c + d_\sigma = 0$$

The optimal solution also satisfies (11) as following: $x^* = \mu W^{-1} B_\sigma - W^{-1} c$.

Proof: According to the KKT condition, equation sets (5) are the optimal solution to problem (4). Equation sets (6) are equivalent to equation sets (5), the solution to equation set (6) is equivalent to the optimal solution to problem (1). They are as following:

$$g(-\mu^* + B_\sigma^T W^{-1} B_\sigma \mu^* - B_\sigma^T W^{-1} c - d_\sigma) - B_\sigma^T W^{-1} B_\sigma \mu^* + B_\sigma^T W^{-1} c + d_\sigma = 0$$

Where $x^* = \mu W^{-1} B_\sigma - W^{-1} c$.
  Now, we are on the stage to state the convergence result of the proposed model.

Theorem 3.1: The proposed (10) converges to the equilibrium point from any start point $\mu \in \mathbb{R}$, and the Eq. (11) is the optimal solution to Eqs. (1).

Proof: Based on the contraction theory for analyzing the convergence of (10):

$$\dot{\mu} = f(\mu, t) = \frac{1}{\varepsilon}(g(-\mu + B_\sigma^T W^{-1} B_\sigma \mu - B_\sigma^T W^{-1} c - d_\sigma) - B_\sigma^T W^{-1} B_\sigma \mu + B_\sigma^T W^{-1} c + d_\sigma) \quad (13)$$

The partial derivative $\frac{\partial f}{\partial \mu}$ is as following:

$$\frac{\partial f}{\partial \mu} = \begin{cases} -\frac{B_\sigma^T W^{-1} B_\sigma}{\varepsilon} & \text{if } \upsilon \geq 0 \\ -\frac{1}{\varepsilon} & \upsilon < 0 \end{cases} \quad (14)$$

Where $\upsilon = -\mu + B_\sigma^T W^{-1} B_\sigma \mu - B_\sigma^T W^{-1} c - d_\sigma$. Note that the derivate $g'(x) = 0$ if $x \geq 0$ and $g'(x) = 1$ if $x < 0$ (more exactly, it should be the upper right-hand derivative since the function $g(x)$ is non-smooth at $x = 0$.) is used in the derivation of (14). By choose $M = 1$, we have

$$\frac{\partial f^T}{\partial \mu} M + M \frac{\partial f}{\partial \mu} + \dot{M} = 2 \frac{\partial f}{\partial \mu} \leq -\frac{2\min(1, B_\sigma^T W^{-1} B_\sigma)}{\varepsilon} \quad (15)$$

Since W^{-1} is positive definite and B_σ is not equal to 0 (otherwise, the corresponding inequality in (2) does not include x), we conclude that $B_\sigma^T W^{-1} B_\sigma > 0$ and $\frac{\min(1, B_\sigma^T W^{-1} B_\sigma)}{\varepsilon} > 0$.

By defining $\beta = \frac{\min\{1, \min_i(1, B_i^T W^{-1} B_i)\}}{\varepsilon}$, with B_i^T denoting the ith row of B and the set $S = \{1, 2, \ldots, p\}$, we have the following inequality for all $\mu \in \mathbb{R}$ and all $\sigma \in S$ under the same metric M,

$$\frac{\partial f^T}{\partial x} M + M \frac{\partial f}{\partial x} + \dot{M} \leq -\beta M \quad (16)$$

The whole state space of μ is a contraction region.

Based on Lemma 3.2, μ^* and x^*, is the optimal solution to problem (1), and it is also an equilibrium point to Eq. (10).

4 Extensions

4.1 Discrete-Time Model

In this part, we propose the discrete-time model to solve the problem (1) and give conditions for global convergence

Replacing μ, $\dot{\mu}$ with μ_n, $\frac{\mu_{n+1} - \mu_n}{\Delta t}$ in (10), respectively, and denoting $\Upsilon = \frac{\Delta t}{\varepsilon}$, we get the following state equation and output equation of the discrete model as following with some trivial manipulations. state equation:

$$\mu_{n+1} = \Upsilon g(-\mu_n + B_\sigma^T W^{-1} B_\sigma \mu_n - B_\sigma^T W^{-1} c - d_\sigma) + (1 \\ - \Upsilon B_\sigma^T W^{-1} B_\sigma) \mu_n + \Upsilon B_\sigma^T W^{-1} c + \Upsilon d_\sigma \quad (17)$$

The output equation:

$$x_n = \mu_n W^{-1} B_\sigma - W^{-1} c \qquad (18)$$

Where σ is the row No. of the largest element of $Bx_n - d$, B and d are as equation sets (3).

In the discrete-time case, the contraction theory is an extension of the well-known contraction mapping theorem. We still use the contraction theory to analyze the convergence. For discrete-time systems, the definition of a contraction region and the condition for contraction are stated as below.

Definition 4.1 ([22]): The discrete-time model is as following:

Equation $x_{n+1} = f_n(x_n, n)$, is a contraction region. Positive definite metric $M_n(x_n, n) = \Theta_n^T \Theta_n$, given in the region $\exists \beta > 0$, $F_n^T F_n - I \leq -\beta I < 0$, where $F_n = \Theta_{n+1} \frac{\partial f_n}{\partial x_n} \Theta_n^{-1}$.

Lemma 4.1 ([22]): If the whole state space is a contraction region, The global convergence to the given trajectory is made sure.

Theorem 4.1: The discrete model (17) will converge to the equilibrium point after initialization $\mu \in \mathbb{R}$ and the output of equilibrium point, as shown in (18), is also the optimal solution to problem (1) under:

$$0 < \gamma < 2, \quad \gamma < \frac{2}{B_i^T W^{-1} B_i} \text{ for all } i \in \mathbf{S} \qquad (19)$$

where B_i^T denotes the ith row of B and $\mathbf{S} = \{1, 2, 3, \ldots, p\}$

Proof: Firstly, solve the equilibrium point for the discrete model (17), and get the output of the Eq. (18). Secondly, prove the global contraction to the equilibrium point.

Step 1: Comparing the discrete-time model (17) and its output (18) with the continuous-time model (10) and its output (11), we can find that they have the same equilibrium and the same output at this equilibrium. According to the Lemma 3.2, the solution to equation set (6) is an equilibrium to the discrete model (17) too, and the output at the point is the optimal solution to the problem (1).

Step 2: To the discrete model (17),

$$\mu_{n+1} = f_n(x_n, n)$$
$$= \Upsilon g(-\mu_n + B_\sigma^T W^{-1} B_\sigma \mu_n - B_\sigma^T W^{-1} c - d_\sigma) + (1 - \Upsilon B_\sigma^T W^{-1} B_\sigma) \mu_n + \Upsilon B_\sigma^T W^{-1} c + \Upsilon d_\sigma \qquad (20)$$

Calculate $\frac{\partial f_n}{\partial \mu_n}$ as follows:

$$\frac{\partial f_n}{\partial \mu_n} = \begin{cases} 1 - \Upsilon B_\sigma^T W^{-1} B_\sigma & \text{if } v_n \geq 0 \\ 1 - \Upsilon & \text{if } v_n < 0 \end{cases} \qquad (21)$$

Where $v_n = -\mu_n + B_\sigma^T W^{-1} B_\sigma \mu_n - B_\sigma^T W^{-1} c - d_\sigma$. Choosing $\Theta_n = 1$ for $n = 0, 1, 2, \ldots, n$ then

$$F_n = \Theta_{n+1} \frac{\partial f_n}{\partial \mu_n} \Theta_n^{-1} \tag{22}$$

Based on (21) and (22),

$$
\begin{aligned}
F_n^T F_n &\leq (max\{|1-\gamma|, |1-\gamma B_\sigma^T W^{-1} B_\sigma|\})^2 \\
&\leq (max\{|1-\gamma|, max_{i \in S}\{|1-\gamma B_\sigma^T W^{-1} B_i|\}\})^2
\end{aligned}
\tag{23}
$$

Set $\beta = 1 - (max\{|1-\gamma|, max_{i \in S}\{|1-\gamma B_\sigma^T W^{-1} B_i|\}\})^2$. To all $\mu_n \in \mathbb{R}$, $\sigma \in S$, the inequality as following (24) is valid.

$$F_n^T F_n - I \leq \beta I \tag{24}$$

Since $0 < \gamma < 2$, $0 < \gamma B_i^T W^{-1} B_i < 2$ for $i \in S$, we get $|1-\gamma| < 1, |1 - \gamma B_i^T W^{-1} B_i| < 1$ for $i \in S$. Therefore, $(max\{|1-\gamma|, max_{i \in S}\{|1-\gamma B_\sigma^T W^{-1} B_i|\}\})^2 \leq 1$ and $\beta > 0$.

Then the whole state space is within contraction range. Let the equilibrium point of (17) to be the given trajectory, according to Lemma 4.1, the discrete model (17) will convergence to the equilibrium point.

4.2   Irredundant Equality Constraint

Based on that the equality constraint exists without redundancy, there is another way to tackle the problem (1). This is to say, the matrix $A \in \mathbb{R}^{m \times n}$ in (1) problem meets the requirement of rank $(A) = m$ and $m < n$.

$Ex \leq e$ in problem (1) is replaced by $max(Ex - e) \leq 0$, then

$$
\begin{aligned}
&\min(\frac{1}{2} x^T W x + C^T x) \\
&\text{s.t. } Ax = b \\
&E_\sigma^T x - e_\sigma \leq 0
\end{aligned}
\tag{25}
$$

Based on KKT conditions and the upper saturation function $g(\cdot)$, the solution to problem (25) meets,

$$
\begin{aligned}
Wx + c - A^T y - \mu B_\sigma^T &= 0 \\
Ax &= b \\
E_\sigma^T x - e_\sigma &= g(E_\sigma^T x - e_\sigma - \mu)
\end{aligned}
\tag{26}
$$

Where $y \in \mathbb{R}^m$, $\mu \in \mathbb{R}$. x and y can be explicitly solved in terms of μ under the above three equations:

$$
\begin{aligned}
x &= \mu P E_\sigma + s \\
y &= (AW^{-1}A^T)^{-1}(-AW^{-1}B_\sigma \mu + AW^{-1}c + b)
\end{aligned}
\tag{27}
$$

Where $P = -W^{-1}A^T(AW^{-1}A^T)^{-1}AW^{-1} + W^{-1}$, $S = W^{-1}(A^T(AW^{-1}A^T)^{-1}(AW^{-1}$ $c+b) - c)$ $P \in \mathbb{R}^{n \times n}$. The third equation in (26) is solved as following:

$$\varepsilon\dot{\mu} = g(E_\sigma^T x - e_\sigma - \mu) - E_\sigma^T x + e_\sigma \tag{28}$$

Where $\varepsilon > 0$ is a scaling parameter. The state equation of the neural network is as following:

$$\varepsilon\dot{\mu} = g(E_\sigma^T P E_\sigma \mu - \mu + E_\sigma^T s - e_\sigma) - E_\sigma^T P E_\sigma \mu - E_\sigma^T s + e_\sigma \tag{29}$$

Before stating the convergence result about the neural network (29), the following lemma is presented, which is used in the convergence proof.

Lemma 4.2: The symmetric matrix $P \in \mathbb{R}^{n \times n}$, $P = -W^{-1}A^T(AW^{-1}A^T)^{-1}AW^{-1} + W^{-1}$ is semi-positive definite, i.e., $z^T P z \geq 0$ for all $z \in \mathbb{R}^n$ and $z^T(W^{-1}A^T(AW^{-1}A^T)^{-1}$ $A - I) \neq 0$.

Proof: Since $W^{-1} \in \mathbb{R}^{n \times n}$ is positive definite, it can be factorized into $W^{-1} = QQ^T$ with Q positive definite. Defining $G \in \mathbb{R}^{m \times n}$, $G = AQ$, then G is also row full rank (since A is row full rank and Q is positive definite) and therefore G can be decomposed into $G = U[\wedge \ 0]V$ via singular value decomposition, where $U \in \mathbb{R}^{m \times m}$, $V \in \mathbb{R}^{n \times n}$ are both unitary matrices, $[\wedge \ 0] \in \mathbb{R}^{m \times n}$, $\wedge \in \mathbb{R}^{m \times m}$ is a diagonal matrix with all positive elements on the diagonal. Bringing

$$W^{-1} = QQ^T, \ G = AQ, \text{ and } G = U[\wedge \ 0]V \text{ into } P$$
$$= -W^{-1}A^T(AW^{-1}A^T)^{-1}AW^{-1} + W^{-1}$$

We get the equations as (30) as below. The stability of the neural network (29) and its global convergence to the optimal solution of (1) is guaranteed by the following theorem.

Theorem 4.2: The neural network (29) exponentially converges to its equilibrium from any initial point $\mu \in \mathbb{R}$, and its output at this equilibrium by following the output Eq. (27) is the optimal solution to problem (1), if $rank(A) = m$ in (1) and $(E_i^T(W^{-1}A^T(AW^{-1}A^T)^{-1}A - I) \neq 0$

For $i = 1, 2, \ldots, q$, with E_i^T denoting the ith row of the matrix E.

Proof: This theorem can be proven in a two-step procedure similar to the proof of Theorem 4.1. The first step is to solve the equilibrium and show that the output at this equilibrium is the optimal solution to the problem (1). The second step is to prove the global contraction to the equilibrium. Note that the condition $(E_i^T(W^{-1}A^T(AW^{-1}$ $A^T)^{-1}A - I) \neq 0$ For $i = 1, 2, \ldots, q$, according to Lemma 4.2, guarantees $E_\sigma^T P E_\sigma > 0$ for all possible switching signal σ. Based on this result, global contraction of the neural network (29) can be proven. Detailed proof process for the two steps is omitted.

$$P = -QQ^T A^T (AQQ^T A^T)^{-1} AQQ^T + QQ^T$$
$$= -QG^T (GG^T)^{-1} GQ^T + QQ^T$$
$$= -QV^T \begin{bmatrix} \Lambda \\ 0 \end{bmatrix} U^T (U\Lambda^{-2}U^T) U [\Lambda \quad 0] VQ^T + QQ^T$$
$$= -QV^T \begin{bmatrix} I_{m \times m} & 0 \\ 0 & 0 \end{bmatrix} VQ + QV^T VQ^T$$
$$= (VQ^T)^T \begin{bmatrix} 0 & 0 \\ 0 & I_{(n-m) \times (n-m)} \end{bmatrix} VQ^T \geq 0 \tag{30}$$

5 Simulations

5.1 Continuous-Time Neural Network Model

The following problem will be solved for showing the convergence of the continuous-time neural network modelled with (10) and (11).

$$\min(3x_1^2 + 3x_2^2 + 4x_3^2 + 5x_4^2 + 3x_1x_2 + 5x_1x_3 + x_2x_4 - 11x_1 - 5x_4)$$

$$\text{s.t.} \quad \begin{aligned} -3x_1 + 3x_2 + 2x_3 - x_4 &\leq 0 \\ -4x_1 - x_2 + x_3 + 2x_4 &\leq 0 \\ -x_1 + x_2 &\leq -1 \\ -2 \leq 3x_1 + x_3 &\leq 4 \end{aligned} \tag{31}$$

To the model, the parameters are as following:

$$W = \begin{bmatrix} 6 & 3 & 5 & 0 \\ 3 & 6 & 0 & 1 \\ 5 & 0 & 8 & 0 \\ 0 & 1 & 0 & 10 \end{bmatrix}, c = \begin{bmatrix} -11 \\ 0 \\ 0 \\ -5 \end{bmatrix}$$

$$B = \begin{bmatrix} -1 & 1 & 0 & 0 \\ 3 & 0 & 1 & 0 \\ -3 & 0 & -1 & 0 \\ -3 & 3 & 2 & -4 \\ -4 & -1 & 1 & 2 \end{bmatrix}, D = \begin{bmatrix} -1 \\ 4 \\ 2 \\ 0 \\ 0 \end{bmatrix} \tag{32}$$

We run the neural network model in simulation with a random initialization of the state variable μ and we choose $\epsilon = 10^{-8}$. Simulations is run for 5×10^{-8} s and the simulation result of μ and x are as shown in the Figs. 2 and 3. The optimal solution is about -19.1794310722, the proposed output $x = [1.8774617068, -1.0393873085, -1.6323851203, 0.6039387308]$, the optimum point is -19.179. The differences are both less than 1.0×10^{-10} for x and the optimum point.

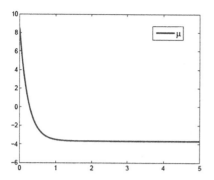

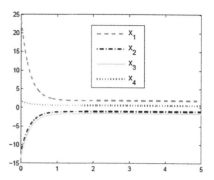

Fig. 2. Transient behavior of μ in Example V-A.

Fig. 3. Transient behavior of x in Example V-A.

5.2 Discrete-Time Model

We also consider the problem studied in Example V-A and solve it with the discrete model based on (17) and (18). In the simulation, $\gamma = 0.05$, randomly initialization μ. The evolution of μ and x are as shown in Figs. 4 and 5. The proposed network output is $x = [1.8774617068, -1.0393873085, -1.6323851203, 0.6039387308]$. The optimal value is -19.179. The difference is less than 5.85×10^{-6} for x and 4.2×10^{-5} for the optimal value.

5.3 Problem with Irredundant Equality Constraint

To show the convergence of the neural network, we solve a quadratic programming problem with irredundant equality constraints as following

$$\min(3x_1^2 + 3x_2^2 + 4x_3^2 + 5x_4^2 + 3x_1x_2 + 5x_1x_3 + x_2x_4 - 11x_1 - 5x_4)$$

$$\text{s.t.} \quad \begin{aligned} -3x_1 + 3x_2 + 2x_3 - x_4 &= 0 \\ -4x_1 - x_2 + x_3 + 2x_4 &= 0 \\ -x_1 + x_2 &\leq -1 \\ -2 \leq 3x_1 + x_3 &\leq 4 \end{aligned} \qquad (33)$$

As the problem (34), A simulation is performed with a random initialization of the state variable μ and the scaling parameter $\epsilon = 10^{-8}$. Simulation result is as shown in Figs. 6, 7 and 8, from which we can observe that the neural network converges very fast after a short transient period shorter after than 10^{-6} s. The error of solution at the final time of the simulation is less than 0.9×10^{-10} in l_2 norm sense, and the optimal solution is $x = [0.5, -0.5, -1.5, 0]$.

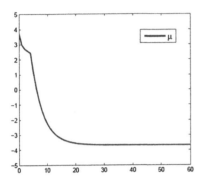

Fig. 4. Transient behavior of μ in Example V-B.

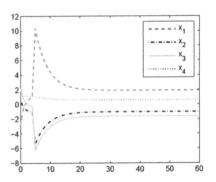

Fig. 5. Transient behavior of x in Example V-B.

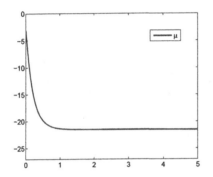

Fig. 6. Transient behavior of μ in Example V-C.

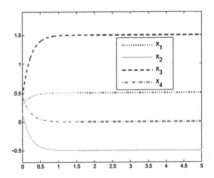

Fig. 7. Transient behavior of x in Example V-C.

$$W = \begin{bmatrix} 6 & 3 & 5 & 0 \\ 3 & 6 & 0 & 1 \\ 5 & 0 & 8 & 0 \\ 0 & 1 & 0 & 10 \end{bmatrix}, \; c = \begin{bmatrix} -11 \\ 0 \\ 0 \\ -5 \end{bmatrix}$$

$$A = \begin{bmatrix} 3 & -3 & -2 & 1 \\ 4 & 1 & -1 & -2 \end{bmatrix}, \; b = \begin{bmatrix} 0 \\ 0 \end{bmatrix} \qquad (34)$$

$$E = \begin{bmatrix} -1 & 1 & 0 & 0 \\ 3 & 0 & 1 & 0 \\ -3 & 0 & -1 & 0 \end{bmatrix}, \; e = \begin{bmatrix} -1 \\ 4 \\ 2 \end{bmatrix}$$

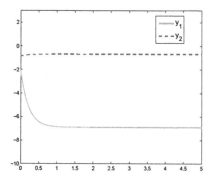

Fig. 8. Transient behavior of y in Example V-C.

6 Conclusions

For solving the general quadratic programming problems, a new recurrent neural network is proposed. The proposed model is with fewer neurons and simple structure. The neural network are shown global convergence based on contraction analysis. The discrete time counterpart to the proposed neural network is studied and an alternative recurrent neural network to solve the problem under irredundant equality constraint is also studied. Finally, simulation results demonstrate the models are efficient and accurate.

Acknowledgements. The authors would like to acknowledge the support of Guangdong Science Foundation of China under Grant No. S2011010006116 and No. 2015A030313587, Shenzhen Science Technology Project No. JCYJ20150417094158025, No. JCY20160307100530069 and GRCK20170424095924228,Shenzhen Institute of Information Technology Scientific Research Platform Cultivation Project (PT201704).

References

1. Hughes, T., Mierle, K.: Recurrent neural networks for voice activity detection. In: IEEE International Conference on Acoustics, pp. 7378–7382 (2013)
2. Schuster, M., Paliwal, K.K.: Bidirectional recurrent neural networks. IEEE Trans. Signal Process. **45**(11), 2673–2681 (2002)
3. Scardino, L., Niess, R.: Performance of recurrent neural networks applied to a simplified pattern recognition problem. J. Comput. Sci. Colleges **15**(3), 251–258 (2000)
4. Nan, B., Fukuda, O.: EMG-based motion discrimination using a novel recurrent neural network. J. Intell. Inf. Syst. **21**(2), 113–126 (2003)
5. Zhang, Y.N., Tan, Z.G.: Repetitive motion of redundant robots planned by three kinds of recurrent neural networks and illustrated with a four-link planar manipulator's straight-line example. Robot. Auton. Syst. **57**(6–7), 645–651 (2009)
6. Vázquez, L.A., Jurado, F.: Decentralized identification and control in real-time of a robot manipulator via recurrent wavelet first-order neural network. In: Mathematical models in Engineering, pp. 1–12 (2015)

7. Xu, R., Wunsch, D.: Inference of genetic regulatory networks with recurrent neural network models using particle swarm optimization. IEEE Trans. Comput. Biol. **4**(4), 681–692 (2007)
8. Hopfield, J.J.: Neurons with graded response have collective computational properties like those of two-state neurons. Proc. Nat. Acad. Sci. **81**(10), 3088–3092 (1984)
9. Liu, Q., Wang, J.: A one-layer recurrent neural network with a discontinuous hard-limiting activation function for quadratic programming. IEEE Trans. Neural Netw. **19**(4), 558–570 (2008)
10. Mestari, M., Namir, A.: Switched capacitor neural networks for optimal control of nonlinear dynamic systems: design and stability analysis. Syst. Anal. Model. Simul. **41**(3), 11–20 (2001)
11. Malek, A., Alipour, M.: Numerical solution for linear and quadratic programming problems using a recurrent neural network. Appl. Math. Comput. **192**(1), 27–39 (2007)
12. Liu, Q., Wang, J.: A one-layer recurrent neural network with a discontinuous activation function for linear programming. Neural Comput. **20**(5), 1366–1383 (2008)
13. Hu, X., Wang, J.: An improved dual neural network for solving a class of quadratic programming problems and its k-winners-take-all application. IEEE Trans. Neural Netw. **19**(12), 2022–2031 (2008)
14. Xia, Y., Han, Y.W.: A mixed-binary convex quadratic reformulation for box-constrained nonconvex quadratic integer program. Mathematics **10**(12), 7897–7905 (2014)
15. Li, S., Chen, S.: Decentralized kinematic control of a class of collaborative redundant manipulators via recurrent neural networks. Neurocomputing **91**(9), 1–10 (2012)
16. Cheng, L., Hou, Z.: A Simplified Neural Network for Linear Matrix Inequality Problems. Neural Process. Lett. **29**(3), 213–230 (2009)
17. Hu, X., Zhang, B.: A simplified dual neural network for quadratic programming with its KWTA application. IEEE Trans. Neural Netw. **17**(6), 1500–1510 (2006)
18. Xia, Y., Feng, G., Wang, J.: A novel recurrent neural network for solving nonlinear optimization problems with inequality constraints. IEEE Trans. Neural Netw. **19**, 1340–1353 (2008)
19. Tymoshchuk, P.: A discrete-time dynamic K-winners-take-all neural circuit. Neurocomputing **72**(13–15), 3191–3202 (2009)
20. Xiao, Y., Liu, Y.: Analysis on the convergence time of dual neural network-based KWTA. IEEE Trans. Neural Netw. Learn. Syst. **23**(4), 676–682 (2012)
21. Hu, F., Zhang, Z.: Contraction theory-based adaptive dynamic surface control for a class of nonlinear systems. Control Decis. **31**(5), 769–775 (2016)
22. Wang, W., Slotine, J.J.E.: Contraction analysis of time-delayed communications and group cooperation. IEEE Trans. Autom. Control **51**(4), 712–717 (2006)

Mutual-Information-SMOTE: A Cost-Free Learning Method for Imbalanced Data Classification

Ying Chen, Yufei Chen$^{(\boxtimes)}$, Xianhui Liu, and Weidong Zhao

College of Electronic and Information Engineering, Tongji University,
Shanghai 201804, China
april337@163.com

Abstract. In recent years, classification of imbalanced data get into trouble due to the imbalanced class distribution. Cost-Sensitive Learning (CSL) is one of the solutions at algorithm level, but this kind of method needs to provide cost information in advance. Mutual-Information Classification (MIC) is a Cost-Free Learning (CFL) method, which could summarize cost information automatically. But this method pays too much attention to minority class and ignores the accuracy of majority class data. Based on MIC, this paper proposed a CFL method for imbalanced data classification, called Mutual-Information-SMOTE Classification (MISC). Firstly, we uses mutual-information classifiers to generate the abstaining samples, which is difficult to be classified. Secondly, we use these abstention samples to synthetize new samples. Thirdly, we construct mutual-information-SMOTE classifiers using MIC and SMOTE. Finally, we uses our classifiers to obtain the final results. Numerical examples are given for indicating that MISC is more effective for imbalanced data classification through comparing with MIC.

Keywords: Imbalanced data · Cost-Free Learning (CFL) · Abstaining
Mutual-Information Classification (MIC)
Mutual-Information-SMOTE Classification (MISC)

1 Introduction

Imbalanced data sets [1, 2] appears frequently in many real-world applications, such as medicine, biology, finance, and computer vision. However, most traditional classification algorithm assume that the class distributions are balanced or the cost of misclassification are equally, so they aim to maximize the overall classification accuracy. In recent years, class-imbalanced learning have received a lot of attention in classifications [3–6]. He and Garcia [4] provide a comprehensive review of learning in the class imbalanced problem and emphasize the importance of the opportunities and challenges in imbalanced data learning. More and more researchers found that the pursuit of accuracy maximization will ignore the importance of minority samples and tend to bias the classification of majority class in the case of imbalance distribution data. Therefore, learning in imbalanced data is of great importance in data mining and machine learning.

© Springer Nature Singapore Pte Ltd. 2018
K. Li et al. (Eds.): ISICA 2017, CCIS 873, pp. 17–30, 2018.
https://doi.org/10.1007/978-981-13-1648-7_2

In order to solve the class imbalance problem, various methods are developed called cost-sensitive learning [8] (CSL) such as cost to sample [9], to weight examples [10], and to find a decision threshold [11, 12]. Researchers usually pay more attention to minority class and deem the cost of misclassifying a minority class is more expensive than that of majority class. Hence, many methods increased the cost of minority class in order to raised concern on the minority class. Although these methods could solve the problem of imbalanced class classification to some extent, they cannot be used without the costs. It means that we need to provide the values of cost in advance when we try to use CSL method. Even then, the price is generally difficult to give in practical application. It may be better to apply abstaining classification [13] when there are some uncertainties in the decision, because it can reduce the probability of potential misclassification. The optimal rejection threshold could be found by minimizing a loss function in a cost-sensitive learning [14, 15]. However, in the case of abstaining classification, the existing CSL methods also require the cost terms associated to rejects. Synthetic Minority Oversampling Technique (SMOTE) [16] is one of over-sampling solutions of the class imbalance problem at data level, but this kind of method doesn't consider the distribution of data sets, thus the results are not satisfied.

One of the method to solve these problem is applying mutual information theory into classification because these kind of method don't need to provide cost terms in advance. Mutual information is an significant definition in entropy theory [17]. Entropy is a measure of uncertainty within random variables while mutual information describes the relative entropy between two random variables [18]. Principe et al. [19] pointed out that the decision outputs could be as relevant as possible to the targets when maximizing mutual information as the target function and then he presented a schematic diagram of information theory learning (ITL). Mackey [18] suggested that mutual information can be used as an evaluation criterion, which was more signification than error rate. A study [20] researched theoretically on both error type and reject type in binary classifications, then it considered that information theory was an effective way to provide an objective classification evaluation in imbalanced classification. On this basis, Hu [22] proposed a method to design classifier based on the mutual-information concept called Mutual-Information Classifier (MIC). Mutual-information classifiers maximize the relative entropy for their learning target and obtain the cost terms automatically. However, he only presented the theoretical formulas and have not learning approaches and results for the real-world data sets. Zhang [8] extended Hu's study on mutual-information classifier and proved the feasibility of Hu's theory on real-world data sets. She defined CFL formally that the approach seeks optimal classification results without requiring any cost information, even in the class imbalance problem. But MIC reduces the error rate of minority class at the cost of the majority class's accuracy. This paper integrate SMOTE methods into MIC and creates a classifier called Mutual-information-SMOTE Classifier. We hope that the mutual-information classifier can consider both minority and majority classes by changing the distribution of data. Experimental results show that the proposed method improves the performance of previous classifier.

The rest of this paper is organized as follows: Sect. 2 provides a brief introduce of mutual-information classifier. Section 3 provides the basic formulas for SMOTE algorithm. Section 4 explains the feasibility reason of mutual-information classifier and presents the improved mutual-information classifier. The experimental results are presented in Sect. 5 and we conclude this work in Sect. 6.

2   Mutual Information Classifier

2.1   Normalized Mutual Information

Normalized mutual information is an evaluation criterion to measure the similarity of the targets variable T and the decision outputs variable Y, we denote $NI(T = t, Y = y) = NI(T, Y)$ as the NI in the following form [21]

$$NI(T; Y) = \frac{I(T; Y)}{H(T)} \tag{1}$$

where t and y are the values of T and Y, m is the total number of class in T.

$I(T, Y)$ is the mutual information of two discrete variables T and Y [18] can be defined as

$$I(T; Y) = \sum_{i=1}^{m} \sum_{j=1}^{m} p(t_i, y_j) \log_2 \frac{p(t_i, y_j)}{p(t_i)p(y_j)} \tag{2}$$

and $H(T)$ is the Shannon entropy defined as

$$H(T) = -\sum_{i=1}^{m} p(t_i) \log_2 p(t_i) \tag{3}$$

$p(t, y)$ is the joint probability distribution of t and y, $p(t)$ and $p(y)$ are the marginal probability distributions functions of T and Y respectively [22]

$$p(t) = \sum_{y} p(t, y), p(y) = \sum_{t} p(t, y) \tag{4}$$

The joint distribution $p(t, y)$ contains the reject option in binary classifications is defined as:

$$p(t, y) = \begin{bmatrix} \int_{R_1} p(t_1)p(x|t_1)dx & \int_{R_2} p(t_1)p(x|t_1)dx & \int_{R_3} p(t_1)p(x|t_1)dx \\ \int_{R_1} p(t_2)p(x|t_2)dx & \int_{R_2} p(t_2)p(x|t_2)dx & \int_{R_3} p(t_2)p(x|t_2)dx \end{bmatrix} \tag{5}$$

where R_j is the region in which the classification result is j. In the case of binary classifications with reject option, R_3 is the region of patterns identified as rejected.

In practical application, the *NI* value may not be effected when rejects appears only in one class. It seems that *NI* cannot distinguish rejects. So Hu and Yuan [20] proposed that turn the range of Y into 1 to m could have good features of original *NI*. The modified form of *NI* represents both non-abstaining and abstaining classification.

$$
NI(T,Y) = \frac{\sum_{i=1}^{m}\sum_{j=1}^{m} P_e(T=i, Y=j) \log_2 \frac{P_e(T=i,Y=j)}{P_e(T=i)P_e(Y=j)}}{-\sum_i P_e(T=i)\log_2 P_e(T=i)}
$$

$$
= -\frac{\sum_{i=1}^{m}\sum_{j=1}^{m} c_{ij}\log_2\left(\frac{c_{ij}}{C_i\sum_{i=1}^{m}\left(\frac{c_{ij}}{n}\right)}\right)}{\sum_{i=1}^{m} C_i \log_2\left(\frac{C_i}{n}\right)}
$$

(6)

where m is the number of classes and Y is counted from 1 to m. c_{ij} is the number of the i_{th} class that is classified as the j_{th} class, $i = 1, 2, \ldots, m, j = 1, 2, \ldots, m+1$.

C_i is the total number of samples in the i_{th} class, which indicates the constrains and relations of an confusion matrix. n is the total number in the confusion matrix.

$$
C_i = \sum_{j=1}^{m+1} c_{ij}, C_i > 0, c_{ij} \geq 0, i = 1, 2, \ldots, m
$$

(7)

$$
n = \sum_{i=1}^{m}\sum_{j=1}^{m+1} c_{ij}, n > 0, c_{ij} \geq 0, i = 1, 2, \ldots, m, j = 1, 2, \ldots, m, m+1
$$

(8)

2.2   Mutual-Information Classifier in Binary Classifications

A mutual-information classifier is the classifier that is obtained from the maximization of mutual information over all patterns [22]

$$
y^+ = \arg\max_{y} NI(T=t, Y=y)
$$

(9)

where y^+ is the optimal classifier in terms of maximal mutual information.

We denote $x = [x_1, x_2, \ldots, x_n]^T$ as a data matrix with n samples to be classified, $x_l \in \Re^d, l = 1, 2, .., n$ is the input feature vector. In the case of m-class abstaining classification, $T_r = [T_{r1}, T_{r2}, \ldots, T_{rm}]^T \in \Re^m$ is defined as the rejection threshold vector, T_{ri} is $[0, 1), i = 1, 2, \ldots, m$. For both abstaining and non-abstaining classifications, the generalize formula to calculate *NI* and the decision thresholds as follow:

$$NI = \max NI(t, y = f(\boldsymbol{\phi}(\boldsymbol{x}), \boldsymbol{T}_r))$$

$$y_l = \begin{cases} \arg\max\limits_{i} \left(\frac{\phi_i(x_l)}{1-T_{ri}} \right) & if \quad \max \left(\frac{\phi_i(x_l)}{1-T_{ri}} \right) \geq 1 \\ m+1 & otherwise \end{cases} \quad (10)$$

$$0 \leq T_{ri} < 1 \, , \; 0 \leq \sum_{i=1}^{m} T_{ri} < m - 1$$

$$i = 1, 2, \ldots, m, \; l = 1, 2, \ldots, n$$

where $\boldsymbol{\phi}(\boldsymbol{x}) \in \Re^{n \times m}$ denotes the output matrix of a probabilistic classifier for n instances, $\phi_i(x_l)$ is the probabilistic output of class i or $\boldsymbol{x}_l$ and meet the conditions $\sum_{i=1}^{m} \phi_i(x_l) = 1 \, , \; 0 \leq \phi_i(x_l) \leq 1$.

Pay attention to formula (10) that $\sum_{i=1}^{m} T_{ri} < m - 1$. Because if $\sum_{i=1}^{m} T_{ri} \geq m - 1$, then $\sum_{i=1}^{m} (1 - T_{ri}) = m - \sum_{i=1}^{m} T_{ri} \leq 1$, we can launch that $\sum_{i=1}^{m} \phi_i(x_l) \geq \sum_{i=1}^{m} (1 - T_{ri})$. One situation to satisfy this conditions is that $\frac{\phi_i(x_l)}{1-T_{ri}} \geq 1$ for all probabilistic outputs. In other words, every sample can be classified into class 1 to m. It means that the rejection would never be met and the abstaining classification is invalid. Therefore, abstaining classification should satisfy the condition $\sum_{i=1}^{m} T_{ri} < m - 1$.

According to maximize NI, we can obtain the optimal threshold vector T_r^*.

$$T_r^* = \arg\max\limits_{T_r} NI(t, y = f(\boldsymbol{\phi}(\boldsymbol{x}), \boldsymbol{T}_r)) \quad (11)$$

3   SMOTE

SMOTE is a common method of over-sampling in imbalanced data and typically used to synthetic minority samples. This approach increases the features available to each class and makes the samples more general. SMOTE works by generating new instances from existing cases. The new instances are created by taking samples of the feature space for each target class and its nearest neighbors, and generating new examples that combine features of the target case with features of its neighbors. SMOTE takes the entire dataset as an input and balanced the data distribution. Hence, SMOTE is a better way of increasing the number of rare cases than simply duplicating existing cases.

However, SMOTE method couldn't consider about the data distribution, so it usually combines with other classification methods in the class imbalance problem. For example, Dai [23] used SMOTE method to process the misclassification data. They synthesized new data and then re-classified them so as to achieve a better classification effect.

SMOTE algorithm searches the K nearest neighbor samples for each sample of the pending data, and then randomly selects N samples from these nearest neighbor samples. Finally, we interpret the original sample data and its nearest neighbor samples.

$$X_{new} = X + rand(0, 1) * (M_i - X), i = 1, 2, \ldots, N \tag{12}$$

where X_{new} is the synthetic samples and X is the original samples. M_i represents the N samples selected from the nearest neighbor samples.

4   Mutual-Information Classifier Based on SMOTE

In this work, we combine MIC and SMOTE to construct a mutual-information-SMOTE classifier. In the previous section, we have introduced the mutual information classification method and SMOTE. Both of them are used to deal with the class imbalance problem.

Algorithm 1. Algorithm for Mutual-Information-SMOTE Classifier

Input: Training set and testing set, the nearest neighbors K;

Output: The thresholds T_{r1}, T_{r2}, the cost term $\lambda_{12}, \lambda_{21}$, normalized mutual information(NI) and classification results of the testing set.

Step 1. Input the training set and testing set into Mutual-Information classifier and obtain the confusion matrix.

Step 2. Search Rej of both positive and negative samples to construct the abstention samples.

Step 3. For each abstention sample, we calculate its nearest neighbors K and compound $K-1$ new samples.

Step 4. Add the new samples to balance the training set, then use the new training set to train our classifier and get the results of the test set.

Our work only discuss the case of binary classifications with reject option. The rejection threshold vector is $T_r = [T_{r1}, T_{r2}]^T, T_r \in [0, 1)$. t is the targets variable and y is the decision outputs variable. The range of decision output y is 1 to 3 class.

$\phi_i(\mathbf{x}_l) = \prod_{k=1}^{m} \phi_i(x_{lk})$ is the posterior probability of the l_{th} pattern belongs to the i_{th} class, where n is the number of patterns and m is the number of attributes.

$$NI = maxNI(t, y = f(\phi(\mathbf{x}), T_r)),$$

$$y_l = \begin{cases} y_1, & \frac{\phi_1(\mathbf{x}_l)}{1-T_1} > \frac{\phi_2(\mathbf{x}_l)}{1-T_2} \text{ and } \frac{\phi_1(\mathbf{x}_l)}{1-T_1} \geq 1 \\ y_2, & \frac{\phi_2(\mathbf{x}_l)}{1-T_2} > \frac{\phi_1(\mathbf{x}_l)}{1-T_1} \text{ and } \frac{\phi_2(\mathbf{x}_l)}{1-T_2} \geq 1 \\ y_3, & otherwise \end{cases} \tag{13}$$

$$0 \leq T_{ri} < 1, l = 1, 2, .., n$$

According to maximize *NI*, we can get the optimal rejection threshold vector T_r^* and the confusion matrix C.

$$T_r^* = \arg \max_{T_r} NI(t, y = f(\phi(x), T_r)) \tag{14}$$

Table 1 shows the confusion matrix in the case of binary classifications with reject option. Where c_{ij} is the number of the i_{th} class that is classified as the j_{th} class, $i = 1, 2, j = 1, 2, 3$.

Table 1. Confusion matrix c in binary class abstaining classification.

T	Y		
	1	2	3
1	c_{11}	c_{12}	c_{13}
2	c_{21}	c_{22}	c_{23}

We synthetize new training data by the rejection data. In order to find K nearest neighbors, we calculate the standard deviation of each pending data. For the multi-attributes binary classification, the following formula is used to synthesize the new data:

$$\begin{cases} D_{new}(aid) = D(aid) + weight(aid) * (T(aid) - D(aid)) \\ D_{new}(aid\_id) = D(aid\_id) \end{cases} \tag{15}$$

where *aid* is the attribute vectors and *aid_id* is the category vector. D_{new} is the synthetic samples and D the category of the original data. $T(aid)$ is the nearest neighbors. Added the processed data to the original training sets and calculate the maximize *NI*. In order to formula (11), we can obtain the optimal classification threshold T_r^*.

Mutual-information classification method can automatically summarize the cost values. Hu [22] has proposed a conversion formula between thresholds and cost terms. In the case of binary classification with reject option, the formula as follow:

$$T_{r1}^* = \frac{\lambda_{13} - \lambda_{11}}{\lambda_{13} - \lambda_{11} + \lambda_{21} - \lambda_{22}}$$
$$T_{r2}^* = \frac{\lambda_{23} - \lambda_{22}}{\lambda_{12} - \lambda_{13} + \lambda_{23} - \lambda_{22}} \tag{16}$$

where λ_{ij} is the cost term that a pattern belongs to t_i but decided as y_j. Generally speaking, there is no cost for correct classification and the cost of rejection classification are equal. So we can set $\lambda_{11} = \lambda_{22} = 0, \lambda_{13} = \lambda_{23} = 1$ to calculate the cost of misclassification. And then the cost can be definition as

$$\lambda_{12} = \frac{1}{T_{r2}^*}, \lambda_{21} = \frac{1}{T_{r1}^*} \tag{17}$$

5   Experimental Results and Conclusions

The previous section introduces the framework of the present method. It can be regarded as an improvement method of mutual-information classifier. The cost information can be automatically summarized by the degree of bias implied in the optimal threshold. In this section, we focus on binary classification with reject option and validate the feasibility of MISC.

5.1   Evaluation Criteria

In the class imbalance class problem, the number of data between categories are generally different. It means that if we go for high accuracy, the minority class may be ignored. We use other evaluation criteria for imbalanced class. Let N denote the negative class and P denote the positive class. In general, we consider the minority class as the positive class and the majority class as the negative class. In order to show the change of each class, E_i and Rej_i are applied as the error rate and the reject rate respectively within the i_{th} class. We also calculate the total error rate and reject rate is denoted as E and Rej respectively. Acc denotes the total accuracy rate and Acc_j are applied as the accuracy rate within the i_{th} class. At the same time, we also calculate the normalized mutual information (NI).

5.2   Experiment on Data Sets

These experiments use three binary class data sets with imbalanced class distributions. All of the data sets are obtained from the UCI Machine Learning Repository, and the data sets are shown in Table 2. All the attributes of them are continuous and are normalized to range of $[0, 1]$ in advance.

Table 2. Description of the data sets.

Data sets	#Inst	#Attr	Class distribution
Letter	20,000	17	19,211/789(=24.35)
Vehicle	846	19	634/212(=2.99)
Yeast	1,484	10	1,055/429(=2.46)
Breast	683	10	444/239(=1.86)
Heart	270	14	150/120(=1.25)

(*#Inst: number of instances, #Attr: number of attributions*).

The following numerical examples are specifically designed for demonstrating the intrinsic differences between MIC and MISC. All experiments are conducted with 10-flod stratified cross validation. Layered cross validation can ensure that each cross validation subset has the same category distribution and is widely used in imbalanced data learning. The results of the experiment are recorded in Table 3.

Table 3. Results of experiment data sets.

| Data sets | Classifier types | CR_1 | E_1 | E | Rej_1 | Rej | NI |
		CR_2	E_2		Rej_2		
Letter	MIC	0.7554	0.0875	0.0892	0.1572	0.4873	0.6335
		0.6681	0.0018		0.3302		
	MISC	0.7845	0.1103	0.1131	0.1052	0.2518	**0.6956**
		0.8506	0.0028		0.1466		
Vehicle	MIC	0.9623	0.0377	0.1056	0	0	0.6711
		0.9322	0.0678		0		
	MISC	1	0	0.1009	0.0016	0.0016	**0.6856**
		0.8975	0.1009		0		
Yeast	MIC	0.5350	0.2336	0.4251	0.2313	0.4522	0.1131
		0.5877	0.1915		0.2209		
	MISC	0.4977	0.2991	0.4526	0.2033	0.3947	**0.1153**
		0.655	0.1536		0.1915		
Breast	MIC	0.9916	0.0402	0.0402	0.0042	0.0042	0.8467
		0.9640	0.036		0		
	MISC	0.9958	0.0042	0.0357	0	0	**0.8617**
		0.9685	0.0315		0		
Heart	MIC	0.9667	0.0267	0.1100	0.0067	0.0233	0.6951
		0.9000	0.0833		0.0167		
	MISC	0.9800	0.0200	0.1033	0	0	**0.7247**
		0.9167	0.0833		0		

Figures 1, 2, 3, 4 and 5 presents the experimental results. Both classifiers have low error rates of the positive class and high overall accuracy. However, MIC is not friendly to negative class, especially when the class distribution differs greatly. We find out that the accuracy of the negative class is low. On *Letter*, the error rate of the positive class is 0.0892 but the accuracy of the negative class only 0.6681 in MIC. By comparison, MISC significantly raises the accuracy of negative class to 0.8506 and the error rate increases only 0.0239. At the same time, the reject rate reduce by 0.2355. In the case of maintaining the accuracy of positive class, the accuracy of negative class may be increased and the reject rate may be reduced. These results show that MISC performs better when dealing with extremely imbalanced data. On all of the five data sets, MIC achieves high accuracy and low error rate, at the price of considerably high reject rate. By comparison, we think that MISC perform better on the imbalanced data. Because we can obtain from Table 3 that although MISC increases a little of error rate, but it improves the accuracy of negative class obviously without reducing the accuracy of the positive class. Moreover, MISC can handle the rejection data better. From Fig. 5 we can see that MISC receives higher value of *NI* than MIC in all of five experimental data sets. It means that the prediction results of MISC classifier have lower uncertainty than MIC, that is, the effect of MISC is better. In a word, MISC performs better than MIC in class imbalance problem because MISC has lower reject rate, higher accuracy, a certain amount of error rate and higher *NI*.

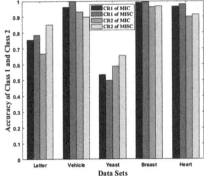

Fig. 1. Accuracy on binary class data sets.

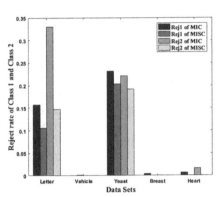

Fig. 2. Reject rates on binary class data sets.

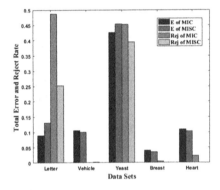

Fig. 3. Error rates on binary class data sets.

Fig. 4. Total error and reject rates on binary class data sets

Table 4 lists the values of rejection threshold T_r and the cost term λ. These values are purely determined by the data sets. We can see that misclassification cost of positive samples are usually greater than negative samples. So the classification is biased towards positive samples. Our method improves the overall value of misclassification cost and the cost of misclassifying negative samples increases to a greater degree or extent. This narrows the difference between λ_{12} and λ_{21}. It means that the classification pays more attention to negative samples while bias to positive samples.

We can come to several conclusions. Firstly, mutual-information classifiers can sum up the cost terms automatically by optimizing decision threshold. It shows that the mutual-information classifier is suitable for the class imbalance problem. Secondly, from the experiment results, we can draw that the performance of mutual information classifier depends on the data distribution rather than the number of data. The degree of imbalanced data had little effect on the results. Therefore, we believe that the classification method based on mutual information theory can be applied to small sample data sets. Finally, our method is able to modify the data boundaries, thereby changes the determine threshold. From the cost term we can find that our classifier can balance the relation between positive and negative classes in imbalanced data.

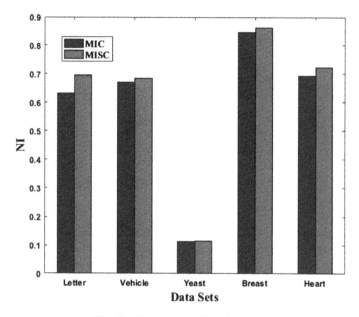

Fig. 5. *NI* on binary class data sets.

Table 4. The "equivalent" costs and the optimal rejection thresholds.

Data sets	Classifier	T_{r1}	T_{r2}	λ_{12}	λ_{21}
Letter	MIC	1.17e−14	5.37e−15	85.575	1.8606
	MISC	1.20e−18	1.17e−18	8.5447	8.3403
Vehicle	MIC	1.03e−38	4.22e−36	971.27	2.3713
	MISC	4.57e−40	1.21e−36	218.91	8.285
Yeast	MIC	2.37e−10	3.43e−10	3.06	3.18
	MISC	1.72e−10	1.55e−10	6.456	5.803
Breast	MIC	1.85e−11	1.07e−09	540.43	9.3452
	MISC	1.42e−12	1.66e−11	70.378	6.0134
Heart	MIC	9.64e−13	5.15e−12	10.372	1.9423
	MISC	6.39e−14	3.97e−13	15.639	2.5204

In summary, mutual information is a very effective method for imbalanced data. MISC can get better results than MIC, because our method enables classifier to increase the attention of majority class and improves the overall classification performance.

5.3   Experiment on Medical Images

The imbalanced data set used in this experiment is collected in the focal liver lesion case. This data set includes 160 samples, of which 60 for the liver patients and 100 for healthy people. We use MIC and MISC respectively in the liver data set. We repeat the experiment three times and calculate the average. Table 5 lists the result of both classifiers. The best experimental result is marked by black print.

Table 5. Results of focal liver lesion data set.

| Classifier | CR_1 | E_1 | E | Rej_1 | Rej | T_{r1} | λ_{12} | NI |
	CR_2	E_2		Rej_2		T_{r2}	λ_{21}	
MIC	0.917	0.017	**0.437**	0.067	0.238	8.11e−09	123.236	0.2159
	0.41	0.42		0.17		2.64e−07	3.782	
MISC	**0.95**	0.05	0.47	0	**0**	1.81e−10	552.606	**0.2495**
	0.58	0.42		0		5.33e−09	18.759	

Compared the proposed classifier with MIC, MISC can achieve better classification results. From Table 5 we can obtain that MISC has greater values of NI. It means that the classification results are more relevant to the actual results and the overall accuracy of the classifier is higher. In our classifier, the accuracy of both positive and negative classes has been improved with a little raise of positive error rates, and all the rejection samples are classified. While CR_1 and CR_2 improve by 0.033 and 0.17 respectively, E only increases by 0.033. Our classifier increases accuracy at the cost of small error rates. Moreover, our method classifies all the rejected samples and the overall accuracy is improved. It shows that most of the rejected samples are classified correctly. T_r represents the classification threshold, which is the classification boundary of data. Among them, T_{r1} and T_{r2} are the discriminant threshold of positive class and negative class respectively. From T_{r1} and T_{r2}, we can see that the boundary width is narrowed and the boundary moves down after SMOTE. It indicates that SMOTE changes the boundary of samples and MISC reduce the rejection threshold scope. λ_{ij} represents the cost of judging i class as j class. From the experiment results we can found that MISC gets greater costs of misclassification than MIC. MIC pays overmuch attention to the minority class and give a high cost to the minority class. When we look at the MIC in Table 5 we can find the ratio of λ_{12} and λ_{21} close to 32. It results in negative class being ignored and its accuracy reduces to 0.41. MISC obviously improves this defect. MISC added value of λ_{12} and λ_{21}, this results indicate that the classifier more focus on misclassification data. Moreover, the increase of λ_{12} is more than λ_{21}, which makes their ratio smaller. It means that MISC not only pays attention to minority class but also considers the majority class. Our method reduces the accuracy of positive class, but it reduces the error rate to a greater degree or extent. It should be noted that all rejection data is classified by our approach, and in this case the classifier can achieve the best result in our data sets. We use the binary classification method which still includes the reject option. Generally speaking, we can draw a conclusion that MISC can achieve better classification results in the focal liver lesion data sets than MIC and MISC can be applied to imbalanced classification well.

6 Conclusion

In this paper, we present a CFL method to deal with the binary-class imbalance problem. Firstly, we acquire abstention samples through mutual-information classifier. Secondly, we use these abstention samples to synthetize new samples through SMOTE method and make up a new data sets. Finally, we construct mutual-information-SMOTE classifier, use the new data sets to train our classifiers and obtain the classification results. We apply our method to six datasets, and compare the results with mutual-information classification method. The experiment results indicate that our method behaves better than mutual-information classification method.

Acknowledgements. This work was supported by the National High Technology Research, Development Program of China (No. 2015IM03030), the National Natural Science Foundation of China (No. 61573235), the Shanghai Innovation Action Project of Science and Technology (No. 17511103502), and the Fundamental Research Funds for the Central Universities.

References

1. Zhang, S., Sadaoui, S., Mouhoub, M.: An empirical analysis of imbalanced data classification. Comput. Inf. Sci. **8**(1), 151–162 (2015)
2. Raeder, T., Forman, G., Chawla, N.V.: Learning from Imbalanced data: evaluation matters. In: Holmes, D.E., Jain, L.C. (eds.) Data Mining: Foundations and Intelligent Paradigms. ISRL, vol. 23, pp. 315–331. Springer, Heidelberg (2012). https://doi.org/10.1007/978-3-642-23166-7_12
3. Beyan, C., Fisher, R.: Classifying imbalanced data sets using similarity based hierarchical decomposition. Pattern Recogn. **48**, 1653–1672 (2015)
4. Díez-Pastor, J.F., Rodríguez, J.J., García-Osorio, C., et al.: Random balance: ensembles of variable priors classifiers for imbalanced data. Knowl. Based Syst. **85**, 96–111 (2015)
5. Gao, M., Hong, X., Harris, C.J.: Construction of neurofuzzy models for imbalanced data classification. IEEE Trans. Fuzzy Syst. **22**(6), 1472–1488 (2014)
6. Li, H., Zou, P., Han, W., et al.: A combination method for multi-class imbalanced data classification. In: Web Information System and Application Conference, pp. 365–368. IEEE (2013)
7. Li, Y., Liu, Z.D., Zhang, H.J.: Review on ensemble algorithms for imbalanced data classification. Appl. Res. Comput. 1001–3695 (2014)
8. Zhang, X., Hu, B.: A new strategy of cost-free learning in the class imbalance problem. IEEE Trans. Knowl. Data Eng. **26**(12), 2872–2885 (2013)
9. Schaefer, G., Krawczyk, B., Doshi, N.P., et al.: Cost-sensitive texture classification. In: IEEE Congress on Evolutionary Computation, pp. 105–108. IEEE (2014)
10. Bahnsen, A.C., Aouada, D.: Example-dependent cost-sensitive decision trees. Expert Syst. Appl. **42**, 6609–6619 (2015)
11. Yu, H., Mu, C., Sun, C., et al.: Support vector machine-based optimized decision threshold adjustment strategy for classifying imbalanced data. Knowl. Based Syst. **76**(1), 67–78 (2015)
12. Lin, W.J., Chen, J.J.: Class-imbalanced classifiers for high-dimensional data. Brief. Bioinf. **14**(1), 13 (2013)

13. Napierala, K., Stefanowski, J.: Abstaining in rule set bagging for imbalanced data. Log. J. IGPL **23**(3), 421 (2015)
14. Zhao, Z., Wang, X.: A research of optimal rejection thresholds based on ROC curve. In: International Conference on Signal Processing, pp. 1403–1407. IEEE (2015)
15. Blankenburg, M., Bloch, C., Krüger, J.: Computation of a rejection threshold used for the bayes classifier. In: International Conference on Machine Learning and Applications, pp. 342–349. IEEE (2015)
16. Hu, B.G.: What are the differences between bayesian classifiers and mutual-information classifiers. IEEE Trans. Neural Netw. Learn. Syst. **25**(2), 249 (2014)
17. Cover, T.M., Thomas, J.A.: Elements of Information Theory, 2nd edn. Wiley, Hoboken (2006). J. Am. Stat. Assoc. **39**(7), 1600–1601 (2006)
18. Zhen, Z., Xue-Gang, H.U.: Classification model based on mutual information. J. Comput. Appl. **31**(6), 1678–1680 (2011)
19. Principe, J.C., Xu, D., Zhao, Q.: Learning from examples with information theoretic criteria. J. VLSI Signal Process. Syst. **26**(1–2), 61–77 (2000)
20. Hu, B.G., He, R., Yuan, X.T.: Information-theoretic measures for objective evaluation of classifications. Acta Autom. Sinica **38**(7), 1169–1182 (2012)
21. Mi, Y.: Imbalanced classification based on active learning SMOTE. Res. J. Appl. Sci. Eng. Technol. **5**(3), 944–949 (2013)
22. Hu, B.G., Wang, Y.: Evaluation criteria based on mutual information for classifications including rejected class. Acta Autom. Sinica **34**(11), 1396–1403 (2008)
23. Dai, H.L.: Imbalanced protein data classification using ensemble FTM-SVM. IEEE Trans. Nanobiosci. **14**(4), 350–359 (2015)

Ontology Sparse Vector Learning Algorithm

Xin Xin Huang$^{(\boxtimes)}$ and Shu Gong

Guangdong University Science and Technology, Dongguan 523083, China
hxx2128@163.com

Abstract. Ontology, as an efficient semantic model, is widely used in various fields of engineering science and semantic similarity. Calculation is the core of the ontology algorithm. Ontology similarity calculation via sparse vector, which can be used for high-dimensional data and big data processing. This paper two experiment results showed algorithm by using squared loss function to express the error term, Finally, two experiment results showed the effective of our ontology sparse vector learning algorithm for ontology similarity computation in specific engineering applications.

Keywords: Ontology · Similarity measure · Ontology mapping
Sparse vector

1 Introduction

Ontology, as a model of structured representation and storage of data, has been widely used in various fields of computer engineering. For example: ontology in information retrieval is often used for query expansion, by computing the similarity between ontology concepts corresponding to the vertex, return the related information and the concept of high similarity was the query in the query process, in order to improve the search coverage. Ontology has been widely used in other fields of computer science, such as medicine, biological computing, geographic information system, educational technology and so on. At present, ontology applications can be classified into two categories: ontology similarity computation and ontology mapping. The common point is to use Ontology graph to represent the Ontology structure, to calculate the similarity between vertices by certain rules; the difference is that the former is to calculate the same Ontology graph concept similarity between the corresponding vertices, the latter which is calculated by the similarity from different Ontology graph between the vertex to determine the similarity between concepts from different ontologies of graphs, and obtained according to the similarity of ontology mapping strategy. The goal of the algorithm from the point of view, the nature of computation of ontology similarity calculation and ontology mapping can be regarded as the similarity Ontology graph vertices, which is based on the known Ontology graph to find the optimal similarity calculation method. In the big data storage, management and calculation, in order to express the relationship between data, structured storage characteristics of data are required for corresponding data model, and the data model require easy to calculation, statistics and analysis. Therefore, as a structured concept of shared storage model, ontology has been paid more and more attention of by data and information managers,

© Springer Nature Singapore Pte Ltd. 2018
K. Li et al. (Eds.): ISICA 2017, CCIS 873, pp. 31–40, 2018.
https://doi.org/10.1007/978-981-13-1648-7_3

and become a hot research issue in the field of information and data gradually in recent years. In general, structured storage of ontology concepts can be used to represent by the hierarchical graph, and it can be used as a graph to represent an ontology, each vertex in the conceptual graph corresponds to a concept or a piece of information, the edges between vertexes represent relationship of concepts or information, such as implicit affiliation, etc.

Ontology, is derived from philosophy to describe the natural connection of things and the inherent hidden connections of their components. In information and computer science, ontology is often taken as a model for knowledge storage and representation. It has shown extensive applications in a variety of fields, such as: knowledge management, machine learning, information systems, image retrieval, information retrieval search extension, collaboration and intelligent information integration. Since a few years ago, because of its efficiency as a conceptually semantic model and an analysis tool, ontology has been favored by researchers from pharmacology science, biology science, medical science, geographic information system and social sciences (for instance, see Koehler et al. [1], Hristoskova et al. [2], Przydzial et al. [3], Kabir et al. [4] and Ivanovic and Budimac [5]).

In addition to structured storage, data model requires to help scholars to calculate, statistical, verify the data and information, and finally come to the conclusion. Therefore, the ontology algorithm generally revolves around the data stored in the ontology, which is to find out their mutual relations. From this perspective, similarity calculation algorithm between ontology concepts between the vertex of the Ontology graph becomes the core ontology engineering algorithm. In the context of large data, an ontology graph often stores vast amounts of information, which leads to the traditional method of heuristic design ontology formula, has been unable to do ontology framework of big data processing. Calculating similarity of ontology concepts through machine learning has become the mainstream in recent ontology algorithm research.

From the point of matrix theory, the strategy of similarity matrix learning for ontology is proposed [6]. The corresponding ontology similarity computation and ontology mapping algorithm are obtained by using the pairwise ranking learning method [7]. The original ontology learning algorithm based on regularization model is improved, and a new ontology learning algorithm with noise condition is obtained [8]. From the other point of view, the ontology learning algorithm based on regularization model is improved, so that it can be applied to the semi supervised learning framework [9]. An ontology learning algorithm based on BMRM iterative sorting method is obtained [10]. Mahalanobis matrix learning is integrated into the ontology algorithm to get the corresponding ontology similarity calculation and ontology mapping algorithm [11]. A new ontology optimization framework is proposed under the framework of multiple segmentation [12]. An ontology learning algorithm based on infinite promotion strategy is obtained under the framework of multiple segmentation [13]. The kernel function is used as the function of similarity function, and the most similar core is obtained by the method of learning, then get the ontology similarity calculation function [14]. The sparse vector learning algorithm based on the ADAL method is obtained to calculate the similarity between vertices by sparse vector [15].

In this paper, which consider in a special frame ontology sparse vector learning, that is through ontology sparse vector get corresponding numerical calculation formula exists nonlinear function, and the effectiveness of the algorithm is verified by experiment.

2 The Framework of Ontology Algorithm Based on Sparse Vector

In order to make the ontology model to adapt learning algorithm framework, numerical treatment of the concept of information is needed, namely for each vertex, a vector is used to represent the all information of this vertex. For the sake of convenience, v *is set to both* represent vertex and its corresponding vector, the v below and can be understood as a vertex and its corresponding vector without causing confusion, while standard vector bold is no longer used.

Let $\beta = (\beta_1, \cdots, \beta_d)^T \in \mathbb{R}^d$ ontology sparse vector, especially $\beta^* = (\beta_1^*, \cdots, \beta_d^*)^T \in \mathbb{R}^d$ for optimal ontology sparse vector. The sparse vector is characterized by the vast majority of the value of 0. For vertex of ontology $v = (v_1, \cdots, v_d)$, the calculation method of the corresponding real value is obtained by using the sparse vector of the Ontology as follows:

$$y = v^T \beta^* + \varepsilon, \tag{1}$$

Where ε is the symbol of the offset or error values. The Ontology algorithm is based on sparse vector, the basic idea of which is to obtain the optimal ontology sparse vector β^*, and then by (1) to calculate the real number of each vertex in the Ontology graph, then the similarity between the ontology concepts is determined by the one dimension distance between the real number corresponding to the vertices: the smaller the distance is, the greater the similarity. The core idea of this algorithm is also a kind of dimension reduction, firstly, the semantic information of each ontology vertex is represented by a d-dimension vector, by reducing the dimension of the sparse vector, the original d dimensional vector is transformed into one dimensional real number y. From this sense, the ontology sparse vector is used as a bridge in the computation model, and the advantages and disadvantages of β^* have a direct impact on the results of similarity calculation. Therefore, ontology sparse vector is based ontology algorithm, the core of which is the learning of sparse vector.

In recent years, because of ontology's wide application in biology, medicine and other field, which involves the representation of ontology information more and more, algorithm of ontology is increasingly demanding. The ontology algorithm based on the sparse vector ontology, while its advantage lying in a specific application, can effectively shield the feature information which is irrelevant to this kind of application, and highlight the valuable information. For example, in genetics, a genetic disease is only related to a number of genes, has nothing to do with most of the genes, and sparse vector can effectively find the target gene, and achieve the desired goals. Due to the magical effect that the sparse algorithm for high dimensional data dimensional reduction has, and for low dimensional data, the algorithm can increase the complexity of the algorithm, therefore, in this paper, the dimension of the vector is much larger than that of the bulk sample, that is $d \gg n$.

This paper considers a change model (1) as follows:

$$y = f(v^T \beta^*) + \varepsilon, \tag{2}$$

Where f can be understood as a kind of conversion functions, when f is the identity function, (2) is degenerate to (1). Therefore, calculation model (2) can be understood as an expansion of (1). Obey $\{v_i, y_i\}_{i=1}^n$ ontology sample set which obey some kind of independent and identically distributed, the optimal sparse vector can be obtained by the following regularization model:

$$\beta^* = \min_{\beta \in \mathbb{R}^d} \frac{1}{n} \sum_{i=1}^n (y_i - f(v_i^T \beta))^2 + \lambda \|\beta\|_1, \tag{3}$$

Where λ is the balance parameter, $\|\beta\|_1$ is used to control the sparse degree of the ontology sparse vector, $\lambda \|\beta\|_1$ is called the balance term. As the main part, $\frac{1}{n} \sum_{i=1}^n (y_i - f(v_i^T \beta))^2$ is the error term. The essence of (3) is to use the square loss as the error representation of the loss function.

3 New Algorithm Description

In this paper, we will discuss the problem of the optimal ontology sparse vector under the nonlinear representation, f as a nonlinear function. The sparse vector of the optimal body. The following are assumed that the given ontology sample is satisfied =, the function f is monotone and continuous.

$$L(\beta) = \frac{1}{n} \sum_{i=1}^n (y_i - f(v_i^T \beta))^2. \tag{4}$$

Prior assumption that A is sparse, and the value is estimated by (3).

Because f is a nonlinear function, $L(\beta)$ may be non convex. Need to find a stagnation point (also known as the equilibrium point) $\hat{\beta}$ satisfy $\lambda \xi + \nabla L(\hat{\beta}) = 0$, where $\nabla L(\hat{\beta})$ denotes the gradient of $L(\hat{\beta})$ and $\partial \|\hat{\beta}\|_1$. Using proximal gradient method to get the stagnation below. The method can get an iterative sequence $\{\beta^{(t)}, t \geq 0\}$, etc.

$$\beta^{(t+1)} = \underset{\beta \in \mathbb{R}^d}{\arg\min} \{ \langle \nabla L(\beta^{(t)}), \beta - \beta^{(t)} \rangle + \frac{\alpha_t}{2} \|\beta - \beta^{(t)}\|_2^2 + \lambda \|\beta\|_1 \}. \tag{5}$$

$$\nabla L(\beta^{(t)}) = -\frac{1}{n} \sum_{i=1}^n (y_i - f(v_i^T \beta^{(t)})) f'(v_i^T \beta^{(t)}) v_i.$$

In (5), $\alpha_t > 0$, $\frac{1}{\alpha_t}$ represents the step size of the t iteration, and the value of $\nabla L(\boldsymbol{\beta}^{(t)})$ can be calculated by the following method:

Denoted by A, then the solution of (5) can be expressed as:

$$\boldsymbol{\beta}_i^{(t+1)} = S(u_i^{(t)}, \frac{\lambda}{\alpha_t}), \tag{6}$$

When $1 \leq i \leq d$, $S(\cdot, \cdot)$ is called soft boundary operator, which is defined as $S(u, a) = \text{sign}(u) \max\{|u| - a, 0\}$.

In this paper, we give a sparse vector learning algorithm based on the proximal gradient calculation:

Input: balance parameter $\lambda > 0$, update factor $\eta > 0$, parameter $\zeta > 0$, $\alpha_{\min}$ and $\alpha_{\max}$ meet $0 < \alpha_{\min} < 1 < \alpha_{\max}$, integer $M > 0$, function $\phi(\boldsymbol{\beta}) = L(\boldsymbol{\beta}) + \lambda \|\boldsymbol{\beta}\|_1$.

Initialization: $t \leftarrow 0$ and select $\boldsymbol{\beta}^{(0)} \in \mathbb{R}^d$

Step1: repeat

Step 2: use the following method to calculate the value of the step size α_t. Enter the iteration count value t, $\delta^{(t)} = \boldsymbol{\beta}^{(t)} - \boldsymbol{\beta}^{(t-1)}$ and $g^{(t)} = \nabla L(\boldsymbol{\beta}^{(t)}) - \nabla L(\boldsymbol{\beta}^{(t-1)})$. if $t = 0$, then $\alpha_t = 1$; otherwise $\alpha_t = \frac{\langle \delta^{(t)}, g^{(t)} \rangle}{\langle \delta^{(t)}, \delta^{(t)} \rangle}$ or $\alpha_t = \frac{\langle g^{(t)}, g^{(t)} \rangle}{\langle \delta^{(t)}, g^{(t)} \rangle}$;

Step 3: repeat the update $u^{(t)} \leftarrow \boldsymbol{\beta}^{(t)} + \frac{1}{n\alpha_t} \sum_{i=1}^{n} (y_i - f(v_i^T \boldsymbol{\beta}^{(t)})) f'(v_i^T \boldsymbol{\beta}^{(t)}) v_i$, $\boldsymbol{\beta}_i^{(t+1)} \leftarrow S(u_i^{(t)}, \frac{\lambda}{\alpha_t})$, $\alpha_t \leftarrow \eta \alpha_t$, during the period if found $\boldsymbol{\beta}^{(t+1)}$ to meet $\phi(\boldsymbol{\beta}^{(t+1)}) \leq \max\{\phi(\boldsymbol{\beta}^{(j)}) - \zeta \frac{\alpha_t}{2} \|\boldsymbol{\beta}^{(t+1)} - \boldsymbol{\beta}^{(t)}\|_2^2 : \max(t - M, 0) \leq j \leq t\}$, then the end of the update;

Step 4: update the value of the iteration count $t \leftarrow t + 1$

Step 5: if $\frac{\|\boldsymbol{\beta}^{(t)} - \boldsymbol{\beta}^{(t-1)}\|_2}{\|\boldsymbol{\beta}^{(t)}\|_2}$ is small enough, then the output $\hat{\boldsymbol{\beta}} \leftarrow \boldsymbol{\beta}^{(t)}$, otherwise return to step 1.

The approximate solution of the optimal sparse vector $\boldsymbol{\beta}^*$ is obtained by the iterative algorithm, and through (2) to get the corresponding y value of each vertex. Let y_i and y_j are corresponding real number of ontology vertex v_i and v_j, then the similarity between v_i and v_j corresponds to the concept of ontology is measured by the value of $|y_i - y_j|$. The smaller the value is, the greater the similarity; the greater the value is, the smaller the similarity.

4 Experiment

The following experiment verifies the efficiency of ontology similarity computation and ontology mapping by the new ontology learning algorithm.

4.1 Ontology Similarity Computation Experiment

First of all, the data derived from the A site construction of the botanical PO ontology O1 (its basic structure can refer to Fig. 1). The ontology can be regarded as a botanical database. We used to test the efficiency of the new algorithm of this paper to calculate the similarity. In order to compare algorithm efficiency, the following three classes of classical ontology learning algorithm also act on the PO ontology: ontology algorithm based on general ordering learning method [17], based on fast sort of learning ontology algorithm [16] and based on NDCG measure calculation ontology algorithm. All the experimental results using the P@N [19] average accuracy rate to judge. The P@N accuracy of the three kinds of classical ontology learning algorithm is compared with the P@N accuracy obtained by the new algorithm, the contrast data for N = 3, 5, 10 is shown in Table 1.

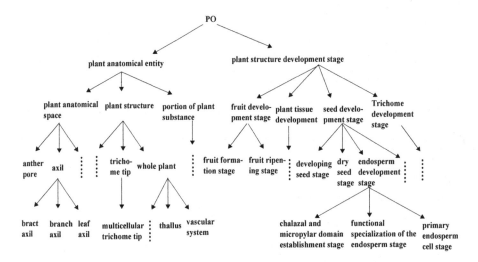

Fig. 1. PO ontology O1

Table 1. Parts of experimental 1 data

Algorithm name	P@3 average accuracy	P@5 average accuracy	P@10 average accuracy
Algorithm in this paper	49.52%	66.99%	89.67%
General sorting algorithm	45.49%	51.17%	58.59%
Fast sorting algorithm	42.82%	48.49%	56.32%
NDCG ontology algorithm	48.31%	56.35%	68.71%

The comparison of P@N accuracy which N = 3, 5, 10 in the above Table 1 shows that, the efficiency of the ontology learning algorithm that this paper proposes for Botany PO ontology similarity calculation, was significantly higher than that of the other three classical ontology learning algorithm.

4.2   Ontology Mapping Experiment

Next, use the following two "bionic robot body" O2 and O3 to verify the efficiency of computing the similarity of ontology mapping of the new ontology learning algorithm, i.e., for a vertex, to find the relatively large similarity vertex in another ontology, as the mapping results returned to the user. In order to allow the data to be compared, we will also use the algorithm just as Ontology mapping based on k-partite ranking learning method [20], Ontology learning method based on similarity measure by optimizing NDCG measure [18] and Ontology learning method based on hypergraph harmonic analysis [21] that act on the "bionic robot" ontology O2 and O3. The accuracy of P@N obtained from these three kinds of learning algorithms is compared with that obtained

Table 2. Parts of experimental 1 data

Algorithm name	P@1 average accuracy	P@3 average accuracy	P@5 average accuracy
Algorithm in this paper	27.78%	53.70%	80.00%
Ontology mapping based on k-partite ranking learning method	27.78%	48.15%	54.44%
Ontology learning method based on NDCG measure	22.22%	40.74%	48.89%
Harmonic analysis ontology method	27.78%	46.30%	53.33%

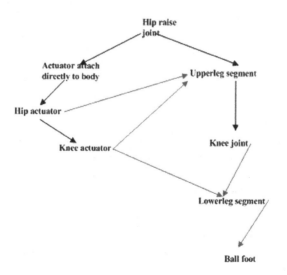

Fig. 2. "Bionic Robot" ontology O2

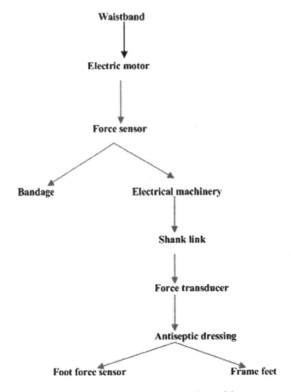

Fig. 3. "Bionic Robot" ontology O3

by the new ontology learning algorithm in this paper. The data is shown in Table 2 when N = 1,3,5 (Figs. 2 and 3).

According to the above P@N accuracy comparison of the data, we can see, in this paper, efficiency based on ontology mapping for the similarity establish between the bionic robot body O2 and O3 of a new ontology learning algorithm, is higher than the other three kinds of algorithm.

5   Conclusion

Ontology is a data management model which tackles data structure storage, analysis, calculation, statistics and reasoning in one. In practical engineering applications, the core of the ontology algorithm is the similarity calculation. Because of its powerful data management and application functions, the ontology has already been applied to pharmaceutical, social science, GIS, management and so on. In this paper, by using the proximal gradient method, an iterative algorithm is obtained for the sparse vector of the optimal ontology under the nonlinear framework, and the similarity calculation method is obtained. The data of the last experiment shows that the similarity computation of the new ontology learning methods for the field of Botany, has higher efficiency.

Acknowledgements. This work was supported by the General project of social science and technology development in Dongguan 2017, number: 2017507154412, Scientific research project of Guangdong University Science & Technology, number: GKY-2016KYYB-10.

References

1. Koehler, S., Doelken, S.C., Mungall, C.J.: The human phenotype ontology project: linking molecular biology and disease through phenotype data. Nucl. Acids Res. **42**(D1), 966–974 (2014)
2. Hristoskova, A., Sakkalis, V., Zacharioudakis, G.: Ontology-driven monitoring of patient's vital signs enabling personalized medical detection and alert. Sensors **14**(1), 1598–1628 (2014)
3. Przydzial, J.M., Bhhatarai, B., Koleti, A.: GPCR ontology: development and application of a G protein-coupled receptor pharmacology knowledge framework. Bioinformatics **29**(24), 3211–3219 (2013)
4. Kabir, M.A., Han, J., Yu, J.: User-centric social context information management: an ontology-based approach and platform. Pers. Ubiquit. Comput. **18**(5), 1061–1083 (2014)
5. Ivanovic, M., Budimac, Z.: An overview of ontologies and data resources in medical domains. Expert Syst. Appl. **41**(11), 5158–15166 (2014)
6. Wu, J.J., Zhu, L.L., Gao, W.: Learning of similarity matrix in ontology algorithm. J. Chin. Comput. Syst. **36**(4), 773–777 (2015)
7. Zhu, L.L., Dai, H.G., Gao, W.: Pairwise ranking ontology learning algorithm. J. Southwest Chin. Norm. Univ. **38**(12), 101–106 (2013). Natural Science Edition
8. Zhu, L.L., Wu, S.F., Ye, Y.F., Gao, W.: Ontology learning algorithm based on regularization model under noise condition. J. Northwest Norm. Univ. **50**(6), 41–45 (2014). Natural Science Edition
9. Zhu, L.L., Dai, H.G., Gao, W.: Semi-supervised ontology algorithm in regularization setting. Microelectron. Comput. **31**(3), 126–129 (2014)
10. Zhu, L.L., Gao, W.: Ontology learning algorithm based on BMRM iterative sorting method. Sci. Technol. Eng. **13**(13), 3653–3657 (2013)
11. Wu, J.J., Yu, X., Gao, W.: Ontology algorithm for learning based on Mahalanobis matrix. J. Southwest Chin. Univ. **37**(2), 117–122 (2015). Natural Science Edition
12. Zhu, L.L., Gao, W.: Ontology similarity measuring and ontology mapping based on new optimization model in multi-dividing setting. J. Comput. Inf. Syst. **11**(1), 377–386 (2015)
13. Gao, W., Zhu, L.L., Guo, Y.: Multi-dividing infinite push ontology algorithm. Eng. Lett. **23**(3), 132–139 (2015)
14. Zhu, L.L., Min, X.Z., Gao, W.: Algorithm of ontology similarity measure based on similarity kernel learning. Int. J. Comput. Technol. **14**(12), 6304–6309 (2015)
15. Gao, W., Zhu, L.L., Wang, K.Y.: Ontology sparse vector learning algorithm for ontology similarity measuring and ontology mapping via ADAL technology. Int. J. Bifurc. Chaos **25** (14), 1540032, 12 p. (2015). https://doi.org/10.1142/s0218127415400349
16. Huang, X., Xu, T., Gao, W., Jia, Z.: Ontology similarity measure and ontology mapping via fast ranking method. Int. J. Appl. Phys. Math. **1**(1), 54–59 (2011)
17. Wang, Y., Gao, W., Zhang, Y., Gao, Y.: Ontology similarity computation use ranking learning method. In: The 3rd International Conference on Computational Intelligence and Industrial Application, Wuhan, China, pp. 20–22 (2010)

18. Gao, W., Liang, L.: Ontology similarity measure by optimizing NDCG measure and application in physics education. Future Commun. Comput. Control Manag. **142**, 415–421 (2011)
19. Gong, S., Gao, W.: Ontology learning algorithm via WMW optimization model. In: 2016 International Conference Computational Intelligence and Security, Xian, China, pp. 431–434 (2016)
20. Lan, H.M., Ren, J.Y., Xu, J., Gao, W.: Ontology similarity computation using k-partite ranking method. J. Comput. Appl. **32**(4), 1094–1096 (2012)
21. Sellami, B., Laskri, Y., Benzine, R.: A new two-parameter family of nonlinear conjugate gradient methods. Optimization **64**(4), 993–1009 (2015)
22. Khodja, D.E., Simard, S., Beguenan, R.: Implementation of optimized approximate sigmoid function on FPGA circuit to use in ANN for control and monitoring. Control Eng. Appl. Inform. **17**(2), 64–72 (2015)
23. Liu, D., Yang, Y., Chen, Y., Zhu, H., Bayley, I., Aldea, A.: Evaluating the ontological semantic description of web services generated from algebraic specifications. In: Proceedings of the 10th IEEE International Conference on Service Oriented System Engineering (SOSE 2016), Oxford, UK (2016)

Bacterial Foraging Algorithm Based on Reinforcement Learning for Continuous Optimizations

Huiyan Jiang, Wanpeng Dong, Lianbo Ma$^{(\boxtimes)}$, and Rui Wang

College of Software, Northeastern University, Shenyang, China
malb@swc.neu.edu.cn

Abstract. Inspired by the adaptive learning and foraging behaviors of the bacteria colony, this paper proposes a re-structured bacterial colony foraging optimizer (RBCFO) based on reinforcement learning (RL) and self-adaptive search strategy for continuous optimizations. The algorithm aims to enhance the individual search efficiency during the evolution process via exploiting the RL mechanism in the multi-operations decision level and the adaptive search strategy in the single-operation level. Specifically, in the multi-operations decision level, the operation of each bacterium is determined by RL in an optimal manner. In the single-operation level (i.e., chemotaxis), each bacterium adaptively varies its own run-length unit and exchange information (i.e., cell-to-cell crossover) to dynamically balance exploration and exploitation during the search process. Then the proposed algorithm is evaluated against several state-of-the-art algorithms on a set of continuous benchmark instances. Experimental results verify the significant superiority of the proposed algorithm.

Keywords: Bacterial colony foraging optimizer · Reinforcement learning
Self-adaptive search strategy

1 Introduction

Swarm-intelligence (SI) optimization algorithms have been increasingly utilized by scientists in many applications [1–4]. The primary point of designing a nature-inspired algorithm is to exploit the potential balance of both global and local search strategies and to incorporate them in an efficient manner. As a promising example, the bacterial foraging algorithm (BFA) is an effective local optimizer, derived from bacterial social cooperation in foraging and communicating process [5, 6]. Until now, a number of BFA variants have been developed with different global search strategies [5–7]. It has been shown in [8] that efficient switching of the local search operations in terms of time and frequency of invoking plays a vital role on the optimization performance. Accordingly, a recent trend is to combine the promising exploitation of local search and global exploration search optimizer into a single model. However, those BFAs still suffer from two major drawbacks: (1) due to the fixed sequential implementation of bacteria behavior, the algorithm mechanism cannot fully adjust the potential balance of exploring operations in unknown region and exploiting operations in promising area.

© Springer Nature Singapore Pte Ltd. 2018
K. Li et al. (Eds.): ISICA 2017, CCIS 873, pp. 41–52, 2018.
https://doi.org/10.1007/978-981-13-1648-7_4

(2) Along with the complex landscapes of the optimization problems, the fixed run-length unit of each bacterium would cause them to be trapped in local optimum easily or oscillated around the optimal point. As a result, with the problem complexity increasing, the performance of the BFA decreases heavily with a poor convergence due to its simple searching strategy [9]. These limitations of BFA have been reported in [9, 10].

In order to address these key issues, we propose a re-structured bacterial colony foraging optimizer (RBCFO) based on reinforcement learning (RL) for the complex problems. RL has been adopted by some SI paradigms, such as PSO, DE [8, 11, 12]. An integration of RL and PSO was proposed by Piperagkas et al. [11], and then developed by Samma et al. [8] to enhance the performance of PSO. Another recent study [12] employed RL for parameter tuning in differential evolution algorithm. Comparing with the existing work in the literature, the RBCFO is to utilize RL method that harnesses the potential of a set of candidate strategies towards the maximum rewards. Specifically, each bacterium is subject to four action, including chemotaxis, reproduction, elimination-dispersal and crossover. And the RL is to regulate the operation sequence of each bacterium.

Intuitively, the main novelties and contributions of our work can lie in the following aspects:

(1) The proposed RBCFO is a more bacterially intelligent model, which incorporates the RL-based bacterial decision mechanism in the multi-operation decision level and the bacterial operations in the single-operation decision level including chemotaxis, reproduction, elimination-dispersal and crossover. This more bacterially intelligent model successfully cast the prime BFA into an adaptive, cooperative and intelligent model.

(2) In order to further balance the exploration and exploitation operation in searching process, the bacterial self-adaptive adjusting strategy of individual run-length unit is designed in the single-operation (i.e., chemotaxis) level. And the bacterial cell-to-cell crossover mechanism is employed to enable the information sharing among bacterial colony.

The rest of this paper is organized as follows. In Sect. 2 the proposed bacterial foraging model based on Reinforcement learning is presented. Section 3 provides a series of experiments to evaluate effectiveness of the new method. Section 4 concludes the paper.

2   Reinforcement Learning Bacterial Foraging Model

2.1   The Principle of the Proposed Model

As reported in previous work [5–9], the key challenge in devising a powerful SI algorithm lies in how to appropriately balance the exploration and exploitation operations during the optimization process. Theoretically, the solution to this issue can be ultimately attributed to the efficient management of the local search operation in terms of time and frequency of its calling [8]. In addition, the standard BFA also suffers from several weaknesses which are primarily the premature convergence and high computational cost problems [9, 10].

Aiming to solve this issue, the Q-learning method from RL is incorporated into RBCFO to manage the operations (i.e., time and frequency of invoking) of each bacterium in the colony. RL provides an effective method to manage the operation decision of the optimization paradigms, which are devised by using a combination of different global and local search operations. In RBCFO, the Q-learning method is used to adaptively switch the individual from one operation to another one according to its current achievement. Positive rewards are assigned to the well-performing bacteria, while penalties are imposed to worse individuals.

Generally, the main idea of RBCFO is to develop an adaptive and self-learning evolutionary model by combining bacterial chemotaxis, elimination-dispersal, cell-to-cell crossover and self-adaptive searching strategies under the RL-based decision-making framework. The proposed RBCFO is a more bacterially-realistic model that the bacteria grows by adaptively varying its run-length unit and its corresponding operations can be managed optimally to obtain an appropriate balance between exploration and exploitation during the searching process, as shown in Fig. 1.

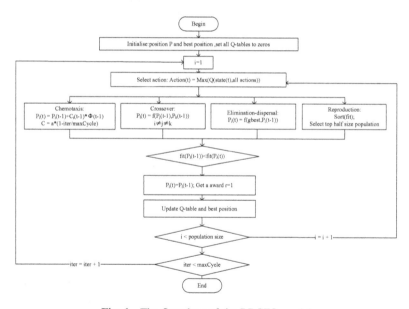

Fig. 1. The flowchart of the RBCFO model

2.2 Main Procedures of the Algorithm

Self-adaptive Chemotaxis
This process simulates the movement of an E. coli by its rotating flagella, which makes it swim toward more suitable areas.

$$P_i(t) = P_i(t-1) + C_i(t-1) * \phi(t-1) \tag{1}$$

Where P_i is the position of ith bacteria, ϕ represents a unit vector in random direction to describe tumble.

In general, a significant task to devise an effective SI algorithm is how to adaptively maintain an appropriate balance between exploration and exploitation in the searching process. Here, the exploration state drives the population towards distant or strange region for potential global optima and the exploitation state is employed to search thoroughly around promising areas obtained currently. In order to achieve this goal, each bacterium in the RBCFO uses a self-adaptive strategy that dynamically balances two conflicting behaviors during the searching process. The run-length unit can be formulated as:

$$C_i = a * (1 - iter/maxCycle) \tag{2}$$

Where a is a coefficient, and is empirically set to 0.01 in this experiment, $iter$ represents the current number of iterations, and $maxCycle$ is the number of total iterations. Note that $iter/maxCycle$ is computed to determine the search stage of current population. Generally, at the beginning of search process, a larger run-length unit is imposed to the agent in order to explore the search space. On the contrary, the agent will take a small step for a local region when $iter/maxCycle$ closes to 1 in convergence phase.

Reproduction

In this action, we first sort fitness values for each bacterium, then the former $N/2$ individuals divide into two (N represents the size of the population); the after $N/2$ individuals can be considered to die or out of the environment. Based on the theory of reinforcement learning, the overall fitness will be improved in reproduction operation, and we set accomplishing this action will get a positive reward value. But, the fitness will not be improved if an agent executes this behavior in the initial stage and it causes to fall into the "premature". Thus execution of reproduction must be delayed until meet the conditions of "$iter > (maxCycle/2)$".

Enhanced Elimination-Dispersal

For consumption of nutrient and some unknown reasons, bacteria are likely to disperse into a new area. In the proposed model, bacteria should take into account the current position and the entire population information when they are scattered. Hence, a learning item (i.e., gbest) is designed, and the individual i updates another location according to the global optimal position.

$$P_{id} = rand1 * (X_{max} - X_{min}) + rand2 * (gbest_d - P_d) \tag{3}$$

Where X_{max}, X_{min} respectively represent the upper and low boundaries of the search space, rand1 and rand2 are random numbers belonging to (0, 1), gbest denotes the optimal location of the entire population. This operation can increase the diversity of the population as well as improves the convergence rate.

Cell-to-Cell Crossover Operation

In the light of the analogy between bacterial quorum sensing and genetic pattern of GA, a principle of cell-to-cell communication for RBCFO model is used: a bacterial in RBCFO needs to exchange information with its neighbor on some dimensions of the problem, which can be identified as the crossover operation to improve the efficiency of information exchange among individuals [13], defined as

$$u_{ij} = \begin{cases} v_{ij} & if\ rand \leq P_{CR}\ or\ j = k \\ x_{ij} & otherwise \end{cases} \tag{4}$$

$$v_{ij} = gbest_j + beta * (P_{aj} - P_{bj}) \tag{5}$$

Where P_{CR} is a crossover probability in [0, 1], k is a random selected dimension from the current individual i, P_a and P_b represent the position of individual a and individual b respectively, beta is a scale factor lie in $[X_{max}, X_{min}]$. This method decreases the probability of the dimension curse in the complex optimization.

RL-Based decision

In classic BFA algorithm, there are four operations (i.e., chemotaxis, swimming & tumbling, reproduction, and elimination-dispersal) that are performed sequentially during the searching process. In every iterative evolution, each individual has experienced $N_c * N_{re}$ times of chemotaxis and swimming & tumbling, N_{re} times of reproduction, until N_{ed} times of elimination-dispersal behavior has completed. Moreover, it is indispensable to set the number of times for each operation in advance. Accordingly, the result will change largely due to the different number of executions. Aiming at this problem, our algorithm employs the RL mechanism to manage the operations of each individual. Specifically, RL adaptively transforms the individual from one state to another one according to its current achievement. The regulation mechanism of RL is shown in Fig. 2.

First, the state space of bacterial foraging colony is mapped to M × M matrix, where M is the number of states. Accordingly, each individual in the colony has its own Q-table. To minimize the computational cost of managing the Q-tables, we only use 4 individuals in the RBCFO. The detailed procedure of the RL algorithm is shown in Table 1. For the sake of clarity, a numerical example of the Q-table of i-th individual is illustrated in Tables 2 and 3. Then, each individual will undergo the transition from current state to next optimal state. As we can see from the example, according to its current state crossover (Cr), the optimal action of i-th individual is determined optimally as chemotaxis (C) based on the principle of potential reward-maximization of RL. Then the individual finishes its state transition from Cr to C.

After action C is finished, the individual will receive a penalty due to failing in further improving its fitness. Mathematically, according to Eqs. (6) and (7), the data item (Cr, C) in the Q-table is updated as

$$Q(s_t, a_t) = Q(s_t, a_t) + \alpha[r_{t+1} + \gamma \max_a Q(s_{t+1}, a) - Q(s_t, a_t)] \tag{6}$$

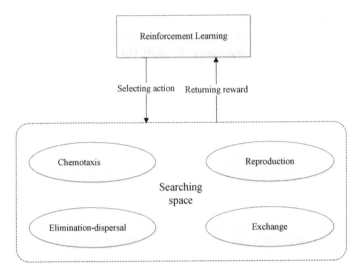

Fig. 2. The proposed RBCFO structure

Table 1. The pseudo code of RL

Step 1 **Initialize:**

The initial state $s_t \in S = \{s_1, s_2, ... s_n\}$ and initial action $a_t \in A = \{a_1, a_2, ..., a_n\}$ for each agent;

Set all Q-tables to zeros.

Step 2 **While** *iter < maxCycle*

Step 2.1 Select the best action a_t for the current state s_t according to Q-table;

Step 2.2 Execute the action a_t, get a reward r;

Step 2.3 Update the data item (s_t, a_t) in the Q-table;

Step 2.4 Update the current state $s_{t+1} = s_t$;

iter = iter + 1;

End

Table 2. The Q-table of an individual after executing 17 iterations

	State/action	C	Re	Ed	Cr
Next state:	C	−0.99982	0	0	−0.0337516
	Re	0	0	0	0
	Ed	0	0	0	0
Cur state:	Cr	**1.001062**	0	0	0

Table 3. The Q-table of an individual after executing 21 iterations

	State/action	C	Re	Ed	Cr
Cur state:	C	−0.99982	0	0	−0.0337516
	Re	0	0	0	0
Next state:	Ed	0	0	0	0
	Cr	−0.99693	0	0	0

$$\alpha(t) = 1 - (0.9 * \frac{iter}{maxCycle}) \tag{7}$$

Where, γ denotes a discount factor within $[0, 1]$, r_{t+1} is the immediate reward for an agent in its current state s_t by executing an action a_t, α is a learning rate falling in $[0, 1]$ to balance the exploration and exploitation. The phenomenon that the value of α is closer to 1 means the corresponding agent has a greater priority to explore unknown states. On the contrary, when the value of α is close to 0, the agent needs to learn more information to exploit its previous state.

It is worth noting that the feedback value of an individual is determined according to the change of its fitness. That is, the new fitness value of the individual should be calculated and compared with the old one if the executing action is not reproduction. Then the corresponding agent can get a reward as following:

$$r = \begin{cases} 1 & \text{if fitness is improved} \\ -1 & \text{otherwise} \end{cases} \tag{8}$$

According to RL, the overall fitness of the individual or agent will be improved remarkable after executing the operation of reproduction (Re). Accordingly, this action will get a positive reward. However, it is noted that, due to the fact that the reproduction operation at the initial phase would easily cause the "premature" of algorithm, this operation should be implemented at the medium-range evolution, such as that the iteration numbers reach to *maxCycle/2*.

3 Experiment Results

3.1 Test Function and Experimental Configure

To evaluate the performance of the proposed algorithm, a suit of benchmark functions are used, including 3 unimodal functions (f1 to f3), 3 multimodal functions (f4 to f6). The mathematical description of test functions is provided in Table 4. We employ three classical SI algorithms, namely PSO [14], BFO [15] and ABC [16], to compare with RBCFO. The related experiment setting is given as below: for the common parameters, the number of fitness evaluation is set to 10000, the population size is set to 20 and the independent run times are set to 30. Other parameters for each algorithm are listed as:

RBCFO: Population size is 4, $\gamma = 0.8$, crossover probability $P_{CR} = 0.9$.
PSO Settings: Acceleration factors $c1 = c2 = 2$, the inertia weight $w = 0.9$.

48	H. Jiang et al.

Table 4. Mathematical formula of the test functions

Function	Formula	Search range
Schwefel	$f_1(x) = \sum\limits_{i=1}^{D} (\sum\limits_{i=1}^{i} x_i)^2$	$x_n \in [-100, 100]$
Rosenbrock	$f_2(x) = \sum\limits_{i=1}^{D} \left[100(x_{i+1} - x_i^2)^2 + (x_i - 1)^2 \right]$	$x_n \in [-2.048, 2.048]$
Quadric-Noise	$f_3(x) = \sum\limits_{i=1}^{D} i x_i^4 + random[0,1]$	$x_n \in [-1.28, 1.28]$
Rastrigin	$f_4(x) = \sum\limits_{i=1}^{D} \left[x_i^2 - 10\cos(2\pi x_i) + 10 \right]$	$x_n \in [-5.12, 5.12]$
Griewank	$f_5(x) = \frac{1}{4000} \sum\limits_{i=1}^{D} x_i^2 - \prod\limits_{i=1}^{D} \cos(\frac{x_i}{\sqrt{i}}) + 1$	$x_n \in [-600, 600]$
Ackley	$f_6(x) = (-20\exp(-0.2\sqrt{\frac{1}{D}\sum\limits_{i=1}^{D} x_i^2}) - \exp(\frac{1}{D}\sum\limits_{i=1}^{D} \cos(2\pi x_i))$ $+ 20 + e$	$x_n \in [-32, 32]$

BFO Settings: Step size C = 0.1, P_{ed} = 0.25, the swimming & tumbling maximum step size N_s = 4, the number of reproduction N_{re} = 5 and number of elimination-dispersal N_{ed} = 2.
ABC settings: Limit = 100.

3.2 Comparison with Other Algorithm in Numerical Value

Table 5 shows the average fitness values and standard deviation values on 30-D benchmark functions from 30 runs. As we can see in Table 5, RBCFO outperforms significantly its compared algorithms, namely PSO [14], BFO [15] and ABC [16], on all the benchmark functions. In addition, RBCFO also gets great improvement on

Table 5. Comparison of the experimental results on four optimization algorithms

Function		BFO	PSO	ABC	RBCFO
Schwefel	mean	1.27e+02	5.67e+01	1.42e+03	**5.79e−03**
	std	2.1e+01	1.3e+01	2.3e+03	8.0e−03
Rosenbrock	mean	9.35e+01	3.55e+01	4.85e+01	**2.77e+01**
	std	8.1e+00	2.3e+01	2.1e+01	1.0e−01
Quadric	mean	1.02e+00	7.36e+00	5.93e−01	**2.11e−03**
	std	2.6e−01	4.9e+00	1.3e−01	2.0e−03
Rastrigin	mean	1.57e+02	2.69e+02	1.44e+01	**3.51e−05**
	std	9.8e+00	3.2e+01	4.7e+00	3.7e−06
Griewank	mean	2.71e−02	1.83e+01	7.32e−01	**1.14e−03**
	std	2.9e−03	3.8e+01	4.4e−01	2.7e−03
Ackley	mean	1.62e+00	1.59e+01	1.44e+00	**4.14e−04**
	std	1.8e−01	1.0e+00	1.1e+00	3.6e−04

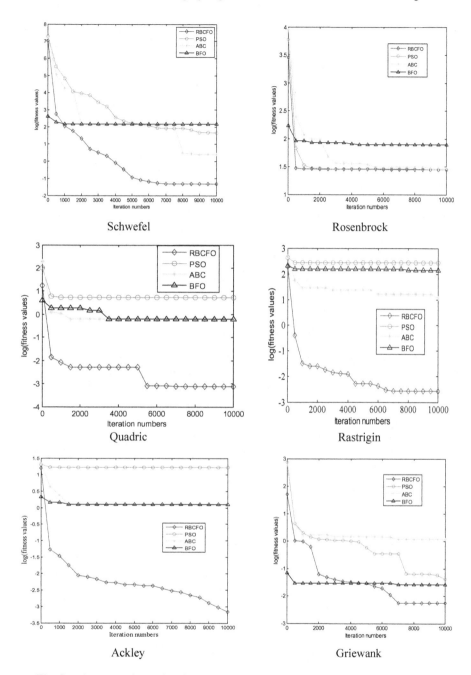

Fig. 3. The comparison of 4 algorithms in terms of convergence rate on 6 functions

Schwefel. Furthermore, RBCFO can obtain nearly seven orders of magnitude better than BFO and PSO, six orders of magnitude better than ABC on Rastrigin function. From these computation results, the performance superiority of RBCFO can be mainly attributed to the RL mechanism and adaptive search strategy that appropriately balance the exploration and exploitation during the search process.

3.3   Comparison with Other Algorithm in Convergence

We compare convergence performance with four different algorithms on 30-dimension functions in 10000 evaluations numbers. It can be observed from Fig. 3 that, RBCFO obtains the fastest convergence profile on all the test functions. The original BFO without RL mechanism performs obviously worse than the enhanced RBCFO, which further verifies the effectiveness and efficiency of the proposed mechanism in RBCFO. In addition, by comparing with other algorithms, the RBCFO also exhibits largely superior performance on all the benchmarks except on Rosenbrock. And only on Ackley, RBCFO does not get satisfactory convergence, which is mainly due to the fact that the adaptive step size makes the convergence results fast in the exploiting stage with the number of iterations approaching *maxCycle*.

3.4   Bacterial Movement Trajectory

In order to further investigate the effect of the proposed mechanism, we make the simulation to show the movement trajectory of populations under the combined mechasmin of RL and adaptive search strategy on 2-D Griewank and 2-D Rosenbrock, as shown in Fig. 4. We set the *maxCycle* to 1000, and other parameter configuration is the same to Sect. 3.1. At initial phase, 4 individuals are initialized distribute randomly in the space region. It can be observed from Fig. 4 that, under the experimental pressure of our proposed mechanism, all bacterial tend to approach the global optimal point along the specific searching path. Thus it also proves the validity of our algorithm in finding the best solution of the optimization problem.

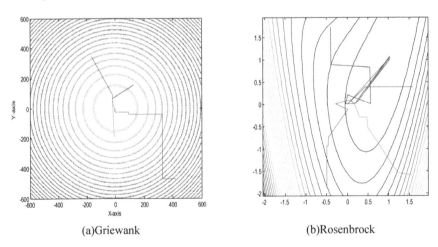

(a)Griewank   (b)Rosenbrock

Fig. 4. The movement trajectory of RBCFO on 2-D Griewank and 2-D Rosenbrock

4 Summary

In order to obtain an appropriate balance between exploration and exploitation of algorithm, this paper proposes a re-structured bacterial colony foraging optimizer (RBCFO) based on reinforcement learning (RL) and self-adaptive search strategy. In the multi-operations decision level, the RL mechanism is exploited to enhance the management efficiency of various foraging operations of each individual during the evolution process. In the single-operation level, the chemotaxis, the self-adaptive strategy and the cell-to-cell crossover are used to maintain dynamical balance between exploration and exploitation. The simulation experiments are conducted on a set of benchmark functions. Computation results in terms of numerical values, convergence rate and diversity show the significant superiority of the proposed RBCFO. We also simulate the movement trajectory of the individual in RBCFO. Simulation results validate the effectiveness of the proposed mechanism.

Acknowledgements. This work is supported by the National Natural Science Foundation of China under Grant No. 61503373 and No. 61572123; Fundamental Research Funds for the Central Universities No. N161705001 and Natural Science Foundation of Liaoning Province under Grand No. 2015020002.

References

1. Hanmandlu, M., Verma, O.P., Kumar, N.K., Kulkarni, M.: A novel optimal fuzzy system for color image enhancement using bacterial foraging. IEEE Trans. Instrum. Meas. **58**(8), 2867–2879 (2009)
2. Verma, O.P., Parihar, A.S.: An optimal fuzzy system for edge detection in color images using bacterial foraging algorithm. IEEE Trans. Fuzzy Syst. **25**(1), 114–127 (2017)
3. Hossain, M.S., Moniruzzaman, M., Muhammad, G., Ghoneim, A., Alamri, A.: Big Data-driven service composition using parallel clustered particle swarm optimization in mobile environment. IEEE Trans. Serv. Comput. **9**(5), 806–817 (2016)
4. Zhang, X., Zhang, X., Yuen, S.Y., Ho, S.L., Fu, W.N.: An improved artificial bee colony algorithm for optimal design of electromagnetic devices. IEEE Trans. Magn. **49**(8), 4811–4816 (2013)
5. Zhao, W., Wang, L.: An effective bacterial foraging optimizer for global optimization. Inf. Sci. **329**, 719–735 (2016)
6. Awadallah, M.A., Venkatesh, B.: Bacterial foraging algorithm guided by particle swarm optimization for parameter identification of photovoltaic modules. Can. J. Electr. Comput. Eng. **39**(2), 150–157 (2016)
7. Abd-Elazim, S.M., Ali, E.S.: A hybrid particle swarm optimization and bacterial foraging for power system stability enhancement. Complexity **21**(2), 245–255 (2015)
8. Samma, H., Lim, C.P., Saleh, J.M.: A new reinforcement learning-based memetic particle swarm optimizer. Appl. Soft Comput. **43**, 276–297 (2016)
9. Chen, H., Zhu, Y., Hu, K., Ma, L.: Bacterial colony foraging algorithm: combining chemotaxis, cell-to-cell communication, and self-adaptive strategy. Inf. Sci. **273**, 73–100 (2014)
10. Ma, L., Hu, K., Zhu, Y., Chen, H.: Computational evolution of social behavior in bacterial colony optimization model. J. pure Appl. Microbiol. **7**, 487–493 (2013)

11. Piperagkas, G.S., Georgoulas, G., Parsopoulos, K.E., Stylios, C.D., Likas, A.C.: Integrating particle swarm optimization with reinforcement learning in noisy problems. In: Proceedings of the 14th Annual Conference on Genetic and Evolutionary Computation. ACM, Philadelphia (2012)

12. Rakshit, P., Konar, A., Bhowmik, P., Goswami, I., Das, S., Jain, L.C., Nagar, A.K.: Realization of an adaptive memetic algorithm using differential evolution and q-learning: a case study in multirobot path planning. IEEE Trans. Syst. Man Cybern. Syst. 43(4), 814–831 (2013)

13. Jiao, L., Gong, M., Wang, S., Hou, B., Zheng, Z., Wu, Q.: Natural and remote sensing image segmentation using memetic computing. IEEE Comput. Intell. Mag. 5(2), 78–91 (2010)

14. Zhan, Z.H., Zhang, J., Li, Y., et al.: Adaptive particle swarm optimization. Prog. Nat. Sci. 39 (6), 1362–1381 (2008)

15. Passino, K.M.: Biomimicry of bacterial foraging for distributed optimization and control. IEEE Control Syst. Mag. 22(3), 52–67 (2002)

16. Horng, S.C.: Combining artificial bee colony with ordinal optimization for stochastic economic lot scheduling problem. IEEE Trans. Syst. Man Cybern. Syst. 45(3), 373–384 (2015)

A Novel Attribute Reduction Approach Based on Improved Attribute Significance

Jun Ye[(✉)] and Lei Wang

School of Information Engineering,
Nanchang Institute of Technology, Nanchang 330099, China
1208561815@qq.com

Abstract. Aiming at the limitations of attribute reduction approach based on Pawlak attribute significance and conditional entropy, an efficient attribute reduction algorithm based on distinguish matrix and the improved attribute significance is put forward. Firstly, the deficiencies of two kinds of classical attribute reduction are analyzed. Furthermore, an improved attribute significance definition is given according to ability to distinguish one object from another in the universe, and then it is used to calculate the significance of attribute in discernibility matrix; Finally, a minimum attribute reduction can be gained by adding attribute to the core attribute set one by one according to descending order of attribute significance. Analysis on numerical example shows that the proposed algorithm can find the minimal attribute reduction effectively. Compared with the previous algorithm, the proposed algorithm can reduce the calculation amount on reduction greatly in the decision table which has more conditional attributes.

Keywords: Attribute significance · Discernibility matrix · Core attribute set
Attribute reduction

1 Introduction

The attribute significance is one of the most important concepts in rough sets theory. It is widely used in attribute reduction (i.e. feature selection) [1–5]. Accurate measurement on attribute significance in the decision table is very important, which affects the efficiency and the implementation of attribute reduction algorithm directly. In rough sets theory, there exists two kinds of classical definition of attribute significance, one is Pawlak-based attribute significance [6] and another is based on conditional entropy [7]. Researchers have got a lot of heuristic attribute reduction algorithms on the basis of these two kinds of attribute significance. References [8–10] proposed an attribute reduction algorithm based on the Pawlak attribute significance. References [11, 12] discussed a reduction algorithm based on conditional entropy. In fact, there are some limitations in attribute reduction algorithm based on these two kinds of attributes significance. On the process of measurement on attribute significance, the changes of classification need to be considered when a certain attribute is deleted or inserted. The prime principle is as follows. The more changes the classification has, the more important the attribute is as an attribute is removed from or added to the condition

© Springer Nature Singapore Pte Ltd. 2018
K. Li et al. (Eds.): ISICA 2017, CCIS 873, pp. 53–66, 2018.
https://doi.org/10.1007/978-981-13-1648-7_5

attribute set and the attribute is of less significance on the contrary. The Pawlak-based attribute significance reflects the variances of classification of objects in the domain, the variation of classification of objects results from the core attribute's removal only. Therefore, the above method can be used to measure core attribute's importance. The non-core attribute's significance is equal to zero according to the above method, so the order cannot be determined, in which non-core attribute add to the core attribute set, and the reduction cannot be completed. Another attribute significance definition, which can reflect the variances of the uncertain objects in the domain, is based on conditional entropy. In a consistent decision table, only the removal of core attribute can cause the change of classification of the object, therefore, the core attribute's significance can be measured, but the significance of non-core attribute is also equal to zero. In the incompatible decision table, the non-core attribute's significance avoid being equal to zero. However, the significance of redundant attribute is sometimes bigger than that of the core attribute. The above deficiency not only contradicts the definition of the core attribute, but also leads to redundant attribute's reservation in the reduction on the process of searching the reductions [13, 14] so it can add space and time costs undoubtedly. On the basis of the previous studies and discernibility matrix, an improved attribute significance definition is presented in this paper, and then a new attribute reduction algorithm is put forward, which provides a new way to find the attribute reduction.

2 Analysis on the Original Attribute Significance

In the decision table, different attribute may have different importance. In order to measure the significance of an attribute (or attribute set), the definitions of Pawlak-based attribute significance and conditional entropy-based attribute significance are given in Definitions 2.1 and 2.2 respectively. When a certain attribute is removed from or added to condition attribute subset, the change of classification should be considered. And if the change is large, the attribute is more important, on the contrary, the attribute is of low significance.

Definition 2.1 (Definition of Pawlak-based attribute significance) [15]. Decision table $DT = (U, C \cup D, V, f)$. U is the universe, C is the condition attribute set and D is decision attribute set. The dependency between C and D is defined as:

$$r(C,D) = \frac{|POS_C(D)|}{|U|} \tag{1}$$

$\forall c \in C$, the attribute significance of C relative to D is (adopting the method of an attribute's removal):

$$Sig(c, C, D) = r(C, D) - r(C - \{c\}, D) \tag{2}$$

$\forall c \in C - A$, $A \subseteq C$, The significance of C relative to D (adopting the method of an attribute's addition)

$$Sig(c, A, D) = r(A \cup \{c\}, D) - r(A, D) \tag{3}$$

$|\cdot|$ indicates the number of elements of set (the same as below).

Definition 2.2 (Definition of conditional entropy-based attribute significance) [15]. Decision table $DT = (U, C \cup D, V, f)$. C is condition attribute set and D is decision attribute set. The conditional entropy for D relative to C is defined as:

$$H(D|C) = -\sum_{i=1}^{n} p(X_i) \sum_{j=1}^{m} p(Y_j|X_i) \log_2 p(Y_j|X_i) \tag{4}$$

$\forall c \in C$. The attribute significance of C relative to D is (adopting the method of an attribute's removal):

$$Sigf(c, C, D) = H(D|C - \{c\}) - H(D|C) \tag{5}$$

$\forall c \in C - A$, $A \subseteq C$. The attribute significance of C relative to D is (adopting the method of an attribute's addition):

$$Sigf(c, A, D) = H(D|A) - H(D|A \cup \{c\}) \tag{6}$$

In rough sets theory, the core idea is to gain the attribute reduction under the premise of keeping the classification ability unchanged.

According to the function of attribute in the classification, all the attributes can be divided into three categories: the necessary (core attribute), the relative necessary attribute, and the non-necessary attribute, the related attributes of three categories can be obtained from Definition 2.3.

Definition 2.3 [16, 17]. Decision table $DT = (U, C \cup D, V, f)$. The set $U/D = \{Y_1, Y_2, ..., Y_n\}$ and the set $U/C = \{X_1, X_2, ..., X_m\}$ are the partitions of universe U by D and C respectively. The positive region of C relative to D is denoted as:

$$POS_C(D) = \bigcup_{Y_i \in U/D} \underline{P}(Y_i), \forall c \in A \subseteq C \tag{7}$$

- If $POS_{C-\{c\}}(D) \neq POS_C(D)$, we call the attribute c is the necessary attribute of C relative to D, else the attribute c is not necessary. The core attribute set is composed of all the necessary attributes, denoted by CORE(C).
- $\subseteq C$, A is a reduction of DT if and only if $POS_A(D) = POS_C(D)$ and $POS_{A-\{c\}}(D) \neq POS_A(D)$. All the reduction sets are denoted as $\cup$ RED(C), and CORE $(C) = \cap$ RED(C), $\cup$ RED(C) $-$ CORE(C) set is called relative necessary set, denoted as R.
- The set which is composed of all non-necessary attributes is denoted as N, i.e. $N = C - \cup$ RED (C).
- From the above Definition 2.3, the followings can be concluded:

- The attribute in the set CORE(C) is the first most important attribute, for any attribute's removal from CORE(C) will affect the classification of objects.
- The attribute in the set R is the second most important attribute, its attribute significance is less than that of the core attribute.
- The attribute significance of attribute in the set N is minimal, it is the third most important attribute.

Namely: $Sig_{core(c)} > Sig_R > Sig_N \geq 0$.

Two examples are given to illustrate the above mentioned definitions of Pawlak-based attribute significance and conditional entropy-based attribute significance, they are employed to measure the attribute significance of attribute in attribute set CORE(C), R and N respectively.

Example 2.1. $DT = (U, C \cup D, V, f)$ is a consistent decision table, as shown in Table 1, where $C = \{c_1, c_2, c_3, c_4\}$ is condition attribute set and $D = \{d\}$ is decision attribute set. According to the method in the Ref. [18], the core attribute set is $\{c_1\}$, and the two attribute reductions are $\{c_1, c_2\}$ and $\{c_1, c_3\}$.

Table 1. A compatible decision table.

U	c_1	c_2	c_3	c_4	D	U	c_1	c_2	c_3	c_4	D
x_1	2	2	0	1	1	x_5	1	0	1	1	0
x_2	1	2	0	1	0	x_6	2	0	1	1	1
x_3	1	1	2	1	1	x_7	2	1	0	0	0
x_4	0	0	0	1	0						

Then necessary attribute set (i.e. core attribute set) can be obtained according to the Definition 2.3: CORE(C) = $\{c_1\}$.

Relative necessary attribute set: $R = \cup$ RED(C) − CORE(C) = $\{c_2, c_3\}$.

Unnecessary attribute set: $N = C - \cup$ RED(C) = $\{c_4\}$.

According to the Definition 2.1, the Pawlak-based attribute significance of each attribute can be calculated:

$$Sig(c_1, C, D) = \frac{|POS_C(D) - POS_{C-\{c1\}}|}{|U|} = \frac{4}{7}$$

In the same way:

$$Sig(c_2, C, D) = Sig(c_3, C, D) = Sig(c_4, C, D) = 0$$

So we can get:

$$Sig(c_1, C, D) > Sig(c_2, C, D) = Sig(c_3, C, D) = Sig(c_4, C, D) = 0$$

The above calculation indicates that the Pawlak metric is only valid to the significance of core attribute c_1 in the compatible decision table for the significance of the non-core attribute (c_2, c_3, c_4) are all equal to zero. c_1 belongs to the necessary attribute set which is composed of the first most important attributes and its attribute significance is equal to 4/7. It just right reflects the nature of the core attribute. c_2, c_3 belongs to the relative necessary attribute set, which is composed of the second most important attributes, but its attribute significance is equal to zero. c_4 belongs to the unnecessary attribute set, its significance is also equal to zero and it reflects the role of the unnecessary attribute basically. However, the difference of these two kinds of attribute in classification do not reflected from the perspective of attribute importance for their attribute significance are all equal to zero. Obviously, the results are in contradiction with the conclusion of $Sig_{core(c)} > Sig_R > Sig_N \geq 0$, which referred from the Definition 2.3. The attribute significance in R and in N are all equal to zero, so which attribute would be added to the core-attribute set firstly cannot be decided while the above method is used to search the relative attribute reduction.

Therefore, the attribute reduction algorithm in the Ref. [9] cannot be realized.

The attribute significance of each attribute based on conditional entropy can be calculated according to the Definition 2.2.

$$Sigf(c_{1,C,D}) = H(D|C - \{c_1\}) - H(D|C)$$
$$= -\frac{2}{7}\left(\frac{1}{2}\log_2\frac{1}{2} + \frac{1}{2}\log_2\frac{1}{2}\right) - \frac{2}{7}\left(\frac{1}{2}\log_2\frac{1}{2} + \frac{1}{2}\log_2\frac{1}{2}\right)$$
$$= \frac{4}{7}$$

In the same way:

$$Sigf(c_2, C, D) = Sigf(c_3, C, D) = Sigf(c_4, C, D) = 0$$

The above results are consistent with the results by Pawlak reduction. For the consistent decision table, the conditional entropy-based attribute significance is the same as the significance of core attribute, the significance of non-core attributes is equal to zero. Therefore, this method is used to find the relative attribute reduction, the order in which the attribute is added to the core attribute set cannot be determined too, leading inefficiency of the corresponding algorithm.

Example 2.2. $DT = (U, C \cup D, V, f)$ is an incompatible decision table, shown in Table 2. Where $C = \{c_1, c_2, c_3\}$ is the condition attribute set and $D = \{d\}$ is the decision attribute set. In the Ref. [14] the core attributes in the Table 2 are c_2 and c_3, and the attribute reduction are $\{c_2, c_3\}$.

The followings can be obtained according to the Definition 2.3.
Necessary attribute set: CORE(C) = $\{c_{2,} c_3\}$.
Relative necessary attribute set: $R = \cup$ RED(C) − CORE(C) = ϕ.
Unnecessary attribute set: $N = C - \cup$ RED(C) = $\{c_1\}$.

Table 2. An incompatible decision table.

U	c_1	c_2	c_3	D	U	c_1	c_2	c_3	D
x_1	1	0	2	2	x_6	1	1	0	2
x_2	1	1	0	2	x_7	0	1	1	1
x_3	1	1	0	0	x_8	0	1	0	1
x_4	1	0	2	2	x_9	0	1	0	0
x_5	1	1	2	0	x_{10}	0	1	1	1

According to the Definition 2.1, the Pawlak-based significance of each attribute can be figured out.

$$Sig(c_1, C, D) = 0, Sig(c_2, C, D) = 0.3, Sig(c_3, C, D) = 0.3.$$

According to the Definition 2.2, the conditional entropy-based attribute significance of each attribute can be figured out:

$$Sigf(c_1, C, D) = 0.09, Sigf(c_2, C, D) = 0.03, Sigf(c_3, C, D) = 0.08.$$

As for the conditional entropy-based attribute significance, it can also measure the attribute significance of core attribute as well as the attribute significance of non-core attribute in the incompatible decision table and the attribute significance of non-core attribute is also equal to zero. By the above calculations, we can see that the attribute significance of c_1 (0.09) is bigger than that of c_2 (0.03) and c_3 (0.08), which is inconsistent with the nature of core attributes. Therefore, the reduction result is $\{c_1, c_2, c_3\}$ which contain the unnecessary attribute c_1. Obviously, the reduction under conditional entropy-based attribute significance contradicts with the general knowledge of rough sets.

3   An Improved Attribute Significance

The above two examples show that the two kinds of attribute significance (Pawlak-based attribute significance and conditional entropy-based attribute significance) cannot reflect the importance of each attribute sufficiently/fully. The Pawlak-based attribute significance emphasizes the qualitative analysis on the attributes, and the conditional entropy-based attribute significance emphasizes the quantitative analysis on the attributes, in order to measure the attributes in the object classification more accurately, we give definition of a new kind of significance of attribute on the basis of discernible matrix.

The discernibility matrix, which is first proposed by Skowron, is used to represent knowledge. The definition of discernible matrix and some related theorems are briefly introduced as follows before the improved attribute significance is discussed.

Definition 3.1. $DT = (U, C \cup D, V, f)$ is a decision information system. Among them, C is the condition attribute set and D is the decision attribute set. The domain is a non-empty finite object set, denoted as $U = \{x_1, x_2, x_3, \ldots, x_n\}$ and $|U| = n$. Then the discernibility matrix of DT decision table is defined as follows:

$$M_{n \times n} = (C_{ij}) = \begin{bmatrix} c_{11} & c_{12} & \cdots & c_{1n} \\ c_{21} & c_{22} & \cdots & c_{2n} \\ \vdots & \vdots & \ddots & \vdots \\ c_{n1} & c_{n2} & \cdots & c_{nn} \end{bmatrix} \tag{8}$$

Among them, $i, j = 1, 2, \ldots, n$, and

$$c_{ij} = \begin{cases} \{a | (a \in C \wedge (f(x_i) \neq f(x_j)\} & f_D(x_i) \neq f_D(x_j) \\ \phi & f_D(x_i) = f_D(x_j) \end{cases} \tag{9}$$

It is obvious that the discernibility matrix is symmetric with respect to the main diagonal of the matrix from the formula (9). The upper or lower triangle matrix is usually used in discernibility matrix for the practical application. The following two theorems can be used to calculate the core attribute set and relative reduction from the discernibility matrix.

Theorem 3.1 [18]. Decision table $DT = (U, C \cup D, V, f)$. Among them, C is the condition attribute set and D is the decision attribute set. The core attribute set in decision table DT is composed of all the individual attribute elements of discernibility matrix, that is:

$$CORE_c D = \{a | (a \in C) \wedge (\exists c_{ij}, ((c_{ij} \in M) \wedge (c_{ij} = \{a\})))\} \tag{10}$$

Theorem 3.2 [18]. In the decision table $DT = (U, C \cup D, V, f)$, $\forall B \subseteq C$, B is a relative reduction of DT if the following two conditions are satisfied.

- $\forall c_{ij} \in M_{nx\ n}$, if $c_{ij} \neq \phi$, then $B \cap c_{ij} \neq \phi$.
- The attribute set B is independent of decision attribute set D.

From the above Definition 3.1, that two objects (x_i, x_j) in the universe have the same values of decision attribute means object x_i and object x_j cannot be distinguished from each other. Object x_i and x_j will fall into the same equivalence class of decision attributes and the differences between them need not to be considered, so $c_{ij} = \phi$. If object x_i can be told from object x_j, then c_{ij} is an attribute set which is composed of all the condition attributes by which object x_i and object x_j can be distinguished from each other. The attribute in c_{ij} may belong to the core attribute set CORE(C), the relative necessary R or the unnecessary attribute set N. However, the attribute has its contribution to tell object x_i from object x_j, regardless of which set it belong to. That is to say all attributes in c_{ij} are important, only the values of attribute significance are not the same. In order to evaluate the significance of attribute in c_{ij} factually, the significance of attribute can be defined by the number of times it appears in all non-empty elements of the discernibility matrix M and the weight it appears per time. For one attribute, the more the number of times and the weight, the more objects it can distinguish, the more contribution it has and the greater significance it has. On the contrary, the smaller significance it has. The definition of improved attribute significance is as follows.

Definition 3.2. $DT = (U, C \cup D, V, f)$ is a consistent decision information system. Among them, C is the condition attribute set, D is the decision attribute set, c_{ij} denotes non-empty elements in discernibility matrix M and K is the total number of all the non-empty element c_{ij} in discernibility matrix M. $\forall c \in C$, the attribute significance of C relative to D can be defined as:

$$Nsig(c) = \frac{\displaystyle\sum_{i,j=1,2\cdots n} r_c \cap c_{ij}}{K} \tag{11}$$

Among them,

$$r_c \cap c_{ij} = \begin{cases} \frac{1}{|c_{ij}|} & c \cap c_{ij} \neq \phi, i, j = 1, 2, \cdots n \\ 0 & c \cap c_{ij} = \phi, i, j = 1, 2, \cdots n \end{cases} \tag{12}$$

The following 3 properties can be gotten from the Definition 3.2.

Property 3.1. $DT = (U, C \cup D, V, f)$ is a consistent decision information system, c_{ij} is a non-empty element of discernibility matrix M. $\forall c \in C$, if $c \in c_{ij}$, then $0 < r_c \cap c_{ij} \leq 1$.

Proof. $\forall c \in C$ and $c \in c_{ij}$, since c_{ij} is a non-empty element of discernibility matrix M, $r_c \cap c_{ij} = \frac{1}{|c_{ij}|} > 0$ can be directly gained by the Definition 3.2. For $c \in c_{ij}$, we have $|c_{ij}| \geq 1$, so $\frac{1}{|c_{ij}|} \leq 1$.

To sum up, $0 < r_c \cap c_{ij} \leq 1$.

Furthermore, we have:

$$r_c \cap c_{ij} < 1 \quad c \in C, c \in c_{ij} \text{ and } c \notin CORE(C) \quad i, j = 1, 2 \cdots n.$$

According to the Property 3.1, while $r_c \cap c_{ij} = 1$, it demonstrates that the distinguishable ability of single core attribute is the strongest in all attributes. This is completely consistent with the properties of core attribute in rough set theory.

Property 3.2. $DT = (U, C \cup D, V, f)$ is a consistent decision information system, c_{ij} is a non-empty element of discernibility matrix M. $\forall c \in C$, if $c \in c_{ij}$, then

$$0 < Nsig(c, C, D) \leq 1 \tag{13}$$

Proof. $\forall c \in C$ and $c \in c_{ij}$, $Nsig(c, C, D) > 0$ can be gotten by Property 3.1.

$\forall c \in C$ and $c \in c_{ij}$, if all the non-empty elements c_{ij} are single attribute in discernibility matrix M. According to theorem 1, all these attributes are core attributes, $|c_{ij}| = 1$ can be obtained by Property 3.1, that is $r_c \cap c_{ij} = \frac{1}{|c_{ij}|} = 1$. So there is:

$$Nsig(c, C, D) = \frac{\sum\limits_{i,j=1,2,\cdots n} r_c \cap c_{ij}}{K} = \frac{1 + 1 + \cdots 1}{K} = \frac{K}{K} = 1.$$

$\forall c \in C$ and $c \in c_{ij}$, If all the non-empty elements c_{ij} are multiple attribute set in discernibility matrix M, there must be $c \notin CORE(C)$ and $c \in R$ or $c \in N$, then $0 < r_c \cap c_{ij} < 1$ (by the Definition 3.2 and Theorem 3.1). So there is:

$$Nsig(c, C, D) = \frac{\sum\limits_{i,j=1,2,\cdots n} r_c \cap c_{ij}}{K} < \frac{1 + 1 + \cdots 1}{K} = \frac{K}{K} = 1.$$

$\forall c \in C$ and $c \in c_{ij}$, in discernibility matrix M, if some non-empty elements c_{ij} are composed of single attribute and some other non-empty elements are composed of multiple attributes, there must exist $c \notin CORE(C)$ and $c \in R$ or $c \in N$. $0 < r_c \cap c_{ij} < 1$ (by the Definition 3.2 and Theorem 3.1). So there is: $0 < \sum\limits_{i,j=1,2\cdots n} r_c \cap c_{ij} < K$.

Therefore, the following formula can be gotten:

$$Nsig(c, C, D) = \frac{\sum\limits_{i,j=1,2,\cdots n} r_c \cap c_{ij}}{K} < \frac{K}{K} = 1.$$

To sum up, $0 < Nsig(b, C, D) \leq 1$ is proved.

In Property 3.2, $\forall c \in c_{ij}$, so long as the attribute c appears in elements c_{ij} of discernibility matrix, its significance is not equal to zero regardless of the core attribute or non-core attribute it belongs to. The improvement overcomes the limitation that the attribute significance of all the non-core attributes is equal to zero.

Property 3.3. Let $DT = (U, C \cup D, V, f)$ is a consistent decision information system, c_{ij} denotes non-empty element of discernibility matrix M. $\forall a, b, c \in C$ and $a, b, c \in c_{ij}$, if $a \in CORE(C)$, $b \in R$, $c \in N$, then

$$Nsig_a > Nsig_b > Nsig_c > 0. \tag{14}$$

Proof. For $a \in CORE(C)$, $b \in R$, $c \in N$, there must be $a \in RED(C)$, $b \in RED(C)$ and $b \notin CORE(C)$, $c \notin CORE(C)$ and $c \notin RED(C)$ by Definition 2.3 and Theorem 3.2. Attribute a appears more times than Attribute b and Attribute b more times than Attribute c in the discernibility matrix, namely:

$$\sum\limits_{i,j=1,2\cdots n} r_a \cap c_{ij} > \sum\limits_{i,j=1,2\cdots n} r_b \cap c_{ij} > \sum\limits_{i,j=1,2\cdots n} r_c \cap c_{ij} > 0.$$

then:

$$\frac{\sum\limits_{i,j=1,2,\cdots n} r_a \cap c_{ij}}{K} > \frac{\sum\limits_{i,j=1,2,\cdots n} r_b \cap c_{ij}}{K} > \frac{\sum\limits_{i,j=1,2,\cdots n} r_c \cap c_{ij}}{K} > 0.$$

So we can get: $Nsig_a > Nsig_b > Nsig_c > 0$.

Property 3.3 is proved.

Property 3.3 shows that the core attribute is the most important attribute, the significance of relative necessary attribute is smaller than that of the core attribute and unnecessary attribute is the least important. This conclusion is consistent with the properties of the Definition 2.3 in the rough sets theory.

4   Reduction Algorithm Based on Improved Attribute Significance and Discernibility Matrix

Three properties from the Definition 3.2 show the definition of attribute significance based on matrix discernibility would more accurately measure the attribute in the role of classification. The improved definition of attribute significance can not only avoid the attribute significance of non-core attribute is equal to zero in Pawlak-based attribute significance, but also avoid the attribute significance of unnecessary attribute is bigger than that of core attribute in conditional entropy-based attribute significance. Next, the reduction algorithm based on the Definition 3.2 can be designed. The idea of corresponding reduction algorithm is to calculate the core attribute set by the discernibility matrix firstly. Secondly, the significance of non-core attributes in the discernibility matrix is needed to be figured out. Thirdly, each non-core attribute need to be added to the core attribute set one by one in the descend order of its attribute significance. Lastly the minimal reduction of the decision table can be obtained.

The proposed attribute reduction algorithm can be described as follows.

Algorithm 1. Attribute Reduction Algorithm Based on Attribute Significance and Discernibility Matrix.

Input: Decision information system $DT = (U, C \cup D, V, f)$.

Output: An attribute reduction set B of decision information system.

Step 1 Let $B = \phi$, CORE(C) $= \phi$.

Step 2 Find all the single attribute element c_{ij} from the discernibility matrix and make CORE(C) = CORE(C) $\cup \{c_i\}$.

Step 3 Let B = CORE(C). If IND (B) is equals to IND (C), then go to step 7.

Step 4 Calculate the attribute significance of $c(\forall c \in (C-B))$ by the discernibility matrix and sort them in the descending order.

Step 5 From the core attribute set B on, the attribute is added to the core attribute set one by one according to the descending order of attribute significance. Let $B = B \cup \{c\}$, if two attribute have the equivalent amount of attribute significance, then take any one.

Step 6 If IND (B) is equal to IND (C), then go to step 7, otherwise go to step 5.

Step 7 Output set B (one of the reduction of C) and the algorithm is over.

In terms of the complexity of the algorithm, the time complexity in calculating the discernibility matrix is equal to $O(|C||U|^2)$. The time complexity of the algorithm 1 is $O((|C| + K)|U|^2)$. Among them, $|C|$ is the number of conditional attributes, $|U|$ is the number of objects in the universe and K denotes the total number of all non-empty elements in distinguish matrix M. The proposed algorithm can greatly reduce the computational workload, and the proposed attribute reduction is the optimal reduction especially in the case of condition attribute set is relatively large.

5   Numerical Example

In order to verify the practicability of the proposed Algorithm 1, the reduction of decision table in the Ref. [17] (as shown in Table 3) is taken as an example. Three methods on attribute reduction are used to find the relative reduction based on attribute significance, and then the results are analyzed.

$DT = (U, C \cup D, V, f)$ is a consistent decision table. Its universe is a non-empty and finite set of object, i.e. $U = \{x_1, x_2, x_3, x_4, x_5, x_6, x_7, x_8\}$.

Table 3. Another compatible decision table.

U	c_1	c_2	c_3	c_4	D	U	c_1	c_2	c_3	c_4	D
x_1	1	1	1	0	1	x_5	2	1	1	2	0
x_2	1	0	1	2	0	x_6	2	2	0	0	1
x_3	0	0	1	1	0	x_7	1	2	0	1	1
x_4	2	2	1	2	1	x_8	0	0	0	0	0

The followings can be gained from the decision Table 3.

$U/D = \{\{x_1, x_4, x_6, x_7\}, \{x_2, x_3, x_5, x_8\}\}$.
$U/C = \{\{x_1\}, \{x_2\}, \{x_3\}, \{x_4\}, \{x_5\}, \{x_6\}, \{x_7\}, \{x_8\}\}$.
$U/(C - \{c_1\}) = \{\{x_1\}, \{x_2\}, \{x_3\}, \{x_4\}, \{x_5\}, \{x_6\}, \{x_7\}, \{x_8\}\}$.
$U/(C - \{c_2\}) = \{\{x_1\}, \{x_2\}, \{x_3\}, \{x_4, x_5\}, \{x_6\}, \{x_7\}, \{x_8\}\}$.
$U/(C - \{c_3\}) = \{\{x_1\}, \{x_2\}, \{x_3\}, \{x_4\}, \{x_5\}, \{x_6\}, \{x_7\}, \{x_8\}\}$.
$U/(C - \{c_4\}) = \{\{x_1\}, \{x_2\}, \{x_3\}, \{x_4\}, \{x_5\}, \{x_6\}, \{x_7\}, \{x_8\}\}$.

According to the Definition 2.1, the Pawlak-based significance of each attribute can be calculated:

$$Sig(c_2, C, D) = \frac{|POS_C(D) - POS_{C-\{c2\}}(D)|}{|U|} = \frac{1}{4}.$$

In the same way:

$$Sig(c_1, C, D) = Sig(c_3, C, D) = Sig(c_4, C, D) = 0.$$

According to the Definition 2.2, the conditional entropy-based significance of each attribute can be calculated:

$$Sigf(c_2) = H(D|C - \{c_1\}) - H(D|C) = 0.09.$$

In the same way:

$$Sigf(c_1, C, D) = Sigf(c_3, C, D) = Sigf(c_4, C, D) = 0.$$

By the above two methods, the core attribute set and the non-core attribute set can all be figured out, they are equal to $\{c_2\}$ and $\{c_1, c_3, c_4\}$ respectively and the significance of non-core attribute is also equal to zero. So the same problem holds: the order in which remained attribute (except for the core-attribute) is joined to the core-attribute set cannot be decided when searching for the attribute reduction, leading algorithm's inefficiency and infeasibility.

The proposed algorithm can be applied to search the attribute reduction by the following procedures.

Step1. According to the Definition 2.1, the discernibility matrix can be gained from decision Table 3, shown in formula (15).

$$M = \begin{bmatrix} \phi & & & & & & & & \\ c_2c_4 & \phi & & & & & & & \\ c_1c_2c_4 & \phi & \phi & & & & & & \\ \phi & c_1c_2 & c_1c_2c_4 & \phi & & & & & \\ c_1c_4 & \phi & \phi & c_2 & \phi & & & & \\ \phi & c_1c_2c_4 & c_1c_2c_4 & \phi & c_2c_4 & \phi & & & \\ \phi & c_2c_3c_4 & c_1c_2c_3 & \phi & c_1c_2c_3c_4 & \phi & \phi & & \\ c_1c_2c_3 & \phi & \phi & c_1c_2c_3c_4 & \phi & c_1c_2c_3 & c_1c_2c_4 & \phi \end{bmatrix} \tag{15}$$

Step2. Find the entire single attribute from discernibility matrix, so the core attribute set CORE(C) = $\{c_2\}$ can be gained, and make $B = \{c_2\}$.

Step3. To calculate the significance of non–core attribute from the discernibility matrix.

According to Definition 3.2, formula (15) shows that the number of all the non-empty elements is $K = 16$ in the discernibility matrix.

Here take the attribute c_1 as an example, it appears 12 times totally in the discernibility matrix. From top to bottom, the first time to appear in the element $\{c_1, c_2, c_4\}$, then there is: $r_c \cap c_{ij} = \frac{1}{|\{c_1,c_2,c_4\}|} = \frac{1}{3}$, the second time to appear in the element $\{c_1, c_2\}$, then $r_c \cap c_{ij} = \frac{1}{|\{c_1,c_2\}|} = \frac{1}{2}$, The third time is in $\{c_1, c_2, c_4\}$, that is: $r_c \cap c_{ij} = \frac{1}{|\{c_1,c_2,c_4\}|} = \frac{1}{3}$, The fourth time is in $\{c_1, c_4\}$, that is: $r_c \cap c_{ij} = \frac{1}{|\{c_1c_4\}|} = \frac{1}{2}$, and so on. Then the significance of the attribute c_1 can be got as follow.

$$Nsig(c_1) = \frac{\sum\limits_{i,j=1,2\cdots n} r_c \cap c_{ij}}{K} = \frac{\frac{1}{3}+\frac{1}{2}+\frac{1}{3}+\frac{1}{2}+\frac{1}{3}+\frac{1}{3}+\frac{1}{3}+\frac{1}{4}+\frac{1}{3}+\frac{1}{4}+\frac{1}{3}+\frac{1}{3}}{16}$$
$$= 0.27.$$

In the same way:

$$Nsig(c_3) = \frac{\frac{1}{3}+\frac{1}{3}+\frac{1}{4}+\frac{1}{3}+\frac{1}{4}+\frac{1}{3}}{16} = 0.11.$$

$$Nsig(c_4) = \frac{\frac{1}{2}+\frac{1}{3}+\frac{1}{3}+\frac{1}{2}+\frac{1}{3}+\frac{1}{3}+\frac{1}{2}+\frac{1}{3}+\frac{1}{4}+\frac{1}{4}+\frac{1}{3}}{16} = 0.25.$$

The results sorted in descending order are as follows:

$$Nsig(c_1) > Nsig(c_4) > Nsig(c_3).$$

Step4. From the core attribute set B on, the most important attribute c_1 is added to the B, then $B = \{c_2\} \cup \{c_1\} = \{c_1, c_2\}$.

Step5. Since IND (B) = IND (C), $B = \{c_1, c_2\}$ is the minimum reduction of decision Table 3.

Repeat step 4 until IND (B) = IND (C), and the reduction set $B = \{c_1, c_2\}$ is the only reduction set from decision Table 3, the reduction is consistent with the Ref. [12]. In addition, the significance of the core attribute c_2 ($Nsig(c_2)$) is equal to 0.38 can be calculated by the Definition 3.2. The following can be obtained compared with the significance of the non-core attribute: $Nsig(c_2) > Nsig(c_1) > Nsig(c4) > Nsig(c3) > 0$. The results are in accordance with the conclusions obtained from the Property 3.3, it is also consistent with the conclusion of the Definition 2.3. Using the proposed algorithm to calculate the decision table of Table 1, the results are in agreement with the results in the Ref. [17]. It proves the validity of the proposed algorithm.

6 Conclusions

A new discernibility matrix-based attribute significance was presented in this paper firstly. Then the improved attribute significance can not only measure the significance of core attribute, but also the significance of non-core attributes. It is reasonable to evaluate the function of conditional attribute on distinguishing object by the improved attribute significance. Furthermore, an improved attribute reduction algorithm was proposed on the basis of the above discussed, it can find the minimum reduction which is in consistent with the decision table. The proposed algorithm can greatly reduce the computational workload, and it is the optimal reduction algorithm, especially in the case of condition attribute set is relatively large. In the future, reliability of the proposed algorithm needs to be studied further when it applied to the incompatible decision table.

Acknowledgements. The research is supported by the Natural Science Foundation of China under grant Nos. 61461032, 61363047, 61562061 and 51669014. The Natural Science Foundation of Jiangxi Province of China under grant Nos. 20151BAB207067 and 20151BAB207032.

References

1. Dai, J., Han, H., Hu, Q., et al.: Discrete particle swarm optimization approach for cost sensitive attribute reduction. Knowl. Based Syst. **102**(6), 116–126 (2016)
2. Teng, S., Lu, M., Yang, A., et al.: Efficient attribute reduction from the viewpoint of discernibility. Inf. Sci. **326**(1), 297–314 (2016)
3. Qian, Y., Liang, J., Pedrycz, W., et al.: Positive approximation: an accelerator for attribute reduction in rough set theory. Artif. Intell. **174**(9–10), 597–618 (2010)
4. Yao, Y., Zhao, Y.: Attribute reduction in decision-theoretic rough set models. Inf. Sci. **178** (17), 3356–3373 (2013)
5. Bing, X., Liam, C., Lin, S., et al.: Binary PSO and rough set theory for feature selection: a multi-objective filter based approach. Int. J. Comput. Intell. Appl. **13**(2), 1157–1948 (2014)
6. Pawlak, Z.: Rough set theory and its applications to data analysis. J. Cybern. **29**(7), 661–688 (1998)
7. Yang, C., Ge, H., Wang, Z.: Overview of attribute reduction on rough set. Appl. Res. Comput. **29**(1), 16–20 (2012)
8. Qian, W., Yang, B., Xie, Y., et al.: A quick algorithm for attribute reduction based on attribute measure. J. Chin. Comput. Syst. **35**(6), 1407–1411 (2014)
9. Zhang, T., Xiao, J., Wang, X.: Algorithms of attribute relative reduction in rough set theory. Acta Electronica Sinica **33**(11), 2080–2083 (2005)
10. Liao, Q., Long, P.: Rough set reduction method of attribute based on importance of attribute. Comput. Eng. Appl. **49**(15), 130–132 (2013)
11. Yang, M.: Approximate reduction based on conditional information entropy in decision tables. Acta Electronica Sinica **35**(11), 2156–2160 (2007)
12. Cao, F., Liang, J., Qian, Y.: Decision table reduction based on information entropy. Comput. Appl. **25**(11), 2630–2631 (2005)
13. Liu, Q., Li, F., Min, F., et al.: An efficient knowledge reduction algorithm based on new conditional information entropy. Control Decis. **20**(8), 878–882 (2005)
14. Li, M., Huang, W., Liu, Z.: Improved CEBARKNC on decision table reduction. J. Comput. Appl. **26**(4), 864–866 (2006)
15. Zhang, W.: Rough Set Theory and Methods, 1st edn. Science Press, Beijing (2001)
16. Liu, Q.: Rough Set and Rough Reasoning, 1st edn. Science Press, Beijing (2001)
17. Ye, J., Wang, L.: An approach of ascertaining the combinatorial attribute weight based on distinguish matrix. J. Comput. Sci. **41**(11), 273–277 (2014)
18. Miao, D., Li, D.: Rough Sets Theory Algorithms and Application, 1st edn. Tsinghua University Press, Beijing (2008)

Neural Networks and Statistical Learning – Transfer of knowledge

Traffic Condition Assessment Based on Support Vectors Machine Using Intelligent Transportation System Data

Deng Lei[1](✉) and Weihua Zhong[2]

[1] College of Information Engineering,
Jiangxi College of Applied Technology, Ganzhou 341000, Jiangxi, China
qjdenglei@163.com
[2] College of Automobile,
Jiangxi College of Applied Technology, Ganzhou 341000, Jiangxi, China
zhongwh168@163.com

Abstract. It is the aim of this research to adopt the data dig technology in the respect of prediction of traffic circumstance and condition by making use of ITS information. There were study results as examples by means of the tool of inductive loop detectors, which were collected to be used as study materials, including 1-37, I-10 and 1-410 in San Antonio. It's usual to use Support Vectors Machine that is one kind of new data dig technology to absorb different knowledge existed in concealment pattern, potential relations and tendencies between variable quantities. Then one compared analysis was made between the prediction results and another two variables, namely BP neural system and Response Surface Methodology, or say RSM which is known as a traditional means in statistics. According to the result, the SVM model is superior to BP neural system and RSM model as to the aspects of MAPE and the RMSE.

Keywords: Traffic condition assessment · SVM · MAPE · RMSE

1 Introduction

Recently, the predictions of traffic circumstance and congestion are gradually emphasized in Intelligent Transportation. The conclusion is that traffic condition prediction is more important in the several systems that are real-time intelligent traffic system existed in traffic signal control, traffic distribution, route and automatism navigation. The method of prediction of traffic volume on the premise of event detection is of high significance. The datas in terms of volume, speed and density are collected by intelligent traffic detectors Intelligent throughout different directions and areas. The three traffic variables are used in many intelligent transportation conditions and there is no exception of this study.

During the past several years, there were a lot of experts and scholars who have been focused on the prediction of traffic volume for short time, and so they established different kinds of models concerned with the prediction of traffic flow. Some researchers discovered that there are five prediction means for short-term which are often used, including mean value method, Auto Regression-Moving Average, neural network

© Springer Nature Singapore Pte Ltd. 2018
K. Li et al. (Eds.): ISICA 2017, CCIS 873, pp. 69–83, 2018.
https://doi.org/10.1007/978-981-13-1648-7_6

system, linear and nonparametric regression. But when these models are used to predict the traffic circumstance, the accuracy and reliability of its results are not pleasing. In the process of traffic condition prediction, since the real-time processing is necessary, the complexity and accuracy of predictions will be affected and some unexpected noise is also to be produced if there are too many variables.

In this paper, we will propose a new data mining algorithms based on SVM to forecast traffic circumstance and congestion condition through the data obtained by San Antonio ITS. By this unnormal method, the "knowledge" existed in potential patterns, tendencies and relationships will be grabbed. Then, we will realize the quality of existing forecast system based on the accuracy level of prediction and calculation efficiency. We compare the new method with neural network system pattern and the other means associated with traditional statistics. Finally, some concrete examples by using this model in practical life and the future develop prospects are recommended.

2 Support Vector Machines

SVM is the short name of support vector machine that is about one method used to monitor study and statistical study theory is its basis that can be used for data classification, regression and pattern recognition. SVMs can be applied to solve many complicated problems that cannot be solved by classical programming techniques due to the absence of a mathematical model (i.e. hand writing character recognition, speech recognition, data mining and knowledge discovery, image classification and several biomedical applications). The basic idea behind SVMs is to find the best hyperplane(s) that can separate n-dimensional data into a number (two in case of binary SVMs) of categories or classes. To make it clear, this section is dedicated to define some technical terms and explain the mathematical background of support vector machines.

2.1 Supervised and Unsupervised Learning

Machine learning is classified as supervised or unsupervised learning. Supervised learning is the type of learning in which training data are available and labeled with the correct result (i.e. the input data and the corresponding output is known). In other words, supervised learning aims to classify objects to one of a pre-specified set of categories or classes. Generally, supervised learning consists of two steps: training step, to learn classifier from training data, and testing step (known as generalization) to enable unseen objects to be identified as belonging to one of the classes. Examples of supervised study strategy are made up of three things that are about making decisions, neural network system and support vector machines. However, this is not always the case, and there is another type of learning in which the training data with pre-defined labels are not available. This is known as unsupervised learning or clustering. In this case, the program search for the similarity between samples of data in order to decide which objects should be grouped together without any prior information. This technique can be used in image segmentation, and speech coding. As a supervised learning technique, SVM is one of the powerful techniques that can be used in classification and pattern recognition. In SVMs, linear and nonlinear classes are the two varieties. For the linear class, there are two situations. First, under the circumstance of binary SVM, the linear class can just divide the linear datas into different classes that above one hyperplane. Second, if SVM is multi-classes, hyperplanes are formed. While for a nonlinear SVM, it has the advantage of dividing more complicated datas in their structure over linear class of SVM. In our case, we use non-linear SVM.

2.2   Nonlinear Support Vector Machines

Linear support vector machines cannot handle many of complicated classification tasks due to computational power limitation. For binary class data that are only nonlinearly separable, a nonlinear binary SVM is required. It first maps, via a function ϕ, the data points in the input space X to a higher (can be infinite) dimensional space so that the mapped data in the new space H (called the feature space) become linearly separable. Figure 1 shows how the input data is mapped to the feature space. The feature space is a vector space where the dot product is applicable. The binary SVM's decision function is $X \to H$. For example, $x_j \to \phi(x_j)$

$$x_j \bullet x \to \phi(x_j)\phi(x) \tag{1}$$

And the decision function becomes

$$D(x) = sign\left[\sum_{j=1}^{l} \alpha_j y_j (\phi(x_j)\phi(x)) + b\right] \tag{2}$$

It is quite clear that the input data appear in the decision function (2) in the form of inner product $x_j \bullet x$ and as the inner product of $\phi(x_j)\phi(x)$. Thus, rather than explicitly mapping the input data into a higher dimension a space and performing a linear SVM classification, we can operate in the input space using the so called Kernel function $K(x_j, x)$ which implicitly represents the inner product of the mapped data in the feature space.

$$K(x_j, x) = \phi(x_j)\phi(x) \tag{3}$$

Mathematically, the valid Kernel must satisfy the Mercer's theorem: for any $g(x)$, $x \in c$ such that,

$$\int_c g^2(x)dx < \infty \tag{4}$$

there must be the case where,

$$\int\int_c K(x_j, x)g(x)g(x')dxdx' \geq 0 \tag{5}$$

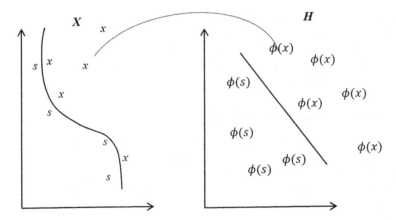

Fig. 1. Mapping input data into a higher dimension feature space

Thus, the Kernel should be a positive semi-definite. In non-linear SVMs we need first to choose a suitable kernel function, and then maximize the function
Maximize α,

$$w(\alpha) = \sum_{i=1}^{l} \alpha_i - \frac{1}{2} \sum_{j,i=1}^{l} \alpha_i \alpha_j y_i y_j K(x_j, x) \tag{6}$$

subject to, $0 \le \alpha_i \le c, i = 1, 2, \ldots, l$

$$\sum_{i=1}^{l} \alpha_i y_i = 0 \tag{7}$$

And substitute the optimum in the decision function (1.27)

$$D(x) = sign\left[\sum_{j=1}^{l} \alpha_j y_j K(x_j, x) + b\right] \tag{8}$$

2.3   Kernel Functions

In the field of literature, it is usual to possess a lot of kernel functions. The commonly used ones include the kernels of linear, polynomial, exponential, and radial basis function or say RBF for short and so on.

(1) *Linear Kernel*: is the simplest Kernel function that manages by the inner product as shown in the followings.

$$K(x, x') = x^T x' + k \qquad (9)$$

where the constant, k is optional.

(2) *Polynomial Kernel*: is a non-stationary Kernel and data should be normalized when using this Kernel. A polynomial of degree, d can be written as:

$$K(x, x') = \left(x^T x' + k\right)^d \qquad (10)$$

where, k is a constant. Under the conditions of the value of 'd' is '1', the linear kernel appears.

(3) Gaussian (radial basis function) kernel:

$$K(x, x') = \exp\left(-\frac{||x - x'||^2}{\sigma^2}\right) \qquad (11)$$

where σ is a design parameter.

(4) *Hyperbolic tangent (Sigmoid) Kernel:* the sigmoid function is the activation function which used in neural networks.

$$K(x, x') = \tanh\left(ax^T x' + k\right) \qquad (12)$$

where, a is the slop, and k is a constant.

(5) *Exponential Kernel*: is similar to a radial basis Kernel.

$$K(x, x') = \exp\left(-\frac{||x - x'||}{\sigma^2}\right) \qquad (13)$$

In terms of any implementation, the key stage is to learn to choose the most suitable kernel function. However, at present, we need to explore a general method to find it. Since the kernel function is closely related to the application of procedures, it is necessary to make efforts to attempt when we prepare to select the former. And the results of the SVM is closely associated with its design variables, let alone that of the kernel.

3   Data Selection Design

In this section, the procedure of modeling will be given. On the whole, the layers of prediction, SVM Data training and Data selection make up the three layers of the process of data processing. Based on differ aims, the first layer is dedicated to collect the suitable data from ITS Data Warehouse. The databases of history, road network and time serial are all used to store the processed datas. Then these datas will be trained by SVM from the three databases one by one. By doing so, with each step of machine study, SVM absorbs the potential knowledge. Finally, by making use of processed input traffic datas, SVM will have the ability to computer and output the result of prediction, traffic situation and congestion automatically.

3.1   Historical Database Selection

The trend of development of the diversity of speed and its distribution is similar to that of previous quality and values. And also, the distribution of them according to week exists. It will helpful to the average speed and the prediction of level of service by using SVM, namely, support vector machine to research this relational pattern. There are some relational expressions about the choice of data in the following and among them, 'DOW means the 'Day of Week'. In the content of the text book, different symbols represent different meanings. For example, 'n' means 'Projection', 'cr' refers to 'Selection', and 'x' points to 'Join':

$$Q_1 = \pi_{speed}(\sigma_{DOW='n'}(DOW \bowtie date)) \tag{14}$$

$$Q_2 = \pi_{vol}(\sigma_{DOW='n'}(DOW \bowtie date)) \tag{15}$$

$$Q_3 = \pi_{Occ}(\sigma_{DOW='n'}(DOW \bowtie date)) \tag{16}$$

When the process of data selection is finished, the database of history will play an important part in which the data is to be choiced and organized. We can present the data structure in the following column (Table 1):

Table 1. Historical database structure

	1st DOW Input Column	2nd DOW Input Column	......	Dth DOW Input Column	Location	Time
Speed	S(i)	S(i + 1)	......	S(i + w)	l(i)	t(i)
Volume	V(i)	V(i + 1)	......	V(i + w)	l(i)	t(i)
Occupancy	O(i)	O(i + 1)	......	O(i + w)	l(i)	t(i)

From this above table, we can know the train process of every point during one day of a week. In this process, (i) and (i + w) shows the meaning of the ith and (i + w)th illustrates the significance of the value of the week in the interval of 5 min, and the symbol of 'w' expounds the length of the input size. In terms of each of factor, l(i) and t(i) will be the considered as the symbols of time and location respectively.

3.2   Road Network Database Selection

As for the concrete traits of vehicle parameters and the distribution in the area of mathe, they have been researched by many researchers. After their study, they discover that whether the speed diversity or its distribution, they always share one kind of relationship and tendency that is similar to that of upstream and downstream traffic circumstance. It is significant to make use of SVM, namely support vector machine, to study the relational pattern and know the average speed and the prediction of traffic congestion. There is the relational expression of data choice as follows:

$$Q_{1.1} = \pi_{speed1}(\sigma_{roadname='r'}(roadname \rhd\lhd location(l)) \tag{17}$$

$$Q_{2.1} = \pi_{vol1}(\sigma_{roadname='r'}(roadname \rhd\lhd location(l)) \tag{18}$$

$$Q_{3.1} = \pi_{Occ1}(\sigma_{roadname='r'}(roadname \rhd\lhd location(l)) \tag{19}$$

The data structure is similar to that Historical Database.

3.3   Time Serial Database Selection

The diversity of speed is related to the distribution of it and the trend of traffic flow and the change of speed of 5 min share the similarities with those of 10 min. Since the obtainment of the evaluation of traffic circumstance and the prediction of congestion is the core point of this study, the interval of 5 min is right. So it is advocated to adopt the method of aggregation. It is significant to make use of SVM, namely support vector machine, to study the relational pattern and know the average speed and the prediction of traffic congestion. There is the relational expression of data choice as follows:

$$Q_{1.1} = \pi_{speed1}(\sigma_{time='t'}(time(t) \rhd\lhd location) \tag{20}$$

$$Q_{2.1} = \pi_{vol1}(\sigma_{time='t'}(time(t) \rhd\lhd location) \tag{21}$$

$$Q_{3.1} = \pi_{Occ1}(\sigma_{time='t'}(time(t) \rhd\lhd location) \tag{22}$$

The data structure is also similar to that Historical Database.

4   SVM Modeling and Prediction

4.1   SVM Modeling Procedure

In general, statistical theory is primarily a machine learning that Vapnik regards as a unique theory that studies the required sample specifications. For SVM, it is mainly based on statistical theory, and then gradually grow from the new universal learning. In the current process of machine learning, support vector machine (SVM) in which there has been a significant increase, mainly due to its access to the confidence interval and the best combination of training errors. In order to effectively improve the regulatory capacity, then the structural risk must be minimized. In addition, vector machines are also needed to effectively handle the problem of local minima, such as the required sample size, nonlinearity and high-dimensional data. Support vector machines that can be used during pattern recognition, require vector machines to be able to process signals and to study and predict time series. Through the above research we can find that there are many advantages of vector machine technology, its structure is relatively simple, in this process, the statistical theory also laid the foundation for its application. This is mainly a part of LibSVM data training, and can predict the accuracy of the calculation. There are five main aspects: First, scaling, the second is cross-validation, the third is to conduct a grid search, the fourth is a coaching model, the fifth is the calculation of the higher accuracy. Its main run in python, the main use of the Python library. The following is a detailed introduction of each function.

4.2 Cross Validation

The following is a simple program in the LibSVM cross-validation program:

1. The training data after the conversion is selected in random form.
2. Select the initial impact factor to begin training, then select other data sets for accuracy testing.
3. Continue to correct the impact factor until the highest accuracy.

With some good influence, it can be automatically stored in the train model. Under normal circumstances LibSVM mainly use the last train model to perform the final prediction (mainly in the unknown test data).

4.3 Index of Evaluating Traffic Condition—Level of Service (LOS)

The transport service level (LOS) system refers to the assessment criteria for traffic conditions and service levels at this stage. Even though most drivers only have a certain interest in the speed of journeys, the level of service is mostly assessed by other factors as well. For road handbooks and the AASHTO Geometrical Design Road,

Table 2. Levels of service

LOS	Traffic condition
A (best)	Free flow
B	Reasonably free flow
C	Stable flow
D	Approaching unstable flow
E	Unstable flow
F (worst)	Forced or breakdown flow

which mainly uses the letters A to F, A is the best condition, F is the worst, as can be seen from the table below:

As can be seen in the Handbook on Road Capacity, the quality of service for expressways is based primarily on the density of vehicles and is mainly expressed in pcpmpl. Used to make specific LOS assessment of the highway [14].

For SAN Antonio ITS data, the main vehicle for various sections of the average velocity (u) and flow (q) value, so, for it shall be selected LOS method based on density (k) = flow (q), that is, every five minutes of each direction of LOS. Based on the city performance evaluation of SAN Antonio in 2015, it mainly defined the free flow velocity of downtown SAN Antonio and expressway at 60 miles per hour. Through Table 2 FHWA service level, the data of the SAN Antonio ITS should be calculated using the following Table 3.

Table 3. Levels of service (LOS)

LOS	Density (vphpm)
A (best)	0–10
B	11–15
C	16–24
D	25–32
E	33–46
F (worst)	47—

4.4 Traffic Condition Forecasting and Congestion Prediction

ITS is the short name of Intelligent Transportation System which can be looked as one kind of transportation management system which is real-time, accurate, and has high efficiency. As to the concrete technologies applied in this system, they contain the technologies of advanced information, electronic control and integration. Besides, with the development of road, vehicles, drivers and passengers, ITS adopts modern advanced science and technology, so it is similar to the intelligent system. So far, the instrument of inductive loop detector has been popular throughout the world. In addition, the equipment of sensor, one of the inseparable part in ITS, is often used to test the speed, volume and occupancy of vehicles when they along one sole way.

According to dashes, three elements which include the symbols of "L", "EN" and "EX" make up the detector location and they are related to the position and illustration of detectors. And concretely speaking, main lane, entrance ramp and exit lane are their meanings respectively. In addition, the number stands for the lane number begins with the lane that is next to the middle line, and the number of expressway and its direction, the point of miles are also indicated. There is one instance. Although the location of the line 10 of the Fig. 2 is 'EN1-0035S-153.017', the loop detector will think and indicate the current speed is '51' while the size and the rate of occupancy are both '0'. Due to indication from such equation, namely Density (k) = flow(q) I speed (U), the final result and value will be '0' if the speed of traffic doesn't corresponds to the count, and if this kind of situation appears, it will also bring huge risks to the prediction of traffic condition by the software.

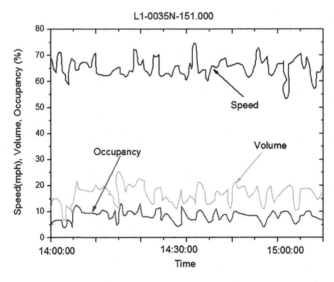

Fig. 2. The relationship of speed, volume and occupancy at uncongested condition

In the following graphs, the separate relationships between four variables are illustrated include time, speed, volume and the rate of occupancy. The graphs put their emphasises on the three traffic conditions which include common situations, the situations where traffic is busy and normal congestion. These graphs provide the materials to discover that the three variables, namely the speed, volume and occupancy have the same trend if the traffic condition is uncongested and when the condition is the opposite, the trend of variable of speed is different from the other two variables which are volume and occupancy and they approach to "1".

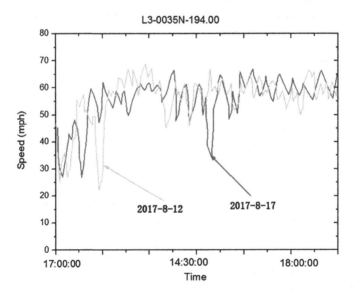

Fig. 3. Speed distribution at different times

The three variables, namely the speed, volume and occupancy have the same trend if the traffic condition is uncongested and it is very appropriate at the place point of dotted line. The develop trends of the diversity and distribution of speed are the same and the conclusion is quite suitable when they have weekly distribution. In the following Fig. 3, the concrete relations of the above mode are illustrated:

It can be seen from the above Fig. 3 that different from one week and two weeks ago, 'L3-0035N-194.00' is the direction and place of the distribution of speed. According to FHWA (http://www.transguide.dot 2025), when the speed is below 45 mph on the expressways, the traffic condition will be considered as the state of congestion. From the Fig. 3, we can obviously see that during the period of 17:00–17:12 pm every day, the traffic condition of the place of section I is usually congested and the speeds are all reduced to below 40 mph. Then when the time goes after 17:12, the phenomenon of traffic congestion doesn't appear during these three days at the location of section II, so the speeds are accumulating to 70 mph and 50 mph. And the traits and distribution in the aspect of mathe of vehicles have been researched by two researchers: A. Galstyan and K. Lerman (2001). After the study made by them, it comes to the conclusion that the diversity and distribution of speed share a very close trend with two traffic circumstances: upstream and downstream. There is no exception of The San Antonio ITS data in terms of this trait; So the following Fig. 4 is listed in order to explain the relationship of the above mode:

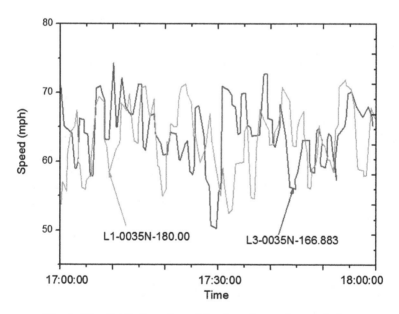

Fig. 4. The distribution of speed by observing road network data

It is transparent that the total five sections, namely the sections I, II, III, IV and V possess the particular same relation of pattern between the different traffic conditions of upstream and downstream. There is no doubt that it's will be useful to predict the speed on average and traffic congestion by taking advantage of support vector machine, or say SVM for short, to research the relationship exists in this mode and pattern.

4.5   Prediction Result

When we decide to input all the datas into the model of the Learn and Prediction, we'd better take advantage of the mode of transformation to train and test the SVM machine firstly in order to study the prediction pattern. When the datas are all transformed, the information contained by these datas is explained by the following Table 4 where the numbers from 1 to 6 refer to respectively the Level of Servers (LOS) 'A', 'B', ' C, 'D', 'E' and 'F'.

Table 4. Transformation data table

LOS	Day of the week	Time 5 min Int	N-2 LOS	N-1 LOS
1	1:03	2:00	3:01	4:01
1	1:03	2:01	3:01	4:01
1	1:03	2:02	3:01	4:01
1	1:03	2:03	3:01	4:01
1	1:03	2:04	3:01	4:01
1	1:03	2:05	3:01	4:01
2	1:03	2:06	3:01	4:02
2	1:03	2:07	3:01	4:02
3	1:03	2:08	3:02	4:02
1	1:03	2:09	3:01	4:01

When we input the current data into the SVM machine, the above table is made by us. The different meanings of the above two columns are expressed and they refer to the latest level of service and someday of one week. For example, for the meaning of the symbol '1:03', the first column means 'Wednesday' while the second column means number '1'. If the same as the third, forth and fifth columns. The symbol of 'Time which is separated by 5 min' stands for the horizontal line's data, and therefore, '2:03' shows us the third time interval; Finally, as for the level of service in the forth and fifth columns, 'N-2 LOS' refers to ten minutes which is same to two time intervals before and 'N-1 LOS' represents five minute that is one time interval before. From this, it can be seen the symbol of '3:01' points to 10 min ago and the T (A) direction and place. while the meaning of another symbol of '4:02' means the same, namely ten minutes ago but '2' (B) is its location; the other two symbols include '3' and '4' refer to the corresponding numbers of their columns.

At 5:00 pm, it is the peak hour of the traffic, we make such one following Table 5 to compare the data we observed and the other data that we predicted.

And Fig. 5 shows the results of prediction that is different from observation value.

Table 5. LOS prediction result table

Time	L1-0035S-166.833		L20035N-188.000		L3-0281N-147.300		L2-0410N-026.076	
	Observed	Predicted	Observed	Predicted	Observed	Predicted	Observed	Predicted
1	A	A	A	A	A	A	A	A
2	A	A	A	A	A	A	A	A
3	A	A	A	A	A	A	A	A
4	A	A	C	C	A	A	B	B
5	C	B	C	C	A	A	B	B
6	C	C	C	C	A	A	B	B
7	C	C	C	C	B	B	D	E
8	C	C	C	C	B	B	C	E
9	C	C	C	C	D	C	C	E
10	C	C	C	C	D	D	D	E
11	C	C	C	C	D	D	D	E
12	C	C	C	C	D	D	E	D

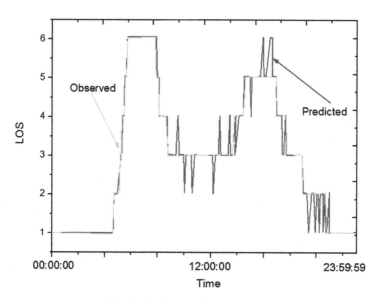

Fig. 5. Traffic condition prediction

5   Conclusion

As one kind of new machine study method, SVM is tested in this research as materials to forecast the situation of real-time traffic circumstance. Besides, the comparison between BP neural network, traditional means concerned with statistics, namely RSM and SVM is also made. One related calculation method is used on the basis of LibSVM package, which provides one example of application of SVM and the parameter and deviation can be adjusted by using the software python 2.6. In addition, in this research, as for the applications of BP neural network and RSM in practical life, they are also mentioned.

It's seen that the traffic datas from five real expressways make their polymerization activity from different directions and places on 1–37, and 1–40 were five separate databases which were about history, road network and time serial. It can be useful for the datas because they can be applied to enhance the development of LibSVM, BP neural network system and at the same time to evaluate the prediction effect. In this process, there will be three ways to make such evaluation, and they are T-test, MAPE and RMSE and through each set of data, the evaluation results will be demonstrated. By the analysis of results, SVM is superior to RSM and BP system through the data in the aspect of values presented in MAPE and RMSE.

Concretely speaking, by using MAPE and RMSE methods, SVM outperforms BP neural network 1.12 and 2.55 on average. As for the set of personal information, the largest improvement of MAPE and RMSE is different. The former is 1.52 (L2-0035N-188.000) and the latter is 4.23 (L2-0410W-026.076).

Generally speaking, the SVM model is better than RSM model in the respect of prediction whether in the level of MAPE or RMSE. Under the way of MAPE, it is 5.43 on average and if the means is RMSE, it is 23.08 on average. While the largest improvement of MAPE and RMSE is both obviously different for personal location and it is 6.64 (L2-0035N-188.000) and 32.52 (L2-0035N-188.000) respectively.

There have also been some other means to be adopted and also make improvement and according to the introduction and comparison, the improvement achieved by the LibSVM is more important than that by B. Park (1998) and Yin, H. (2002). Therefore, the method of SVM has obvious advantages both in the level of theory and practice from the analysis and so this means is expected to become the alternative tool in the prediction of traffic congestion and real-time expressway traffic circunstance. Besides, there is no need for SVM to have more runs, which can cost less model exercising time. So it is benefit when the size of training data is large. With the development of computers, its calculation abilities have also been improved, so the real-time application of SVM, neural network system and RSM models can be realized on individual computers. But in this research, when SVM runs during the single time, the speed of it is slow; the absorbed knowledge is to be collected that exists the database. It would be the main point to study the method of storing the knowledge of former researches and make the period of prediction shorter than before.

References

1. Wang, Y.: Research on the traffic flow data mining in road network. Fudan University, December 2007
2. Cui, Z.: Model and analysis of automatic incident detection based on support vector machine. Beijing Jiaotong University, June 2008
3. Li, B.: On the recursive estimation of vehicular speed using data from a single inductance loop detector: a Bayesian approach. Transp. Res. Part B **43**, 391–402 (2009)

4. Nguyen, M.N., Shi, D., Quek, C., Ng, G.S.: Traffic prediction using Ying-Yang fuzzy cerebellar model articulation controller. In: ICPR 2006 (2006)
5. Xia, H., Bi, J., Li, Y.: Support vector machine approach for retained introns prediction using sequence features. In: Wang, J., Yi, Z., Zurada, J.M., Lu, B.-L., Yin, H. (eds.) ISNN 2006. LNCS, vol. 3973, pp. 654–659. Springer, Heidelberg (2006). https://doi.org/10.1007/11760191_96
6. Vert, J.-P.: Support vector machine prediction of signal peptide cleavage site using a new class of kernels for strings. Pac. Symp. Biocomput. **7**, 649–660 (2002)
7. Smith, B.L., Williams, B.M., Oswalsd, R.K.: Comparison of parametric and nonparametric models for traffic flow forecasting. Transp. Res. Part C **10**(4), 303–321 (2002)
8. Agrawal, R., Kiernan, J.: Watermarking relational databases. In: Proceedings of the 28th VLDB Conference, Hong Kong, China (2002)
9. Sion, R., Atallah, M., Prabhakar, S.: Watermarking relational databases. CERIAS TR 2002-28 (2002)
10. Sion, R., Atallah, M., Prabhakar, S.: On watermarking numeric sets. In: Kim, H.J. (ed.) IWDW 2002. LNCS, vol. 2613, pp. 130–146. Springer, Heidelberg (2003). https://doi.org/10.1007/3-540-36617-2_12

Bidirectional Negative Correlation Learning

Yong Liu[✉]

The University of Aizu, Aizu-Wakamatsu, Fukushima 965-8580, Japan
yliu@u-aizu.ac.jp

Abstract. Negative correlation learning method is to create different individual learners for building a committee machine. In the original version of negative correlation learning, the learning target on a give data point was set to be the same for all the individual learners in the committee. The same learning target could lead the individual learners to become similar to each other if the learning process would be conducted for long. In order to create more different and cooperative individual learners for a committee machine, different learning targets should be set on each learning data for different individual learners in negative correlation learning. In this paper, negative correlation learning with two different learning targets was implemented. On learning each training data, the individual learners could go to the two different learning directions so that there would be little chance for them to become similar even if a long learning process would be performed. Experimental results would show how the two different learning targets would allow the individual learners to become both weak and different in negative correlation learning.

1 Introduction

Supervised learning can be formulated as an optimization problem by minimizing an error function. One example of such error functions is the mean squared error between the outputs of a trained model on a set of training points and the given target values. During the minimization process, the model weights will be adjusted. If the error function would be fixed in the whole optimization process, the absolute values of the weights might be so big that the learned models would become too complex to generalize the unseen data. One way to deal with this problem is to follow the basic principle of Occam's Razor that implies to use the simplest models among all the models well fitting the given data [1–4].

Two practical ways of early stopping and complexity penalty are often applied in learning for controlling the complexity of the learned models. Early stopping is to stop the learning when the performance reaches the best one on another selected validation set. Two problems exist in early stopping. One problem is that the selected validation set should be able to represent the unseen data well. Such well-represented validation set is not easy to be selected from

© Springer Nature Singapore Pte Ltd. 2018
K. Li et al. (Eds.): ISICA 2017, CCIS 873, pp. 84–92, 2018.
https://doi.org/10.1007/978-981-13-1648-7_7

the giving data. The other problem is that less data could be used in training when the selected data in the validation set cannot be used in training anymore.

Complexity penalty is often to limit the ranges of the adjustable weights. The idea is actually to learn less from some data when those data might lead the weights to change too much. However, a good balance between learning the given data and preventing the weights from changing too much is hard to be reached. This good balance still depends on another selected validation set. Therefore, the complexity penalty also faces the same problems in early stopping.

Different to the single model learning approaches, multiple model learning approaches use the combined outputs of a set of learned models to make the predictions. Such combined system should still cooperate the training points with a low cost in complexity. If the combined system had too higher complexity, it might overfit the training data. There are normally two ways to control the complexity of the combined system. One is to use small number of learning models with moderate complexity. The other is to combine large number of learning models with low complexity. Learning models with low complexity might just learn better than random guessing, and become weak learners [5]. The combined system with large number of such weak learners could become a strong learner that could generate high-accuracy predictions. Therefore, even if each individual learner had low complexity, the combined system with large number of such individual learners could have high complexity. It suggests that the complexity of the combined system by the weak learners could be mainly decided by the number of individuals in the combined system. When the number of individuals is not so hard to be decided as the complexity itself, the next problem is how to create such weak learners.

Negative correlation learning is able to create weak learners by encouraging each individual to learn to be different to the rest of the individuals in the same ensemble [6–12]. It had been observed that the differences among the learned individuals by negative correlation learning could go down if the learning process would be conducted for long. The reason is that the learning targets are set to either the upper limit or the lower limit for some applications such as two-class classification problems. When the outputs of the individuals would be limited between the lower limit to the upper limit, all the individuals would be driven to give the same output on the same data. After all the individuals had learned too much, they would be neither weak nor different. Surely, early stopping could be adopted in negative correlation learning to prevent the individuals from becoming too similar. However, it would unavoidably remove some data from the given training data. Alternatively, this paper proposed to set different learning targets for the different individuals in learning the same data. It would not need another validation set so that the whole given data could be used for training. It is helpful when the number of the given data would be few in some real applications. Meanwhile, no matter how long the learning would be conducted, all the individuals would be kept to be rather different after the different learning targets are introduced. Particularly, two different learning targets are set for the same learning data point in negative correlation learning in this paper. Each individual would randomly decide to learn which targets with the different probabilities. Experimental results would show how the two different learning targets would allow the individual learners to become both weak and different in negative correlation learning.

The rest of this paper contains the following three sections. Section 2 describes how to set up the different learning targets in negative correlation learning. Section 3 displays how both the learning performance and the learning behaviors would change when the learning targets are set in the different ways. Finally, Sect. 4 gives a short summary, and point out the possible improvements.

2   Negative Correlation Learning with Two Learning Targets

Negative correlation learning [13] was proposed to create negatively correlated neural networks (NNs) for neural network ensemble (NNE) learning system. In negative correlation learning, the output y of a NNE is defined by a simple average of outputs F_i among all the NNs in the NNE.

$$F(n) = \frac{1}{M} \Sigma_{i=1}^{M} F_i(n) \tag{1}$$

where $F_i(n)$ is the output of the i-th NN on the n-th training point $\mathbf{x}(n)$, $F(n)$ is the output of the NNE on the n-th training point, and M is the number of individual NNs in the NNE. The training data set D consists of N input-output pairs of $(\mathbf{x}(n), y(n))$ where $\mathbf{x}$ and $y(n)$ are the given input vector and output value, respectively. It should be pointed that the output could be vectors as well. The index value of n is from 1 to N.

In negative correlation learning, the i-th NN will be trained on the whole training set D where its error function E_i on D is given by two items:

$$
\begin{aligned}
E_i &= \frac{1}{N} \Sigma_{n=1}^{N} E_i(n) \\
&= \frac{1}{N} \Sigma_{n=1}^{N} \frac{1}{2} \left[(F_i(n) - y(n))^2 - \lambda (F_i(n) - F(n))^2 \right]
\end{aligned}
\tag{2}
$$

where the first item is the sum of the mean-squared error $E_i(n)$ on the whole training points. The second term is the sum of the mean-squared error between the output of the n-th NN and the output of the NNE. The first item will be minimized so that each NN will be trained to generate the desired output. The second item with the minus sign will be maximized so that each NN will be different to the NNE. The weight λ is set to a positive value no bigger than 1 so that all the individual NNs could be trained cooperatively in negative correlation learning.

Learning direction for the i-th NN in negative correlation learning is determined by the partial derivative of its error function E_i with respect to its output on the n-th pattern:

$$\frac{\partial E_i(n)}{\partial F_i(n)} = (1 - \lambda)(F_i(n) - y(n)) + \lambda(F(n) - y(n)) \tag{3}$$

At $\lambda = 1$, the derivative of E_i will be decided by the difference between $F(n)$ and $y(n)$ only:

$$\frac{\partial E_i(n)}{\partial F_i(n)} = F(n) - y(n) \tag{4}$$

Based on Eq. (4), learning directions would be the same for all the individual NNs in a NNE by negative correlation learning with $\lambda = 1$. Negative correlation learning by Eq. (4) is still able to create the different NNs in the early learning stage. However, the individual NNs would turn to be the similar if the Negative correlation learning process would be conducted for long. One way to prevent them from being similar is to let the NNs in the NNE have the different learning directions on the same learning data. It can be implemented by setting two learning directions as follows. On the n-th data, the i-th NN could be trained in the error direction decided by either Eq. (5) with the probability α, or Eq. (6) with the probability $(1 - \alpha)$:

$$\frac{\partial E_i(n)}{\partial F_i(n)} = F(n) - |y(n) - \beta| \tag{5}$$

$$\frac{\partial E_i(n)}{\partial F_i(n)} = F(n) - |y(n) - \gamma| \tag{6}$$

where β and γ are the small positive values between 0 and 0.5 when the target value of $y(n)$ is 1.0 or 0.0 for a two-class problem. $|y(n) - \beta|$ and $|y(n) - \gamma|$ can be regarded as the two different targets. In the case of the desired target value at 1.0, any values between 0.5 to 1.0 can serve as a new target. Correspondingly, when the desired target value at 0.0, any values between 0.0 to 0.5 can be chosen as a new target value as well.

In the simulations described in the next section, the values of β and γ were fixed to 0.4 and 0.1 although other values could be used as well. If the desired output were 1.0, the two new targets would be 0.6 and 0.9. Once the out of the NNE falls between 0.6 and 0.9, Eqs. (5) and (6) would have the opposite learning directions. Therefore, negative correlation learning with such two learning targets is named as bidirectional negative correlation learning. Actually, more than two new targets could be introduced in negative correlation learning. The probability of α would determine which learning target each NN should go to. For example, at $\alpha = 0.5$, half of the NNs in an ensemble would go to the target of $|y(n) - \beta|$ while the other half would learn the target of $|y(n) - \gamma|$.

3 Learning Performance

Three items have been checked in the learning process of bidirectional negative correlation learning (BNCL). The first item is the average error rates of the learned ensembles by BNCL. If the ensembles' error rates were nearly zero on the training set, they would become the strong learners. The second item is the average error rates of the learned individuals by BNCL. If one individual would have the error rate near to 0.5, it would become a weak learner. The first two items are to confirm the relationship between the weak learners and the strong learners. The third item is the average output overlapping rates that indicate how similar every pair of individuals in the learned ensembles are. If the two individuals would have the same output classifications on the whole measured set, their overlapping rate would be 1. On the contrary, if the two individuals would have the different output classifications on every data in the measured set, their overlapping rate would be 0. The third item would suggest how to create the different individuals in order to achieve the better performance of the learned ensembles.

10 runs of 10-fold cross-validation were conducted on the Australian credit card assessment data set from the UCI machine learning bench-mark repository. This data set is on assessing applications for credit cards based on a number of attributes, which has 14 attributes and 690 examples. The number of individual NNs was set to 50. Each individual NN has one hidden layer where 10 and 20 hidden nodes were tested. The number of training epochs was set to 2000. $\lambda = 1$, $\beta = 0.4$ and $\gamma = 0.1$ were used in BNCL. Eleven values of α between 0 and 1 were tested with the step size of 0.1.

3.1   Results of NNEs with Two Architectures

Tables 1 and 2 described both the training and the testing error rates of the two NNE architectures by BNCL from the 100 runs. Results on each value of α were obtained by 100 runs. It was surprised to see that the NNE with small NNs learned better than the NNE with large NNs at the same training epochs. For an example, the NNE with the 10-hidden-node NNs reached to a very low training error rate 0.0081 by BNCL with $\alpha = 1$. In this case, there was only one learning direction for each NN in BNCL. It was noticed that BNCL with $\alpha = 0$ had one learning direction, and also achieved the lower training error rates. Although the lower training error rates could be achieved by BNCL with only one learning

Table 1. Average error rates of the ensemble with 50 10-hidden-node NNs by BNCL.

No. of epochs	250		500		1000		2000	
α	Training	Testing	Training	Testing	Training	Testing	Training	Testing
0	0.0618	0.1368	0.0454	0.1387	0.0310	0.1409	0.0190	0.1417
0.1	0.0690	0.1380	0.0571	0.1372	0.0469	0.1377	0.0349	0.1390
0.2	0.0698	0.1359	0.0589	0.1354	0.0496	0.1380	0.0400	0.1378
0.3	0.0723	0.1401	0.0592	0.1410	0.0504	0.1397	0.0419	0.1380
0.4	0.0720	0.1393	0.0593	0.1384	0.0500	0.1388	0.0422	0.1399
0.5	0.0723	0.1388	0.0584	0.1377	0.0488	0.1367	0.0408	0.1409
0.6	0.0721	0.1364	0.0575	0.1381	0.0470	0.1401	0.0394	0.1403
0.7	0.0698	0.1399	0.0553	0.1396	0.0460	0.1413	0.0385	0.1419
0.8	0.0662	0.1394	0.0529	0.1420	0.0445	0.1419	0.0373	0.1422
0.9	0.0600	0.1403	0.0468	0.1432	0.0396	0.1420	0.0336	0.1428
1	0.0854	0.1362	0.0536	0.1368	0.0232	0.1415	0.0081	0.1486

Table 2. Average error rates of the ensemble with 50 20-hidden-node NNs by BNCL.

No. of epochs	250		500		1000		2000	
α	Training	Testing	Training	Testing	Training	Testing	Training	Testing
0	0.0644	0.1341	0.0486	0.1333	0.0345	0.1371	0.0193	0.1406
0.1	0.0708	0.1394	0.0598	0.1358	0.0500	0.1387	0.0361	0.1399
0.2	0.0729	0.1371	0.0612	0.1358	0.0524	0.1365	0.0411	0.1354
0.3	0.0736	0.1391	0.0615	0.1381	0.0527	0.1393	0.0420	0.1387
0.4	0.0751	0.1367	0.0614	0.1387	0.0522	0.1370	0.0423	0.1394
0.5	0.0749	0.1396	0.0604	0.1378	0.0509	0.1365	0.0417	0.1377
0.6	0.0747	0.1374	0.0599	0.1391	0.0500	0.1390	0.0407	0.1386
0.7	0.0746	0.1396	0.0590	0.1380	0.0497	0.1380	0.0409	0.1370
0.8	0.0712	0.1387	0.0566	0.1409	0.0482	0.1410	0.0406	0.1435
0.9	0.0655	0.1406	0.0514	0.1396	0.0435	0.1413	0.0370	0.1438
1	0.0876	0.1352	0.0624	0.1359	0.0345	0.1377	0.0127	0.1461

direction at $\alpha = 0$ or $\alpha = 1$, the worse overfitting after the longer training was observed in such cases. The testing error rates rose from 0.1352 to 0.1461 for the NNEs with the 20-hidden-node NNs by BNCL with $\alpha = 1$ after 2000-epoch learning.

For BNCL with more than one learning direction at α from 0.1 to 0.9 in Tables 1 and 2, the testing error rates changed in the smaller ranges. The largest differences observed among the testing error rates were less than 0.005 so that overfitting was less severe no matter how long the learning were conducted. When α is near to 0 or 1, the training error rates were lower than those obtained around $\alpha = 0.5$. It suggests that two different learning directions could prevent the ensembles from learning too well to overfit the training data.

Table 3. Average error rates of the individual NNs with 10 hidden nodes by BNCL.

No. of epochs	250		500		1000		2000	
α	Training	Testing	Training	Testing	Training	Testing	Training	Testing
0	0.1844	0.2380	0.1705	0.2416	0.1577	0.2452	0.1455	0.2499
0.1	0.2233	0.2628	0.2124	0.2627	0.2036	0.2645	0.1945	0.2679
0.2	0.2518	0.2850	0.2433	0.2851	0.2368	0.2864	0.2312	0.2885
0.3	0.2776	0.3072	0.2724	0.3080	0.2687	0.3091	0.2657	0.3111
0.4	0.3023	0.3277	0.2995	0.3301	0.2982	0.3311	0.2977	0.3328
0.5	0.3261	0.3481	0.3259	0.3506	0.3265	0.3528	0.3272	0.3544
0.6	0.3484	0.3677	0.3500	0.3720	0.3519	0.3733	0.3531	0.3740
0.7	0.3697	0.3872	0.3723	0.3902	0.3747	0.3913	0.3758	0.3914
0.8	0.3899	0.4053	0.3911	0.4068	0.3921	0.4070	0.3930	0.4072
0.9	0.4069	0.4222	0.4063	0.4231	0.4056	0.4232	0.4052	0.4228
1	0.4224	0.4322	0.4131	0.4298	0.4069	0.4321	0.4022	0.4345

Table 4. Average error rates of the individual NNs with 20 hidden nodes by BNCL.

No. of epochs	250		500		1000		2000	
α	Training	Testing	Training	Testing	Training	Testing	Training	Testing
0	0.1901	0.2387	0.1752	0.2399	0.1617	0.2431	0.1456	0.2473
0.1	0.2267	0.2630	0.2159	0.2624	0.2060	0.2634	0.1951	0.2663
0.2	0.2545	0.2849	0.2459	0.2843	0.2387	0.2845	0.2314	0.2871
0.3	0.2800	0.3074	0.2740	0.3078	0.2696	0.3080	0.2659	0.3100
0.4	0.3043	0.3285	0.3010	0.3294	0.2989	0.3308	0.2981	0.3326
0.5	0.3268	0.3479	0.3259	0.3499	0.3263	0.3520	0.3270	0.3535
0.6	0.3492	0.3673	0.3502	0.3704	0.3516	0.3718	0.3533	0.3731
0.7	0.3705	0.3871	0.3720	0.3887	0.3738	0.3904	0.3753	0.3911
0.8	0.3900	0.4047	0.3909	0.4069	0.3922	0.4075	0.3929	0.4077
0.9	0.4074	0.4215	0.4064	0.4226	0.4056	0.4227	0.4053	0.4227
1	0.4266	0.4356	0.4204	0.4344	0.4118	0.4333	0.4025	0.4314

3.2   Results of Small and Large NNs

Tables 3 and 4 compared both the training and the testing error rates of the individual NNs with 10 and 20 hidden nodes by BNCL from the 100 runs. Although the NNEs by these two different structures of NNs had rather different performances, the error rates reached by these NNs were nearly the same. With the increased α, both the training and testing error rates increased as well except for few cases. At $\alpha > 0.5$, the individual NNs were rather weak while their error rates were over 0.35. By altering the value of α, the NNs with different level of weakness could be created by BNCL.

Table 5. Average similarity ratios between every pair of two individual NNs with 10 hidden nodes by BNCL.

No. of epochs	250		500		1000		2000	
α	Training	Testing	Training	Testing	Training	Testing	Training	Testing
0	0.7386	0.7220	0.7468	0.7180	0.7558	0.7147	0.7653	0.7092
0.1	0.6884	0.6784	0.6958	0.6790	0.7012	0.6778	0.7068	0.6741
0.2	0.6515	0.6451	0.6565	0.6456	0.6601	0.6446	0.6626	0.6428
0.3	0.6207	0.6156	0.6221	0.6150	0.6231	0.6142	0.6233	0.6125
0.4	0.5937	0.5914	0.5931	0.5889	0.5919	0.5877	0.5903	0.5865
0.5	0.5706	0.5691	0.5679	0.5667	0.5654	0.5651	0.5632	0.5639
0.6	0.5512	0.5499	0.5476	0.5475	0.5446	0.5466	0.5427	0.5463
0.7	0.5350	0.5350	0.5313	0.5330	0.5286	0.5321	0.5270	0.5324
0.8	0.5218	0.5220	0.5195	0.5212	0.5180	0.5212	0.5167	0.5214
0.9	0.5119	0.5123	0.5110	0.5120	0.5105	0.5121	0.5102	0.5124
1	0.5073	0.5068	0.5090	0.5078	0.5100	0.5074	0.5107	0.5068

3.3   Results of Similarity Ratios

It is expected that the weaker the individual NNs are, the lower the similarity ratios will be. As shown in Tables 5 and 6, the similarity ratios fell from the highest value 0.7653 to the lowest value 0.5062. Even if the individual NNs had overlapped each other for just around 50%, the NNEs by these NNs could still learn the training data well. Although both BNCL with $\alpha = 0$ and BNCL with $\alpha = 1$ had one learning direction, the individual NNs by BNCL with $\alpha = 1$ were the most different while the individual NNs by BNCL with $\alpha = 0$ were the least different.

Table 6. Average similarity ratios between every pair of two individual NNs with 20 hidden nodes by BNCL.

No. of epochs	250		500		1000		2000	
α	Training	Testing	Training	Testing	Training	Testing	Training	Testing
0	0.7331	0.7188	0.7434	0.7188	0.7527	0.7164	0.7653	0.7128
0.1	0.6853	0.6772	0.6930	0.6788	0.7001	0.6790	0.7066	0.6767
0.2	0.6494	0.6444	0.6547	0.6459	0.6592	0.6463	0.6627	0.6448
0.3	0.6189	0.6154	0.6213	0.6153	0.6229	0.6155	0.6231	0.6143
0.4	0.5926	0.5899	0.5923	0.5890	0.5918	0.5880	0.5899	0.5867
0.5	0.5705	0.5685	0.5685	0.5671	0.5661	0.5655	0.5636	0.5648
0.6	0.5513	0.5506	0.5481	0.5482	0.5454	0.5473	0.5426	0.5468
0.7	0.5350	0.5343	0.5321	0.5332	0.5296	0.5323	0.5276	0.5322
0.8	0.5223	0.5224	0.5201	0.5213	0.5182	0.5211	0.5171	0.5213
0.9	0.5122	0.5124	0.5113	0.5119	0.5109	0.5119	0.5104	0.5122
1	0.5066	0.5062	0.5081	0.5072	0.5107	0.5084	0.5142	0.5111

4   Conclusions

In order for the individual NNs in a NNE to become different and cooperative, it is important for them to have different learning directions in training. BNCL is able to generate such different learning directions by setting different learning targets for the individual NNs in the NNE. Not only could the two different learning directions in BNCL let the NNE learn the given data, but also make the individual NNs learn the data differently. Although NCL with one learning target close to 0.5 could also force the individual NNs learn the given data very differently, the whole NNE would be trained in one direction only. Therefore, NNC with one learning target might overfit the learning data if the learning would be run for long.

The output of the NNE was generated by the simple average in the given simulation results. Although rather different individual NNs were created by BNCL or NCL with targets defined near to 0.5, these different NNs still could not show good generation by voting even if the majority voting from these NNs could perform well on the training data. It suggests that some redundant NNs might be needed in a NNE for having the consistent performances on both the training and testing data. Therefore, the size of NNE should be enlarged if the majority voting would be implemented. From the view of complexity, when the size of NNEs is increased, the size of the individual NNs would be reduced. It would be interesting to see how to combine large number of NNs with few hidden nodes into a strong NNE.

References

1. Fahlman, S.E., Lebiere, C.: The cascade-correlation learning architecture. In: Touretzky, D.S. (ed.) Advances in Neural Information Processing Systems 2, pp. 524–532. Morgan Kaufmann, San Mateo (1990)
2. Śmieja, F.J.: Neural network constructive algorithms trading generalization for learning efficiency? Circ. Syst. Signal Process. 12(2), 331–374 (1993)
3. Tolstrup, N.: Pruning of a large network by optimal brain damage and surgeon: an example from biological sequence analysis. Int. J. Neural Syst. 6(1), 31–42 (1995)
4. Kwok, T.-Y., Yeung, D.-Y.: Constructive algorithms for structure learning in feed-forward neural networks for regression problems. IEEE Trans. Neural Netw. 8(3), 630–645 (1997)
5. Schapire, R.E.: The strength of weak learnability. Mach. Learn. 5, 197–227 (1990)
6. Liu, Y.: Error awareness by lower and upper bounds in ensemble learning. Int. J. Pattern Recogn. Artif. Intell. 30(9), 14 (2016)
7. Liu, Y., Zhao, Q., Pei, Y.: Error awareness by lower and upper bounds in ensemble learning. In: Proceedings of 2015 11th International Conference on Natural Computation (2015)
8. Liu, Y.: Negative selection in negative correlation learning. In: Proceedings of 2016 12th International Conference on Natural Computation, Fuzzy Systems and Knowledge Discovery (2016)
9. Liu, Y.: Build correlation awareness in negative correlation learning. In: Proceedings of 2017 13th International Conference on Natural Computation, Fuzzy Systems and Knowledge Discovery (2017)
10. Liu, Y., Zhao, Q., Pei, Y.: Bounded learning for neural network ensembles. In: Proceedings of the 2015 IEEE International Conference on Information and Automation (2015)
11. Liu, Y.: Enforcing negativity in negative correlation learning. In: Proceedings of the 2016 IEEE International Conference on Information and Automation (2016)
12. Liu, Y.: Computational awareness for learning neural network ensembles. In: Proceedings of the 2017 IEEE International Conference on Information and Automation (2017)
13. Liu, Y., Yao, X.: Simultaneous training of negatively correlated neural networks in an ensemble. IEEE Trans. Syst. Man Cybern. Part B Cybern. 29(6), 716–725 (1999)

Reflectance Estimation Based on Locally Weighted Linear Regression Methods

Dejun Lu, Weifeng Zhang[(⊠)], Kaixuan Cuan, and Pengfei Liu

Department of Mathematics, South China Agricultural University,
483 Wushan Road, Guangzhou, China
zhangwf@scau.edu.cn

Abstract. Regression methods have been successfully applied to the area of reflectance estimation. The local linear methods show better generalization performance than the global nonlinear methods in this problem. However, the local linear models which treat every neighbor equally would lose some nonlinear information. To improve the learning ability for the nonlinear structure and reserve the generalization ability of the linear method at the same time, we propose the locally weighted linear regression method for reflectance estimation. The proposed method assigns weights to the neighbors with kernel functions and solves the weighted least squares problem to reconstruct the spectral reflectance. Experiment results show that our approach has better recovery precision and generalization performance than both the global kernel methods and the local linear methods.

Keywords: Locally weighted linear regression · Reflectance estimation
Kernel function

1 Introduction

The commonly used color measurements changes along with the illumination condition and imaging system. Thus, color values represented by RGB signals or CIE XYZ values do not have a comprehensive description of the characteristic of the objects. Spectral reflectance, which is independent of device and illumination, represents the intrinsic physical property of a natural object. With the reflectance information, one can generate color values under arbitrary illumination condition and imaging system. Accurate knowledge of spectral reflectance is needed in many practical applications, such as computer aided design (CAD) [1], art painting archiving [2], image segmentation [3], object classification [4], and hyperspectral image analysis [5]. Spectral reflectance can be measured by spectrophotometer directly, but this specific equipment is expensive and heavy. Also, the slow measure procedure and low resolution makes it impossible to measure a high-resolution image pixel by pixel. As a result, the problem of reflectance estimation from color measurements arouses widespread concern.

The reflectance estimation methods can be roughly divided into two categories: using the camera response to reconstruct the reflectance spectra with the prior knowledge of the spectral sensitivities of the sensors and the spectral power distribution of the

© Springer Nature Singapore Pte Ltd. 2018
K. Li et al. (Eds.): ISICA 2017, CCIS 873, pp. 93–101, 2018.
https://doi.org/10.1007/978-981-13-1648-7_8

illumination; or using only the camera response. Both two categories of methods aim to transform a low dimensional camera response into a high dimensional reflectance vector and this is an underdetermined problem. Many researches showed that reflectance functions tend to be smooth functions of wavelength, which are bounded between 0 and 1, and they fall within a small subset of all such possible functions [4]. According to this property, lots of effective estimation methods were proposed. The Wiener estimation methods use the spectral responsivity, autocorrelation matrix of spectral reflectance, and system noise variance to construct a transformation matrix which minimizes the mean square errors [2, 6]. The finite-dimensional methods are based on the dependency of the reflectance functions [7] and estimate the reflectance through a weighted combination of set of basic functions which can be calculated by principal component analysis (PCA) or others [8, 9].

The above two kinds of methods both need to know the prior spectral information of the sensors and the illumination, which are difficult to obtain. The pseudoinverse methods, however, build models without knowing the prior knowledge and use the regression analysis directly on the camera response and the reflectance spectra, thus is more applicable. The early proposed pseudoinverse methods compute the transformation matrix by applying pseudoinverse on the data matrix, which is equal to a multivariate linear regression problem [10, 11]. Since the output dimension is higher than the input dimension, the linear models are not flexible enough to represent the nonlinear relationship between the camera response and the spectral reflectance. Connah *et al.* proposed the polynomial models with Tikhonov regularization for this problem [12]. Heikkinen *et al.* consider a regularized learning framework based on the reproducing kernel Hilbert spaces (RKHS) [13]. They show that the nonlinear models can obtain better recovery than the linear models.

The pseudoinverse methods above use all the training samples to build the estimation functions. These types of methods can be called global methods. Contrast to the global methods, the local methods use only a small subset of the training dataset according to every test sample and build different estimation functions. For reflectance estimation problem, the training data is usually sparse in the global view, thus nonlinear methods are prone to be overfitting. In such case, local linear learning methods which use local statistics are more feasible than the global models [14]. Lansel introduced a local, linear, learned (L3) pipeline for image and reflectance estimation [15]. Zhang *et al.* proposed a regularized local linear model (RLLM) for reflectance estimation [16] and proved that it is equal to the local ridge regression [17]. They also compared the performances of different local regression methods and the global regression methods [18].

The local linear methods that treat every sample equally cannot gain the local nonlinear relationship between the camera responses and the reflectance spectra. In this paper, we propose to use locally weighted linear regression (LWLR) to estimate the reflectance. By putting weights on the locally selected training samples, LWLR can not only avoid overfitting but also learn the nonlinear structure. A comparative study of three weighted strategies to three former estimation methods is conducted on multiple reflectance datasets. Experimental results demonstrate the effective of the new method.

2   Background and Methods

The reflectance functions of natural objects are usually discretized by uniformly sampling in the visible wavelengths from 400–700 nm. If the sampling interval is 10 nm, the reflectance can be represented as a 31-dimensional vector $\mathbf{r} \in \mathbb{R}^{31}$. The camera response $\mathbf{x} \in \mathbb{R}^3$ can be modeled as

$$\mathbf{x} = \mathbf{Mr} + \mathbf{e} \tag{1}$$

where $\mathbf{M} = \mathbf{SL}$ denotes the spectral responsivity matrix, $\mathbf{S}$ is a 3×31 matrix of spectral sensitivities of the camera sensors, $\mathbf{L}$ is a 31×31 diagonal matrix having the illuminant's spectral power distribution in the diagonal, and $\mathbf{e}$ represents the system noise vector.

The problem of reflectance estimation is that for a given $\mathbf{x}$, to reconstruct the reflectance $\mathbf{r}$. This paper focuses on the regression methods which don't need to know the spectral responsivity matrix.

2.1   Global Regression Methods

By given the training set $D = \{(\mathbf{x}_i, \mathbf{r}_i), \mathbf{x}_i \in \mathbb{R}^3, \mathbf{r}_i \in \mathbb{R}^{31}, i = 1, \ldots, l\}$ of known camera response and reflectance vector, the regression methods solve the reflectance estimation problem by learning a set of functional relationships

$$r_j = f_j(\mathbf{x}), \quad j = 1, \ldots, 31 \tag{2}$$

from D, where r_j is the jth component of the reflectance vector.

The simplest assumption for f_j is the linear model

$$f_j(\mathbf{x}) = \mathbf{w}_j^T \mathbf{x} \tag{3}$$

The coefficient vector $\mathbf{w}_j$ can be found by minimizing the least squares error on the training set D

$$\min \sum_{i=1}^{l} (\mathbf{w}_j^T \mathbf{x}_i - r_{ij})^2 \tag{4}$$

Although simple, the linear model is not flexible enough to represent the nonlinear relationship between the camera response and the reflectance, so that many nonlinear regression methods are proposed. Typically, Heikkinen et al. proposed the kernel regression based estimation method [13]. They modeled the functions f_j in a reproducing kernel Hilbert space (RKHS) as

$$f_j(\mathbf{x}) = \sum_{i=1}^{l} c_i K(\mathbf{x}_i, \mathbf{x}) \tag{5}$$

The choice of kernel function $K(,)$ is important to the kernel regression method. The most widely used kernel function is the classical Gaussian kernel

$$K(\mathbf{x}, \mathbf{y}) = \exp\left(-\frac{\|\mathbf{x} - \mathbf{y}\|^2}{2\sigma^2}\right) \tag{6}$$

Another well-performed kernel proposed in [13] is the Duchon spline kernel

$$K(\mathbf{x}, \mathbf{y}) = \|\mathbf{x} - \mathbf{y}\|_2^{2m-n} \tag{7}$$

2.2 Regularized Local Linear Model

The conventional regression based reflectance estimation methods use all the training samples to build the functions in Eq. (3). Recent studies have shown that reflectance vectors usually locate on a lower dimensional manifold in the embedded high-dimensional space [15, 16]. The training data rarely are evenly distributed in the input space. Thus, the global regression methods which attempt to discover the global structure of the data space, is unreasonable with the limited training data. The local learning methods, which use only small subset of the training dataset, can not only exploit the local manifold structure of data but also perform more efficiently than the global learning models. Zhang $et\ al.$ proposed a regularized local linear model and achieved better results than the traditional global nonlinear model [16].

For a given camera response $\hat{\mathbf{x}}$, Let $\{\mathbf{x}_1, \ldots, \mathbf{x}_k\}$ be its k-nearest neighbors by Euclidean distance or other kind of distance measure to the training set, $\{\mathbf{r}_1, \ldots, \mathbf{r}_k\}$ be the corresponding reflectance. The regularized local linear model sets

$$\hat{\mathbf{r}} = \sum_{i=1}^{k} w_i \mathbf{r}_i \tag{8}$$

The coefficient $\mathbf{w} = (w_1, \ldots, w_k)^T$ can be learned by solving the regularized linear regression problem

$$\arg\min_{\mathbf{w}} \left\|\mathbf{x} - \sum_{i=1}^{k} w_i \mathbf{x}_i\right\| + \lambda\|\mathbf{w}\|^2 \tag{9}$$

In their following work [17], Zhang $et\ al.$ proved that the regularized local linear model is equal to the local ridge regression which builds the functions by Eq. (4) from the local k-nearest neighbors.

2.3 Locally Weighted Linear Regression

The conventional global and local regression methods treat every training sample equally. Since their aims are to build the universal models and it is hard to decide which training sample is more important than the others. However, for the problem of estimating spectral reflectance, the training data is always sparse and non-evenly distributed. The training sample which is more like the test sample should be trustier than the farther ones. It is necessary to treat the training samples differently. Here we propose to use the locally weighted linear regression (LWLR) method to model the functions in Eq. (3).

LWLR method minimizes the following weighted squared loss to get Eq. (3)

$$L(f_j, D) = \sum_{i=1}^{l} v_i (\mathbf{w}_j^T \mathbf{x}_i - r_{ij})^2 \tag{10}$$

where v_i is the weight of the ith training sample. We hope that the closer the training sample $\mathbf{x}_i$ to the test sample $\mathbf{x}$, the bigger the weight v_i should be. This can be achieved by using a kernel function. A commonly used kernel is the Gaussian kernel in Eq. (6) and the weights can be calculated as

$$v_i = \exp\left(-\frac{\|\mathbf{x}_i - \mathbf{x}\|}{2\sigma^2}\right) \tag{11}$$

The parameter σ controls the local weighted scope of $\mathbf{x}$. When σ is set to be small, the weighted scope is small, which means the model focus on the local training samples near $\mathbf{x}$. When σ is set larger, the models consider a bigger area of training samples.

Note that when using weight function Eq. (11) into Eq. (10), although training samples far away from $\mathbf{x}$ would be given small weights, the weights would not be zero. This non-compact form would result in a large amount of calculation. To tackle this problem, we consider using only the k-nearest neighbors of $\mathbf{x}$ into Eq. (10).

The choice of the weight function is crucial for LWLR. Except for the Gaussian weight function in Eq. (11), another popular compact kernel weight function is tri-cube function [19]:

$$v_i = \left(1 - \left(\frac{\|\mathbf{x}_i - \mathbf{x}\|}{\max\|\mathbf{x}_i - \mathbf{x}\|}\right)^3\right)^3 \tag{12}$$

3   Experimental Analysis

3.1   Datasets and Procedure

Three spectral reflectance datasets are used for evaluation: the Munsell Matte color spectra dataset, the paper spectra dataset, Agfa IT8.7/2 spectra dataset [20]. The Munsell dataset includes 1269 reflectance spectra from different color chips. The paper dataset contains 426 reflectance spectra in which there are 216 reflectance spectra of 72 colored paper samples and 210 reflectance spectra of 70 cardboard samples. The Agfa dataset includes 288 reflectance spectra. All the reflectance spectra in the datasets are sampled at 10 nm intervals from 400 to 700 nm. The corresponding camera responses are simulated on a real camera system in which the spectral sensitivity curves are from Sony

DXC-930 3CCD camera measured by Barnard et al. [21] and the illumination light is assumed to be the CIE illuminant D65. Signal-independent Gaussian noise with zero mean and standard deviation $\sigma = 0.0032$ is added to each of the RGB channel. Experiments are conducted on datasets both without noise and with noise.

The datasets are used respectively as training set, test set and validation set. The training set is used to build the models for reflectance estimation, the validation set is used to choose the best hyper parameters of the models and the test set is used to estimate the generalization error under the best hyper parameters. Four kinds of training set sizes: 100, 300, 600, and 900, are randomly selected from the Munsell Matte color spectra dataset. Corresponding data composition is show in Table 1.

Table 1. Data composition of the experiments.

Type	Name	Size
Training set	Munsell matte	100, 300, 600, 900
Validation set	Agfa IT8.7/2	288
Test set	Paper	426

The estimation error is measured by root mean square error (RMSE). For each training set size, we sample 50 times and conduct 50 trials to guarantee the experimental results to be statistically unbiased.

3.2 Experiments Between Different Methods

We compare the performance of the methods mentioned in Sect. 2: kernel regression with the Gaussian kernel (KR-G) and the Duchon spline kernel (KR-D), regularized local linear model (RLLM), locally weighted linear model with Gaussian kernel (LWLR-G), locally weighted linear regression within k-nearest neighbors using Gaussian kernel (LWLR-KG) and using tri-cube kernel (LWLR-KC). Each method has several hyper parameters, which have vital influence on the performance. The hyper parameters which minimize the RMSE on the validation set would be chosen.

In the experiments, the best parameters of KR-G are $\lambda = 0.001, \sigma = 0.7$ and the best parameters of KR-D are $\lambda = 0.001, m = 4$. RLLM performed best when $\lambda = 0.0001$ and $k = 20$. The best parameter σ of LWLM-G lies in the interval [0.2, 0.25]. For LWLM-KG method, k changed to 30 and σ lies in the interval [0.5, 0.7]. The parameter k of LWLM-KC should be chosen from 35 to 50, but is sensitive to the size of the training set. When the training set size is bigger, the best k should also be bigger.

With the best parameters, the average RMSE on the test set of different methods is showed in Table 2. As can be seen, locally weighted linear regression within k-nearest neighbors provides the lowest reflectance estimation errors. The conventional global regression methods of KR-G and KR-D perform the worst. This can be due to the training set; validation set and test set are differently distributed. Although global kernel regression methods have strong nonlinear learning ability to learn the structure of training set but it fail to generalize well to other dataset. Our LWLM-G method performs better than

KR-G and KR-D but worse than RLLM. This is because LWLM-G works with a non-compact kernel, so that they use all the samples in the training set to build models. The models they built subject to the structure of the whole training set, thus generalize worse than RLLM which use only a small subset of the training set. This verifies that local methods that use only the samples near the test sample would have better generalization capability in the reflectance estimation problem.

Table 2. Average RMSE on the test set of different methods.

Noise	Training size	KR-G	KR-D	RLLM	LWLM-G	LWLM-KG	LWLM-KC
No	100	0.0505	0.0513	0.0475	0.0483	0.0468	**0.0465**
	300	0.0489	0.0501	0.0457	0.0466	0.0448	**0.0444**
	600	0.0480	0.0496	0.0444	0.0458	0.0435	**0.0431**
	900	0.0478	0.0497	0.0444	0.0458	**0.0432**	0.0434
Yes	100	0.0506	0.0516	0.0479	0.0485	0.0471	**0.0467**
	300	0.0487	0.0500	0.0456	0.0463	0.0448	**0.0442**
	600	0.0483	0.0499	0.0446	0.0461	0.0436	**0.0434**
	900	0.0479	0.0499	0.0445	0.0459	**0.0432**	**0.0432**

LWLM-KG and LWLM-KC have smaller averaged RMSE than RLLM independent of the noise level and the training set size. LWLM-KC performs a little better than LWLM-KG. This result shows that the locally weighted linear models are more effective than the regularized linear models in the local and could estimate the reflectance more accurately.

4   Conclusion

In this paper, we propose a novel spectral reflectance estimation method based on locally weighted linear regression. We propose three weighted strategies: Gaussian kernel weighted on the whole training samples, Gaussian kernel weighted on the k-nearest neighbors and tri-cube kernel weighted on the k-nearest neighbors. We compared the three weighted strategies with three conventional methods on multiple reflectance datasets. The performance of the locally weighted linear regression is better than the other methods regardless of the training set size and with or without noise. Our future work will focus on how to learn better local structure for the problem of spectral reflectance estimation.

Acknowledgements. This work is supported in part by the National Natural Science Foundation of China (NSFC, grants 61375006) and the Outstanding Young College Teacher Program of Guangdong Province (grant Yq2013032).

References

1. Hardeberg, J.Y., Schmitt, F., Brettel, H., Crettez, J.P., Maitre, H.: Multispectral image acquisition and simulation of illuminant changes. In: Colour Imaging: Vision and Technology, pp. 145–164 (1999)
2. Haneishi, H., Hasegawa, T., Hosoi, A., Yokoyama, Y., Tsumura, N., Miyake, Y.: System design for accurately estimating the spectral reflectance of art paintings. Appl. Opt. **39**(35), 6621–6632 (2000)
3. Li, C., Li, F., Kao, C.Y., Xu, C.: Image segmentation with simultaneous illumination and reflectance estimation: an energy minimization approach. In: 2009 IEEE 12th International Conference on Computer Vision, pp. 702–708. IEEE, September 2009
4. Qi, B., Zhao, C., Youn, E., Nansen, C.: Use of weighting algorithms to improve traditional support vector machine based classifications of reflectance data. Opt. Express **19**(27), 26816–26826 (2011)
5. Khan, Z., Shafait, F., Mian, A.: Adaptive spectral reflectance recovery using spatio-spectral support from hyperspectral images. In: 2014 IEEE International Conference on Image Processing, ICIP, pp. 664–668. IEEE, October 2014
6. Shimano, N.: Recovery of spectral reflectances of objects being imaged without prior knowledge. IEEE Trans. Image Process. **15**(7), 1848–1856 (2006)
7. Cohen, J.: Dependency of the spectral reflectance curves of the Munsell color chips. Psychon. Sci. **1**(1–12), 369–370 (1964)
8. Fairman, H.S., Brill, M.H.: The principal components of reflectances. Color Res. Appl. **29**(2), 104–110 (2004)
9. Schettini, R., Zuffi, S.: A computational strategy exploiting genetic algorithms to recover color surface reflectance functions. Neural Comput. Appl. **16**(1), 69–79 (2007)
10. Ribes, A., Schmitt, F.: Linear inverse problems in imaging. IEEE Sig. Proc. Mag. **25**(4), 84–99 (2008)
11. Zhao, Y., Berns, R.S.: Image-based spectral reflectance reconstruction using the matrix R method. Color Res. Appl. **32**(5), 343–351 (2007)
12. Connah, D.R., Hardeberg, J.Y.: Spectral recovery using polynomial models. In: Proceedings of SPIE, vol. 5667, pp. 65–75, January 2005
13. Heikkinen, V., Jetsu, T., Parkkinen, J., Hauta-Kasari, M., Jaaskelainen, T., Lee, S.D.: Regularized learning framework in the estimation of reflectance spectra from camera responses. J. Opt. Soc. Am. A **24**(9), 2673–2683 (2007)
14. Shen, H.L., Xin, J.H.: Colorimetric and spectral characterization of a color scanner using local statistics. J. Imaging Sci. Technol. **48**(4), 342–346 (2004)
15. Lansel, S.P.: Local linear learned method for image and reflectance estimation. Doctoral dissertation, Stanford University (2011)
16. Zhang, W.F., Tang, G., Dai, D.Q., Nehorai, A.: Estimation of reflectance from camera responses by the regularized local linear model. Opt. Lett. **36**(19), 3933–3935 (2011)
17. Zhang, W.F., Lu, D.J.: Regularized local linear model with core neighbors for reflectance estimation. In: 2015 IEEE International Conference on Systems, Man, and Cybernetics, SMC, pp. 2996–2999. IEEE, October 2015
18. Zhang, W.F., Yang, P., Dai, D.Q., Nehorai, A.: Reflectance estimation using local regression methods. In: Wang, J., Yen, G.G., Polycarpou, M.M. (eds.) ISNN 2012. LNCS, vol. 7367, pp. 116–122. Springer, Heidelberg (2012). https://doi.org/10.1007/978-3-642-31346-2_14

19. Friedman, J., Hastie, T., Tibshirani, R.: The Elements of Statistical Learning. Springer Series in Statistics. Springer, New York (2001)
20. Spectral Database: University of Joensuu Color Group. http://spectral.joensuu.fi/
21. http://www.cs.sfu.ca/colour/data/

A Multi-task Learning Approach
for Mandarin-English Code-Switching
Conversational Speech Recognition

Xiao Song[1,2], Yi Liu[2], Daming Yang[3], and Yuexian Zou[1(✉)]

[1] ADSPLAB/Intelligent Lab, SECE, Peking University, Shenzhen, China
zouyx@pkusz.edu.cn
[2] IMSL Shenzhen Key Lab, PKU-HKUST Shenzhen Hong Kong Institution, Shenzhen, China
[3] PKU Shenzhen Institute, Shenzhen, China

Abstract. We propose a new approach based on Deep Neural Network via Multi-task Learning (MTL-DNN) for simultaneous Mandarin-English code-switching conversational speech recognition (MECS-CSR) (primary task) and language identification (LID) (auxiliary task). In our approach, the hidden layers of the DNNs for primary task fuse with ones of the DNN for auxiliary task by sharing weights/bias parameters. Extensive experiments are carried out on LDC2015S04 and Mixed Error Rate (MER) is used as performance metric for the code-switching speech recognition. Compared with the baseline and the first MECS-CSR system [1] on LDC2015S04, MER of proposed approach is relatively reduced by 4.57% and 4.07%, respectively. Results show that the proposed approach is able to capture more language switching information from the auxiliary task and significantly outperforms the competitive algorithms for the single tasks.

Keywords: Multi-task learning · Deep Neural Network
Mandarin-English code-switching · Speech recognition
Language identification

1 Introduction

Code-switching is a common linguistic phenomenon that results from different languages coexisting in an utterance among multilingual speakers. It is particularly prevalent among the large population of speakers who alternate between two or more languages, such as Chinese/English, Spanish/English, Arabic/French, etc. [2].

The challenges exist in language modeling and acoustic modeling when we build an automatic speech recognition (ASR) system on a code-switching dataset. Especially, due to the sparsity of code-switching data [3], the prediction of the switching languages in an utterance makes it difficult to model the acoustic units at the code-switching points. For tackling the problem, on one hand, phone sharing and speaker adaptation between languages were investigated in paper [4, 5]. On the other hand, combinating language identification (LID) task with monolingual acoustic models were used for code-switching speech recognition [6, 7]. Besides, a LID system was integrated into the

© Springer Nature Singapore Pte Ltd. 2018
K. Li et al. (Eds.): ISICA 2017, CCIS 873, pp. 102–111, 2018.
https://doi.org/10.1007/978-981-13-1648-7_9

decoding process by using multi stream approach [1]. However, even if adding the speaker adaptation information or combining LID, the code-switching problem is still taken as a single task to solve and still tackled faultily.

Recently, multi-task learning with deep neural network (MTL-DNN) has been shown that learning correlated tasks simultaneously can boost the performance of individual tasks. This ability of the knowledge integration makes it suitable for many complex tasks, for example low resource recognition [8], multilingual recognition [9], recognition in reverberant [10], and so on. A novel use of MTL-DNN was proposed for the code-switching speech recognition [3]. Three schemes of the auxiliary tasks were used to provide language information for MTL-DNN. These three schemes were phoneme language classification, phoneme prediction and combination of the formers. These auxiliary tasks provided more language information to enhance the recognition of Mandarin-English code-switching speech. However, except [3], there is few studies use the MTL-DNN for the code-switching problem.

Inspired by the success of single task learning and the correlated tasks learning simultaneously, we present a new approach based on MTL-DNN for simultaneous Mandarin-English code-switching conversational speech recognition (MECS-CSR) and LID task, denoted as CSR-LID-MTL. The proposed method performs the MECS-CSR by fusing with the hidden layers which leverage the knowledge of linguistics obtained from the DNN for LID task to provide more language switching information.

Besides, we study choosing two different acoustic units in acoustic modeling to build two types of primary tasks, which leverage the semantic content information obtained from the DNN for MECS-CSR task to improve the modeling ability of acoustic units at the code-switching points. On one hand, we investigate that which types of the acoustic units are more suitable for the Mandarin-English code-switching speech recognition. And on the other hand, the effectiveness of the proposed CSR-LID-MTL approach with the further study will be proven for the MECS-CSR task.

This paper is organized as follows. Section 2 presents the proposed CSR-LID-MTL approach. The introduction of the experimental settings and results are given in Sect. 3. Finally, Sect. 4 gives the conclusion.

2 Proposed CSR-LID-MTL Approach

We propose a CSR-LID-MTL approach for simultaneous MECS-CSR and LID task via MTL-DNN. The framework of the proposed approach is shown in Fig. 1, where the primary task and auxiliary task share the hidden representations. It contains two phases: pre-selection and multi-task learning.

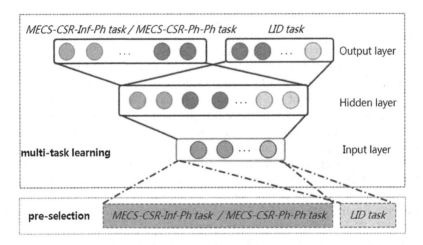

Fig. 1. The framework of the proposed CSR-LID-MTL approach.

2.1 Pre-Selection Phase for the Primary Tasks

As shown in the bottom of Fig. 1, the pre-selection phase is choosing two types of MECS-CSR tasks which contain different information in two types of acoustic units in acoustic modeling to overcome the challenge of poor modeling ability of the acoustic units at the code-switching points [1].

In the code-switching speech recognition, we adopt International Phonetic Alphabet (IPA) to build the universal phone set for the MECS-CSR task. Then, propose two types of primary tasks according to the choice of acoustic units for Mandarin in acoustic model. The motivation is that the modeling units can be used to describe the salient acoustic and phonetic information for Mandarin in our speech recognition system. Therefore, this pre-selection phase is to select the primary task for the MTL-DNN.

***MECS-CSR-Inf-Ph* Task:** As known to all, the phoneme and the initial-and-final are the best acoustic units for English and Mandarin in acoustic modeling, respectively. Besides, the initial-and-final as acoustic units can reflect the knowledge and characteristics of Chinese phonetics. Therefore, we adopt the initial-and-final for Mandarin and phoneme for English as the acoustic units in the first acoustic modeling for MTL-DNN (*Inf-Ph*).

***MECS-CSR-Ph-Ph* Task:** Although the phoneme as acoustic units is not the best choice for Mandarin, the acoustic modeling can give comparable recognition performance with the initial-and-final as in acoustic model [11]. Besides, the number of phoneme is less than the initial-and-final for Mandarin. The parameters of the phoneme modeling units can be more fully and accurately estimated in the case of the same amount of training data. Therefore, we adopt the phoneme for both languages as the acoustic units in the second acoustic modeling for MTL-DNN (*Ph-Ph*).

Based on the discussion above, two types of primary tasks are chosen (*MECS-CSR-Inf-Ph* and *MECS-CSR-Ph-Ph* task). Thus, we adopt the *MECS-CSR-Inf-Ph* or *MECS-CSR-Ph-Ph* task fused with auxiliary task respectively to investigate the effectiveness of the choice of acoustic units in acoustic modeling.

Besides, in order to make full use of the language information, LID task is chosen as the auxiliary task for the MTL-DNN. Here, we use 'CH' for Mandarin and 'EN' for English as the acoustic units in acoustic model for a simple LID task. While the effectiveness of the CSR-LID-MTL approach with the further study will be proven for the MECS-CSR task.

2.2 Multi-task Learning Phase for the CSR-LID-MTL

As shown in the top of Fig. 1, the *MECS-CSR-Inf-Ph* (*TASK-P₁*) or *MECS-CSR-Ph-Ph* (*TASK-P₂*) task and *LID* (*TASK-A*) share the hidden representations, which enables these related tasks to be learned together and transfer the knowledge from LID task to the primary task. As a result, better high-level features can be obtained compared to one with a single task model.

In the proposed CSR-LID-MTL approach, the hidden layers of the MTL-DNN for primary task fuse with ones of the DNN for auxiliary task by sharing weights/bias parameters. Here, we assume there are K tasks $T \equiv \{T_1, T_2..., T_k, ..., T_K\}$ to learn via the MTL-DNN. The model parameters are represented by $\Lambda \equiv \{\lambda_0\} \cup \{\lambda_1, \lambda_2, ..., \lambda_k, ..., \lambda_K\}$. The output layers of the tasks are trained to model the posterior probabilities of the triphone senones (tied states). That is, given x as an input acoustic vector, the posterior probabilities of the i-th senone y_i^k at the output layer is computed using the softmax function as following:

$$p(y_i^k|x; \lambda_0, \lambda_k) = \frac{\exp(h_i^k)}{\sum_{j=1}^{N_k} \exp(h_j^k)}, \forall i = 1, ..., N_k \tag{1}$$

where h_i^k is activation of senone y_1^k, N_k is the total number of senones (in task T_k). For each training frame, the error function of task T_k is to minimize the following per-frame cross entropy:

$$CE_k(x; \lambda_0, \lambda_k) = -\sum_{i=1}^{N_k} r_i^k \log p(y_i^k|x; \lambda_0, \lambda_k) \tag{2}$$

where $CE_k(x)$ is the entropy of task T_k. r_i^k is target value of the i-th senone in task T_k. Finally, training objective function $CE(D, \Lambda)$ is formulated as the weighted sum of the cross-entropies of all the tasks. The optimization of minimizing $CE(D, \Lambda)$ in MTL-DNN can be formulated as follows:

$$CE(D, \Lambda) = \sum_{x \in D} (\sum_{k=1}^{K} \beta_k CE_k(x; \lambda_0, \lambda_k)) \tag{3}$$

where β_k is the task weight of T_k and $\sum_{k=1}^{K} \beta_k$. D is the training set. The parameters Λ are learned ultimately in the MTL-DNN, where λ_0 is the model parameters that are shared by K tasks and λ_k is the model parameters associated with task T_k. Without loss of generality, T_1 is taken as primary task, and the rest are auxiliary tasks.

In the proposed CSR-LID-MTL approach, to control the contribution of each task during the process of sharing information, the training objective function $CE(D, \Lambda)$ for each primary task is designed as follows:

$$CE_i(D, \Lambda) = \sum_{x \in D} (\beta_{TASK-P_i} CE_{TASK-P_i}(x) + \beta_{TASK-A} CE_{TASK-A}(x)), \; i = 1, 2 \qquad (4)$$

where $CE_i(D,\Lambda)$ represent the final objective function for $TASK-P_1$ and $TASK-P_2$. $CE_{TASK-Pi}(x)$, and $CE_{TASK-A}(x)$ represent the entropy of $TASK-P_i$ ($TASK-P_1$, $TASK-P_2$) and $TASK-A$, respectively. β_{TASK-P_i} (β_{TASK-P_1}, β_{TASK-P_2}) and β_{TASK-A} are the task weight of $TASK-P_i$, and $TASK-A$ ($\beta_{TASK-Pi} \geq \beta_{TASK-A}$), which control the tasks of different entropy proportion that impacts the back-propagations to learn different high-level features.

Generally, after training, only the model parameters (λ_0 and λ_1) will be kept and the other parameters ($\lambda_2, \ldots, \lambda_K$) will be discarded. Therefore, the kept λ_0 and λ_1 are learned by the CSR-LID-MTL approach for the primary task.

2.3   The Proposed CSR-LID-MTL Approach

The CSR-LID-MTL approach is realized as follows:

First, preparing the input features for the MTL-DNN: Gaussian Mixture Model (GMM) of the $TASK-P_1$, $TASK-P_2$ and $TASK-A$ task are trained respectively. Then, the context dependent decision tree, the audio alignment and the feature transform are adopted from the GMM systems, which are the input of the MTL-DNN.

Second, sharing information from the auxiliary task to the primary task: the MTL-DNN is trained with the back-propagation through time (BPTT) algorithm. Via the BPTT algorithm, the parameters (weights and bias of each hidden layer) between each task are shared, which means that all tasks are sharing their information to each other.

Third, decoding for the primary task: the final layer is separated for the primary task [12]. Accordingly, the primary task can use its specific language model to decode individually, which can be addressed efficiently for speech using weighted finite state transducers (WFSTs).

3   Experimental Settings and Results

3.1   Dataset

LDC2015S04, a public South East Asia Mandarin-English (SEAME) corpus [13], is employed as our experimental dataset. It contains 63 h spontaneous Mandarin–English code-switching transcribed speech, which are recorded with the form of unscripted interviews and conversation from 157 Singaporean and Malaysian speakers. About 82%

of the transcribed utterances contain one or more intra-sentential code-switches. We divide the corpus into three sets (training, development and evaluation set) the same proportion as [1], where the details are shown in Table 1.

Table 1. Details of the datasets

	Train set	Dev. set	Eval. set	Total
Speakers	139	8	10	157
Duration (hours)	53.65	4.00	4.27	61.92
Utterances	48040	1943	2162	52145

3.2 Setting of the Baseline ASR System

We use the Kaldi speech recognition toolkit to build our ASR system for each single-task, where each system employs the single-task learning with Kaldi recipe 'nnet2' (DNN, denoted as STL-DNN). As shown in Table 2, the tasks and models are denoted as their abbreviations simply. To build the STL-DNNs, we train the networks using a forced alignment given by GMM systems, which are trained by using Kaldi recipe 's5' (model 'tri5a'). Accordingly, GMM for each basic task is trained firstly with the standard 13 dimensional Mel-frequency cepstral coefficients (MFCC) features, which is calculated by the 25 ms Hanning windows with 10 ms shift for neighboring frames. Each DNN has four hidden layers with 1024 units per layer. Both of them are trained using stochastic gradient descent (SGD) with one weight update per utterance, and the truncated BPTT learning algorithm. At the same time, the learning rate declines from 5×10^{-3} to 5×10^{-4} that is exponentially decayed during training.

Table 2. Correspondence between the tasks and their models

Task	Model
Single $TASK\text{-}P_1$	STL-DNN-$P1$
Single $TASK\text{-}P_2$	STL-DNN-$P2$
Single $TASK\text{-}A$	STL-DNN-A
Jointly training $TASK\text{-}P_1$ and $TASK\text{-}A$	MTL-DNN-$P1A$
Jointly training $TASK\text{-}P_2$ and $TASK\text{-}A$	MTL-DNN-$P2A$

In our baseline ASR systems for basic single-tasks, N-gram language models are trained with the SRILM toolkit by using the training data transcription. As a result, 3-gram back-off models smoothened using Kneser-Ney discounting are obtained.

3.3 Settings of the CSR-LID-MTL Approach

We implement the proposed approach by utilizing Kaldi recipe 'nnet2/train_multi-lang2.sh'[1].

---

[1] https://github.com/xiaosdawn/Kaldi-multi-task/blob/master/egs/wsj/s5/local/online/run_multitask2.sh

To fully investigate the performance of the proposed approach, we design the following experiments:

Experiment (I): We choose either $TASK\text{-}P_1$ or $TASK\text{-}P_2$ as the basic primary task to fuse with $TASK\text{-}A$, which results in two MTL-DNN models respectively, shown in Table 2.

Experiment (II): We investigate various task weights of the jointed tasks during learning to evaluate their effects on the final performance.

Every MTL-DNN has the same hidden layers with the same units as the STL-DNNs. As for the training of MTL-DNNs, the model is initialized by the well-trained single primary task (when jointly training $TASK\text{-}P_1$/$TASK\text{-}P_2$ and $TASK\text{-}A$, the well-trained STL-DNN-$P1$/STL-DNN-$P2$ is chosen as the initial model for MTL-DNN). It is noted that the choice of the initial model doesn't influence the final results, due to the fact that the final model will converge by two or more iterations. During each iteration, the parameters of each hidden layer between each task are weighted (as description in Sect. 2.2) via the BPTT algorithm with the same learning rate of STL-DNNs training.

3.4 Experimental Results

The performance of recognition results is measured by mix error rate (MER) which applies character error rate (CER) for Mandarin and word error rate (WER) for English.

The baseline results of two single-tasks for the MECS-CSR are shown in the first row of Tables 3 and 4, respectively. The result of STL-DNN-A has been improved but is not shown in this paper. In every table, '*Pure Eng./Man./CS. Sen.*' represents the single statistics for pure English/Mandarin/code-switching sentences, and '*Overall*' is the overall performance for all sentences.

Table 3. Recognition results with different weights of jointly training the $TASK\text{-}P_1$ and $TASK\text{-}A$ (MTL-DNN-$P1A$)

Weights of Tasks $(\beta_{TASK\text{-}P1} : \beta_{TASK\text{-}A})$	Evaluation (%)			
	Pure Eng. Sen. (WER)	Pure Man. Sen. (CER)	Pure CS. Sen. (MER)	Overall SEN, JFFF(MER)
1:0 (STL-DNN-$P1$)	50.09	39.90	34.21	36.79
0.9:0.1	49.01	37.74	32.38	35.71
0.8:0.2	49.23	**37.58**	**32.34**	35.66
0.7:0.3	**48.36**	37.85	32.65	**35.11**
0.6:0.4	49.95	37.31	32.93	35.47
0.5:0.5	49.19	37.15	33.08	35.83

Table 4. Recognition results with different weights of jointly training the *TASK-P₂* and *TASK-A* (MTL-DNN-*P2A*)

Weights of Tasks ($\beta_{TASK-P2}$:β_{TASK-A})	Evaluation (%)			
	Pure Eng. Sen. (WER)	Pure Man. Sen. (CER)	Pure CS. Sen. (MER)	Overall SEN, JFFF(MER)
1:0 (STL-DNN-*P2*)	51.88	42.88	35.18	37.24
0.9:0.1	51.63	41.44	34.88	36.66
0.8:0.2	**50.01**	**41.08**	34.74	**36.63**
0.7:0.3	51.25	41.27	**34.45**	36.76
0.6:0.4	51.58	42.50	35.05	36.90
0.5:0.5	51.82	43.45	35.16	37.09

Comparison Tables 3 and 4 (the different task weights of jointly training *TASK-P₁*/ *TASK-P₂* and *TASK-A*), we can observe that the STL-DNN-*P1* has a better ability to recognize the code-switching speech. When observing single statistics for pure English/ Mandarin/code-switching sentences, the performance of STL-DNN-*P1* relatively reduces WER/CER/MER by 3.45%/6.95%/2.76% than STL-DNN-*P2*. It is shows that the STL-DNN-*P1* is benefits learning of the primary task. The improvement on Mandarin is obvious. We can conclude that the initial-and-final-based acoustic model is more suitable for the code-switching speech recognition system.

Besides, when jointly training each primary task and LID task with different task weights respectively, we can observer that the *TASK-A* has a greater contribution to the *TASK-P₁* than *TASK-P₂* task in all terms. When observing single statistics in Table 3 for pure English/Mandarin/code-switching sentences, the best performance relatively reduces WER/CER/MER by 3.45%/5.81%/5.47% than the baseline (STL-DNN-*P1*). When observing single statistics in Table 4 for pure English/Mandarin/code-switching sentences, the best performance relatively reduces WER/CER/MER by 3.60%/4.20%/ 2.08% than the baseline (STL-DNN-*P2*). It is shown that jointly training any primary task and *TASK-A* can improve the final performance. While we can observer that jointly training *TASK-P₁* and *TASK-A* is benefits learning pure Mandarin and code-switching sentences obviously. It is verified that the information of the *TASK-A* is indeed transferable to the *TASK-P₁* or *TASK-P₂* by the hidden layers of MTL-DNN in different degree. Besides, it is showed that our proposed method provides more language switching information, by fusing with the LID task, and then outperforms the primary tasks. From all the results, we can observer that the best overall performance (MER) is 35.11% when jointly training *TASK-P₁* and *TASK-A* with the task weight of 0.7:0.3.

Finally, results with optimal weight on the proposed CSR-LID-MTL approach and the best results on the first MECS-CSR system [1] are shown in Table 5, when jointly training *TASK-P₁* and *TASK-A*, the best overall performance relatively reduces MER by 4.57% than the first baseline (STL-DNN-*P1*), and 4.07% than the first MECS-CSR system on the same corpus [1].

Table 5. Recognition results with the optimal weights

Model	Evaluation (%)			
	Pure Eng. Sen. (WER)	Pure Man. Sen. (CER)	Pure CS. Sen. (MER)	Overall SEN, JFFF(MER)
IPA-based (H.Li [1] -Dev)	–	–	–	37.10
LID + IPA-based (H.Li [1] - Dev)	–	–	–	36.60
STL-DNN-*P1*	50.09	39.90	34.21	36.79
STL-DNN-*P2*	51.88	42.88	35.18	37.24
MTL-DNN-*P1A* (0.7: 0.3)	**48.36**	**37.85**	**32.65**	**35.11**
MTL-DNN-*P2A* (0.8: 0.2)	50.01	41.08	34.74	36.63

4 Conclusions

In this paper, we proposed CSR-LID-MTL, a new approach based on Multi-task Learning with DNN (MTL-DNN) for simultaneous primary task, Mandarin-English code-switching conversational speech recognition (MECS-CSR) and auxiliary task, language identification (LID). In CSR-LID-MTL, the primary task fused with the auxiliary task by sharing weights/bias parameters during the hidden layers of the MTL-DNN. The experiments were carried out on LDC2015S04 and showed the effectiveness of the CSR-LID-MTL in enhancing the recognition of Mandarin-English code-switching speech. The best performance was obtained by jointly training the *MECS-CSR-Inf-Ph* task with the auxiliary task. A relative improvement of 4.57% and 4.07% were obtained in terms of MER on the evaluation set than the single *MECS-CSR-Inf-Ph* task learning and the first speech recognition system for Mandarin-English code-switch speech [1]. In the future, we will investigate jointly training more related tasks together and focus on the works of language model to build a better MECS-CSR system.

Acknowledgements. This work is partially supported by Shenzhen Science & Research projects. (No: JCYJ20160331104524983, JSGG20160229121006579). The authors also gratefully acknowledge the helpful comments and suggestions of the reviewers, which have improved the presentation.

References

1. Vu, N.T., Lyu, D.-C., Weiner, J., Telaar, D., Schlippe, T., Blaicher, F., et al.: A first speech recognition system for Mandarin-English code-switch conversational speech. In: 2012 IEEE International Conference on Acoustics, Speech and Signal Processing (ICASSP), pp. 4889–4892 (2012)
2. Li, Y., Fung, P.: Code switching language model with translation constraint for mixed language speech recognition. In: Proceedings of COLING, pp. 1671–1680 (2012)

3. Chen, M., et al.: Multi-Task Learning in Deep Neural Networks for Mandarin-English Code-Mixing Speech Recognition. IEICE Trans. Inf. Syst. **99**(10), 2554–2557 (2016)
4. Yeh, C.F., Huang, C.Y., Sun, L.C., Lee, L.S.: An integrated framework for transcribing Mandarin-English code-mixed lectures with improved acoustic and language modeling. In: 7th International Symposium on Chinese Spoken Language Processing (ISCSLP), pp. 214–219 (2010)
5. Yu, S., Zhang, S., Xu, B.: Chinese-English bilingual phone modeling for cross-language speech recognition. In: Proceedings of IEEE International Conference on Acoustics, Speech, and Signal Processing (ICASSP 2004), pp. I–917 (2004)
6. Bhuvanagiri, K., Kopparapu, S.: An approach to mixed language automatic speech recognition. In: Oriental COCOSDA, Kathmandu, Nepal (2010)
7. Lyu, D.-C., Lyu, R.-Y., Chiang, Y.-C., Hsu, C.-N.: Speech recognition on code-switching among the Chinese dialects. In: Proceedings of IEEE International Conference on Acoustics, Speech and Signal Processing, ICASSP 2006, p. I (2006)
8. Chen, D., Mak, B.K.-W.: Multitask learning of deep neural networks for low-resource speech recognition. IEEE/ACM Trans. Audio Speech Lang. Process. **23**, 1172–1183 (2015)
9. Huang, J.-T., Li, J., Yu, D., Deng, L., Gong, Y.: Cross-language knowledge transfer using multilingual deep neural network with shared hidden layers. In: IEEE International Conference on Acoustics, Speech and Signal Processing (ICASSP), pp. 7304–7308 (2013)
10. Giri, R., Seltzer, M.L., Droppo, J., Yu, D.: Improving speech recognition in reverberation using a room-aware deep neural network and multi-task learning. In IEEE International Conference on Acoustics, Speech and Signal Processing (ICASSP), pp. 5014–5018 (2015)
11. Davis, K., Biddulph, R., Balashek, S.: Automatic recognition of spoken digits. J. Acoust. Soc. Am. **24**, 637–642 (1952)
12. Chen, D., Mak, B., Leung, C.-C., Sivadas, S.: Joint acoustic modeling of triphones and trigraphemes by multi-task learning deep neural networks for low-resource speech recognition. In IEEE International Conference on Acoustics, Speech and Signal Processing (ICASSP), pp. 5592–5596 (2014)
13. Lyu, D.-C., Tan, T.-P., Chng, E.-S., Li, H.: Mandarin–English code-switching speech corpus in South-East Asia: SEAME. Lang. Resour. Eval. **49**, 581–600 (2015)

Feature Selection of Network Flow Based on Machine Learning

Taian Xu[✉]

Zaozhuang University, Zaozhuang 277160, Shandong, China
tyanxu@163.com

Abstract. This paper studies stages of feature selection in machine learning. CSF algorithm selects feature based on the correlation between features as well as features and categories, but information gain algorithm only considers the contribution values of individual feature to classification. It combines the subsequent learning algorithm to set appropriate threshold and complete feature selection, thus the consideration of redundancy between features are neglected. In view of this, we introduce symmetrical imbalance to improve information gain algorithm. We hope the improved information gain algorithm can increase classification accuracy through further removing its redundant features. Finally, under three kinds of learning algorithms, we compare the classification effect of three feature selection methods on WEKA platform respectively.

Keywords: Machine learning · Flow identification · Feature selection

1 Introduction

With the growing of the Internet scale and its scope of application, services such as flow monitoring, network management and billing have received much attention. IP flow classification or application identification has become an important basis for network research and network applications. In general, the rapid development of network technique mainly caused two problems: network security threats and unreasonable network resources, which posed a great challenge to the quality of network services [1]. Based on data flow and message information, the network flow identification technique determines which upper application category that flow belongs to. To network monitoring, network security and the implementation of access control as well as content auditing, the above process plays a fundamental role. Flow identification technique has undergone three stages of development: port detection, deep packet inspection and deep flow inspection based on machine learning method, whose technique becomes the current research hotspot.

The precondition of applying machine learning method in the field of network flow identification is to extract statistic characteristics of the network flow, then build a multi-sample data set (including flow category labels) and take it as prior knowledge of machine learning. By learning this priori knowledge, machine learning constructs a classification model. In this process, it uses the built-up model to classify unknown flows. However, some of the flow's statistic feature in the data set are irrelevant to the

© Springer Nature Singapore Pte Ltd. 2018
K. Li et al. (Eds.): ISICA 2017, CCIS 873, pp. 112–124, 2018.
https://doi.org/10.1007/978-981-13-1648-7_10

category, and there may be some redundancy between features at the same time. The irrelevance and redundant features will seriously affect the speed and precision of classification. Therefore, it is necessary to perform feature selection to characteristics of the dataset before building the classification model of machine learning [2].

2 Flow Identification and Machine Learning

2.1 Flow Identification Technique

From the perspective of the development of technical history, the Internet flow identification has experienced three different stages of technical research: (1) stage of port-based identification research; (2) stage of deep packet inspection (DPI) identification search; (3) stage of identification research based on machine learning method [3]. These three stages correspond to three different flow classification techniques respectively.

2.1.1 Detection Based on Port Number

In traditional network environment, the identification of network flow is not very complicated, since each protocol and each network application follows the rules of custom. They use specific well-known ports. For example, Web application based on HTTP protocol adopts the server port 80 whiling the file transfer application based on FTP using port 20 and 21, and telnet remote terminal using port 23. Thus in the early development of flow classification technology, people naturally think of using port mapping method to identify the network flow.

The advantages and disadvantages of the flow identification technique based on port mapping are quite obvious. This original technology is simple and intuitive, and easy to implement. Without the need for additional hardware and software support, it can be achieved by using simple rules on basic network devices. Moreover, in traditional network environment, its recognition efficiency and accuracy is very high. However, with the development of Internet applications, especially the subsequent development of P2P applications, the vast majority of applications are adopting dynamic port technology to keep up with objective requirements. Namely, they no longer use the well-known standard port to provide network services. For example, many mainstream Web servers and FTP server software allow users to manually specify server port while building Web server and FTP server, instead of using fixed port 80 (Web) and port 20, 21 (FTP). This not only improves server's flexibility, but also provides a possibility for multi-purpose server and reusing of single application. However, it also has made port-based flow identification technique gradually ineffective. What's more, a lot of Internet applications deliberately adopted dynamic port and camouflage port technology in order to evade detection, which has increased the difficulty of identification network flow based on port mapping.

2.1.2 DPI (Deep Packet Inspection)

Due to the fact that port-based flow identification technique cannot effectively identify Internet applications with dynamic and camouflage port, another analytical identification technique, Deep Packet Inspection (DPI), came into being. The basic principle of DPI technique is as follows: Firstly, it conducts a protocol or application feature analysis of the target flow to be identified, analyzing their carried signatures sent in the packet load when they are in network communication [4]. This kind of signature can be binary data at specific bit in load, some feature string in the load or digital signature of load after hash transformation. After obtaining the signature of application or protocol, they are applied to flow identification. When flow-generated data packets passing through the identification system, they will be unpacked by the system, checking whether the packet carried the signature of the target flow type or not. If so, it indicates that the flow matches the target flow type [5]. This method has the advantages of high speed and high accuracy, but its scalability is low and has a certain hysteretic quality. Since basing on the application layer, it may violate personal privacy and there is a certain security risks. To make matters worse, it cannot identify the encrypted flow.

2.1.3   Flow Identification Based on Machine Learning

Port-based flow identification and DPI flow identification are essentially analytical methods, which require the identification of flow in accordance with man-made laws. Thus, in addition to above-mentioned disadvantages, these two techniques also have a common flaw: they are incapable of intelligent identification. Under an increasingly complex and changeable modern Internet environment, the intelligence of flow identification model is particularly important. So in recent years, flow identification technique based on ML (Machine Learning) has emerged in the field of flow identification research.

Unlike the first two analytical methods, the machine learning method is not focused on flow's local and analytic features, but identifies from its macroscopic feature and according to its statistical behaviors feature. This makes machine learning method have inherent advantages compared with the first two techniques: it can adapt to the changes of flow's behavior feature to some extent. In other words, the machine learning method has a certain degree of intelligence. Besides this, it can also overcome other inherent defects of the first two techniques, e.g. the identification of encrypted flow.

The identification method based on machine learning first extracts a series of statistical features from the network flow, such as the number of packets, the size of data packet, the duration of stream and the average time interval at which data packets arrived, and so on. These statistical features are payload-independent, thus they are easy to be extracted and with a low computational complexity. Then, using the extracted feature data, this method trains an identification model based on machine learning algorithms. Once the identification model is built, it can predict the type of unknown flows, thus completing the flow identification task.

A typical machine learning can be divided into supervised and unsupervised learning. From the perspective of data mining, they correspond to classification and clustering. For supervised learning, a set of known data of target problem is required at first. This set of known data should have a certain scale, known as the training dataset. The training dataset is used to train a classification model, such as artificial neural network (ANN), support vector machine (SVM), decision tree (DT), etc. The training process is generally an iterative process. It constantly adjust the parameters of theoretical model by random optimization or analytic methods, so as to maximally approach the actual circumstances of the training dataset [6]. When the model training is completed, it can be used to identify unknown samples. This process is called 'testing' or 'actual classification'. The whole classification process is shown in Fig. 1.

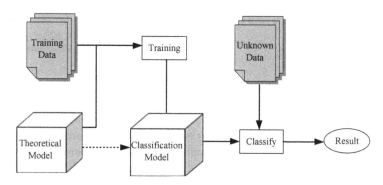

Fig. 1. Supervised learning

In contrast, the objective and task of unsupervised learning are relatively simple. Unsupervised learning finds samples with same or similar behavior in sample space from a certain angle (e.g. proximity to the distance). From the viewpoint of data-mining, it is the clustering of data samples. By analyzing and comparing interrelationships between the samples, unsupervised learning searches the type of sample. It does not require known training data, and this is clearly different from supervised learning. Therefore, unsupervised learning is widely applied in many practical problems [7].

Because unsupervised learning just clusters samples, it cannot truly identify the type of sample, so it cannot be directly applied to flow identification. However, with the guidance of a certain number of known samples, unsupervised learning could achieve the purpose of classification [8]. Such a process is known as a semi-supervised learning. The following is its basic process:

(1) Clustering sample set through unsupervised learning algorithm;
(2) Calibrating each cluster's type with a small number of known samples;
(3) Determining categories of the unknown sample according to its clustering afflation.

2.2 Concept of Feature Selection

Feature selection refers to finding out the feature subset with optimal flow classification effect from the original feature data set. It can eliminate irrelevant and redundant features, and reduce the feature dimension of dataset. Besides, feature selection is able to simplify the classifier model while improving classifier accuracy. The process of a typical feature selection is shown in Fig. 2.

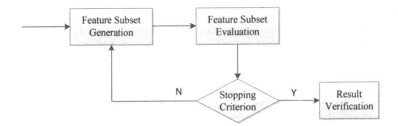

Fig. 2. Feature selection process

In Fig. 2, feature subset generation refers to the search of original dataset according to specific search strategy (search strategy is divided into full search, random search, and heuristic search). Feature subset evaluation refers to the evaluation of the generated feature subset according to evaluation function, which can be divided into independent and dependent evaluation function according to its relationship with learning algorithm. Stopping criteria refers to the set threshold or the number of iterations, which circulates until meet stopping criteria and then outputs the optimal feature subset. Result verification refers to the verification of the classification effect of the output feature subset. By analyzing the modeling time and accuracy of subsequent classifier, it verifies the validity of the feature subset [9].

According to its relationship with classification algorithm, feature selection algorithm can be divided into three types: Filter, Wrapper and Embedded.

Filter feature selection algorithm considers that the process of feature selection is independent of machine learning. It uses the inherent independent information of the feature to select feature subset. To this selection, the commonly-used evaluation criterion includes distance measurement, consistency measurement, information measurement and correlation measurement. This method is universal, but the selected feature subset may not be the optimal one that matches the subsequent machine learning algorithm.

Wrapper feature selection algorithm directly trains the selected feature subsets by using subsequent machine learning algorithm. It evaluates the merits of the feature subset according to classifier's result. The advantage of this method is that feature subsets obtained by different classification algorithms are more targeted, but it also has a disadvantage, viz. a large number of calculations is required in screening process, thus it is unsuitable for high-dimensional feature selection. The universality of this algorithm is also relatively low.

Embedded feature selection algorithm refers to the embedding the feature selection into machine learning algorithm. In this algorithm, feature selection and training process are carried out at the same time, and the classification model determines whether using the feature (e.g. artificial neural network) or not.

2.3   Algorithm Evaluation Criteria

2.3.1   Confusion Matrix

Now turning to the research of flow classification based on machine learning. After learning training samples getting the classifier, the type of all testing samples can be predicted by running the classifier on testing sample. According to the number of samples correctly and incorrectly predicted, we can also build the model of evaluating classifier's classification capacity. For example, there are N network flow samples in the test set, which respectively belong to m types of network flow. Using the trained classifier, the prediction results on the test set can be represented in the matrix form of Table 1:

Table 1. Confusion matrix

Predicted class				Actual class
Class 1	Class 2	...	Class n	
N_{11}	N_{12}	...	N_{1n}	Class 1
N_{21}	N_{22}	...	N_{2n}	Class 2
...	...	...	...	...
N_{n1}	N_{n2}	...	N_{nn}	Class n

In this table, each N_{ij} represents the number of samples which its actual class is i, but is identified as class j by the classifier, and this matrix is called as a Confusion Matrix. We can easily calculate classification results of the classifier from the Confusion Matrix.

2.3.2   Evaluation Method

The following concepts can be defined according to confusion matrix:

(1)   TP (true positive): the number of samples whose actual class is i and is correctly classified as i by the classifier.

$$TP_i = N_{ii} \tag{1}$$

(2)   FN (false negative): the number of samples whose actual class is i but is incorrectly classified as non i by the classifier.

$$FN_i = \sum_{j \neq i} N_{ij} \tag{2}$$

(3)   FP (false positive): the number of samples whose actual class is non i but is incorrectly classified as i by the classifier.

$$FP_i = \sum_{j \neq i} N_{ij} \tag{3}$$

Based on above concepts, commonly-used indicators for measuring the accuracy of the classification model can be given:

$$\text{Accuracy} = \frac{\sum TP}{\sum (TP + FN)} \tag{4}$$

$$\text{Recall (class i): Recall} = \frac{TP_i}{TP_i + FN_i} \tag{5}$$

$$\text{Precision (accuracy of class i): Precision} = \frac{TP_i}{TP_i + FP_i} \tag{6}$$

$$\text{FP\%} = \frac{\sum FP}{\sum (FP + TP)} \tag{7}$$

$$\text{FN\%} = \frac{\sum FN}{\sum (FN + TP)} \tag{8}$$

The above indicators evaluate the classification algorithm only from the perspective of classification accuracy. In fact, the evaluation of algorithm should be considered from a variety of considerations, such as real-time, scalability and so on.

3   Feature Selection Algorithm and Its Improvement

3.1   CFS Algorithm

CFS (Correlation-based Feature Subset) feature selection algorithm searches the feature subset based on the redundancy between features. Its purpose is to find features highly correlated with categories and with a low correlation each other, and to build a feature subset. This method effectively eliminates redundant and irrelevant features on the basis of finding features highly related to categories. The evaluation function of CFS for the feature subset s is expressed as:

$$R_s = k.\bar{r}_{cf} / \sqrt{k + k.(k - 1).\bar{r}_{ff}} \tag{9}$$

where, s represents the feature subset containing k features; R_s is a whole evaluation on feature subset s; k is the number of features that feature subset contains; $\bar{r}_{cf}$ is the average correlation measure between class and feature; $\bar{r}_{ff}$ is the average correlation measure between two features in the feature set.

CFS comprehensively considers the prediction capacity of each feature and the existence of redundant features in the feature set. Its main idea is to first calculate the correlation between each feature and category feature, and then perform a cycle assessment on feature subset according to some sort of search strategies, namely, keep trying to delete the individual feature of the current feature subset, so that the correlation between each feature and category characteristic is high and there is almost little correlation between the features. By this way, irrelevant and redundant features are removed.

According to different search strategies, CFS algorithm can be divided into BestFirst-CFS algorithm and GA-CFS algorithm.

BestFirst search algorithm is also known as best priority search. Its basic idea is to extend the current optimal solution. If there is a better solution in the extension node, it will replace the current optimal solution and extend it by repeating the above processes. If not, the current optimal solution is taken as the final one. The expansion mode of the current optimal solution is determined according to searching direction. For example, forward search constantly adds features to current set in order to perform the expansion, whereas backward search keeps on removing features from the current set. The evaluation function of the feature subset adopts CFS function. In Table 2, with forward search direction as an example, BestFirst-CFS algorithm process are described.

Table 2. Feature selection process based on association rules

Input: $X = \{x_1, x_2, \cdots x_k\}$//containing k features
Output: Feature subset X_i
Step 1: X_i is null initially. Take a feature from X and add it to X_i, $C(X_i)$ is up to the maximum, then record its evaluation value C_{max};
Step 2: If there are features in X which not be added to X_i, then perform step 3;
Step 3: Take a new feature from X and add it to X_i, $C(X_i)$ is up to the maximum, then record its value C_{max}. If $C_{max} \geq C_{max}$, then $C_{max} = C_{max}$, return to step 2; otherwise finish the algorithm, C_{max} is the optimal evaluation value. Now the obtained X_i is the optimal feature subset;
Step 4: If X is null, then return directly to subset with X_i is the optimal feature.

The algorithm is easy to be implemented. Its time complexity is O(nm2), in which m is the number of features, and n is the number of samples. Since it is easy to fall into local optimum during the running process and jump out the cycle ahead of time, the algorithm usually runs for a short time.

3.2 Information Gain Algorithm

Information gain (info gain, IG) algorithm is based on information metrics. By calculating the size of each feature on the classified information, this algorithm determines the importance of the features. It belongs to the filter feature selection classification, whose results are as follows: based on the contribution to the classification, each feature is sorted from large to small order. By combining the subsequent learning algorithm, IG algorithm selects the appropriate threshold and then get the feature subset. Its calculation formula is $I(X, Y) = H(X) - H(X|Y)$. Where, $H(X)$ represents the information entropy of random variable X, $H(X|Y)$ represents the condition information entropy of X after random variable Y is known. The mathematical expressions of IG algorithm are as follows.

$$H(X) = - \sum_i P(x_i) \log_2 P(x_i) \tag{10}$$

$$H(X|Y) = - \sum_j P(y_j) \sum_i P(x_i|y_j) \log_2 P(x_i, y_j) \tag{11}$$

The larger of the information gain value, the greater contribution of the feature to classification. Namely, this feature is more important to the classification. The algorithm flow is described in Table 3.

Table 3. Process of information gain algorithm

Input: Sample data set
Output: Sorting of the feature information gain value
Step 1: enter the data set, initialize the null matrix I for depositing feature weights, calculate the number of sample features n;
Step 2: calculate the information entropy of class H(Y);
Step 3: for i=1 to n calculates the condition information entropy of Y H(Y|X$_i$)and its information gain I(X$_i$, Y) = H(Y) − H(Y|X$_i$) as feature X$_i$ is known.
Step 4: sort the feature in descending order according to the information gain value;
Step 5: end.

From its process, we can see that information gain algorithm only considers the correlation between features and categories when it carries out feature selection, thus ignores the redundancy between the features.

3.3 Improved Information Gain Algorithm Based on Symmetric Uncertainty

The above-mentioned CFS feature selection algorithm not only considers the correlation between the features and categories, but also calculates the redundancy between features. However, the IG feature selection algorithm only sorts the information gain value of the category based on calculation characteristics. Combining with learning algorithms, it select the threshold and then determines the feature subset. Features filtered out by this method are likely to be redundant between each other. In the light of its various disadvantages, we improved the information gain algorithm on the basis of symmetrical imbalance. Then we further calculated the redundancy between the features. After this, we eliminated features with high redundancy and low information gain. Based on three machine learning algorithms, we also done experiments on WEKA platform in order to get more compact feature subsets and higher classification accuracy.

SU is the correlation between the information gain described earlier and the computational features of information entropy. Assuming that X and Y represent two features, the correlation formula is as follows:

$$SU(X, Y) = \frac{2I(X, Y)}{H(X) + H(Y)} \tag{12}$$

The value range of SU (X, Y) is [0, 1]. The greater the value, the higher correlation between feature X and Y. When the value is greater than the setting threshold, it means there is redundancy between feature X and Y. At this point, we can delete the feature with lower information gain. Then, at the basis of symmetric uncertainty, we further screen out redundant features from the feature subset got by the information gain feature selection algorithm. Finally, we assume that the obtained feature subset is the optimal one. The process of the improved feature selection algorithm is shown in Fig. 3.

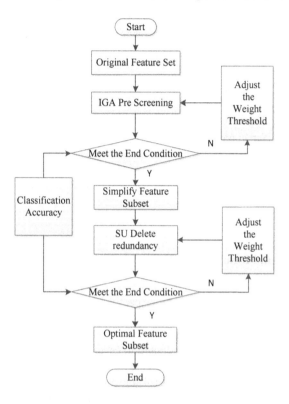

Fig. 3. Process of information gain feature selection algorithm based on symmetric uncertainty

4 Research on Classification Performance of Improved Feature Selection Algorithm

4.1 Data Mining Tools WEKA

WEKA (Waikato Environment for Knowledge Analysis) is an open source platform for data mining developed by Professor Witten at the University of Waikato. It combined a large number of machine learning algorithms which can undertake the task of data mining, including data pre-processing, classification, regression, clustering, association rules, and visualization on new interactive interfaces.

4.2 Moore Dataset

Moore dataset is collected by Prof. Moore of Cambridge University from a network center at several time periods. With the complete TCP bi-directional flow as its research object, Moore collected a total of 377,526 flow samples. This dataset is divided into 10 set (from Entry01 to Entry10), and it defines 248 flow features. These features were defined in the form of 'serial number, abbreviation, detailed description'. Item 249 indicates the application category of the flow corresponds. It covers 10 categories of common network applications, such as WWW, FTP, P2P, etc. Some of the features and their description are shown in Table 4.

Table 4. Description of Moore dataset

Serial number	Abbreviation	Detailed description
1	Server port	Server-side port number
2	Client port	Client-side port number
3	Min_IAT	The minimum value in packet arrival time interval
4	Q1_IAT	The first quartile of flow packet arrival time interval
...	...	...
249	Classes	Flow application type

4.3 Experiment and Result Analysis

Entry03 data set is selected from 10 Moore data sets as the training data set, and feature selection is carried out on this data set by respectively using CFS, IG and the improved IG feature selection algorithm based on symmetric uncertainty. To the resulting three feature subsets, machine learning algorithms of Naive Bayes, J48 Decision Tree and Support Vector Machine (SVM) are adopted to establish a classification model. Here, WEKA is set to cross-validation mode, and the remaining nine data sets are regarded as test datasets, their test results are averaged. Table 5 shows features obtained by different methods, and Fig. 4 depicts the overall classification accuracy tested with three kinds of machine learning algorithms.

Table 5. Features obtained by three kinds of feature selection methods

Selection algorithm	Selected feature identification number	Feature number
CFS	4, 72, 78, 108, 113	6
IGA	4, 107, 108, 201, 98, 194, 94, 200, 193, 198, 96	11
IGA-SU	4, 107, 201, 194, 200, 198	6

Through the comparison of feature number in above tables and the comparison of the overall classification accuracy, it is not difficult to find that in J48 and SVM these two machine learning algorithms, the feature subset selected by CFS algorithm has a higher accuracy than feature subset selected by IG algorithm. However, under three machine learning algorithms, the feature subset obtained by the improved IG algorithm based on symmetric uncertainty has higher overall accuracy than the feature subset

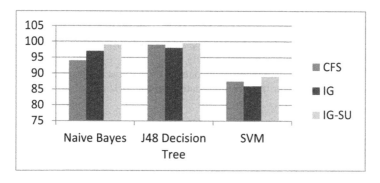

Fig. 4. Overall classification accuracy of three kinds of feature selection method

obtained by CFS algorithm. Moreover, the improved feature subset selection algorithm also has a good performance in reducing the feature dimension, which is of great practical significance in saving algorithm's running time and reducing calculation's complexity when faced with a higher dimensional data set.

5   Conclusion

In this paper, we discussed network flow feature selection based on machine learning. We proposed an improved information gain algorithm based on symmetric uncertainty. For feature subsets obtained by three feature selection methods, we have made comparisons of their overall accuracy respectively on the classifier established by three kinds of machine learning algorithms. The experimental results show that the improved feature selection method has certain feasibility and validity for the overall classification accuracy and the reduction of the feature dimension.

References

1. Lin, C., Li, Y., Wan, J.: Research review of computer network service quality optimization methods. J. Comput. **34**(1), 1–14 (2011)
2. Wang, H.: Design and implementation of network flow classification system based on decision tree. Doctoral dissertation. University of Chinese Academy of Sciences, Beijing (2014, unpublished)
3. Li, T., Wang, J., Wang, L.: Research on abnormal flow detection mechanism of large-scale network based on IPFIX. J. Tianjin Univ. Technol. **31**(3), 1–5, 11 (2015)
4. Pan, Z., Yang, B., Cao, B.: Research on deep packet detection system based on network processor. Microcomput. Inf. **25**(27), 115–116 (2009)
5. Zhang, X.: Research and system implementation of key technology of network flow analysis. Doctoral dissertation. National University of Defense Technology, Changsha (2012, unpublished)
6. Li, R., Cheng, L.: Decision tree construction based on SVM optimal decision surface. J. Electron. Meas. Instrum. **30**(3), 342–351 (2016)

7. Hu, G., Li, N., Xing, Y., et al.: Algorithm of SAR segmentations based on KMM and ultra-pixel. Foreign Electron. Meas. Technol. **35**(6), 101–108 (2016)
8. Seeger, M.: Learning with labeled and unlabeled data, p. 7. Institute for Adaptive and Neural Computation, University of Edinburgh (2001)
9. Xing, H., Lu, C., Zhang, Q.: Frequency modulated weak signal detection based on stochastic resonance and genetic algorithm. Instrumentation **3**(1), 4 (2016)

Evolutionary Multi-objective and Dynamic Optimization – Optimal Control and Design

Multi-objective Optimal Scheduling of Valves and Hydrants for Sudden Drinking Water Pollution Incident

Chengyu Hu, Lu Zou, Xuesong Yan[(✉)], and Wenyin Gong

School of Computer Science, China University of Geosciences, Wuhan, China
yanxs@cug.edu.cn

Abstract. In the last decades, water pollution incidents have occurred frequently, causing severe significant economic losses, and negative social influence. How to establish and improve emergency disposal mechanism for water pollution incident is an important issue, which has become a foremost concern in the world. In this paper, we first make a theoretical analysis on the optimal scheduling of valves and hydrants and prove that it is NP-Complete, then a multi-objective optimization model for contaminant response is established and two conflicting objectives are explored: (1) minimization of the volume of contaminated water exposure to the public, (2) minimization of the costs of operations on hydrants and valves which needed for isolation and flushing of the contaminant in the water distribution network. Finally, a customized multi-objective non-dominated sorted genetic algorithm-II (NSGA-II) linking to EPANET simulation is utilized to trade off the two optimization objectives. A medium size of water distribution network is employed for demonstrating the validity of proposed model and methodology.

Keywords: NSGA-II · NP-complete · Water distribution system

1 Introduction

In recent years, drinking water safety has aroused widespread concern in society, in order to reduce its contamination risk, sensors for water quality monitoring should be deployed in water distribution systems (WDS) to enable realtime contaminant detection. When water quality sensor raise a warning, the first step is to find the contaminant source using the information which is taken from water

Xuesong Yan received him B.E. degree in Computer Science and Technology in 2000 and M.E. degree in Computer Application from China University of Geosciences in 2003. He received his Ph.D. degree in Computer Software and Theory from Wuhan University in 2006. He is currently with School of Computer Science, China University of Geosciences, Wuhan, China and was as a visiting scholar with Department of Computer Science, University of Central Arkansas, Conway, USA. His research interests include evolutionary computation, data mining and computer application.

© Springer Nature Singapore Pte Ltd. 2018
K. Li et al. (Eds.): ISICA 2017, CCIS 873, pp. 127–137, 2018.
https://doi.org/10.1007/978-981-13-1648-7_11

quality sensors. The second step is to take some effective emergency response strategy to avoid much harming on the public.

To identify the source of contaminant injection is one of premises in making further scheduling of valves and hydrants, therefore, it is necessary to find contamination source, at the meanwhile, identify the location, starting time, duration, and mass rate of contaminant injection incident. Several studies have tackled CSI problem when water quality sensors raise an alarm [1]. In our study, we assume that the characteristic of contaminant incident is known in advance by CSI algorithm [2,3].

In this paper, we mainly focus on the second step, that is how to better guide water utilities in preparing for and responding to drinking pollution incident after the source of contaminant is already known, we need to determine which actions should be taken to limit the impact of potential contamination on public health. Generally speaking, an efficient response strategy may include:

- Using appropriate valve closures to isolate the contaminated water;
- Using hydrant to flush and evacuate foul water out of WDS, in some times, coupling flushing actions with valve manipulations to remove pollution;
- Inject edible colored dye at the location of contaminant to make a warning;

However, it is quite a challenging task to choose appropriate valve closures and hydrant in a complex water distribution system, Several studies have applied mixed integer programming, particle swam optimization, and evolution algorithm to find some strategies for operating hydrants and valves to reduce the impacts of public health. For example, Poulin et al. [4] proposed an emergency response strategy to ensure drinking water safety, in which the operational sequence of valves and hydrants was determined with the objective of minimizing consumed contaminated water. Poulin et al. [5] further considered the delay time of reaction, and then used heuristic search algorithm to obtain an optimal strategy. In 2010, Poulin et al. [6] proposed a new operational unidirectional flushing (UDF) strategy for the discharge of contaminated water, taking into account the combination of valves and hydrants.

Considering the multiple impacts on society of water pollution accident, some researchers used multi-objective optimization to trade off multiple goals, i.e., the number of polluted nodes, operation cost. In 2008, Baranowski and LeBoeuf [7] first used genetic algorithm to minimize the concentration of contaminant in WDS, at the meantime, to minimize the increased water demand to open hydrants for flushing. Preis and Ostfeld [8] aimed to minimize contaminated water consumption and the number of valves and hydrants needed to operated by using NSGA-II [9]. Similarly, Alfonso et al. [10] added thresholds for calculating the absorbed dose of contaminants and compared the results of NSGA-II with single objective method.

More specifically, the effort in this study lies in three aspects:

- We first reduce the optimal scheduling of valves and hydrants to a Knapsack problem, then prove that it is NP-Complete.

- We propose a modified multi-objective model for the optimal scheduling of valves and hydrants. To be specific, we further use operation location and time as decision variable. Two goals include: minimization of the volume of contaminated water exposure to public, minimization of the costs of operations on hydrants and valves. These two objectives conflict with each other as minimizing the absorption of polluted water will inevitably result in multiple operations of water valves and hydrants, thus lead to the incasement of operational costs.
- We design a customized NSGA-II to trade off bi-objective optimization problem. By a medium sized water distribution network, we give a comprehensive analysis of the impact of flow rate of hydrant, contaminant source and location of monitoring station on our proposed model and algorithm.

The rest of the paper is organized as follows. Section 2 establish system model and formulates optimal scheduling problem. Sections 3 and 4 propose a multi-objective optimization model and a customized Algorithm respectively. Section 5 validates the efficiency and robustness of our algorithm. Section 6 concludes our work and puts forward some issues which need to be highlighted in the future work.

2   System Model and Formulation

2.1   System Model

water supply network (WSN) consists of thousands of pipes, junctions and hydro-valves. Some WSNs may have a loop or branch network topology, or a combination of both. As one kind of special network, WSN is often modeled as a graph $G = (V, E)$, where vertices in V represent junctions, tanks, or other sources, and edges in E represent pipes, pumps, and valves. In this paper, a typical water distribution network is used as example shown as Fig. 1 [11].

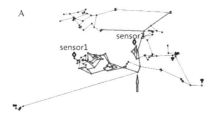
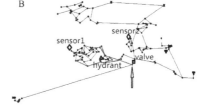

Fig. 1. Contaminant spreads without any response action (Color figure online)

Fig. 2. Contaminant spreads with response actions

As shown in the Fig. 1, the contaminant spread in 5 h with the flow of water without any emergency response after the sensors give a warning. The black arrow represents contaminant injection point, and rhombus represents water quality sensors. We can see that the contaminant (red lines) passed through a lot of nodes with the water flow if we do not take any operation on the valves and hydrants. In the Fig. 2, by opening a hydrant(indicated by □) in one direction for flushing, and closing a valve (△) for isolating contaminated water, we can see that contaminants pass through much less nodes comparing to Fig. 1, thereby proves that it is quite valid to take a series of actions on valves and hydrants.

2.2 Formulation

There are several competing design objectives which are widely used for dispatching valves and hydrants, i.e., minimization of the public health impact, minimization of the number of individuals exposed to a contaminant. In this paper, we use the maximization draining of polluted water as optimization objective:

$$Maximize \quad f_{a,n,t_a}(x_h, x_v) = Mc - \sum_{i=1}^{N} \sum_{t=t_d}^{t_{end}} C_i(x_h, x_v, t) * V_{i(t)} * \Delta t \qquad (1)$$

$$s.t. \quad x_h \in \{0, 1\}, x_v \in \{0, 1\} \qquad (2)$$

$$\sum x_h \leq H_{max}, \sum x_v \leq V_{max} \qquad (3)$$

where Mc is the total mass of injected contaminant when contamination accident a occurred at the node n, time t_a, x_h, x_v are decision variable, $x_h = 1$ means to open the hydrant and $x_h = 0$ means to close the hydrant, $x_v = 1$ means to open the valve and $x_v = 0$ means to close the valve. $t = t_d$ and t_{end} are the first time of detection of contaminant and the end time of our simulation. $C_{i(t)}$ is the contaminant concentration at the node i, time t, $V_{i(t)}$ is the water demand at the node i, time t, so the Eq. 1 is to maximize the draining of the polluted water. The Eq. 3 is the constraints of hydrant and valves, H_{max}, V_{max} means the max number of hydrants and valves. In a general way, most scheduling problem is NP-Complete, here we give a proof that our problem is NP-complete.

Theorem. *Optimal Scheduling of Valves and Hydrants(OSVH) is NP-complete.*

Proof. There are two steps to prove that *OSVH* problem is NP-complete.

First, we argue that this problem is NP. Given a threshold value θ, use the Eq. 1, we can easily to verify that the draining of polluted water is bigger than θ, so this problem is NP.

Second, we will reduce Knapsack Problem (KP) to OSVH, that is $KP \leq_P OSVH$, as Knapsack problem is quite simple to state, and it's known to be NP-complete, if there exist a polynomial time reduction from KP to $OSVH$, then we can prove optimal scheduling of valves and hydrants problem is also NP-complete.

Suppose 0–1 knapsack problem, which restricts the number x_i of copies of each kind of item to zero or one. Given a set of n items numbered from 1 up to n, each with a weight w_i and a value v_i, along with a maximum weight capacity W:

$$Maximize \quad \sum_i^n v_i * x_i \tag{4}$$

$$s.t. \quad \sum_i^n w_i * x_i \leq W, x_i \in \{0, 1\} \tag{5}$$

Based on the Knapsack problem, we will show that there exists $f : KP \rightarrow OSVH$ which is a poly-time reduction of KP to $OSVH$. We are given an instance of Knapsack problem, from which we construct an instance of $OSVH$. Suppose the decision variable $x_h \in \{x_h = x_i | i = 1, \cdots, H_{max}\}$, similarly, $x_v \in \{x_v = x_i | i = H_{max} + 1, \cdots, H_{max} + V_{max}\}$. Let the weight $w_i = 1$, and v_i is the draining of the polluted water when we operate the valves and hydrants in the form of a binary vector $\{x_1, \cdots, x_n, n = V_{max} + Hmax\}$.

We now have shown that Knapsack problem is basically a special case of optimal scheduling of valves and hydrants, since our construction takes polynomial time, we can conclude that optimal scheduling of valves and hydrants is NP-Complete.

3 Multi-objective Optimization Model for Drinking Water Pollution Incident

In order to conform with standard multi-objective optimization problem, we use minimization of the volume of contaminant exposure to public as our first fitness function instead of Eq. 1, the second fitness function is to minimize the cost of the operations on valves and hydrants.

3.1 Objective Function f_1 - Minimizing the Amount of Contaminant Exposure to Public

The first objective function f_1 is defined as the total dose (mg) of contaminant exposure to public during the period from the first intrusion of contaminant to the end time of simulation:

$$Minimize \quad f_{a,n,t_a}(x_h, x_v) = \sum_{i=1}^{N} \sum_{t=t_d}^{t_{end}} C_i(x_h, x_v, t) * V_{i(t)} * \triangle t \tag{6}$$

Where the constraints is as Eqs. 2 and 3. We can compute the Eq. 6 by simulation software EPANET2.0 [12], which can simulation the transmission of water quality and compute the hydraulic process.

3.2 Objective Function f_2 - Minimizing the Cost of Scheduling of Valves and Hydrants

The second objective function f_2 is related to the cost of operation on the valves and hydrants which needed, as the cost includes many factors, such as material, financial resources and the salary for operator, et al., so it is difficult to quantify. In our study, the cost is defined as the sum of the distances from the monitoring station of utility to the places where the hydrants and valves need to be operated. The objective function f_2 is defined as Eq. 7.

$$Minimize \quad f_2 = \sum_{k=1}^{valves} D(valve_k) + \sum_{j=1}^{hydrants} D(hydrant_j) \tag{7}$$

$$valves \leq V_{max}, hydrants \leq H_{max}$$

where k is the valve index, $valves$ is the total number of valves need to be operated in the WDS; in the same way, j is hydrant index, $hydrants$ denotes the total number of valves need to be operated; $D(valve_k)$ and $D(hydrant_j)$ represent the Euclidean distance between the monitoring station of utility and the valves or the hydrants needed respectively.

4 A Customized NSGA-II Approach for Bi-criterion Scheduling Problem

For this Bi-criterion scheduling problem, we propose a customized NSGA-II approach. The algorithm trades off the exposure of contaminant to the public (f_1) with the cost of operations on valves and hydrant (f_2), the steps of the algorithm is as follows.

- Step 1: Initializing population of chromosomes, and each chromosome represent a scheduling operation on a group of hydrants or valves. Set the maximization generation and stopping criteria.
- Step 2: Select, crossover, and mutation operation. Similar to the simple genetic algorithm, generate a new population using GA operators.
- Step 3: Evaluation operations. Combine the old and new populations into one temporary population, then evaluate each chromosomes's fitness value for all objective functions.
- Step 4: Performing a fast non-dominant sorting. Chromosome i dominates chromosome j if the fitness of chromosome i is as good as the fitness of chromosome j for all objectives, or the fitness of chromosome i is better than the fitness of chromosome j for at least one objective.
- Step 5: Crowding distance sorting within each Pareto Front. Each front is sorted from the highest crowding distance to the lowest to guarantee the spreading of solutions at the Pareto fronts, and then go to step 2, until the stop conditions are met.

In the subsections, we further explain the detailed procedures such as encoding and initial population, crossover and mutation operation, fitness function evaluation.

4.1 Encoding and Initialization of Populations

For evolutionary algorithm and its application in WDS, the first step is to encode the hydrants, valves. In our scheme, we use binary coding to describe the status of valves and hydrants, 0 means close and 1 means open. For each valve and hydrant, we also need to determine the open time and duration time, we use a three-dimension gene represents whether and when to open the valve or hydrant, and how long to close them, for instance, a gene V(0, 2:10, 4) means to close the valve at the time 2:10, and open it after 4 h later, a gene H(1, 4:30, 6) means to open the hydrant at the time 4:30 and close it after 6 h later. In addition, if a gene is like V(1, 4:20, 4), we will neglect it and take no operation on the valve because its initial status is open. We assume that all the valves remain opened and all the hydrants remain closed before the contaminant intrusion. Once the sensor raise a warning, we need to know the optimal response actions, that is, which valves should be closed and which hydrant should be opened at an appropriate time.

In our study, there are 5 valves and 5 hydrants in the water distribution network, therefor, a chromosome consist of 10 genes, e.g., a chromosome is represented by H1(1, 4:15, 8), H2(1, 7:10, 6), ..., H5 (0, T, D), V1(0, 3:10, 5), V2(1, 2:10, 6), ...,V5(1, 3:20, 6), which indicates that the hydrant $H1$ is opened at the time 4:15, and closed after 8 h later, the hydrant $H2$ is opened at the time 7:10 and closed after 6 h later, the valve $V1$ is shut off at the time 3:10 and opened after 5 h later.

4.2 Selection, Crossover and Mutation Operators

To reproduce a second generation population of solution, we randomly select one solution from non-dominated set and an individual in the current population as a pair of "parents". By producing a "child" solution using the single-point crossover, a new solution is created which typically shares many of the characteristics of its "parents". In our scheme, We first determine the cross point, then exchange the first part from the beginning of chromosome to the cross point.

For the mutation operator, we mainly use uniform mutation to change one gene in the chromosome, some simulation results show that uniform mutation is as good as Gauss mutation strategy, and better than basic bit mutation strategy.

4.3 Evaluation of Fitness Function

For each chromosome in the population, we first use EPANET simulator to compute the spreading of contaminant at each node in the water distribution network, then we use formula 6 to compute the total dose (mg) of contaminant exposure to public. For the cost of operation on valves and hydrants, we can use the coordinates of nodes to compute the Euclidean distances by the formula 7.

5 Experiment Simulation and Analysis

5.1 Parameter Setting

In our study, a typical water distribution network is used an example to demonstrate the effectiveness of our proposed model and algorithm. The network consists of 126 nodes and 168 pipes. The red triangle is the monitoring station and the red pentacle represent the contaminant injection. There are 5 valves in the pipes No4, No46, No73, No100, No161 and 5 hydrants at the nodes No21, No30, No37, No58, No102. The flow rate of hydrant set as 170 GPM empirically.

For our proposed algorithm, the number of gene is 3, the length of chromosome is 10 and the size of Population is 100, the maximum iteration is 50 and crossover probability is 0.9, mutation probability is 0.02.

5.2 Pareto Front with Different Generation

The Fig. 3 shows the curve of Pareto front when the contaminant accident occurs at node 30. we find that the convergence rate is the fastest during the 10th–20th generation, and gradually slow at 40th–50th generation. At the last generation, we obtain 4 non-dominated solutions.

For our proposed bi-criterion optimization model, there are many factors, e.g., the flushing rate of the hydrants, the contaminant source and the location of monitoring station, may have a great impact on the Pareto Front and solution. Therefore, in the following subsection, we will further discuss their effect on the model and algorithm.

5.3 Impact of Hydrants Flow Rate

In an emergency condition, we can use hydrant flushing as a response action to maintain high water quality in a network. By opening hydrants and flushing, it is easy to discharge the contaminated water, at the meantime, it will not take too much effort to maintain water quality in WDS.

In our study, we assume that contaminated water is driven by demand, when we open hydrant, the pressure will drive the contaminated water to flow out at a steady flow rate which is set a priori, such as 170 GPM. In order to investigate the impact of flow rate, we make a comparison with the flow rate of 70 GMP and 270 GPM.

As showed in the Fig. 4, it is obvious that when the flow rate is 270 GPM, the Pareto front is worst comparing to the flow rate of 70 GPM and 170 GPM. This shows that stronger flushing may not lead to better results because it may change the flow direction of contaminant, thus resulting in more operation on valves.

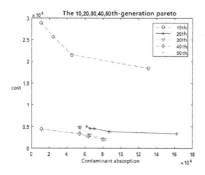

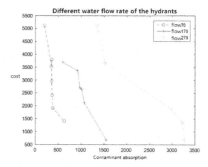

Fig. 3. Pareto front when contaminant accident occur at node 30, monitoring station is at node 20 and the flow of hydrant is 170 GPM

Fig. 4. Pareto front with different flow rates of hydrants

5.4   Impact of Monitoring Station Location

In the objective function f_2, we use the Euclidean distance as the cost of operation action, so the location of monitoring station will have an effect on the fitness function f_2. In this subsection, we assume that the monitoring stations are located at the node 8, 15 and 20.

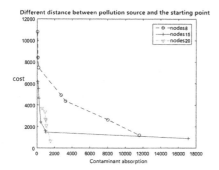

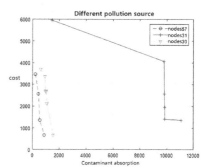

Fig. 5. Pareto fronts with different locations of monitoring station

Fig. 6. Pareto fronts with different contaminant sources

It can be seen from the Fig. 5 that different location of monitoring station will lead to different Pareto front, accordingly, so do the Pareto solutions.

5.5   Impact of Contaminant Source

When the location of contaminant injection is known in advance, the contaminant spreads with a fixed pattern by hydraulic transport model. However, different contaminant source will have an enormous impact on the Pareto front. In most cases, contaminant injection point is unknown, so in this subsection, we further discuss the impact of different contaminant source. We assume that there are three different contaminant accidents at node 30, 31, 57.

It can be seen from the Fig. 6, the optimization result is the worst when the injection of contaminant is at the node 31, and the result is the best when the injection of contaminant is at the node 57. This shows that different contaminant injection point will lead to different impact on the simulation results. By comparing the number of the adjacent edge of the three injection point, we found that the node 31 has 4 adjacent edges, and the node 57 has 2 adjacent edges. This indicates that contaminant injection at the nodes with more adjacent edges will lead to greater impact.

6   Conclusion

In this paper, we first make a theoretical proof that scheduling of valves and hydrants is a NP-Complete problem, then a multi-objective model is proposed to enhance the response to a contamination event. Finally, a customized Non-Dominated Sorted Genetic Algorithm-II (NSGA-II) was employed to tradeoff two conflicting objectives: one is minimization of contaminated water absorbed by consumers, the other is minimization of the cost of operation on the valves and hydrants. In further, we study thee impact of flow rate of hydrant, the source of contaminant injection and location of monitoring station on the Pareto fronts.

Future work will deal with incorporating more conflicting objectives in the model and reducing the computational time of the overall multi-objective process. In addition, we will use a more appropriate objective function to describe the cost of operation on valves and hydrants.

Acknowledgments. This research was partially supported by NSF of China (Grant No. 61673354, 61305087, 61502439, 61501412). This paper has been subjected to Hubei Key Laboratory of Intelligent Geo-Information Processing, China University of Geosciences, Wuhan 430074, China.

References

1. Yang, X., Boccelli, D.L.: Bayesian approach for real-time probabilistic contamination source identification. J. Water Resour. Plann. Manage. **140**(8), 04014019 (2013)
2. Hu, C., Zhao, J., Yan, X., Zeng, D., Guo, S.: A mapreduce based parallel niche genetic algorithm for contaminant source identification in water distribution network. Ad Hoc Netw. **35**, 116–126 (2015)
3. Yan, X., Zhao, J., Hu, C., Wu, Q.: Contaminant source identification in water distribution network based on hybrid encoding. J. Commun. **16**(2), 379–390 (2016)
4. Poulin, A., Mailhot, A., Grondin, P., Delorme, L., Villeneuve, J.-P.: Optimization of operational response to contamination in water networks. In: Proceedings of the Water Distribution System Analysis Symposium (2006)
5. Poulin, A., Mailhot, A., Grondin, P., Delorme, L., Periche, N., Villeneuve, J.-P.: Heuristic approach for operational response to drinking water contamination. J. Water Resour. Plann. Manage. **134**(5), 457–465 (2008)

6. Poulin, A., Mailhot, A., Periche, N., Delorme, L., Villeneuve, J.-P.: Planning uni-directional flushing operations as a response to drinking water distribution system contamination. J. Water Resour. Plann. Manage. **136**(6), 647–657 (2010)
7. Baranowski, T.M., LeBoeuf, E.J.: Consequence management utilizing optimization. J. Water Resour. Plann. Manage. **134**(4), 386–394 (2008)
8. Preis, A., Ostfeld, A.: Multiobjective contaminant response modeling for water distribution systems security. J. Hydroinf. **10**(4), 267–274 (2008)
9. Deb, K., Pratap, A., Agarwal, S., Meyarivan, T.: A fast and elitist multiobjective genetic algorithm: NSGA-II. IEEE Trans. Evol. Comput. **6**(2), 182–197 (2002)
10. Alfonso, L., Jonoski, A., Solomatine, D.: Multiobjective optimization of operational responses for contaminant flushing in water distribution networks. J. Water Resour. Plann. Manage. **136**(1), 48–58 (2009)
11. Preis, A., Ostfeld, A.: Genetic algorithm for contaminant source characterization using imperfect sensors. Civil Eng. Environ. Syst. **25**(1), 29–39 (2008)
12. U.S. Environmental Protection Agency: EPANET 2 programmer's toolkit, us environmental protection agency. Water Supply and Water Resources Division, National Risk Management Research Laboratory, Cincinnati, OH, vol. 45268 (2012)

A Novel Mutation and Crossover Operator for Multi-objective Differential Evolution

Qingxia Li[1] and Wenhong Wei[2(✉)]

[1] Department of Computer,
City College of Dongguan University of Technology,
Dongguan 523419, China
[2] School of Computer, Dongguan University of Technology,
Dongguan 523808, China
weiwh@dgut.edu.cn

Abstract. Differential evolution is a simple evolutionary algorithm by simulating Darwinian evolution principle where the population of individuals are evolved and adapted with some reproduction mechanisms such as mutation, crossover and selection operator in the computer environment. During a mutation in the nature, only a few vectors of the individuals will be mutated instead of a mutation in whole vectors. Also during a crossover operation, there is still a mutation chance for individuals. This work proposed a novel mutation and crossover operation for multi-objective differential evolution inspired by above, named as NMCO-MODE. In the NMCO-MODE, offspring is allowed to mutate during a crossover as it is same for living individuals in nature, or keep some of its vector same after a mutation. At last, the NMCO-MODE is tested with multi-objective optimization problems (MOP), it's found out that the new mechanism significantly improved performance of differential evolution algorithm. It has sharp convergence character and gets stuck in local minima less frequently than other multi-objective evolutionary algorithms.

Keywords: Mutation · Crossover · Multi-objective · Differential evolution

1 Introduction

In real-world cases, such conflicted objectives may hinder experts from gaining an exact and efficient solution (global optimum state of a system). Under such a condition, it is highly recommended to contrive proper frameworks (known as vector optimizers or multi-objective methods) which are capable of unveiling a set of eligible solutions (non-dominated solutions). This lets experts to exert a trade-off among collected solutions to attain a proper compromise about the problem at hand. Even if such paradigms are very useful for engineering design, their implementation is a very exacting task. This is strictly because in a single objective problem, one should find the optimal solution which is a single point in the solution span. But, for a multi-objective optimization problem (MOP), one should devise a method that is able to find multi optimal solutions (known as Pareto front). The problem may even be more tedious if the expert does not have any information regarding the physics of the problem at hand

© Springer Nature Singapore Pte Ltd. 2018
K. Li et al. (Eds.): ISICA 2017, CCIS 873, pp. 138–147, 2018.
https://doi.org/10.1007/978-981-13-1648-7_12

(known as reference point). In such conditions, the obtained Pareto front should be distributed uniformly and un-biased [1]. Metaheuristics have proved their capability for developing Pareto based frameworks.

Differential evolution (DE), which was first proposed by Storn and Price [2], is one of the most powerful metaheuristic algorithms for global numerical optimization. The advantages of DE are its ease of use, simple structure, speed, efficiency, and robustness [3, 4]. Recently, DE has been successfully applied in diverse domains [5, 6]. For multi-objective optimization problems, multi-objective differential evolution (MODE) typically shows better convergence properties and diversity metrics around the true Pareto optimal front compared to the other multi-objective evolutionary algorithms [7, 8].

As we well known, during offspring production, if the individuals are accumulated to parents, crossover operator will not work efficiently to diversity the population. Also, if the individuals are well distributed, mutation is not an efficient diversification method for the population. Based on this, a novel mutation and crossover operator (NMCO) is prepared which allows change probabilities of crossover and mutation operator for each individual. By doing so, if the individuals are accumulated, the mutation rate will able to be increased and this operation will be controlled with a feedback mechanism.

So in this work, the NMCO is developed for MODE by changing the offspring production mechanism, which is called as NMCO-MODE. At last, the performances of NMCO-MODE and other multi-objective evolutionary algorithms are explained and compared.

The rest of this paper is organized as follows. We firstly introduce the related work in Sect. 2. Section 3 discusses the NMCO-MODE algorithm description. Section 4 reports experimental results. Finally, conclusions are drawn in Sect. 5.

2   Relate Work

Since Chang et al. first extended DE to handle MOPs [9] in 1999, many multi-objective differential evolution algorithms have been developed [10–17]. In [10], a new approach to multi-objective optimization based on DE was proposed (DEMO). It combined the advantages of DE with the mechanisms of Pareto based ranking and crowding distance sorting, used by state-of-the-art evolutionary algorithms for multi-objective optimization. In [11], an adaptive DE algorithm (ADEA) was proposed to solve MOPs. In ADEA, the mutation factor was adjusted by the number of non-dominated solutions and crowding distance. The selection operator of ADEA combined the advantages of DE with the mechanisms of non-dominated-based sorting and crowding distance ranking. In [12], a multi-objective self-adaptive DE (MOSaDE) was proposed. In MOSaDE, it automatically used the generation strategies of the trial vectors and their associated parameters based on their previous experience of generating promising or inferior solutions. The work developed in [13] is an extended version of MOSaDE by combining objective-wise learning strategies. In [14], another self-adaptive DE with Pareto dominance was presented to solve MOPs. In this MODE, the two control parameters of DE were self-adapted. Furthermore, the determined non-dominated solutions during the evolutionary process were retained in an external elitist archive, and the crowding entropy diversity measure was used to preserve the diversity of the

Pareto front. In [15], an efficient MODE variant with opposition-based learning (OBL) and random localization strategy (RLS) to solve MOPs was developed, where the OBL was used to generate an initial population of potential candidate, and the RLS was used to select better solutions for mutation. In [16], a new MODE variant with the ranking-based mutation operator (MODE-RMO) was proposed. In MODE-RMO, the ranking-based mutation operator was used to accelerate the convergence speed. In [17], a multi-objective particle swarm-differential evolution algorithm (MOPSDE) is proposed that combined a particle swarm optimization (PSO) with a DE. During consecutive generations, a scale factor is produced by using a proposed mechanism based on the simulated annealing method and is applied to dynamically adjust the percentage of use of PSO and DE.

3   NMCO-MODE Algorithm

The mutant, an organism which is mutated, will have a new but not a completely changed set of vectors. In other words, some vectors will remain same as they were after a mutation. While in conventional mutation and crossover operation, all vectors are performed, in the proposed operator both mutation and crossover operations are applied to individuals' vectors. This allow offspring to mutate during a crossover as it is same for the living organisms in nature, or keep some of its vectors same after a mutation. By combining all these features, the NMCO-MODE is a variant for MODE. Based on the probabilities, the NMCO-MODE tires to imitate the nature closer than the conventional version by giving a chance to mutation during crossover and by keeping some vectors same after mutation in the parent's vectors.

The NMCO-MODE produces 2 offspring after each offspring production and the vectors of the offspring are produced with either crossover or mutations, or it remains same from the randomly selected parent. The Pseudo steps of the new operator are summarized in Table 1. In NMCO-MODE, first, a random number is generated. If the generated number is in mutation range, it performs a mutation. Otherwise, it creates

Table 1. Pseudo code for NMCO-MODE

NMCO-MODE algorithm
Evaluate the initial population P of random individuals.
While stopping criterion not met do
Randomly select two parents P_1, P_2 from P
if rand(1) < Mutation probability
child$_1$ ← performs mutation operator
child$_2$ ← performs mutation operator
else if rand(1) < Crossover probability
child$_1$ ← performs crossover operator
child$_2$ ← performs crossover operator
else
child$_1$ ← P_1
child$_2$ ← P_2
end
Evaluate the children using non-dominated sorting and crowding distance and produce new population P.
end

another random number and checks if the condition for performing a crossover is satisfied. If yes, it performs crossover. If not, the offspring gets the same value for related variable from his parent.

4 Performance Measures and Test Results

In this study, NMCO-MODE is compared with some typical multi-objective evolutionary algorithms by using non-constrained multi-objective optimization problems. These multi-objective evolutionary algorithms include NSGA-II, MOEA/D, MODE-RMO, DEMO and MOPSDE. To compare the performances of both operators, the five two-objective benchmark functions are used (ZDT1, ZDT2, ZDT3, ZDT4 and ZDT6).

All approaches are run with 100 individuals and 250 generations (25000 function evaluations for each test). All problems are repeated 100 times to overcome the randomness nature of the algorithm and their average is assigned as the result of simulations. The probability of mutation p_m is set as 20% and the probability of crossover $p_c = 80\%$. The distribution index for mutation and crossover operators are set as $\eta_m = 20\%$ and $\eta_c = 20\%$ respectively.

In these experiments, two metrics was used. The smaller the value of these metrics, the better the performance of the algorithm. Convergence metric γ measures the distance between the obtained non-dominated front Q and the set P^* of Pareto-optimal solutions. Diversity metric Δ measures the extent of spread achieved among the non-dominated solutions.

Tables 2, 3, 4, 5 and 6 show the mean and variance of the values of the convergence metric and diversity metric obtained using NSGA-II, MOEA/D, MODE-RMO, DEMO, MOPSDE and NMCO-MODE for ZDT1, ZDT2, ZDT3, ZDT4 and ZDT6, averaged over 20 runs.

As it can be seen by Tables 2, 3, 4, 5 and 6, NMCO-MODE shows better performance than other algorithms except in ZDT3 function. For ZDT3 function, such as in Table 3, NMCO-MODE is a litter worse than MOPSDE, but NMCO-MODE is better than NSGA-II, MOEA/D and DEMO.

Table 2. Convergence and diversity comparison between NMCO-MODE and other algorithms for ZDT1.

Algorithms	γ		Δ	
	Mean	Std	Mean	Std
NMCO-MODE	**2.52E−04**	4.12E−05	**2.34E−01**	1.13E−02
NSGA-II	8.94E−04	0.00E+00	4.64E−01	4.16E−02
MOEA/D	7.79E−04	6.50E−04	3.78E−01	1.23E−01
MODE-RMO	6.17E−04	5.51E−04	4.25E−01	2.43E−01
DEMO	1.08E−03	1.13E−04	3.25E−01	4.84E−03
MOPSDE	2.59E−04	3.24E−05	2.57E−01	3.69E−02

Table 3. Convergence and diversity comparison between NMCO-MODE and other algorithms for ZDT2.

Algorithms	γ		Δ	
	Mean	Std	Mean	Std
NMCO-MODE	**1.24E−05**	4.14E−04	**1.41E−01**	1.05E−02
NSGA-II	8.24E−04	0.00E+00	4.35E−01	2.46E−02
MOEA/D	5.28E−04	3.73E−04	3.03E−01	1.68E−01
MODE-RMO	4.55E−04	3.03E−04	2.93E−01	2.54E−01
DEMO	7.55E−04	4.50E−05	3.29E−01	3.24E−02
MOPSDE	4.75E−05	6.68E−05	2.71E−01	1.14E−02

Table 4. Convergence and diversity comparison between NMCO-MODE and other algorithms for ZDT3.

Algorithms	γ		Δ	
	Mean	Std	Mean	Std
NMCO-MODE	1.02E−03	2.54E−06	**2.82E−01**	2.18E−03
NSGA-II	4.34E−02	4.20E−05	5.76E−01	5.08E−03
MOEA/D	5.23E−02	3.45E−03	8.86E−01	1.48E−02
MODE-RMO	4.12E−02	1.45E−04	5.71E−01	1.16E−03
DEMO	1.19E−03	5.90E−05	3.09E−01	1.86E−02
MOPSDE	**5.76E−04**	1.34E−04	5.82E−01	6.93E−02

Table 5. Convergence and diversity comparison between NMCO-MODE and other algorithms for ZDT4.

Algorithms	γ		Δ	
	Mean	Std	Mean	Std
NMCO-MODE	**3.91E−03**	1.89E−01	**2.18E−01**	1.68E−01
NSGA-II	3.23E+00	7.31E+00	4.79E−01	9.84E−03
MOEA/D	3.25E−00	8.49E−05	6.23E−01	3.27E−01
MODE-RMO	3.36E−00	2.19E−02	6.39E−01	2.57E−01
DEMO	1.04E−03	1.34E−04	3.40E−01	3.77E−02
MOPSDE	5.11E−03	8.36E−02	3.28E−01	5.58E−01

Table 6. Convergence and diversity comparison between NMCO-MODE and other algorithms for ZDT6.

Algorithms	γ		Δ	
	Mean	Std	Mean	Std
NMCO-MODE	**1.04E−04**	1.17E−04	**2.26E−01**	2.23E−01
NSGA-II	7.81E+00	1.67E−03	6.44E−01	3.50E−02
MOEA/D	7.32E−04	1.41E−04	2.98E−01	3.10E−02
MODE-RMO	7.12E−03	1.41E−04	5.88E−01	3.10E−02
DEMO	6.29E−04	4.40E−05	4.42E−01	3.93E−02
MOPSDE	1.28E−04	3.22E−06	2.50E−01	3.52E−01

Figures 1, 2, 3, 4 and 5 present the non-dominated fronts obtained by one run of NMCO-MODE.

Another observed from Figs. 1, 2, 3, 4 and 5 promising result of the NMCO-MODE was the better convergence of test functions to final Pareto front.

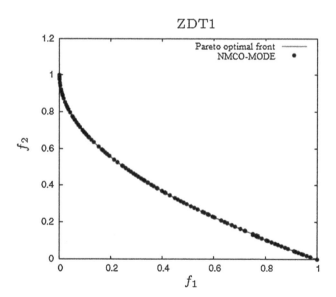

Fig. 1. Computed Pareto front on ZDT1

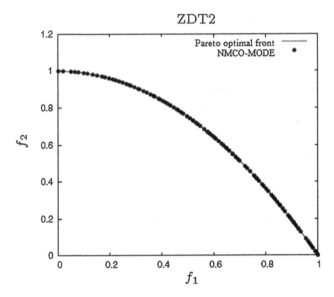

Fig. 2. Computed Pareto front on ZDT2

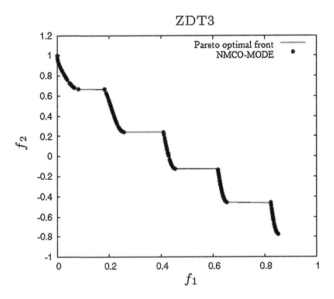

Fig. 3. Computed Pareto front on ZDT3

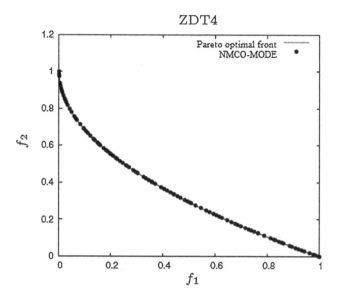

Fig. 4. Computed Pareto front on ZDT4

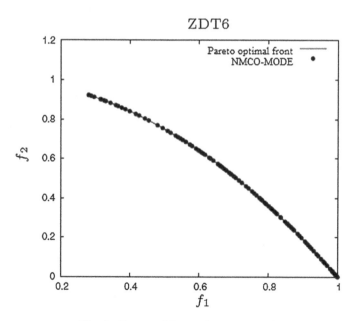

Fig. 5. Computed Pareto front on ZDT6

5   Conclusion

In this study, a novel mutation and crossover operator is developed for multi-objective differential evolution algorithm, which is named by NMCO-MODE. In NMCO-MODE, offspring is allowed to mutate during a crossover as it is same for living individuals in nature, or keep some of its vector same after a mutation. It is tested with selected optimization problems and compared its' performance with other multi-objective evolutionary algorithms. Results proved that NMCO-MODE shows better performance than other algorithms.

Acknowledgement. This work was supported by the National Nature Science Foundation of China (No. 61370185, 61402217), Guangdong Higher School Scientific Innovation Project (No. 2014KTSCX188), the outstanding young teacher training program of the Education Department of Guangdong Province (YQ2015158); and Guangdong Provincial Science and Technology Plan Projects (No. 2016A010101034, 2016A010101035). Guangdong Provincial High School of International and Hong Kong, Macao and Taiwan cooperation and innovation platform and major international cooperation projects (No. 2015KGJHZ027).

References

1. Fleischer, M.: The measure of Pareto optima applications to multi-objective metaheuristics. In: Fonseca, C.M., Fleming, P.J., Zitzler, E., Thiele, L., Deb, K. (eds.) EMO 2003. LNCS, vol. 2632, pp. 519–533. Springer, Heidelberg (2003). https://doi.org/10.1007/3-540-36970-8_37
2. Storn, R., Price, K.: Differential evolution-a simple and efficient heuristic for global optimization over continuous spaces. J. Glob. Optim. **11**, 341–359 (1997)
3. Wei, W., Wang, J., Tao, M.: Constrained differential evolution with multiobjective sorting mutation operators for constrained optimization. Appl. Soft Comput. **33**, 207–222 (2015)
4. Zhou, X., Zhang, G., Hao, X., Yu, L.: A novel differential evolution algorithm using local abstract convex underestimate strategy for global optimization. Comput. Oper. Res. **75**, 132–149 (2016)
5. Rajesh, K., Bhuvanesh, A., Kannan, S., Thangaraj, C.: Least cost generation expansion planning with solar power plant using differential evolution algorithm. Renew. Energy **85**, 677–686 (2016)
6. Malathy, P., Shunmugalatha, A., Marimuthu, T.: Application of differential evolution for maximizing the loadability limit of transmission system during contingency. In: Pant, M., Deep, K., Bansal, J.C., Nagar, A., Das, K.N. (eds.) Proceedings of Fifth International Conference on Soft Computing for Problem Solving. AISC, vol. 437, pp. 51–64. Springer, Singapore (2016). https://doi.org/10.1007/978-981-10-0451-3_6
7. Wei, W., Wang, J., Tao, M., Yuan, H.: Multi-objective constrained differential evolution using generalized opposition-based learning. Comput. Res. Dev. **53**(6), 1410–1421 (2016)
8. Cheng, J., Yen, G.G., Zhang, G.: A grid-based adaptive multi-objective differential evolution algorithm. Inf. Sci. **367–368**, 890–908 (2016)
9. Chang, C.S., Xu, D.Y., Quek, H.B.: Pareto-optimal set based multiobjective tuning of fuzzy automatic train operation for mass transit system. IEE Proc.: Electric Power Appl. **146**, 577–583 (1999)

10. Robič, T., Filipič, B.: DEMO: differential evolution for multiobjective optimization. In: Coello Coello, C.A., Hernández Aguirre, A., Zitzler, E. (eds.) EMO 2005. LNCS, vol. 3410, pp. 520–533. Springer, Heidelberg (2005). https://doi.org/10.1007/978-3-540-31880-4_36

11. Qian, W.Y., Li, A.J.: Adaptive differential evolution algorithm for multiobjective optimization problems. Appl. Math. Comput. **201**(1–2), 431–440 (2008)

12. Huang, V.L., Qin, A.K., Suganthan, P.N., Tasgetiren, M.F.: Multi-objective optimization based on self-adaptive differential evolution algorithm. In: 2007 IEEE Congress on Evolutionary Computation, pp. 3601–3608. IEEE Press, New York (2007)

13. Huang, V.L., Zhao, S.Z., Mallipeddi, R., Suganthan, P.N.: Multi-objective optimization using self-adaptive differential evolution algorithm. In: 2009 IEEE Congress on Evolutionary Computation, pp. 190–194. IEEE Press, New York (2009)

14. Wang, Y.N., Wu, L.H., Yuan, X.F.: Multi-objective self-adaptive differential evolution with elitist archive and crowding entropy-based diversity measure. Soft. Comput. **14**, 193–209 (2010)

15. Ali, M., Siarry, P., Pant, M.: An efficient differential evolution based algorithm for solving multi-objective optimization problems. Eur. J. Oper. Res. **217**(2), 404–416 (2012)

16. Chen, X., Du, W.L., Qian, F.: Multi-objective differential evolution with ranking-based mutation operator and its application in chemical process optimization. Chemom. Intell. Lab. Syst. **136**(16), 85–96 (2014)

17. Su, Y., Chi, R.: Multi-objective particle swarm-differential evolution algorithm. Neural Comput. Appl. **28**(2), 407–418 (2017)

Multi-objective Gene Expression Programming Based Automatic Clustering Method

Ruochen Liu[1(✉)], Jianxia Li[1], and Manman He[2]

[1] Key Laboratory of Intelligent Perception and Image Understanding of Ministry of Education,
Xidian University, Xi'an 710071, China
ruochenliu@xidian.edu.cn
[2] School of Computer, Xi'an Shiyou University, Xi'an 710065, China

Abstract. We have proposed multi-objective Gene Expression Programming (GEP) based automatic clustering method (do not need prior knowledge), which is denoted as MOGEPC. In our algorithm, we adopt GEP based multi-objective optimization, which has a powerful global search ability to optimize the two objective functions, namely, compactness and connectedness simultaneously. We use center-based encoding to generate chromosomal in the encoding phase and expression tree (ET) to decode chromosome to cluster centers. The introduction of multi-objective in GEP is helpful to make the clustering quality of data sets with different structures better. Finally, we apply MOGEPC on artificial data sets, real data sets from UCI. Experiments show that MOGEPC is robust in clustering various data sets without any apriori information.

Keywords: GEP · Multi-objective optimization · Clustering

1 Introduction

A key issue in data mining [1] is data clustering. The purpose of clustering is to partition the data with the same characteristics into one cluster and to separate the data with different characteristics. Scholars have proposed some clustering algorithms [2]. However, these algorithms always need apriori information (such as clustering number [3], data structures) which is difficult to obtain accurately, so the clustering quality is not satisfied.

In recent years, multi-objective [4–7] is widely used in clustering algorithm. Meanwhile, as Gene Expression Programming (GEP) has a powerful search capability, some researchers have combined the advantage of GEP to solve problems, such as function finding [8, 9], data clustering [10, 11]. In our work, we introduce a automatic clustering method based on multi-objective GEP, which optimizes multi objective function in the evolutionary process. GEP is a new evolutionary algorithm which integrates the strength of simple genetic algorithm (SGA) and genetic programming (GP), and GEP was proposed by Ferreira in 2001 [12]. We also introduce the idea of Non-dominated Sorting Genetic Algorithm II (NSGA-II) [13, 14]. Global search capabilities of GEP and the idea learnt from NSGA-II are effective in improving clustering quality.

© Springer Nature Singapore Pte Ltd. 2018
K. Li et al. (Eds.): ISICA 2017, CCIS 873, pp. 148–158, 2018.
https://doi.org/10.1007/978-981-13-1648-7_13

The main contents of each chapter are arranged in this way. Section 2 shows a simple introduction of the GEP. Section 3 depicts NSGA-II. In Sect. 4, we introduces MOGEPC in detail. Section 5 presents experiment results. The last Section makes a conclusion and summarizes the future research.

2   An Overview of GEP

In GEP, firstly, randomly generating initial population which is linear strings with a fixed length. Then the chromosomes are coded as different expression trees (ETs). After decoding to ETs, fitness is calculated according to the given objective function. Then selecting a part of the current chromosome to generate a new population. During the reproduction, a series of operators like mutation, rotation and recombination are applied to chromosomes. This process can introduce new elements, and make the population evolve. It is worth pointing out that elite reserves is used to make sure the best chromosome not lost in the evolution. When the termination condition is reached, the algorithm stops. Figure 1 presents the flowchart [8] of GEP.

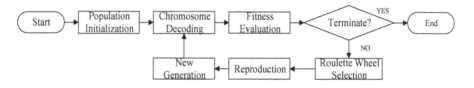

Fig. 1.   The flowchart of GEP

2.1   Initialization

Each chromosome in GEP is fixed-length. Each gene is conducted by head and tail. The composition of head is from Function Set (*FS*) and Terminal Set (*TS*), while the tail is merely from *TS*. And the length of the tail t is $t = h^*(n - 1) + 1$, where h is the length of the head, n is the maximum of function arguments.

Consider *FS* is $\{+, -, *, /\}$ and *TS* is $\{a, b\}$, so $n = 2$, let $h = 4$, so $t = 5$, and gene's length is $t + h = 9$. An example with one-gene is: $+*a/baabb$. And a two-gene chromosome is shown as: $+*a/baabb*ab+babba$.

2.2   Decoding

Although the chromosome is fixed-length, it can be decoded to the ET with different sizes and shapes. In decoding process, open reading frames (ORFs) is used. For the multi-gene chromosome, each ET is linked to form final ET by predefined linking functions. Take chromosome ($+*a/baabb*ab+babba$) as an example, if we choose linking function is '+', it can be decoded as Fig. 2:

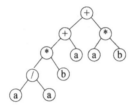

Fig. 2. The decoded ET of the chromosome

2.3   Fitness Evaluation

Fitness function is the direction of the population evolve. Candida Ferreira, the founder of GEP algorithm, proposed eight forms of fitness function aimed at classification, function finding and logic synthesis [15].

2.4   Roulette Wheel Selection (RWS)

RWS is used in GEP to produce the next generation and elitism is also adopted in GEP to prevent losing the best individual in the evolutionary process. Every kind of selection scheme has its own field of use, but for complex problems roulette wheel with elitism is best.

2.5   Reproduction with Modification

The modification is used to introduce new element so that the population evolved, which includes replication, mutation, recombination and inversion.

(1) According to fitness, copying the chromosomes are copied into the next generation.
(2) Mutation introduces new elements, thereby increasing the diversity of the population.
(3) Recombination: randomly selected two parent chromosomes and exchanged some material between them in pairs.
(4) The inversion operator randomly chooses a section of the chromosome, and inverts the start and termination points. Because it has a big impact when it occurs in the heads of genes, so it usually restricted use in this region.

3   NSGA-II

Deb proposed NSGA-II in 2002 [13]. It mainly consists of four parts, crowding-distance, elite reserves, diversity preservation and fast non-dominated sorting. Crowing-distance is used in sorting and elitist reserves, diversity preservation are introduced to ensure evolution toward a good direction. The process of fast non-dominated sorting put solution into different levels. So, we can select out good solutions to ensure the population evolve toward good direction.

The process of fast non-dominated sorting is presented in Fig. 3.

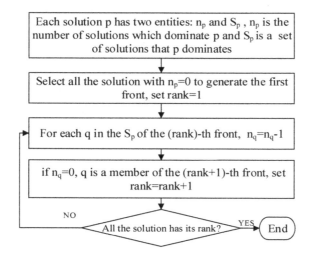

Fig. 3. The process of fast non-dominated sorting

4 Multi-objective GEP Based Automatic Clustering

Figure 4 gives the flowchart of MOGEPC.

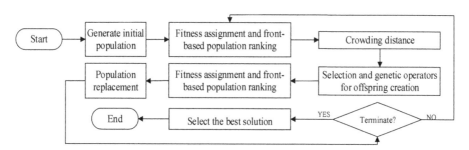

Fig. 4. Flowchart of the proposed MOGEPC

First, randomly generate initial population P. Second, fitness is calculated. According to the fitness, dividing the individuals in the population into different fronts, using a strategy NSGA-II [13] described in Sect. 3. Third, crowding distance is computed. This step is introduced to further subdivide the solutions in the same rank, which is used in the selection. Fourth, after fitness is computed and population is stratified, creating the offspring population according to applying crossover operator and mutation operator on parent individuals [16], then, mixing the parent and offspring population. Fifth, fitness and rank are calculated again. Sixth, a new population is created based on rank and crowing distance. The higher the rank is, the larger the crowing distance is, the greater probability the chromosome is chosen to produce the nest population. When the termination condition is reached, the algorithm stops [17]. At last calculate the correct rate for each individual and select the result with the highest correct rate.

4.1 Initialization

In GEP-cluster we introduce clustering operators, namely 'd' and 'm'. 'd' means split the data into different clustering centers and 'm' means merge the data into one clustering centers. So the FS is $\{d, m\}$, and the TS is $\{1, 2, \ldots, i\}, i \in [1, n]$. In our paper, we forced the first gene element is 'd' and the head is only selected from FS to make chromosome have more cluster center. If the random initialization of the head is ddmm and the tail is 58419, then we get a correct chromosome $CH0$ as: **ddmm58419**.

4.2 Objective Functions

In order to reflect data structure, we choose two types of complementary objectives: compactness and connectedness [4].

We choose the overall deviation as the measurement of compactness. This is a simple and widely used criterion. It is expressed as follows:

$$Deviation(c) = \sum_{i=1}^{n} \sum_{j=1}^{m_i} d(c_i, x_j^i) \tag{1}$$

As to evaluating the connectedness, we choose connectivity which is defined as follows:

$$Connectivity(c) = \frac{1}{N} \sum_{i=1}^{n} \left(\frac{\sum_{j=1}^{h} x_{i,nn_i(j)}}{h} \right) \text{ with } x_{m,n} = \begin{cases} 1 & if \exists c_k : m, n \in c_k \\ 0 & otherwise \end{cases} \tag{2}$$

We choose these two objective functions because they can balance each other. Overall deviation tends to decrease the clustering number, while connectivity devotes to increase the clustering number. For purpose of making these two objective functions both to be maximized, we take the reciprocal of overall deviation as objective functions.

4.3 Decoding the Chromosome

We use pre-order traversal to map the chromosome into an ET. The ET's root node is the first elements of the gene. All the offspring branches are handled left to right and layer-by-layer and the gene elements read from left to right too. From our rulers, the chromosome $CH0$ given above can build a binary tree as Fig. 5.

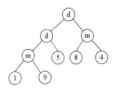

Fig. 5. The ET of $CH0$

Once the ET is established, we use the method of post-order traversal to decode ET to cluster centers. Specifically, the recursive algorithm is used. This step uses the knowledge of data structure and the main steps are given as Algorithm 1.

Algorithm 1: Post order (binary tree, Visit)

1) For each ET do
2) Read the root node of ET and store it as t
3) If t is not NULL, recursive decoding left sub-tree and right sub-tree; if t is NULL, call the function Visit
4) End for

Visit is a function to calculate cluster centers and its main steps are as follows:

Algorithm 2: Visit (int num, char element, bool center)

1) Read the element in the binary tree node and store it as e
2) If e is from TF, store data in Data Center array and Center Number plus one
3) If e is 'm', merge the last two elements in the Data Centers array and Center Number minus 1

In order to understand the decoding process better, we break down the ET into sub-ET according the two algorithms mentioned above. It is shown in Fig. 5.

In Fig. 6, S_1 means sub-ET1 and so on. First from sub-ET1, we get a center number $C_1 = (x_1 + x_9)/2$. Because sub-ET2's root is 'd', so S_1 and 5 represent two cluster centers, then we get a new cluster centers $C_2 = x_5$. From sub-ET3, we get cluster centers $C_3 = (x_4 + x_8)/2$. Because ET's root is specially designed to 'd', so chromosome have more cluster centers, so at last we get three cluster centers in chromosome *CH0*.

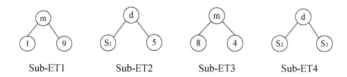

Sub-ET1   Sub-ET2   Sub-ET3   Sub-ET4

Fig. 6. Sub-ETs from chromosome *CH0*

Now, we find out all cluster centers hidden in chromosome *CH0*, they are $C_1 = (x_1 + x_9)/2$, $C_2 = x_5$ and $C_3 = (x_4 + x_8)/2$.

After acquiring the hidden cluster centers, evaluating the fitness. The main steps are as follows: (1) For each data, finding the nearest cluster center; (2) Recalculate the center of each cluster, after replace the old center with the new center; (3) Calculate the objective functions.

4.4 Selection and Variation

We adopt elite reserves and selection strategy based on NSGA-II to create next generation. Elite reserves is to make the best chromosome not lost during evolution, and selection strategy is to ensure the population evolve toward good direction.

Variation operators include mutation and recombination. The law of mutation operator is the first element in head does not participate in mutation, while the others can change to FT freely. The tail element can change to TF freely. The recombination operation is described as: randomly selecting two chromosomes, then exchanging some elements of the two chromosomes to generating two new chromosomes.

4.5 The Choice of Solution

After multi-objective algorithm, we could get a set of solutions. Choosing promising solution from Pareto front is a crucial step. There are mainly two methods [18, 19]. The first kind is based on Gap statistic and the others are using a special index to measure the quality of solution. The first method is time-consuming and the second one has a poor performance when it meets of data sets with different structures. At this step, we calculate the correct rate of all chromosomes and that the largest one as our final solution. This method can quickly and easily select the optimal solution, but it can only handle data with the known label.

5 Experiment and Analysis

5.1 Experimental Data

Table 1 presents all the data set used in the experiment.

Table 1. Data sets

Name	Total	The amount of data in each cluster	Dimension	Clusters
Iris	150	3 * 50	4	3
New_thyroid	215	150,35,30	5	3
Wine	178	5,9,71,48	13	3
Ionosphere	351	126,225	17	2
Glass	214	70,76,17,13,9,29	9	6
Diabetes	768	268,500	8	2
Ad_5_2	250	5 * 50	2	5
Ad_20_2	1000	20 * 50	2	20
Sticks	512	117,123,150,122	2	4
Eyes	300	100,100,100	2	3

5.2 Parameter Setting

The basic parameters in our algorithm are introduced Table 2.

Table 2. Parameters setting

Parameters	Setting
Number of generation	200
Population size	20
Head length	10
Tail length	11
Mutation rate	0.04
Recombination rate	0.77

5.3 Experimental Comparison

We evaluate the performance of MOGEPC, compared with k-means and another single-objective GEP clustering algorithm (SOGEPC) [10], since k-means is a classic clustering algorithm and a comparison with single-objective can show the advantages of multi-objective algorithm. We evaluate solutions of different algorithms using accuracy as shown in Table 3, the higher the accuracy is, the better. We also show average of clustering number and its standard deviation, which is an important measure about correct clustering numbers, as shown in Table 4. We run each algorithm fifty times and take the average as final results.

Table 3. Experimental comparison (accuracy)

Name	Cluster	K-means	SOGEPC	MOGEPC
Iris	3	80.16	82.00	**90.61**
New_thyroid	3	**92.99**	39.89	92.53
Wine	3	81.71	78.37	**87.35**
Ionosphere	2	**68.55**	54.05	66.21
Glass	6	35.52	47.85	**49.63**
Diabetes	2	**66.02**	62.67	64.39
Ad_5_2	5	**96.29**	89.16	94.20
Ad_20_2	20	78.33	90.95	**92.50**
Sticks	4	44.77	65.08	**88.44**
Eyes	3	45.67	70.53	**79.21**

From Table 3, we can see that the accuracy of MOGEPC is better than SOGEPC in all the data sets, however the accuracy of MOGEPC is not as good as K-means in some data sets, because K-means obtain the correct clustering numbers as a priori knowledge. So we can draw the conclusion that MOGEPC show a good performance across our data set than the other two algorithms. From Table 4, we could see that the average of clustering number of MOGEPC is much better than SOGEPC, which shows that MOGEPC has a strong capability in finding the correct clustering number. That is owe to the global search ability of GEP and the introduction of NSGA-II.

Table 4. Experimental comparison (the average of clustering number)

Name	Cluster	K-means	SOGEPC	MOGEPC
Iris	3	3.0 ± 0.0	3.2 ± 0.4	3.0 ± 0.0
New_thyroid	3	3.0 ± 0.0	1.0 ± 0.0	3.0 ± 0.0
Wine	3	3.0 ± 0.0	3.5 ± 0.6	3.0 ± 0.0
Ionosphere	2	2.0 ± 0.0	2.4 ± 0.7	2.3 ± 1.0
Glass	6	6.0 ± 0.0	2.9 ± 0.3	3.0 ± 0.0
Diabetes	2	2.0 ± 0.0	3.9 ± 0.6	2.0 ± 0.0
Ad_5_2	5	5.0 ± 0.0	5.0 ± 0.0	5.0 ± 0.0
Ad_20_2	20	20.0 ± 0.0	18.8 ± 0.9	21.0 ± 1.8
Sticks	4	4.0 ± 0.0	9.1 ± 0.7	4.0 ± 1.0
Eyes	3	3.0 ± 0.0	12.3 ± 0.7	3.2 ± 0.4

From Table 5, we could find that MOGEPC is time-consuming, but considering the above Tables 3 and 4, MOGEPC does not need any prior knowledge and has a better accuracy than K-means and SOGEPC, so time consumption is acceptable. The performance of MOGEPC is mainly due to two aspects: (i) MOGEPC is more effective at optimizing the objective functions; (ii) MOGEPC benefits from the introduction of multiple objectives.

Table 5. Experimental comparison (run time (Unit: s))

Data	K-means	SOGEPC	MOGEPC
Iris	0.44	0.83	5.30
New_thyroid	0.40	1.43	2.84
Wine	0.30	2.62	4.54
Ionosphere	0.29	5.94	11.26
Glass	0.78	1.34	3.12
Diabetes	0.40	8.14	15.24
Ad_5_2	0.40	1.08	2.79
Ad_20_2	2.15	4.24	21.04
Sticks	0.47	2.12	5.42
Eyes	0.27	1.38	3.26

6   Conclusion

We propose automatic clustering method using multi-objective algorithm based on GEP, which simultaneously optimizes two objectives, compactness and connectedness, which uses center-based encoding to generate chromosomal in the encoding phase and ET to decode chromosome to cluster centers. Experiments show that MOGEPC is effective and promising in clustering various data sets without any apriori information when compared with SOGEPC and K-means. In the future, we will primarily focus on encoding of chromosome and the choice of solution from solution set, since encoding determine the number of clustering centers, meanwhile the choice of solution is a challenge in multi-objective clustering faced by all researchers.

Acknowledgments. This work was supported by the National Natural Science Foundation of China (No. 60803098, No. 60872135), Research Fund for the Doctoral Program of Higher Education of China (No. 20070701022); the Provincial Natural Science Foundation of Shaanxi of China (2010JM8030, No. 2010JQ8023), and the Fundamental Research Funds for the Central Universities (No. K50511020014, No. K50510020011).

References

1. Han, J.W., Kambr, M.: Data Mining Concepts and Techniques. Higher Education Press, Beijing (2001)
2. Lee, J., Lee, D.: Dynamic characterization of cluster structures for robust and inductive support vector clustering. IEEE Trans. Pattern Anal. Mach. Intell. **28**(11), 1869–1874 (2006)
3. McQueen, J.: Some methods for classification and analysis of multivariate observations. In: 5th Berkeley Symposium in Mathematics, Statistics and Probability, University of California Press, vol. 1, no. 14, pp. 281–297 (1967)
4. Handl, J., Knowles, J.: An evolutionary approach to multi-objective clustering. IEEE Trans. Evol. Comput. **11**(1), 56–76 (2007)
5. Wikaisuksakul, S.: A multi-objective genetic algorithm with fuzzy c-means for automatic data clustering. Appl. Soft Comput. **24**, 679–691 (2014)
6. Chen, E., Wang, F.: Dynamic clustering using multi-objective evolutionary algorithm. In: Hao, Y., et al. (eds.) CIS 2005. LNCS (LNAI), vol. 3801, pp. 73–80. Springer, Heidelberg (2005). https://doi.org/10.1007/11596448_10
7. Nanda, S.J., Panda, G.: Automatic clustering algorithm based on multi-objective immunized PSO to classify actions of 3D human models. Eng. Appl. Artif. Intell. **26**(5), 1429–1441 (2013)
8. Liu, R., Lei, Q., Liu, J., Jiao, L.: A population diversity-oriented gene expression programming for function finding. In: Deb, K., et al. (eds.) SEAL 2010. LNCS, vol. 6457, pp. 215–219. Springer, Heidelberg (2010). https://doi.org/10.1007/978-3-642-17298-4_22
9. Ferreira, C.: Function finding and the creation of numerical constants in gene expression programming. In: Benítez, J.M., Cordón, O., Hoffmann, F., Roy, R. (eds.) Advances in Soft Computing, pp. 257–265. Springer, London (2003). https://doi.org/10.1007/978-1-4471-3744-3_25
10. Chen, Y., Tang, C., Zhu, J., Li, C., Qiao, S., Li, R.: Clustering without prior knowledge based on gene expression programming. In: International Conference on Natural Computation, vol. 3, pp. 451–455 (2007)
11. Ni, Y., Du, X., Xie, D., Ye, P., Zhang, K.: A multi-objective cluster algorithm based on GEP. In: IEEE International Conference on Cloud Computing and Big Data (CCBD 2014), pp. 33–38 (2014)
12. Ferreira, C.: Gene expression programming: a new adaptive algorithm for solving problem. Complex Syst. **13**(2), 87–129 (2001)
13. Deb, K., Pratap, A., Agarwal, S., Meyrivan, T.: A fast and elitist multi-objective genetic algorithm: NSGA-II. IEEE Trans. Evol. Comput. **6**(2), 182–197 (2002)
14. Deb, K.: Multi-objective evolutionary algorithms: introducing bias among pareto-optimal solutions. In: Ghosh, A., Tsutsui, S. (eds.) Advances in Evolutionary Computing, pp. 263–292. Springer, London (2003). https://doi.org/10.1007/978-3-642-18965-4_10

15. Ferreira, C.: Gene Expression Programming, 2nd edn, pp. 263–292. Springer, Heidelberg (2006). https://doi.org/10.1007/3-540-32849-1
16. Eiben, A.E., Smith, J.E.: Introduction to Evolutionary Computing, 2nd edn. Springer, Heidelberg (2007)
17. Coelho, A.L.V., Fernandes, E., Faceli, K.: Inducing multi-objective clustering ensembles with genetic programming. Neurocomputing 74(1–3), 494–498 (2010)
18. Tibshirani, R., Walther, G., Hastie, T.: Estimating the number of clusters in a data set via the gap statistic. J. Roy. Stat. Soc. 63(2), 411–423 (2001)
19. Tasdemir, K., Merényi, E.: A validity index for prototype-based clustering of data sets with complex cluster structures. IEEE Trans. Syst. Man Cybern. Soc. 41(4), 1039–1053 (2011)

Multi-objective Firefly Algorithm Guided by Elite Particle

Jiayuan Wang, Li Lv$^{(\boxtimes)}$, Zhifeng Xie, Xi Zhang, Hui Wang,
and Jia Zhao

School of Information Engineering, Nanchang Institute of Technology,
Nanchang 330099, China
wangjiayuan1201@163.com, lvli623@163.com,
xiezhifeng_nit@163.com, zhaojia925@163.com,
1023624613@qq.com, huiwang@whu.edu.cn

Abstract. With the diversification and complexity of the social needs, multi-objective optimization problems gradually attract more and more attention. The traditional multi-objective optimization algorithms cannot meet the practical needs. Therefore, it is urgent to improve and develop new multi-objective optimization algorithms to meet the challenges. On the basis of standard firefly algorithm, this paper proposed a multi-objective firefly algorithm based on population evolution guided by elite particle. The algorithm randomly selects a non-inferior solution as the elite particle to participate in the population evolution, extends the detection range of firefly, and improves the diversity and accuracy of the non-inferior solution set. The experimental results show that the proposed algorithm is superior to the MOPSO, MOEA/D, PESA-II, NSGA-III algorithm on the GD, SP, MS and other quantitative indexes for the seven classic test functions, and the proposed algorithm is an effective method for multi-objective optimization.

Keywords: Firefly algorithm · Multi-objective optimization · Pareto optimum
Elite particle

1 Introduction

Optimization problems exist in many engineering fields. With the growth of problem complexity, stronger optimization algorithms are required. In the past decades, several new optimization techniques have been proposed by the inspiration of swarm intelligence, such as particle swarm optimization (PSO) [1, 2], artificial bee colony (ABC) [3–6], cuckoo search (CS) [7, 8], bat algorithm (BA) [9–11], and artificial plant optimization algorithm [12, 13]. The multi-objective optimization problem (MOP) [14] originates from the design, modeling and planning of practical complex systems, it's a class of complex, difficult to solve the optimization problem. Since 1990s, large numbers of experimental results show that the multi-objective evolutionary algorithms (MOEAs) [15] have obvious advantages in solving this kind of problems. Scholars from different countries have also proposed different MOEAs. For example, the multi-objective genetic algorithm (MOGA) [16] proposed by Fonseca and Fleming, the non-dominated sorting

© Springer Nature Singapore Pte Ltd. 2018
K. Li et al. (Eds.): ISICA 2017, CCIS 873, pp. 159–170, 2018.
https://doi.org/10.1007/978-981-13-1648-7_14

genetic algorithm(NSGA) [17] proposed by Srinivas and Deb, the multi-objective particle swarm optimization (MOPSO) [18] proposed by Coello and so on.

The firefly algorithm (FA) [19–21] is a metaheuristic algorithm proposed by Yang in 2008 [22]. It achieves good performance in lots of applications in the field of single objective optimization. Yang proposed the multi-objective firefly algorithm (MOFA) [23] in 2013, the FA is applied to the multi-objective optimization problem, and the results show that the MOFA has obvious advantages in multi-objective evolutionary algorithm. On the basis of MOFA, this paper introduces the mechanism of elite particle guidance, which increases the population diversity and search range, improves the accuracy and uniformity of the solution set for multi-objective optimization, optimizes the fitting performance of Pareto front.

The Sect. 2 in this paper introduces the concept of multi-objective optimization problem and the standard firefly algorithm. The Sect. 3 elaborates the multi-objective firefly algorithm and its improved strategy. The Sect. 4 gives the validity tests of the proposed algorithm and its results. The conclusions are summarized in the Sect. 5.

2   The Multi-objective Optimization Problem and Standard Firefly Algorithm

2.1   The Multi-objective Optimization Problem

In general, the mathematical description of multi-objective optimization problems is

$$
\begin{aligned}
&\boldsymbol{x} = [x_1, x_2, \cdots, x_n] \\
\min \boldsymbol{y} = f(x) &= [f_1(\boldsymbol{x}), f_2(\boldsymbol{x}), \cdots, f_m(\boldsymbol{x})] \\
s.t. &\begin{cases} g_i(x) \le 0, \ i=1,2,\cdots,p \\ h_j(x) = 0, \ j=1,2,\cdots,q \end{cases}
\end{aligned} \tag{1}
$$

Where, n is the number of decision variables, m is the number of objective functions, x is the decision vector, y is the objective vector, $g_i(x)$ is the inequality constraint, and $h_j(x)$ is the equality constraint.

Different with the single objective optimization problem, in the multi-objective optimization problem, it is impossible to find a solution that can make all the objective functions achieving their optimal solutions at the same time. Instead, it can only coordinate the various objective functions. With a compromise, an optimal solution set can be obtained to make all the objective functions as near as possible to their optimal values. This optimal solution set is called the Pareto optimal solution set.

2.2   Standard Firefly Algorithm

The core idea of the firefly algorithm is that a firefly moves toward another firefly who is brighter than itself. This movement depends mainly on two factors: brightness and attraction degree. The brightness determines the location quality of an individual and its moving direction. The attraction degree determines the moving distance. The objective optimization is achieved by the constant updating of brightness and attraction degree.

To reduce the complexity of the problem, we assume that the absolute brightness I_i of a firefly i at the point of $x_i(x_{i1}, x_{i2}, \cdots, x_{id})$ is equal to the target function value at that point, i.e. $I_i = f(x_i)$. According to the literature [22], other relevant definitions of FA are as follows.

Definition 1. The attraction degree between firefly i and j:

$$\beta = \beta_0 e^{-\gamma r_{ij}^2} \tag{2}$$

Where, β_0 is the maximum attraction degree, that is, the attraction degree at $r = 0$, which is generally set to 1. γ is the light absorption coefficient, which represents the changes of attractiveness, and is generally set to $\gamma \in [0.01, 100]$. r_{ij} is the Euclidean distance between firefly i and j.

Definition 2. The position movement equation of firefly i attracted by another brighter firefly j:

$$x_i(t+1) = x_i(t) + \beta(x_j(t) - x_i(t)) + \alpha \cdot \varepsilon_i \tag{3}$$

Where, t is the iteration numbers of the algorithm. x_i and x_j are the locations of the firefly i and j, α is the step size factor, ε_i is the obtained random number vector, which obeys the uniform distribution, Gauss distribution or other distributions.

3 The Multi-objective Firefly Algorithm and Its Improvement

3.1 The Multi-objective Firefly Algorithm

In solving multi-objective optimization problems, MOFA determines the attractive relationship between individual firefly using the concept of Pareto dominant. If firefly $j \prec i$, it indicates that firefly j is brighter, and firefly i will update its position according to Eq. (3). An individual that is not controlled by any firefly updates its position using Eq. (4):

$$x_i(t+1) = \mathbf{g}^* + \alpha \cdot \varepsilon_i \tag{4}$$

$\mathbf{g}^*$ is the obtained optimal value of a single objective function converting from random weighted sum of multiple objective functions using Eq. (5), and each generation of w_k is required to be different. The definition of α and ε_i are the same with Eq. (3), their values can be chose based on specific problems. Here [23] is used as a reference.

$$\psi(x) = \sum_{k=1}^{K} w_k f_k, \sum_{k=1}^{K} w_k = 1 \tag{5}$$

After the evolution of each generation fireflies, all the non-inferior solutions are filed in the external archive (EA) [24], and the EA is maintained to ensure that all the solutions in the EA are not mutually dominant. In order to save computing resources, the maximum capacity of EA is generally set. When the number of non-inferior solutions exceeds the number of EA, the adaptive mesh deletion method is adopted to make the distribution of obtained Pareto non-inferior solutions more uniform [25]. In addition, to increase the population diversity, the mechanism of partial mutation [26] is usually introduced.

3.2   The Multi-objective Firefly Algorithm Guided by Elite Particle

From Sect. 3.1 we found that although each generation of $\mathbf{g}^*$ will change, this method still has great randomness. Not only the convergence is slow, but also the distribution of the obtained Pareto non-inferior solutions is not uniform. It cannot achieve good optimization performance. Here the elite particle is introduced, a non-inferior solution is randomly selected from EA as an elite particle to participate in the evolution of firefly. It is called as the multi-objective firefly algorithm guided by elite particle (EMOFA). As shown in Eq. (6), this algorithm can extend the search range of fireflies, guide the moving direction of fireflies, improve the convergence speed, and achieve better performance.

$$x_i(t+1) = c_1\mathbf{g}^* + c_2\boldsymbol{leader} + \alpha \cdot \boldsymbol{\varepsilon}_i \tag{6}$$

$$\boldsymbol{leader} = \boldsymbol{x}_i', \boldsymbol{x}_i' \in EA \tag{7}$$

where $c_1, c_2 \in [0, 1]$, $\sum c_i = 1$, $\boldsymbol{leader}$ is a randomly selected firefly from EA. The definitions of α and ε_i are the same with Eq. (3). The specific theoretical analysis is shown in Fig. 1.

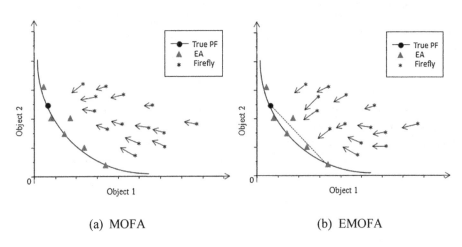

(a) MOFA   (b) EMOFA

Fig. 1.   Theoretical analysis of the algorithm

As can be seen from Fig. 1, the firefly in Fig. 1(a) only moves to a certain point in the front of the theory Pareto, although each generation of guiding points is different, but it still limits the scope of firefly exploration. In Fig. 1(b), guided by the particles in the EA that have been found, the direction of the firefly movement is more comprehensive, the coverage area is wider, and the fitting effect of the algorithm is improved.

The detailed algorithm is shown in Table 1:

Table 1. The multi-objective firefly algorithm guided by elite particle

1	Initialize the population of fireflies and set the parameter.
2	Calculate the fitness values of each firefly.
3	while (t < MAX_GEN) do
4	for $i = 1:n$ do
5	for $j = 1:n$ do
6	if $f(x_j) \prec f(x_i)$ do
7	Move firefly x_i towards x_j according to Eq.(6)
8	end if
9	if no one dominates $f(x_i)$
10	Move firefly x_i using Eq.(9)
11	end if
12	end for
13	end for
14	Update and save Pareto non-dominates solutions in EA
15	Gen++
16	end while
17	Postprocess results and visualisation

4 Experiments and Results

4.1 Experimental Setup

4.1.1 Test Functions

To verify the effectiveness of the proposed algorithm, seven classic MOP problem test functions [27–29] are selected to investigate the performance of the algorithm in solving different types of problems. Table 2 provides detailed information about the name and nature of each test function. The Pareto fronts of these test problems have different characteristics, so that the performance of the algorithm can be investigated more comprehensively.

4.1.2 Algorithm Comparison and Experimental Parameters

In order to test the performance of the algorithm, this paper selects five classic multi-objective optimization algorithms for comparative analysis, including MOPSO [18], NSGA-III [30], MOEA/D [31], PESA-II [32] and MOFA [23]. The parameter settings of these algorithms are all derived from the corresponding literature. The parameters of the proposed algorithm in this paper are set to $\alpha = 0.2$, $\beta_0 = 1$, $\gamma = 1$. In order to ensure the fairness of the algorithm, the iteration number of all the algorithms is 300 times, the population number is 50, and the number of EA is 200. Meanwhile, in order

Table 2. The test function set for multi-objective optimization problem.

Function	Pareto frontier features	Number of decision variables	Number of objective functions	Variable range
SCH	Convex	1	2	$[-10^5, 10^5]$
KUR	Discontinuous	3	2	$[-5, 5]$
ZDT1	Convex	30	2	$[0, 1]$
ZDT2	Concave	30	2	$[0, 1]$
ZDT3	Discontinuous	30	2	$[0, 1]$
ZDT6	Concave, multi-modal, biased	10	2	$[0, 1]$
Viennet3	Mixed, degraded	2	3	$[-3, 3]$

to reduce the interference of random factors, all the algorithms run 30 times independently for each test function. The evaluation index data is the average value of 30 runs, and the performance diagram is the point diagram of one random run.

4.2 Experimental Results and Analysis

The performance of each algorithm is evaluated by comparing the advantages and disadvantages of each algorithm on the three evaluation indexes. Tables 3, 4 and 5 show the mean and variance (Std) of the GD [33], SP [34] and MS [35] on seven test functions for each algorithm. The double tailed t-test with level of 5% is used to judge the significance difference of the results. The data with grey background in the table is the optimal value on the same test function for several algorithms. Table 6 calculates the number of optimal values obtained by each algorithm on the three evaluation functions, so that the performance of the algorithm can be evaluated comprehensively. Table 7 shows the Friedman test results of each algorithm on the three evaluation functions.

Table 3. The average value and variance of GD.

Problem	MOPSO Mean$_{(Std)}$	NSGA-III Mean$_{(Std)}$	MOEA/D Mean$_{(Std)}$	MOFA Mean$_{(Std)}$	PESA-II Mean$_{(Std)}$	EMOFA Mean$_{(Std)}$
SCH	6.74e-04$_{(1.56e-05)}$	6.44e-04$_{(4.40e-05)}$	5.85e-04$_{(4.39e-05)}$	6.70e-04$_{(4.29e-05)}$	6.67e-04$_{(4.01e-05)}$	6.62e-04$_{(2.53e-05)}$
KUR	1.31e-01$_{(3.33e-02)}$	6.01e-02$_{(4.33e-03)}$	9.86e-02$_{(6.01e-02)}$	9.98e-02$_{(3.01e-02)}$	5.17e-02$_{(3.98e-03)}$	8.92e-02$_{(1.76e-02)}$
ZDT1	8.16e-04$_{(1.97e-03)}$	9.84e-03$_{(1.67e-03)}$	2.24e-02$_{(1.87e-02)}$	2.36e-03$_{(8.57e-04)}$	8.25e-03$_{(3.27e-03)}$	2.24e-04$_{(3.65e-04)}$
ZDT2	8.37e-01$_{(2.23)}$	7.30e-03$_{(2.30e-03)}$	1.08e-01$_{(4.95e-02)}$	6.19e-03$_{(5.35e-03)}$	1.45e-02$_{(4.92e-03)}$	4.95e-04$_{(9.10e-04)}$
ZDT3	3.27e-04$_{(1.76e-03)}$	1.97e-02$_{(5.72e-03)}$	3.26e-02$_{(2.79e-02)}$	1.24e-02$_{(6.31e-03)}$	8.88e-03$_{(2.82e-03)}$	4.62e-03$_{(3.71e-04)}$
ZDT6	2.77e-02$_{(5.22e-02)}$	3.61e-05$_{(1.41e-05)}$	3.43e-01$_{(2.09e-01)}$	2.00e-02$_{(1.31e-01)}$	2.27e-03$_{(3.51e-03)}$	7.82e-03$_{(7.56e-03)}$
Viennet3	3.10e-04$_{(8.49e-05)}$	4.35e-04$_{(2.60e-04)}$	4.96e-03$_{(1.16e-02)}$	4.19e-04$_{(1.75e-04)}$	2.62e-04$_{(1.05e-04)}$	3.59e-04$_{(8.42e-05)}$

The performance of each algorithm on GD can be seen from Table 3. The optimization performance of EMOFA on ZDT1 and ZDT2 is obviously better than that of other algorithms, and the differences from the optimal values are not significant on other test functions. Although MOPSO has good performance, its robustness is poor, especially the variance on ZDT2 even reaches 2.23. The optimization performance of NSGA-III on ZDT6 is the best, and is obviously better than other algorithms. As can be

Table 4. The average value and variance of SP.

Problem	MOPSO Mean(Std)	NSGA-III Mean(Std)	MOEA/D Mean(Std)	MOFA Mean(Std)	PESA-II Mean(Std)	EMOFA Mean(Std)
SCH	2.23e-02(1.54e-03)	3.20e-02(9.83e-03)	9.48e-02(1.84e-02)	2.90e-02(6.92e-03)	2.24e-02(2.91e-03)	2.04e-02(1.59e-03)
KUR	1.09e-01(4.59e-03)	7.47e-02(1.41e-02)	2.80e-03(1.19e-02)	9.83e-02(8.81e-02)	8.54e-02(5.65e-03)	1.38e-03(4.74e-03)
ZDT1	5.19e-03(6.87e-04)	1.72e-02(5.63e-03)	3.62e-03(2.02e-03)	1.80e-03(4.74e-03)	5.81e-03(1.59e-03)	6.45e-03(1.07e-03)
ZDT2	3.39e-03(2.50e-03)	1.11e-02(6.36e-03)	3.62e-03(4.31e-03)	2.11e-03(4.67e-03)	6.72e-03(2.62e-03)	5.18e-03(1.43e-03)
ZDT3	4.32e-03(1.76e-03)	2.33e-02(1.15e-02)	1.96e-03(3.09e-03)	1.63e-02(1.78e-02)	8.36e-03(8.97e-03)	6.61e-03(2.64e-03)
ZDT6	5.45e-03(2.88e-03)	4.18e-02(4.01e-02)	1.91e-03(5.63e-03)	4.21e-02(3.18e-02)	6.67e-03(6.41e-03)	7.26e-03(1.38e-03)
Viennet3	4.46e-02(8.24e-03)	2.88e-02(2.05e-02)	2.11e-02(7.23e-02)	3.97e-02(2.77e-02)	4.35e-02(5.79e-03)	1.10e-01(1.14e-01)

Table 5. The average value and variance of MS.

Problem	MOPSO Mean(Std)	NSGA-III Mean(Std)	MOEA/D Mean(Std)	MOFA Mean(Std)	PESA-II Mean(Std)	EMOFA Mean(Std)
SCH	1.00(2.57e-03)	9.69e-01(3.77e-02)	7.33e-01(7.57e-02)	1.00(3.46e-04)	9.87e-01(6.98e-03)	1.00(6.49e-04)
KUR	9.35e-01(4.45e-02)	9.55e-01(4.14e-03)	4.73e-01(2.28e-01)	9.43e-01(1.99e-02)	8.72e-01(8.37e-02)	9.42e-01(2.67e-02)
ZDT1	9.96e-01(1.41e-02)	6.28e-01(9.69e-02)	8.34e-01(8.04e-02)	9.43e-01(1.38e-02)	9.41e-01(1.71e-02)	1.00(0.00)
ZDT2	6.85e-01(3.42e-01)	3.29e-01(9.36e-02)	8.09e-01(1.28e-01)	9.23e-01(8.78e-02)	8.40e-01(7.94e-03)	1.00(0.00)
ZDT3	9.81e-01(1.94e-02)	7.08e-01(1.28e-01)	7.09e-01(1.17e-01)	9.88e-01(9.02e-03)	9.46e-01(1.10e-02)	9.99e-01(3.04e-03)
ZDT6	1.00(2.69e-04)	9.17e-01(1.95e-01)	2.81e-01(1.02)	5.96e-01(1.93e-01)	1.00(1.77e-05)	1.00(1.04e-05)
Viennet3	9.94e-01(4.95e-03)	8.03e-01(2.56e-02)	4.84e-01(2.54e-01)	8.70e-01(7.34e-02)	8.40e-01(7.94e-03)	8.76e-01(6.07e-02)

Table 6. The summary of optimal solutions for each algorithm.

Evaluation index	MOPSO	NSGA-III	MOEA/D	MOFA	PESA-II	EMOFA
GD	1	1	1	0	2	2
SP	0	0	3	2	0	2
MS	3	1	0	1	1	5
Total	4	2	4	3	3	**9**

Table 7. The Friedman tests of three evaluation indexes for each algorithm.

GD Friedman Test		SP Friedman Test		MS Friedman Test	
algorithm	Average ranking	algorithm	Average ranking	algorithm	Average ranking
EMOFA	**2.29**	MOEA/D	1.57	EMOFA	**5.57**
PESA-II	2.86	MOPSO	3.29	MOPSO	4.29
NSGA-III	3.29	EMOFA	3.57	MOFA	4.14
MOFA	3.71	MOFA	3.71	PESA-II	3.14
MOPSO	4.00	PESA-II	4.00	NSGA-III	2.29
MOEA/D	4.86	NSGA-III	4.86	MOEA/D	1.57

seen from Table 4, the performances of all algorithms on the SP do not have obvious difference. Both EMOFA and MOFA achieved the optimal value for twice on the seven test functions, while MOEA/D achieved the optimal value for three times, which shows good dispersibility. As can be seen from Table 5, EMOFA takes advantage of its wide search range and large coverage, it achieved the optimal value for five times on the seven test functions. Table 6 shows that EMOFA achieved the optimal value for nine times in total on the three evaluation indexes. Both MOPSO and MOEA/D achieved the optimal value for four times. Both MOFA and PESA-II achieved the optimal value for three times. However, NSGA-III only achieved the optimal value for twice. These results show that the overall optimization performance of EMOFA is better than that of other algorithms. Compared with MOFA, EMOFA has optimal value for 16 times on the three evaluation indexes, which fully proves the effectiveness of the improved algorithm.

The results of Friedman test in Table 7 also validate the above experimental results. EMOFA ranked the first for both GD and MS, and ranked the third for SP. Based on comprehensive consideration, it can be concluded that EMOFA is an effective and competitive multi-objective optimization algorithm.

Due to space limitations, Figs. 2, 3, 4 and 5 shows the effect of each algorithm on some test functions, which clearly show the distributions of the obtained Pareto non-inferior solution and the real Pareto front. In which the black points represent the real Pareto front, and the red circles represent the obtained Pareto front. The advantages and disadvantages of the proposed algorithm are intuitively presented by comparing all the tables.

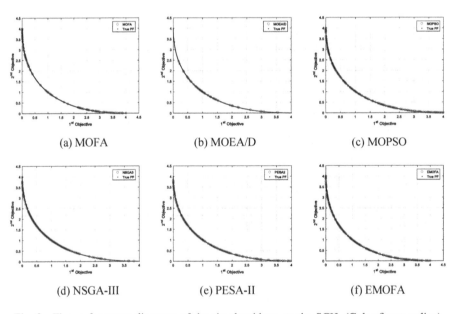

(a) MOFA   (b) MOEA/D   (c) MOPSO

(d) NSGA-III   (e) PESA-II   (f) EMOFA

Fig. 2. The performance diagrams of the six algorithms on the SCH. (Color figure online)

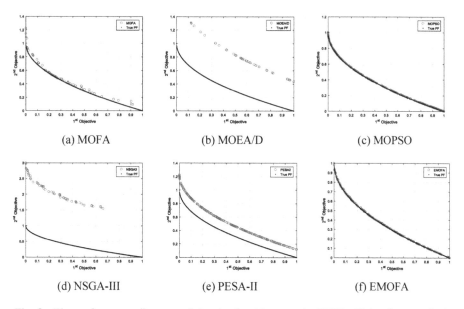

(a) MOFA　　　　　　　(b) MOEA/D　　　　　　　(c) MOPSO

(d) NSGA-III　　　　　　(e) PESA-II　　　　　　　(f) EMOFA

Fig. 3. The performance diagrams of the six algorithms on the ZDT1. (Color figure online)

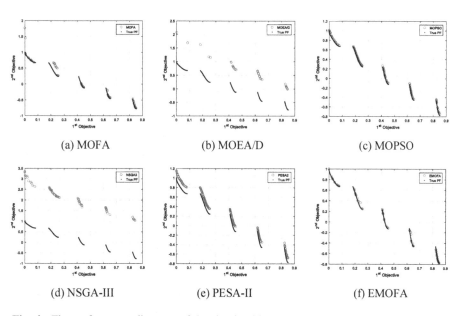

(a) MOFA　　　　　　　(b) MOEA/D　　　　　　　(c) MOPSO

(d) NSGA-III　　　　　　(e) PESA-II　　　　　　　(f) EMOFA

Fig. 4. The performance diagrams of the six algorithms on the ZDT3. (Color figure online)

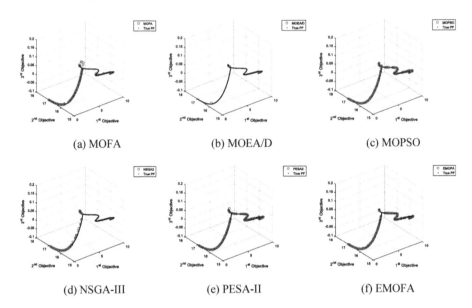

Fig. 5. The performance diagrams of the six algorithms on the Viennet3. (Color figure online)

5   Conclusions

The multi-objective optimization problems are getting more and more complex. Therefore, it is urgent to develop a new and effective multi-objective optimization algorithm to solve the complex MOP problems. As a new heuristic algorithm, the firefly algorithm has gradually shown its advantages. This paper presents a multi-objective firefly algorithm guided by elite particle (EMOFA). On the basis of MOFA, the archive elite particle is introduced to participate in the evolution, the convergence speed and search area of the proposed algorithm are accelerated, so that the obtained Pareto non-inferior solution set is closer to the real Pareto front. The coverage is more extensive, and it is an effective multi-objective optimization algorithm with superior performance. In the future, we will continue our research work on the following three aspects: (1) Apply the algorithm to more test functions to obtain high-quality algorithms that can adapt to various types of problems. (2) The processing method in MOEA/D can be used as a reference for the processing of EA to improve the uniformity of Pareto non-inferior solutions. (3) Further improve the learning mechanism of firefly, and improve its performance in solving MOP problem.

Acknowledgment. This research was supported by the Jiangxi Province Department of Education Science and Technology Project under Grant (No. GJJ161108), the National Natural Science Foundation of China under Grant (Nos. 61663029, 51669014, 61663028), Science Foundation of Jiangxi Province under Grant (Nos. 2016IBAB212037, 20171BAB202035), National Undergraduate Training Programs for Innovation and Entrepreneurship under Grant (No. 201711319001) and the project of Nanchang Institute of Technology's graduate student innovation program under Grant (No. YJSCX2017023).

References

1. Zhao, J., Fan, T., Lv, L., Sun, H., Wang, J.: Adaptive intelligent single particle optimizer based image de-noising in shearlet domain. Intell. Autom. Soft Comput. **23**(4), 661–666 (2017)
2. Zhao, J., Lv, L., Wang, H., Sun, H., Wu, R., Nie, J., Xie, Z.: Particle swarm optimization based on vector gaussian learning. KSII Trans. Internet Inf. Syst. **11**(4), 2038–2057 (2017)
3. Sun, H., Wang, K., Zhao, J., Yu, X.: Artificial bee colony algorithm with improved special centre. Int. J. Comput. Sci. Math. **7**(6), 548–553 (2016)
4. Lv, L., Wu, L.Y., Zhao, J., Wang, H., Wu, R.X., Fan, T.H., Hu, M., Xie, Z.F.: Improved multi-strategy artificial bee colony algorithm. Int. J. Comput. Sci. Math. **7**(5), 467–475 (2016)
5. Lu, Y., Li, R.X., Li, S.M.: Artificial bee colony with bidirectional search. Int. J. Comput. Sci. Math. **7**(6), 586–593 (2016)
6. Yun, G.: A new multi-population-based artificial bee colony for numerical optimization. Int. J. Comput. Sci. Math. **7**(6), 509–515 (2016)
7. Cui, Z., Sun, B., Wang, G., Xue, Y.: A novel oriented cuckoo search algorithm to improve DV-Hop performance for cyber-physical systems. J. Parallel Distrib. Comput. **77**(103), 42–52 (2017)
8. Wang, G.G., Gandomi, A.H., Yang, X.S., Alavi, A.H.: A new hybrid method based on krill herd and cuckoo search for global optimization tasks. Int. J. Bio-Inspir. Comput. **8**(5), 286–299 (2016)
9. Cai, X., Wang, L., Kang, Q., Wu, Q.: Bat algorithm with Gaussian walk. Int. J. Bio-Inspir. Comput. **6**(3), 166–174 (2014)
10. Cai, X., Gao, X.Z., Xue, Y.: Improved bat algorithm with optimal forage strategy and random disturbance strategy. Int. J. Bio-Inspir. Comput. **8**(4), 205–214 (2016)
11. Xue, F., Cai, Y., Cao, Y., Cui, Z., Li, F.: Optimal parameter settings for bat algorithm. Int. J. Bio-Inspir. Comput. **7**(2), 125–128 (2015)
12. Cui, Z., Fan, S., Zeng, J., Shi, Z.Z.: APOA with parabola model for directing orbits of chaotic systems. Int. J. Bio-Inspir. Comput. **5**(1), 67–72 (2013)
13. Cui, Z., Fan, S., Zeng, J., Shi, Z.Z.: Artificial plant optimisation algorithm with three-period photosynthesis. Int. J. Bio-Inspir. Comput. **5**(2), 133–139 (2013)
14. Xiao, X.W., Xiao, D., Lin, J.G., et al.: Overview on multi-objective optimization problem research. Appl. Res. Comput. **28**(3), 805–808 (2011)
15. Zheng, X.W.: Progress of research on multi-objective evolutionary algorithms. Comput. Sci. **34**(7), 187–192 (2007)
16. Fonseca, C.M., Fleming, P.J.: Genetic algorithms for multi-objective optimization: formulation discussion and generalization. In: Proceedings of the 5th International Conference on Genetic Algorithms, pp. 416–423. Morgan Kaufmann Publishers Inc. (1993)
17. Srinivas, N., Deb, K.: Multi-objective optimization using non-dominated sorting in genetic algorithms. Evol. Comput. **2**(3), 221–248 (1994)
18. Coello, C.A.C., Pulido, G.T., Lechuga, M.S.: Handling multiple objectives with particle swarm optimization. IEEE Trans. Evol. Comput. **8**(3), 256–279 (2004)
19. Wang, H., Wang, W.J., Sun, H., Rahnamayan, S.: Firefly algorithm with random attraction. Int. J. Bio-Inspir. Comput. **8**(1), 33–41 (2016)
20. Lv, L., Zhao, J.: The firefly algorithm with Gaussian disturbance and local search. J. Signal Process. Syst. **9**(11), 1–9 (2017)
21. Wang, H., Wang, W.J., Zhou, X.Y., Sun, H., Zhao, J., Yu, X., Cui, Z.: Firefly algorithm with neighborhood attraction. Inf. Sci. **23**(s), 382–383, 374–387 (2017)

22. Yang, X.S.: Nature-Inspired Metaheuristic Algorithms. Luniver Press, Somerset (2008)
23. Yang, X.S.: Multi-objective firefly algorithm for continuous optimization. Eng. Comput. **29** (2), 175–184 (2013)
24. Deb, K., Gupta, H.: Introducing robustness in multi-objective optimization. Evol. Comput. **14**(4), 463–494 (2014)
25. Yang, J.J., Zhou, J.Z., Wan, R.C.: Multi-objective particle swarm optimization based on adaptive grid algorithms. J. Syst. Simul. **20**(21), 5843–5847 (2008)
26. Li, K., Kwong, S., Deb, K.: A dual-population paradigm for evolutionary multi-objective optimization. Inf. Sci. 309(C), 50–72 (2015)
27. Zitzler, E., Deb, K., Thiele, L.: Comparison of multi-objective evolutionary algorithms: empirical results. Evol. Comput. **8**(2), 173–195 (2000)
28. Zitzler, E., Thiele, L.: Multiobjective optimization using evolutionary algorithms - a comparative case study. In: Eiben, A.E., Bäck, T., Schoenauer, M., Schwefel, H.-P. (eds.) PPSN 1998. LNCS, vol. 1498, pp. 292–301. Springer, Heidelberg (1998). https://doi.org/10.1007/BFb0056872
29. Tang, L.X., Wang, X.P.: A hybrid multi-objective evolutionary algorithm for multi-objective optimization problems. IEEE Trans. Evol. Comput. **17**(1), 20–46 (2013)
30. Mkaouer, W., Kessentini, M., Shaout, A.: Many-objective software remodularization using NSGA-III. ACM Trans. Softw. Eng. Method. **24**(3), 1–45 (2015)
31. Zhang, Q.F., Li, H.: MOEA/D: a multi-objective evolutionary algorithm based on decomposition. IEEE Trans. Evol. Comput. **11**(6), 712–731 (2007)
32. Corne, D.W., Jerram, N.R., Knowles, J.D., Oates, M.J.: PESA-II: region-based selection in evolutionary multi-objective optimization. In: Proceedings of the 3rd Annual Conference on Genetic and Evolutionary Computation, pp. 283–290, Morgan Kaufmann Publishers Inc. (2001)
33. Schott, J.R.: Fault tolerant design using single and multi-criteria genetic algorithm optimization. Cell. Immunol. **37**(1), 1–13 (1995)
34. Veldhuizen, D.A.V.: Multi-objective evolutionary algorithms: classifications, analyses, and new innovations. Evol. Comput. **8**(2), 125–147 (1999)
35. Cai, Z., Wang, Y.: A Multi-objective optimization-based evolutionary algorithm for constrained optimization. IEEE Trans. Evol. Comput. **10**(6), 658–675 (2006)

Improving Energy Demand Estimation Using an Adaptive Firefly Algorithm

Hui Wang[1(✉)], Zhangxin Chen[2], Wenjun Wang[3], Zhijian Wu[4],
Keliu Wu[2], and Wei Li[5]

[1] School of Information Engineering, Nanchang Institute of Technology,
Nanchang 330099, China
huiwang@whu.edu.cn
[2] Department of Chemical and Petroleum Engineering, University of Calgary,
Calgary, AB T2N 1N4, Canada
zhachen@ucalgary.ca, wukeliu19850109@163.com
[3] School of Business Administration, Nanchang Institute of Technology,
Nanchang 330099, China
wangwenjun881@126.com
[4] State Key Laboratory of Software Engineering, School of Computer,
Wuhan University, Wuhan 430072, China
zhijianwu@whu.edu.cn
[5] School of Information Engineering,
Jiangxi University of Science and Technology, Ganzhou 341000, China
nhwslw@gmail.com

Abstract. In recent years, energy demand has increased rapidly in developing countries. Energy demand estimation (EDE) plays an important role for policy makers and related organizations. Generally, energy demand can be mathematically modelled by population, economic, and other indicators using various forms of equations. However, it is difficult to choose optimal or near-optimal weighting factors for these models. In this paper, an adaptive firefly algorithm (AFA) is proposed to improve the efficiency of energy demand estimation in Turkey. Two different estimation models, including linear and quadratic forms, are used. Historical data in Turkey from 1979 to 2005 is utilized for training and testing these models. Experimental results show that our approach achieves better relative estimation errors than two other existing algorithms, ant colony (ACO) and particle swarm optimization (PSO).

Keywords: Firefly algorithm · Swarm intelligence · Energy demand estimation
Optimization

1 Introduction

Energy plays a significant role in improving the quality of human life. In the past decades, energy consumption has been increasing all over the world. In 2016, the global primary energy consumption increased by 1%, and the growth in developing countries is much faster. In China, India, and Turkey, the growth in energy consumption was up to 1.3%, 5.4%, and 4.2%, respectively [1]. According to the latest

© Springer Nature Singapore Pte Ltd. 2018
K. Li et al. (Eds.): ISICA 2017, CCIS 873, pp. 171–181, 2018.
https://doi.org/10.1007/978-981-13-1648-7_15

report of U.S. Energy Information Administration (EIA) in 2017 [2], the global energy demand will rise 30% in 2040. Therefore, energy demand estimation (EDE) is important. It can help policy makers to find the underlying imbalance of demand and supply and make appropriate responses.

To estimate the energy demand, several different approaches were proposed [3–9]. Toksarı [3] attempted to use ant colony optimization (ACO) to estimate the energy demand in Turkey. First, two models in linear and quadratic forms were developed based on population, gross domestic product (GDP), import and export. Then, ACO was utilized to optimize the weighting factors of these models. In [4], the performance of PSO was compared with ACO for energy demand estimation in Turkey. Results showed that PSO based on the quadratic model was better than ACO. Yu et al. [5] hybridized PSO and a genetic algorithm (GA), and proposed a PSO-GA model for energy demand estimation in China. Three models including linear, exponential, and quadratic were constructed by GDP, population, economic structure, an urbanization rate, energy structure and an energy price. Experimental results showed that PSO-GA performed better than GA, PSO, and ACO. Canyurt and Ozturk [6] presented an application of GA on forecasting the demand of fossil fuels in Turkey. A quadratic model was applied to the coal, oil and natural gas demand, respectively. Askarzadeh [7] used seven PSO variants to estimate the electricity demand of Iran. Simulation results showed that PSO with velocity control (PSO-vc) achieved higher accuracy than other algorithms. In [8], Wu and Peng presented a hybrid energy demand estimation approach, which combined a bat algorithm, Gaussian perturbations, and simulated annealing (BAG-SA) to optimize the weighting factors of multiple linear and quadratic models. Simulation results showed that the hybrid BAG-SA outperformed the genetic algorithm (GA), PSO, and bat algorithm in terms of the prediction accuracy. In [9], artificial bee colony (ABC) was used to estimate transportation energy demand in Turkey. Three models based on GDP, population, and total annual vehicle-km were designed. Results revealed the effectiveness of ABC for transportation energy planning.

In the literature, energy demand estimation can be divided into three tasks. First, different mathematical estimation models are developed based on the social and economic structure. Then, different methods are employed to find the optimal weighting factors of these models. Finally, the optimized models are used to forecast the future energy demand. However, choosing the optimal weighting factors is a challenging task, because the models are usually nonlinear and multimodal. A firefly algorithm (FA) is a recently proposed optimization technique, which has been successfully applied to various practical optimization problems [10]. In this paper, an adaptive firefly algorithm (AFA) is proposed to improve the efficiency of energy demand estimation in Turkey. Two estimation models in linear and quadratic forms are used. Historical data between 1979 and 2005 is utilized for training and testing the models. In the experiments, the performance of AFA is compared with that of PSO and ACO.

The rest of the paper is organized as follows. In Sect. 2, the standard FA is briefly described. Our approach is proposed in Sect. 3. Results and discussions are presented in Sect. 4. Finally, this work is concluded in Sect. 5.

2 Firefly Algorithm

Swarm intelligence is a new computational paradigm inspired by the social behaviours from the nature [11]. In the past decades, some swarm intelligence algorithms have been proposed, such as particle swarm optimization (PSO) [12, 13], artificial bee colony (ABC) [14–18], firefly algorithm (FA) [19–21], cuckoo search (CS) [22, 23] and bat algorithm (BA) [24]. Recently, some studies proved that FA is a good search algorithm on many optimization problems. FA is a population-based random search algorithm, which starts with a set of initial fireflies. Each firefly in the population is called a candidate solution. The updating of solutions is defined by the attracting behavior among fireflies [10]. In FA, a less bright firefly can move to a brighter one because of attraction. It means that a solution jumps to a new position and tries to find a better one. Recently, FA was successfully used to solve various optimization problems [25–29].

Assume that $X_i = (x_{i1}, x_{i1}, \ldots, x_{iD})$ is the ith firefly (candidate solution) in the current population, where $i = 1, 2, \ldots, N$, N is the population size, and D is the dimension size. For any two fireflies X_i and X_j, their attractiveness is defined by [10]

$$\beta(r_{ij}) = \beta_0 e^{-\gamma r_{ij}^2} \tag{1}$$

where β_0 is the attractiveness at $r = 0$, γ is a light absorption coefficient, and r_{ij} is the distance between X_i and X_j. The distance r_{ij} is calculated by [10]

$$r_{ij} = \left\| X_i - X_j \right\| = \sqrt{\sum_{d=1}^{D} (x_{id} - x_{jd})^2} \tag{2}$$

If X_j is better than X_i, X_i will move to X_j. The movement is defined by [10].

$$x_{id}(t+1) = x_{id}(t) + \beta(r_{ij}) \cdot \left(x_{jd}(t) - x_{id}(t) \right) + \alpha \cdot \left(rand - \frac{1}{2} \right) \tag{3}$$

where x_{id} and x_{jd} are the dth dimensions of X_i and X_j, respectively, $\alpha \in [0,1]$ is called a step factor, and $rand$ is a random value with the range $[0,1]$.

3 Proposed Approach

3.1 Adaptive Firefly Algorithm (AFA)

Like other population based random search algorithms, the performance of FA is sensitive to its control parameters. According to reference [30], the step factor α is related to the convergence of FA. When FA is convergent, the parameter α should satisfy the following condition:

$$\lim_{t \to \infty} \alpha = 0 \tag{4}$$

where t is the generation index.

To avoid manually adjusting the parameter α, an adaptive parameter strategy was proposed in our previous work [30]. The parameter α is dynamically updated as follows:

$$\alpha(t+1) = \alpha(t) \cdot \left(1 - \frac{t}{T_{max}}\right) \tag{5}$$

where T_{max} is the maximum number of generations. It is obvious that the above parameter strategy satisfies Eq. (4). As the generation increases, α gradually decreases to zero.

According to the movement model (described in Eq. (3)), $\alpha \cdot \left(rand - \frac{1}{2}\right)$ is regarded as a step size. At the beginning stage of the search, a large α results in a large step size. This is helpful to enhance the exploration ability. With increasing of t, α becomes smaller. At the last stage of the search, a small α is beneficial for improving the exploitation ability. Therefore, the proposed parameter strategy can adaptively adjust the search and make a good balance between exploration and exploitation. In AFA, the parameter α is updated by using Eq. (5), and the initial $\alpha(0)$ is set to 0.5 [30].

3.2 Estimation Models

In this paper, energy demand estimation in Turkey is used as a case study. According to references [3, 4], the energy demand estimation model is determined by four factors: GDP (X_1), population (X_2), import (X_3), and export (X_4). Table 1 presents the historical data in Turkey on energy demand, GDP, population, import, and export between 1979 and 2005 [3, 4]. As seen, the GDP, population, import and export increased 3.40, 0.68, 22.03, and 31.51 times, respectively, from 1979 to 2005. The energy consumption increased 1.98 times during this period. It seems that the social development and economic growth lead to the increase of energy consumption.

Based on the social and economic indicators, the estimation of energy demand can be modeled by using different forms, such as linear, exponential, and quadratic. By the suggestions in [3, 4], two models in linear and quadratic forms are used as follows:

Linear model:

$$Y_l = w_0 + \sum_{i=1}^{n} w_i X_i \tag{6}$$

Quadratic model:

$$Y_q = w_0 + \sum_{i=1}^{n} w_i X_i + \sum_{i=1}^{n} \sum_{j=i+1}^{n} k_{ij} X_i X_j + \sum_{i=1}^{n} u_i X_i^2 \tag{7}$$

Table 1. Historical data in Turkey on energy demand (MTOE), GDP ($\$10^9$), population ($10^6$), import ($\$10^9$) and export ($\$10^9$) from 1979 to 2005.

Year	Energy demand	GDP	Population	Import	Export
1979	30.71	82.00	43.53	5.07	2.26
1980	31.97	68.00	44.44	7.91	2.91
1981	32.05	72.00	45.54	8.93	4.70
1982	34.39	64.00	46.69	8.84	5.75
1983	35.70	60.00	47.86	9.24	5.73
1984	37.43	59.00	49.07	10.76	7.13
1985	39.40	67.00	50.31	11.34	7.95
1986	42.47	75.00	51.43	11.10	7.46
1987	46.88	86.00	52.56	14.16	10.19
1988	47.91	90.00	53.72	14.34	11.66
1989	50.71	108.00	54.89	15.79	11.62
1990	52.98	151.00	56.10	22.30	12.96
1991	54.27	150.00	57.19	21.05	13.59
1992	56.68	158.00	58.25	22.87	14.72
1993	60.26	179.00	59.32	29.43	15.35
1994	59.12	132.00	60.42	23.27	18.11
1995	63.68	170.00	61.53	35.71	21.64
1996	69.86	184.00	62.67	43.63	23.22
1997	73.78	192.00	63.82	48.56	26.26
1998	74.71	207.00	65.00	45.92	26.97
1999	76.77	187.00	66.43	40.67	26.59
2000	80.50	200.00	67.42	54.50	27.78
2001	75.40	146.00	68.37	41.40	31.33
2002	78.33	181.00	69.30	51.55	36.06
2003	83.84	239.00	70.23	69.34	47.25
2004	87.82	299.00	71.15	97.54	63.17
2005	91.58	361.00	72.97	116.77	73.48

where Y is the energy demand; n is the number of factors affecting the energy demand; X_i and X_j are the ith and jth factors, respectively; and w_0, w_i, k_{ij}, and u_i are the weighting factors.

In the above models, four factors GDP (X_1), population (X_2), import (X_3), and export (X_4) are used. Therefore, n is equal to 4. For the linear model, there are five weighting factors. For the quadratic model, there are 15 weighting factors.

3.3   Fitness Evaluation Function

In this paper, we use the proposed AFA to optimize the weighing factors (w_0, w_i, k_{ij}, and u_i) in the energy demand estimation models. To evaluate the quality of obtained weighting factors, the sum of squared errors (SSE) is employed to construct the fitness evaluation function:

$$f(X) = \sum_{i=1}^{m} (Y_{act} - Y_{est})^2 \tag{8}$$

where Y_{act} and Y_{est} are the actual (observed) and estimated energy demand, respectively, and m is the number of training data.

3.4   Data Normalization

In this paper, the historical data listed in Table 1 is used for training and testing the energy demand estimation models. To remove the effects of different units of data, a normalization method is used. In Table 1, the data of energy demand, GDP, population, import and export is normalized as follows:

$$X^* = \frac{X - X_{min}}{X_{max} - X_{min}} \tag{9}$$

where X is the value to be normalized, X^* is the normalized value, and $[X_{min}, X_{max}]$ is the boundary range. When training a model, the related data (1979–1995) is normalized. When testing the model, the related data (1996–2005) is normalized.

4   Simulation Experiments

4.1   Experimental Setup

In the experiments, the proposed AFA is applied to energy demand estimation of Turkey. In reference [3], all data between 1979 and 2005 is used to train the models (determine the weighting factors), and the data between 1996 and 2005 is used to test the models. This is not fair to use the training data to test the models. In this paper, we only use the partial data (1979–1995) to train the models, and the rest data (1996–2005) is applied to test the models.

In AFA, the population size N is set to 30. We use the maximum number of fitness evaluations (Max_FEs) as the stopping condition, and Max_FEs is set to 1.0E + 06. The initial $\alpha(0)$, β_0, and γ are set to 0.5, 1.0, and $1/\Gamma^2$, respectively, where Γ is the length of a search range.

For each model, AFA is run 10 times. The best, worst and mean results are reported. In the experiments, we use the relative error (RE) and mean relative error (MRE) to measure the performance of FA:

$$RE = \left| \frac{Y_{act} - Y_{est}}{Y_{act}} \right| \tag{10}$$

$$MRE = \frac{1}{M} \cdot \sum_{i=1}^{M} \left| \frac{Y_{act}(i) - Y_{est}(i)}{Y_{act}(i)} \right| \tag{11}$$

where $Y_{act}(i)$ and $Y_{est}(i)$ are the actual and estimated energy demand for the ith test data, respectively, and M is the number of test data.

To check the performance of our approach, we compare it with PSO and ACO on the same data. The involved algorithms are listed as follows:

- AFA with the linear model (AFA-L);
- AFA with the quadratic model (AFA-Q);
- ACO with the linear model (ACO-L) [3];
- ACO with the quadratic model (ACO-Q) [3];
- PSO with the linear model (PSO-L) [4];
- PSO with the quadratic model (PSO-Q) [4].

4.2 Results

Table 2 shows the average results of AFA-L and AFA-Q over 10 runs, where "Best", "Worst", and "Mean" represent the best, worst, and mean MRE, respectively, and "Std" indicates the standard deviation. For AFA-L, the best MRE is 1.15%, and the worst is only 3.21%. AFA-Q achieves more accurate prediction than AFA-L and the best MRE is 0.76%. It demonstrates that the quadratic model is more suitable for fitting the data.

Table 2. Statistical results achieved by AFA-L and AFA-Q between 1996 and 2005.

MRE	AFA-L	AFA-Q
Best	1.15%	**0.76%**
Worst	3.21%	**2.39%**
Mean	2.17%	**1.26%**
Std	6.84E−03	**4.74E−03**

Table 3 presents the best estimation of energy demand achieved by AFA-L and AFA-Q on the test data. The results from ACO-L, ACO-Q, PSO-L, and PSO-Q were taken from [3, 4]. As shown, the largest and smallest relative errors are 2.95% and 0.10% for AFA-L, respectively. For AFA-Q, the largest relative error is 1.98%, and the smallest one is 0.09%. It seems that AFA-L and AFA-Q almost perfectly fit the actual energy demand on some test years.

Figure 1 shows the actual energy demand and the best estimation of energy demand on the training data. Figure 2 illustrates the actual energy demand and the best estimation of energy demand on the test data. From Figs. 1 and 2, AFA-L and AFA-Q can obtain a good fit to the training and test data.

Table 4 gives the comparison of the best estimation of energy demand for different algorithms between 1996 and 2005. For the linear model, AFA-L achieves a smaller RE (1.15%) than PSO-L (1.43%) and ACO-L (1.41%). ACO-L is slightly better than

Table 3. The best estimation of energy demand achieved by AFA-L and AFA-R between 1996 and 2005.

Years	Actual energy demand (MTOE)	Estimated energy demand (MTOE)		RE	
		AFA-L	AFA-Q	AFA-L	AFA-Q
1996	69.86	71.10	69.78	1.78%	0.12%
1997	73.78	73.33	72.57	0.61%	1.65%
1998	74.71	75.00	74.78	0.39%	0.09%
1999	76.77	76.17	75.90	0.78%	1.13%
2000	80.50	80.16	80.02	0.42%	0.60%
2001	75.40	77.62	75.82	2.95%	0.55%
2002	78.33	80.54	79.88	2.82%	1.98%
2003	83.84	83.76	84.36	0.10%	0.62%
2004	87.82	87.70	88.26	0.14%	0.50%
2005	91.58	92.92	91.22	1.47%	0.38%
Total average				**1.15%**	**0.76%**

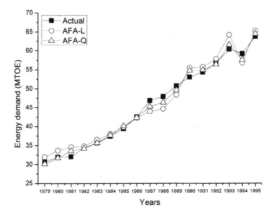

Fig. 1. Actual energy demand and the best estimation of energy demand on the training data.

PSO-L, while PSO-Q is better than ACO-Q. For the quadratic model, AFA-Q still obtains the smallest RE (0.76%) among three algorithms. The comparison results demonstrate that our approaches AFA-L and AFA-Q can effectively improve the prediction accuracy of energy demand. Due to the space limitation, some forecasting results on energy demand between 2018 and 2020 are not presented.

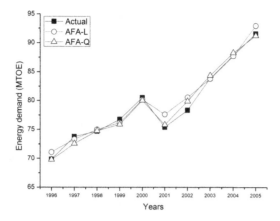

Fig. 2. Actual energy demand and the best estimation of energy demand on the testing data.

Table 4. Comparison of the best estimation of energy demand with different algorithms between 1996 and 2005.

Years	PSO-L	ACO-L	AFA-L	PSO-Q	ACO-Q	AFA-Q
	RE	RE	RE	RE	RE	RE
1996	0.57%	0.54%	1.78%	0.13%	0.94%	0.12%
1997	2.29%	2.33%	0.61%	1.16%	0.15%	1.65%
1998	2.06%	2.10%	0.39%	0.11%	1.28%	0.09%
1999	3.31%	3.87%	0.78%	2.03%	0.89%	1.13%
2000	0.14%	0.50%	0.42%	0.35%	1.20%	0.60%
2001	0.90%	0.61%	2.95%	1.25%	0.89%	0.55%
2002	1.39%	0.28%	2.82%	2.22%	2.83%	1.98%
2003	1.38%	1.90%	0.10%	0.27%	0.64%	0.62%
2004	0.38%	0.32%	0.14%	0.50%	0.32%	0.50%
2005	1.91%	1.66%	1.47%	0.30%	1.58%	0.38%
Total average	1.43%	1.41%	**1.15%**	0.83%	1.07%	**0.76%**

5   Conclusions

Energy demand estimation models are usually complex, nonlinear, and multimodal. Thus, choosing optimal weighting factors for these models is a challenging task. In this paper, an adaptive FA (AFA) algorithm is proposed to improve the prediction accuracy of energy demand. The estimation of energy demand in Turkey is considered as a case study. Two estimation models in linear and quadratic forms are developed. In addition, data normalization is employed to remove the effects of different units of data.

Historical data between 1979 and 1995 is utilized for training the models and the rest data (1996–2005) is used for testing the models. Simulation results show that AFA achieves a smaller relative error than PSO and ACO for the linear and quadratic models. The best prediction accuracy obtained by AFA-Q is up to 99.24%. These results demonstrate that our approach can effectively improve the prediction accuracy of energy demand.

This paper only uses four factors (GDP, population, import, and export) to construct the estimation models. However, there are some other factors, such as social policy, energy structure and an energy price, which may affect the energy demand. Moreover, this paper focuses on comparing AFA with PSO and ACO with the same data, and the data is not new (1979–2005). Some latest data can be added in the experiments. These will be further investigated in the future work.

Acknowledgement. This work was supported by the Science and Technology Plan Project of Jiangxi Provincial Education Department (No. GJJ170994), the National Natural Science Foundation of China (No. 61663028), the Distinguished Young Talents Plan of Jiangxi Province (No. 20171BCB23075), the Natural Science Foundation of Jiangxi Province (No. 20171BAB202035), the China Scholarship Council (No. 201608360029), and the Open Research Fund of Jiangxi Province Key Laboratory of Water Information Cooperative Sensing and Intelligent Processing (No. 2016WICSIP015).

References

1. BP Group: BP statistical review of world energy June 2017. Technical report. BP Company (2017)
2. Energy Information Association: Annual energy outlook 2017 with projections to 2050. US Department of Energy (2017)
3. Toksarı, M.D.: Ant colony optimization approach to estimate energy demand of Turkey. Energy Policy 35(8), 3984–3990 (2007)
4. Ünler, A.: Improvement of energy demand forecasts using swarm intelligence: the case of Turkey with projections to 2025. Energy Policy 36(6), 1937–1944 (2008)
5. Yu, S., Wei, Y.M., Wang, K.: A PSO–GA optimal model to estimate primary energy demand of China. Energy Policy 42, 329–340 (2012)
6. Canyurt, O.E., Ozturk, H.K.: Application of genetic algorithm (GA) technique on demand estimation of fossil fuels in Turkey. Energy Policy 36(7), 2562–2569 (2008)
7. Askarzadeh, A.: Comparison of particle swarm optimization and other metaheuristics on electricity demand estimation: a case study of Iran. Energy 72, 484–491 (2014)
8. Wu, Q., Peng, C.: A hybrid BAG-SA optimal approach to estimate energy demand of China. Energy 120, 985–995 (2017)
9. Sonmez, M., Akgüngör, A.P., Bektaş, S.: Estimating transportation energy demand in Turkey using the artificial bee colony algorithm. Energy 122, 301–310 (2017)
10. Yang, X.S.: Nature-Inspired Metaheuristic Algorithms. Luniver Press, Beckington (2008)
11. Torres-Treviño, L.M.: Let the swarm be: an implicit elitism in swarm intelligence. Int. J. Bio-Inspired Comput. 9(2), 65–76 (2017)
12. Zhao, J., Lv, L., Wang, H., Sun, H., Wu, R., Nie, J., Xie, Z.: Particle swarm optimization based on vector Gaussian learning. KSII Trans. Internet Inf. Syst. 11(4), 2038–2057 (2017)
13. Wang, H., Sun, H., Li, C.H., Rahnamayan, S., Pan, J.S.: Diversity enhanced particle swarm optimization with neighborhood search. Inf. Sci. 223, 119–135 (2013)
14. Sun, H., Wang, K., Zhao, J., Yu, X.: Artificial bee colony algorithm with improved special centre. Int. J. Comput. Sci. Math. 7(6), 548–553 (2016)

15. Yu, G.: A new multi-population-based artificial bee colony for numerical optimization. Int. J. Comput. Sci. Math. **7**(6), 509–515 (2016)
16. Wang, H., Wu, Z.J., Rahnamayan, S., Sun, H., Liu, Y., Pan, J.S.: Multi-strategy ensemble artificial bee colony algorithm. Inf. Sci. **279**, 587–603 (2014)
17. Lv, L., Wu, L.Y., Zhao, J., Wang, H., Wu, R.X., Fan, T.H., Hu, M., Xie, Z.F.: Improved multi-strategy artificial bee colony algorithm. Int. J. Comput. Sci. Math. **7**(5), 467–475 (2016)
18. Cui, L.Z., Li, G.H., Lin, Q.Z., Du, Z.H., Gao, W.F., Chen, J.Y., Lu, N.: A novel artificial bee colony algorithm with depth-first search framework and elite-guided search equation. Inf. Sci. **367–368**, 1012–1044 (2016)
19. Yu, G.: An improved firefly algorithm based on probabilistic attraction. Int. J. Comput. Sci. Math. **7**(6), 530–536 (2016)
20. Wang, H., Cui, Z., Sun, H., Rahnamayan, S., Yang, X.S.: Randomly attracted firefly algorithm with neighborhood search and dynamic parameter adjustment mechanism. Soft. Comput. **21**(18), 5325–5339 (2017)
21. Lv, L., Zhao, J.: The firefly algorithm with Gaussian disturbance and local search. J. Signal Process. Syst. **90**, 1123–1131 (2017). https://doi.org/10.1007/s11265-017-1278-y. (in press)
22. Cui, Z., Sun, B., Wang, G., Xue, Y., Chen, J.: A novel oriented cuckoo search algorithm to improve DV-Hop performance for cyber-physical systems. J. Parallel Distrib. Comput. **103**, 42–52 (2017)
23. Zhang, M., Wang, H., Cui, Z., Chen, J.: Hybrid multi-objective cuckoo search with dynamical local search. Memet. Comput. **10**, 199–208 (2017). https://doi.org/10.1007/s12293-017-0237-2. (in press)
24. Cai, X., Gao, X.Z., Xue, Y.: Improved bat algorithm with optimal forage strategy and random disturbance strategy. Int. J. Bio-Inspired Comput. **8**(4), 205–214 (2016)
25. Marichelvam, M.K., Geetha, M.: A hybrid discrete firefly algorithm to solve flow shop scheduling problems to minimise total flow time. Int. J. Bio-Inspired Comput. **8**(5), 318–325 (2016)
26. Wang, H., Wang, W., Sun, H., Rahnamayan, S.: Firefly algorithm with random attraction. Int. J. Bio-Inspired Comput. **8**(1), 33–41 (2016)
27. Kaur, M., Sharma, P.K.: On solving partition driven standard cell placement problem using firefly-based metaheuristic approach. Int. J. Bio-Inspired Comput. **9**(2), 121–127 (2017)
28. Wang, H., Wang, W.J., Zhou, X.Y., Sun, H., Zhao, J., Yu, X., Cui, Z.: Firefly algorithm with neighborhood attraction. Inf. Sci. **382–383**, 374–387 (2017)
29. dos Santos Coelho, L., Mariani, V.C.: Improved firefly algorithm approach applied to chiller loading for energy conservation. Energy Build. **59**, 273–278 (2013)
30. Wang, H., Zhou, X.Y., Sun, H., Yu, X., Zhao, J., Zhang, H., Cui, L.Z.: Firefly algorithm with adaptive control parameters. Soft. Comput. **21**(17), 5091–5102 (2017)

Evolutionary Multi-objective and Dynamic Optimization – Hybrid Methods

Firefly Algorithm with Elite Attraction

Jing Wang[(⊠)]

School of Software and Internet of Things Engineering,
Jiangxi University of Finance and Economics, Nanchang, China
wj.jxufe@gmail.com

Abstract. Firefly algorithm is a simple and efficient meta-heuristic optimization algorithm which has outstanding performance on many optimization problems. However, in the standard FA, the fireflies will be attracted by all the other bright fireflies, and there is a lot of attraction that does not affect, but will increase the computational time of the algorithm. In addition, all the best firefly information in the search process has not been recorded, which may lead the algorithm to be inefficient. To over these problems, this paper proposed en elite-k attraction firefly algorithm (EkFA), which can not only reduce the no effective attractions between the fireflies but also can make full use of the best firefly's information to guide other nearby fireflies to movement. Thirteen well-known benchmark functions are used to verify the performance of our proposed method. The experimental results show that the accuracy and efficiency of the proposed algorithm are significantly better than those of other FA variants.

Keywords: Firefly algorithm · Meta-heuristic algorithm · Elite-k attraction
Global optimization

1 Introduction

The firefly algorithm is a swarm-based optimization algorithm which is proposed by Yang in 2008 [1]. The firefly algorithm searches for the brighter companion around the region by simulating the firefly luminescence behavior in nature, moving towards the better position in the region, and finally gathering to the brightest and optimal position, so as to realize the function of searching the optimal solution. Compared with other typical intelligent algorithms such as particle swarm optimization algorithm and genetic algorithm, the existing simulation results show that the firefly algorithm has high convergence speed and revenge accuracy. And the firefly algorithm's parameters are simple and effective, so it is popular in a short time, and has been widely used in many fields, such as scheduling problem [2, 3], wireless sensor networks [4, 5], stock forecasting [6], mechanical design optimization problem [7] and so on.

Although the firefly algorithm has been widely used in many optimization problems, the algorithm still has some flaws. In the standard FA algorithm, each firefly will be attracted to all the remaining fireflies, and this behavior will increase the number of invalid objective function evaluations. When faced with high-dimensional problems or objective functions are complex, the complexity of the algorithm will be dramatically increased. To overcome these problems, Wang [8] proposes a neighborhood attraction

© Springer Nature Singapore Pte Ltd. 2018
K. Li et al. (Eds.): ISICA 2017, CCIS 873, pp. 185–194, 2018.
https://doi.org/10.1007/978-981-13-1648-7_16

FA (NaFA), In NaFA, each firefly is attracted by other brighter fireflies selected from a predefined neighborhood, but the predefined neighborhood is a virtual structure, and not the real structure of the fireflies. The essence of the algorithm is to reduce the number of times that fireflies are attracted by other remaining fireflies. In this paper, we proposes a new FA variant, namely elite-k attraction FA (EkFA). EkFA can not only reduce the no effective attractions between the fireflies, but also can make full use of the best firefly's information to guide other nearby fireflies to movement.

The rest of this paper is organized as follows. In Sect. 2, stand FA and its variant are briefly reviewed. Our approach DFA is described in Sect. 3. Experimental results and analysis are presented in Sect. 4. Finally, the work is concluded in Sect. 5.

2 A Brief Review of Firefly Algorithm

In FA, the location of each firefly represents a solution to the problem to be solved. The brightness of the firefly depends on the objective function value of the problem to be solved. The better the objective function is, the stronger the brightness of the firefly. As the iterative process progresses, the weak fireflies in the population are moving closer to their own fireflies, and most of the fireflies will gather near the brightest fireflies, and the brightest fireflies are the optimal solution to the problem.

There are four very important concepts in the firefly algorithm: light intensity, attractiveness, distance and movement.

Light Intensity: The light intensity $I(r)$ is defined by Yang [9]:

$$I(r) = I_0 e^{-\gamma r^2} \tag{1}$$

where I_0 is the initial brightness. The parameter γ is the light absorption coefficient, r is the distance between two fireflies.

Attractiveness: The attractiveness of a firefly is monotonically decreasing as the distance increases, and the attractiveness is as follows [9]:

$$\beta(r) = \beta_0 e^{-\gamma r^2} \tag{2}$$

Where $\beta 0$ is the attractiveness at $r = 0$. The light absorption coefficient γ will determine the variation of attractiveness β and $\gamma \in [0, \infty]$. For most practical implementations, Yang suggest that $\gamma = 1$ and $\beta_0 = 1$.

Distance: For two fireflies x_i and fireflies x_j, r is defined by Yang [9]:

$$r_{ij} = \|x_i - x_j\| = \sqrt{\sum_{d=1}^{D} \left(x_{id} - x_{jd}\right)^2} \tag{3}$$

Movement: The light intensity of the weak firefly will move to another brighter firefly, assuming that a firefly x_j is more brighter than firefly x_i, the position update equation given by the following formula [9]:

$$x_i(t+1) = x_i(t) + \beta_0 e^{-\gamma r^2}\left(x_j(t) - x_i(t)\right) + \alpha \epsilon_i \tag{4}$$

where t is the iterations. The third term of the right is a random disturbance term which contains α and ϵ_i, $\alpha \in [0,1]$ is the step factor, $\epsilon_i \in [0.5, 0.5]$ is a random number vector obtained by Gaussian distribution or Levy flight [10].

The framework of stand FA is listed as follow, FEs is the number of evaluations, MAXits is the maximum number of evaluations, and PS is the population size.

Algorithm1. Framework of FA

1	Randomly initialize N fireflies (solutions) as an initial population $\{X_i \mid i = 1, 2, ..., N\}$;
2	Calculate the fitness v of each firefly X_i;
3	FEs = 0 and PS = N;
4	**while** FEs ≤ $MAXFEs$
5	**for** i= 1 to N
6	**for** j = 1 to N
7	**if**$f(X_j) < f(X_i)$
8	Move X_i towards X_j according to Eq. 4;
9	Calculate the fitness value of the new X_i;
10	FEs++;
11	**end if**
12	**end for**
13	**end for**
14	**end while**

Fister et al. [11] proposed a memetic self-adaptive FA (MFA), where values of control parameters are changed during the run. Experimental results show that MFA were very promising and showed a potential that this algorithm could successfully be applied in near future to the other combinatorial optimization.

MFA makes the dynamic change of the step factor with the evolutionary iteration. It is redefined as follows:

$$\alpha(t+1) = \left(\frac{1}{9000}\right)^{\frac{1}{t}} \alpha(t), \tag{5}$$

where t represents the current iteration. In the above-mentioned MFA [11], Fister also changes the fireflies's movement strategy, which can be defined by the following equation:

$$x_{id}(t+1) = x_{id}(t) + \beta\left(x_{jd}(t) - x_{id}(t)\right) + \alpha(t)s_d \epsilon_i, \tag{6}$$

Where

$$\beta = \beta_{min} + (\beta_0 - \beta_{min})e^{-\gamma r^2}, \tag{7}$$

$$s_d = x_d^{max} - x_d^{min}, \tag{8}$$

x_{id} denotes the d-dimensional variable of the i-th firefly, β_{min} is usually set to 0.2 that is the minimum value in β, s_d is the length of the domain of the initialization variable. x_d^{max} and x_d^{min} are the maximum and minimum boundaries of the variable, respectively. In this paper, our proposed strategies are based on MFA.

3 Our Proposed Firefly Algorithm

Firefly algorithm is a simple and efficient meta-heuristic algorithm. However, it also has two unfavorable factors. First, as shown in Algorithm 1, there are two inner cycles and one outer loop in the firefly algorithm. The number of outer cycles is the maximum number of iterations *MAXits*, and the number of internal cycles is the population number N. Assume that the time spent on the evaluation of the objective function is T, the complexity of the firefly algorithm is $O(N^2 * MAXits * T)$. In the algorithm, the main cost of the firefly algorithm is to spend on the evaluation of the objective function, no matter how far the distance between the fireflies, the fireflies will be attracted by other brighter fireflies. The attraction will cause the fireflies to move, and the movement will cause the algorithm to evaluate the objective function again. Although the attraction will increase the diversity of the algorithm, however it also will increase the complexity of algorithm. Second, the algorithm does not use the history optimal value of the individual, nor does it use the global optimal value. It also may affect the accuracy and convergence speed of the firefly algorithm.

Therefore, based on this phenomenon, this paper proposes a new FA variant with elite-k attraction method, we called it EkFA. We select the best fireflies and then move the K other fireflies closest to the best fireflies, rather than moving to other brighter fireflies in each generation. For a more clear description of elite-k attraction method, Fig. 1 shows the structure of elite-k attraction and full attraction. Assume that all N fireflies are organized into circular topological structures according to their indices. Figure 1(a) and (b) show the state of elite-k attraction and full attraction when N is 12, respectively. It is easy to conclude that the time complexity of the algorithm is $O(N^2 * MAXits * T)$ in full attraction, and the time complexity of the algorithm in elite-k attraction is $O((N^2 - K^2) * MAXits * T)$, and in each generation there are K^2 individuals that are moving according to the global optimal value. In this paper, K is set to 10.

Experiments show that MFA is an efficient FA variant. In fact, the EkFA algorithm not only uses the elite-k attraction method, but also incorporates the advanced technology of MFA. The EkFA algorithm framework is shown in Algorithm 2, FEs is the number of evaluations, MAXits is the maximum number of evaluations, and PS is the population size.

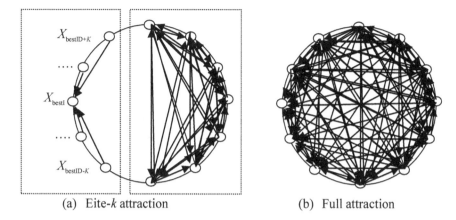

(a) Eite-*k* attraction   (b) Full attraction

Fig. 1. Eite-*k* attraction vs. Full attraction, where $N = 12$

Algorithm 2. Framework of EkFA

1	Randomly initialize N fireflies (solutions) as an initial population $\{X_i \mid i = 1, 2, …, N\}$;
2	Calculate the fitness v of each firefly X_i;
3	FEs = 0 and PS = N;
4	**while** FEs ≤ $MAXits$
5	⎸ Find the best individual's ID;
6	⎸ Update the step factor α according to Eq. 5;
7	⎸ Update the attractiveness β according to Eq. 7
8	⎸ **for** i= 1 to N
9	⎸ ⎸ **if** (abs(i - bestID) < K)
10	⎸ ⎸ ⎸ Move X_i towards X_{best} according to Eq. 6;
11	⎸ ⎸ **else**
12	⎸ ⎸ ⎸ **for** j = 1 to N
13	⎸ ⎸ ⎸ ⎸ **if** $f(X_j) < f(X_i)$
14	⎸ ⎸ ⎸ ⎸ ⎸ Move X_i towards X_j according to Eq. 6;
15	⎸ ⎸ ⎸ ⎸ ⎸ Calculate the fitness value of the new X_i;
16	⎸ ⎸ ⎸ ⎸ ⎸ FEs++;
17	⎸ ⎸ ⎸ ⎸ **end if**
18	⎸ ⎸ ⎸ **end for**
19	⎸ ⎸ **end else if**
20	⎸ **end for**
21	**end while**

4 Experimental Study

4.1 Test Problems

In this experiment, 13 well-known benchmark functions [12] were used to test the performance of the algorithm. In these functions, f_1-f_5 are unimodal functions, f_6 is a step function with a minimum value, and discontinuous, f_7 is a quadratic function with noise, and $f_8 - f_{13}$ are multimodal functions with many local minimums. All functions are minimization problems, and their descriptions are shown in Table 1.

4.2 Experimental Results

In order to ensure the fairness of the algorithm, the following parameters of all algorithms are set to the same, and list as follow:

- Population size: 30
- Max iterations: 5.0E+05
- Run times: 30
- Problem dimension: 30

The computational results of EkFA are recorded in Table 2. As shown in the Table 2, except f_3, f_5 and f_8 fall into the local optimal, most of the other functions find the approximate optimal solution. Especially the f_6 and f_{11}, they finds the global optimal solution every time.

In 2010, Yang [9] suggested $\gamma = \frac{1}{\Gamma^m}, m \geq 1$, Γ is the given length scale for a given problem. So EkFA and the two versions of FA are compared in the following experiment. Table 3 records the computational results between the two versions FA and EkFA. "Mean" represents the average optimal value for 30 runs, and "Std Dev" is the standard deviation. As shown in the Table 3, when $\gamma = 1.0$, the stand FA is hard to get the global optimal solution in most benchmark functions, but when γ change to $1/\Gamma^2$, the performance of stand FA improved significantly. f_1, f_4, f_7 and f_{11} get the approximate optimal solution, and the quality of the rest functions is also improved. EkFA is much better than the two versions of FA except f_3, f_5 and f_7, FA ($\gamma = 1/\Gamma^2$) is better than EkFA on f_5 and f_7, but their optimal values are very close. In other functions, the optimal value of EkFA is significantly improved compared with the two versions of FA.

4.3 Comparison of EkFA with Other FA Variants

In order to more fully test the performance of the EkFA algorithm, in the next experiment, the EkFA algorithm will compare with the several recently proposed FA variants. The details of all the parameters of the involved algorithm set as shown in Table 4.

The comparison results between VSSFA, WSSFA, MFA, NaFA and EkFA are recorded in Table 5. Because we did not implement the code for the other four versions of the FA algorithm, the comparison data in the Table 5 came from [8, 15]. All the adjustment parameters of the algorithms are set as shown in Table 4, and the run parameters are set to the same as shown in Sect. 4.2. As shown in Table 5, VSSFA and

Table 1. Benchmark functions used in the experiments, where D is the problem dimension

Name	Function	Search Range	Global Optimum
Sphere	$f_1(x) = \sum_{i=1}^{D} x_i^2$	[-100, 100]	0
Schwefel 2.22	$f_2(x) = \sum_{i=1}^{D} |x_i| + \prod_{i=1}^{D} x_i$	[-10, 10]	0
Schwefel 1.2	$f_3(x) = \sum_{i=1}^{D} (\sum_{j=1}^{i} x_j)^2$	[-100, 100]	0
Schwefel 2.21	$f_4(x) = \max\{|x_i|, 1 \leq i \leq D\}$	[-100, 100]	0
Rosenbrock	$f_5(x) = \sum_{i=1}^{D} [100(x_{i+1} - x_i^2)^2 + (1 - x_i^2)^2]$	[-30, 30]	0
Step	$f_6(x) = \sum_{i=1}^{D} \lfloor x_i + 0.5 \rfloor$	[-100, 100]	0
Quartic with noise	$f_7(x) = \sum_{i=1}^{D} ix_i^4 + random[0, 1)$	[-1.28, 1.28]	0
Schwefel 2.26	$f_8(x) = \sum_{i=1}^{D} -x_i \sin\left(\sqrt{|x_i|}\right) + 418.9829D$	[-500, 500]	0
Rastrigin	$f_9(x) = \sum_{i=1}^{D} [x_i^2 - 10\cos 2\pi x_i + 10]$	[-5.12, 5.12]	0
Ackley	$f_{10}(x) = -20 \exp\left(-0.2\sqrt{\frac{1}{D}\sum_{i=1}^{D} x_i^2}\right)$ $- \exp\left(\frac{1}{D}\sum_{i=1}^{D} \cos(2\pi x_i)\right) + 20$ $+ e$	[-32, 32]	0
Griewank	$f_{11}(x) = \frac{1}{4000}\sum_{i=1}^{D} x_i^2 - \prod_{i=1}^{D} \cos\left(\frac{x_i}{\sqrt{i}}\right) + 1$	[-600, 600]	0
Penalized 1	$f_{12}(x) = \frac{\pi}{D}\left\{\sum_{i=1}^{D-1}(y_i - 1)^2[1 + \sin(\pi y_{i+1})] + (y_D - 1)^2 \right.$ $\left. + 10\sin^2(\pi y_1)\right\}$ $+ \sum_{i=1}^{D} u(x_i, 10, 100, 4), y_i$ $= 1 + (x_i + 1)/4$ $u(x_i, a, k, m) = \begin{cases} u(x_i, a, k, m), & x_i > a \\ 0, & -a \leq x_i \leq a \\ k(-x_i - a)^m, & x_i < -a \end{cases}$	[-50, 50]	0
Penalized 2	$f_{13}(x) = 0.1\left\{\sin^2(3\pi x_1)\right.$ $+ \sum_{i=1}^{D-1}(x_i - 1)^2[1 + \sin^2(3\pi x_{i+1})]$ $\left. + (x_D - 1)^2[1 + \sin^2(2\pi x_D)]\right\}$ $+ \sum_{i=1}^{D} u(x_i, 5, 100, 4)$	[-50, 50]	0

Table 2. Computational results of EkFA

Function	Worst	Best	Mean	Std dev
f_1	1.41E$-$127	7.02E$-$128	1.09E$-$127	2.86E$-$128
f_2	1.85E$-$64	1.35E$-$64	1.60E$-$64	2.14E$-$65
f_3	5.39E+01	2.75E$-$02	1.83E+01	2.07E+01
f_4	2.27E$-$64	1.45E$-$64	1.76E$-$64	3.29E$-$65
f_5	9.91E+01	2.31E+01	4.06E+01	3.28E+01
f_6	0.00E+00	0.00E+00	0.00E+00	0.00E+00
f_7	1.48E$-$01	4.21E$-$02	9.88E$-$02	4.58E$-$02
f_8	$-$5.66E+03	$-$7.02E+03	$-$6.16E+03	5.66E+02
f_9	5.37E+01	2.69E+01	3.90E+01	1.16E+01
f_{10}	2.18E$-$14	1.47E$-$14	1.75E$-$14	3.89E$-$15
f_{11}	0.00E+00	0.00E+00	0.00E+00	0.00E+00
f_{12}	1.57E$-$32	1.57E$-$32	1.57E$-$32	0.00E+00
f_{13}	5.29E$-$32	2.34E$-$32	3.72E$-$32	1.50E$-$32

Table 3. Computational results between the two versions FA and EkFA

Function	FA ($\gamma = 1.0$)		FA ($\gamma = 1/\Gamma^2$)		EkFA	
	Mean	Std dev	Mean	Std dev	Mean	Std dev
f_1	6.67E+04	1.83E+04	5.14E$-$02	1.36E$-$02	**1.09E$-$127**	**2.86E$-$128**
f_2	5.19E+02	1.42E+02	1.07E+00	2.65E$-$01	**1.60E$-$64**	**2.14E$-$65**
f_3	2.43E+05	4.85E+04	**1.26E$-$01**	**1.86E$-$01**	1.83E+01	2.07E+01
f_4	8.35E+01	3.16E+01	9.98E$-$02	2.34E$-$02	**1.76E$-$64**	**3.29E$-$65**
f_5	2.69E+08	6.21E+07	**3.41E+01**	**6.23E+00**	4.06E+01	3.28E+01
f_6	7.69E+04	3.38E+03	5.24E+03	1.08E+03	**0.00E+00**	**0.00E+00**
f_7	5.16E+01	2.46E+01	**7.55E$-$02**	**1.42E$-$02**	9.88E$-$02	4.58E$-$02
f_8	1.10E+04	3.77E+03	9.16E+03	1.78E+03	$-$**6.16E+03**	**5.66E+02**
f_9	3.33E+02	6.28E+01	4.95E+01	2.39E+01	**3.90E+01**	**1.16E+01**
f_{10}	2.03E+01	2.23E$-$01	1.21E+01	1.96E+00	**1.75E$-$14**	**3.89E$-$15**
f_{11}	6.54E+02	1.69E+02	2.13E$-$02	1.47E$-$02	**0.00E+00**	**0.00E+00**
f_{12}	7.16E+08	1.82E+08	6.24E+00	4.62E+00	**1.57E$-$32**	**0.00E+00**
f_{13}	1.31E+09	4.76E+08	5.11E+01	1.28E+01	**3.72E$-$32**	**1.50E$-$32**

Table 4. The parameters of the algorithms

	α	α_{min}	$\alpha(0)$	γ	β	β_{min}
MFA [11]	-	-	0.5	$1/\Gamma^2$	$\beta_0 = 1.0$	0.2
WSSFA [13]	-	0.04	-	1.0	$\beta_0 = 1.0$	-
VSSFA [14]	-	-	-	1.0	$\beta_0 = 1.0$	-
NaFA [8]	-	-	0.5	$1/\Gamma^2$	$\beta_0 = 1.0$	0.2
EKFA	-	-	0.5	$1/\Gamma^2$	$\beta_0 = 1.0$	0.2

WSSFA seem difficult to find the global optimal solution in many functions. MFA is better than VSSFA and WSSFA, especially for f_6, MFA achieves global optimal solutions. Both NaFA and EkFA obtain global optimal values on functions f_6 and f_{11}. MFA achieves 4 optimal solutions (f_5,f_6,f_7,f_8), NaFA achieves 4 optimal solutions (f_3,f_6,f_9,f_{11}), while EkFA achieves 8 optimal solutions $(f_{1-2},f_4,f_6,f_{10-13})$, the EkFA algorithm is significantly better than the other four algorithms.

Table 5. Computational results of Mean best fitness value by VSSFA, WSSFA, MFA, NaFA and EkFA

Function	VSSFA mean	WSSFA mean	MFA mean	NaFA mean	EkFA mean
f_1	5.84E+04	6.34E+04	1.56E−06	4.43E−29	**1.09E−127**
f_2	1.13E+02	1.35E+02	1.85E−03	2.98E−15	**1.60E−64**
f_3	1.16E+05	1.10E+05	5.89E−05	**2.60E−28**	1.83E+01
f_4	8.18E+01	7.59E+01	1.73E−03	3.43E−15	**1.76E−64**
f_5	2.16E+08	2.49E+08	**2.29E+01**	2.39E+01	4.06E+01
f_6	5.48E+04	6.18E+04	**0.00E+00**	**0.00E+00**	**0.00E+00**
f_7	4.43E+01	3.24E−01	**1.30E−01**	2.91E−02	9.88E−02
f_8	1.07E+04	1.06E+04	**4.94E+03**	6.86E+03	−6.16E+03
f_9	3.12E+02	3.61E+02	6.47E+01	**2.09E+01**	3.90E+01
f_{10}	2.03E+01	2.05E+01	4.23E−04	3.02E−14	**1.75E−14**
f_{11}	5.47E+02	6.09E+02	9.86E−03	**0.00E+00**	**0.00E+00**
f_{12}	3.99E+08	6.18E+08	5.04E−08	1.36E−31	**1.57E−32**
f_{13}	8.12E+08	9.13E+08	6.06E−07	2.13E−30	**3.72E−32**

5   Conclusions

In the standard FA, the fireflies will be attracted by all other bright fireflies, there is a lot of attraction without effect, but it will increase the computational time of the algorithm. In addition, the best firefly information in all the search process have not been recorded, which may cause the algorithm inefficiency. This paper presents an elite-k attraction firefly algorithm; elite-k attraction has two important roles:

- Reduce the number of attraction.
- Make full use of elite information to guide the movement of fireflies near the elite.

Thirteen well-known benchmark functions are used to verify the performance of our proposed method. The experimental results show that the accuracy and efficiency of EkFA are significantly better than the standard FA, VSSFA, WSSFA, MFA and NaFA.

In our proposed EkFA, k plays an important role. In this paper, when the population size is 30, the parameter k is set to 10, and a good solution can be obtained. In the future work, k will be further studied.

Acknowledgments. The authors would like to thank anonymous reviewers for their detailed and constructive comments that help us to increase the quality of this work. This work was supported by the National Natural Science Foundation of Jiangxi Province (No. 20151BAB207023) and Science and Technology project of Jiangxi Provincial Department of Education (No. GJJ150448).

References

1. Yang, X.-S.: Nature-Inspired Metaheuristic Algorithms. Luniver Press, Bristol (2008)
2. SundarRajan, R., Vasudevan, V., Mithya, S.: Workflow scheduling in cloud computing environment using firefly algorithm. In: 2016 International Conference on Electrical, Electronics, and Optimization Techniques (ICEEOT), pp. 955–960 (2016)
3. Marichelvam, M.K., Prabaharan, T., Yang, X.S.: A discrete firefly algorithm for the multi-objective hybrid flowshop scheduling problems. IEEE Trans. Evol. Comput. **18**, 301–305 (2014)
4. Manshahia, M.S., Dave, M., Singh, S.B.: Firefly algorithm based clustering technique for wireless sensor networks. In: 2016 International Conference on Wireless Communications, Signal Processing and Networking (WiSPNET), pp. 1273–1276 (2016)
5. Lalwani, P., Ganguli, I., Banka, H.: FARW: firefly algorithm for routing in wireless sensor networks. In: 2016 3rd International Conference on Recent Advances in Information Technology (RAIT), pp. 248–252 (2016)
6. Kazem, A., Sharifi, E., Hussain, F.K., Saberi, M., Hussain, O.K.: Support vector regression with chaos-based firefly algorithm for stock market price forecasting. Appl. Soft Comput. **13**, 947–958 (2013)
7. Baykasoğlu, A., Ozsoydan, F.B.: Adaptive firefly algorithm with chaos for mechanical design optimization problems. Appl. Soft Comput. **36**, 152–164 (2015)
8. Wang, H., Wang, W., Zhou, X., Sun, H., Zhao, J., Yu, X., Cui, Z.: Firefly algorithm with neighborhood attraction. Inf. Sci. **382–383**, 374–387 (2016)
9. Yang, X.-S.: Engineering Optimization: An Introduction with Metaheuristic Applications. Wiley Publishing, Hoboken (2010)
10. Yang, X.-S.: Firefly algorithm, lévy flights and global optimization. In: Bramer, M., Ellis, R., Petridis, M. (eds.) Research and Development in Intelligent Systems XXVI, pp. 209–218. Springer, Heidelberg (2010). https://doi.org/10.1007/978-1-84882-983-1_15
11. Fister Jr., I., Yang, X.-S., Fister, I., Brest, J.: Memetic firefly algorithm for combinatorial optimization. arXiv preprint arXiv:1204.5165 (2012)
12. Lin, Y., Wang, L., Zhong, Y., Zhang, C.: Control scaling factor of cuckoo search algorithm using learning automata. Int. J. Comput. Sci. Mathematics **7**, 476–484 (2016)
13. Yu, S., Su, S., Lu, Q., Huang, L.: A novel wise step strategy for firefly algorithm. Int. J. Comput. Math. **91**, 2507–2513 (2014)
14. Yu, S., Zhu, S., Ma, Y., Mao, D.: A variable step size firefly algorithm for numerical optimization. Appl. Math. Comput. **263**, 214–220 (2015)
15. Wang, H., Zhou, X., Sun, H., Yu, X., Zhao, J., Zhang, H., Cui, L.: Firefly algorithm with adaptive control parameters. Soft. Comput. **21**, 5091–5102 (2016)

A Hybrid Fireworks Explosion Algorithm

Liping Wang[1,2], Renwen Chen[1(✉)], and Chengwang Xie[3]

[1] State Key Laboratory of Mechanics and Control of Mechanical Structures,
Nanjing University of Aeronautics and Astronautics,
29 Yudao Street, Nanjing 210016, Jiangsu, China
rwchen@nuaa.edu.cn
[2] School of Information and Computer Engineering, Pingxiang University,
Pingxiang 337000, China
[3] School of Computer and Information Engineering,
Guangxi Teachers Education University, Nanning 530023, China

Abstract. In order to improve the abilities of global exploration and local exploitation for the fireworks explosion algorithm (FA), the opposition-based learning method is introduced to generate the opposition-based population and to expand the exploration range of the FA. In addition, a computing method of adaptive adjustment of explosion radius is proposed based on the fitness differences of the individuals in the population. These above strategies are integrated to form an adaptive fireworks explosion optimization algorithm with opposition-based learning. The presented FA is experimented with other four swarm intelligence optimization algorithms, and the results show that our improved FA clearly outperforms the others in the performance of convergence and accuracy.

Keywords: Fireworks algorithm · Opposition-based learning
Explosion radius

1 Introduction

Nowadays, a large number of increasingly complex optimization problems emerge in the field of scientific computing and engineering practice, which generally cannot meet the mathematical prerequisites, making them difficult to be solved effectively with the classical mathematical methods. At the same, many researchers, by abstracting and modeling some natural phenomena, proposed some new stochastic optimization algorithms such as ant colony algorithm [1], particle swarm optimization algorithm [2], simulated annealing algorithm [3] and bee colony algorithm [4] which all belong to the swarm intelligence optimization algorithm. Commonly, the swarm intelligence optimization algorithm is used to generate a population of candidate solutions randomly. Such candidate solutions are developed into better solutions with an iterative process, and the algorithm terminates when solution reaching a satisfactory fitness level is found. With the swarm intelligence optimization algorithm, the problems to be solved don't need to be continuous or differentiable or meet additional conditions. As a result, the swarm intelligence optimization algorithm can be applied to a wider range, especially appropriate to solve problems where only approximate solutions are required.

© Springer Nature Singapore Pte Ltd. 2018
K. Li et al. (Eds.): ISICA 2017, CCIS 873, pp. 195–203, 2018.
https://doi.org/10.1007/978-981-13-1648-7_17

In 2010, Tan and Zhu proposed a novel fireworks explosion algorithm (FEA) by simulating the scattering of sparks in the fireworks explosion. When a firework is set off, a shower of sparks will fill the circular space around the point where the firework is set off (the "explosion point"). If such circular space is viewed as a local space of the problem, and the sparks generated by the explosion are viewed as points in the local space, then one explosion can be viewed as an exploration of the local space, and such exploration can be viewed as a search in the local space around the explosion point. Due to its advantages such as simple execution process, high precision, good convergence and robustness, the FEA has become popular among researchers. So far, researchers have improved the fundamental FEA from different aspects, further bettering the performance of the FEA in solving optimization problems. In literature [6], the parameters of the fundamental FEA, such as explosion point and explosion radius, are modified to improve the performance of the FEA. In literature [7], the exchange operator is introduced by reference to the PSO algorithm to realize the exchange among individuals in the population and guide the population to evolve into the global best solution. In literature [8], the idea of the genetic algorithm is introduced into the FEA, where a certain explosion point is selected randomly to exchange for the information of the current best explosion point, so as to increase the diversity of the populations and overcome the premature convergence of the algorithm. In literature [9], the FEA is combined with the differential evolution(DE) algorithm to propose a combined FEA-DE algorithm and good experimental results are obtained. In literature [10], the radius of the fireworks explosion is determined on whether better sparks can be generated by the firework, and a dynamic firework explosion algorithm, i.e., the dyn FEA, is proposed. In literature [11], the biogeography-based optimization algorithm is introduced to the FEA to enhance the information exchange among the fireworks, and the BBO-FWA, a new combined fireworks explosion algorithm, is put forward. To some extent, all the foresaid algorithms accelerate the convergence of the algorithm increase the accuracy of the algorithm, and improved the performance of the algorithm.

However, it should be noted that fireworks explosion optimization algorithm is a greedy algorithm. With the ongoing of the iterative process, all individuals will gradually draw close to the current best individual, and the explosion radius will gradually shrink, so does the diversity of the populations. This search mechanism makes the FEA give a poor performance in solving the optimization problems in an irregular solution space because maintaining the diversity of the firework populations and promoting the search in a wider space can increase the algorithm's probability of finding the global best solution in a complicated space, thereby avoiding the premature convergence of the algorithm. Moreover, in the existing FEAs, the fireworks explosion radius is generally set to decrease in a nonlinear way, which can ensure that the algorithm can perform a global exploration at the early stage, and a precise local search for the best solution in the final stage. But the preset variation mode of the fireworks explosion radius is applied only to the firework populations, and aims at effectively utilizing the search resources, without considering that the radii of the fireworks in the same generation are set differently according to the different fitness values of fireworks. Given all that, in this paper, the opposition-based learning (OBL) is introduced to the

FEA: first, the opposition-based populations are generated through the opposition-based learning, and better individuals are selected from the original population and the opposition-based populations for evolution, to promote the algorithm to converge to the global best solution; second, in this paper, the explosion radii of different individuals in populations in different generations are not set to decrease in a nonlinear way, while the explosion radii of fireworks in the same generation are assigned according to their respective fitness values, so as to utilize the computing resources efficiently and improve the search efficiency.

Based on the above, an Adaptive Fireworks Explosion Optimization Algorithm Using Opposition-Based Learning (AFEAOL) is proposed in this paper. This algorithm has three characteristics: (1) it gives full play to the relatively stronger local exploration capacity of the basic FEA algorithm and improves the accuracy of the FEA; (2) it the global exploration capacity of the algorithm with the opposition-based learning so as to improve the capacity of the algorithm of finding the global best solution; (3) it adaptively adjusts the explosion radius of the fireworks according to the fitness values of individuals inside the population of the algorithm, so as to effectively utilize the search resources. The organic combination of the foresaid tactics aims at improving the FEA's capability in solving complicated optimization problems.

2 AFAOL Algorithm

2.1 Opposition-Based Learning

In 2005, Tizhoosh [12] proposed the concept of the Opposition-Based Learning (OBL) and explained that the opposite solution is 50% more likely to be closer to the global best solution than the current solution. The main idea is to generate opposite individuals in the space where the current individuals are located and make the opposite individuals to compete with current ones and then lead better individuals to evolve to the next generation.

Definition 1. Given a feasible solution in D-dimensional search space $x_i = (x_{i,1}, x_{i,2}, \cdots, x_{i,D})$, where $x_{i,j} \in [a_j, b_j], j = 1, 2, \ldots, D$, then the opposite solution $x_i' = (x_{i,1}', x_{i,2}', \cdots, x_{i,D}')$ should satisfy the Eq. (1).

$$x_{i,j}' = \left(a_{i,j} + b_{i,j}\right) - x_{i,j} \tag{1}$$

Definition 2. The generalize opposition-based learning (GOBL): Let $x_{i,j}' = k \cdot (a_j + b_j) - x_{i,j}$, where $x_{i,j} = [a_j, b_j], i = 1, 2, \cdots, |Popsize|, j = 1, 2, \ldots, D, |Popsize|$ denotes the size of population, D indicates the dimension of search space, and k is a random number within the range of [0, 1].

Definition 3. Dynamic Generalized Opposition-based Learning (DGOBL): Let $x'_{i,j} = k \cdot (da_j + db_j) - x_{i,j}$, where da_j and db_j are the minimum value and maximum value respectively on the jth dimension of search space of the current population, that is:

$$\begin{cases} da_j = \min(A_j) \\ db_j = \max(A_j) \end{cases} \tag{2}$$

where A_j is the set of all values of individuals in the population on the jth dimension of the search space, and $k \in [0, 1]$ is the generalized coefficient.

To validate the effectiveness of the opposition-based learning, an example is given below to show the strong exploration ability of such mechanism. Let the problem to be optimized min $f(q) = \|q - A\|$, where the dimension of the search space for the optimization problem is 2, and A is the global best location for the optimization problem and set as (14, 15). When the value range of the decision variables x_1 and x_2 is [0, 20], select a point p = (9, 8) in the decision space, whose fitness value f(p) = $\sqrt{74}$. According to Definition 1, it can be obtained that the opposite solution point $p' = (11, 12)$, with its fitness value of $f(p') = 3\sqrt{2}$. Obviously, $f(p') < f(p)$, which means the opposite solution point is closer to the location of the best solution.

In this paper, the operator of DGOBL is adopted to generate the opposite population, and individuals with better fitness values in the union set of the current population and the opposite group are selected to involve the next generation. The strategy of DGOBL can expand the range of exploration and avoid the ineffective search, which results in the improvement of convergence and accuracy of the algorithm.

2.2 Fireworks Algorithm Optimization

The search method of basic fireworks explosion algorithm is as follows. N initial locations (fireworks) are generated randomly in the search space. For example, the ith explosion point in the D-dimensional search space can be denoted as $(x_{i,1}, x_{i,2}, \cdots, x_{i,D})$. For a good firework, the maximum radius can be set as r, which is also denoted as the maximum scattering range of sparks. If the number of the explosion layers is w, the explosion radius at each layer can be expressed as $j \cdot r/w (j = 1, 2, ..., w)$, and r is generally decreased in a nonlinear way, which results in the global exploration at the early stage and local exploitation at the end stage for the algorithm. Due to the limitation of time and space resources, and considering to generating sufficient sparks, the distances between sparks and explosion sparks can be set as $r/4$, $r/2$, $3r/4$ and r. For a firework i, the sparks generated by explosion point i can be denoted as follows.

$$x_i^* = x_i + r_j \cdot \vec{b}_k \tag{3}$$

Where, x_i is the current location of the firework i, x_i^* is the location of sparks generated by the explosion of firework i, r_j ($j = 1, 2, 3, 4$) is the explosion radius, and $r_1 = r$, $r_2 = 3r/4$, $r_3 = r/2$, $r_4 = r/4$, $r = r(t)$ represent the explosion radii of the tth generation of fireworks. $\vec{b}_k (k = 1, 2, ..., m)$ is the explosion direction, and m is the total number of all explosion directions after the explosion of the firework i. For a search space with a low dimension (for example, $D \leq 5$), the algorithm selects the intuitive standard coordinate axis direction as the explosion directions, that is, the number of explosion directions for D-dimensional problems is 2D. However, for search space with higher dimension (for example, $D > 5$), if 2D explosion directions are still adopted, the time and space resources consumed by the algorithm are too large. As a result, when an optimization problem with high dimension is solved, D/3 directions and additional D/3 opposite ones shall be selected randomly from the 2D standard coordinate axis directions as the explosion directions at each layer, thus forming 2D/3 explosion directions.

Furthermore, let $x^l (l = 1, 2, 3, \cdots, D)$ denotes any dimensional decision variable in D-dimensional search space and we set $x^l \in [a_l, b_l]$, if the spark x_j generated by firework x_i falls out of the range $[a_l, b_l]$, it will become an infeasible solution, so we reset x_j on the lth dimension as follows.

$$x_j^l = rand(a_l, b_l) \quad if\ x_j^l < a_l\ or\ x_j^l > b_l \tag{4}$$

where $rand(\cdot)$ denotes a uniform random number in the interval $[a_l, b_l]$.

2.3   Adaptive Explosion Radius

In the basic FA, the radius is identical in each generational population, that is to say, the search scope of better individuals is the same as that of the worse ones in a group. However, for a better individual, it is generally closer to the best solution, so it should search around a smaller range so that converges quickly. In contrast, the worse individuals are generally farther from the optimum to the problem solved, so they should explore in a larger space to search the best solution. It is obvious that the one-cut way of radius assignment exists some defects, which will result in difficulty when assigning the computational resources of the algorithm. Therefore, we assign the adaptive explosion radii to the fireworks based on their fitness values in the paper. The method of assignment radius can be described as follows.

(1) Combining the fireworks and their sparks to form a union set U.
(2) Evaluating the individuals in the set U, and sort them based on their fitness values.
(3) Computing the initial explosion radius $r(t)$ of the tth generational population as follows.

The method of calculating the explosion radius is characterized as follows. (1) The initial explosion radius of each generational population is calculated according Eq. (3), and the radius will decrease gradually with the increase of the iterative number, thus our algorithm can perform an exploration in a wider search space at the early stage and a local exploitation at the end stage. (2) Individuals can assign different explosion radii based on the quality of their fitness values, which can utilize the computational resources effectively, avoid the ineffective search, and improve the efficiency of research in the best solution.

2.4   Generating Explosion Sparks

The initial fireworks population is generated, the fireworks and their sparks are combined to form a union set, a certain number of individuals are selected according to the quality of their fitness values to construct the next generational population. This process continues until the predefined condition is satisfied. The strategy of selection sparks is presented in Algorithm 1.

Algorithm 1. Generating Explosion Sparks

Step1. Combine the tth population with their sparks to form an union
      set $U(t)$.

Step2. Calculate the fitness value of each individual in $U(t)$.

Step3. IF $|U(t)| > |Pop|$

Step4. Select $|Pop|$ better individuals from $U(t)$ based on their
      fitness values to form the next generational population
      $Pop(t+1)$.

Step5. END IF

Step6. Output $Pop(t+1)$.

It is important to point out that after combining each generation of fireworks and the sparks they produce, the size of the combination is larger than that of the population $|Pop|$, so there is no "ELSE" bifurcation in the IF sentence.

3   Experiments Study

To verify the effectiveness of the proposed AFAOL, four other peering algorithms are used to perform comparison experiments. The four algorithms include (1) basic FA [5], (2) basic PSO [2], (3) elite opposition-based PSO (EOPSO) [13], and (4) artificial bee colony algorithm based on orthogonal experimental design (ABC-OED) [14].

For the sake of fairness, the size of population for each peering algorithm is 20, and the maximum iterative number is 1,000. Some other parameters involved the compared algorithms can refer to the corresponding literature. The literature [13] indicates that it is helpful to improve the performance when the generalization coefficient $k = 1$, so we similarly set the k value of AFAOL to be 1 in the paper.

In order to reduce the negative effects derived from the random factors, each algorithm is independently experimented 30 times on each test function, so as to obtain a sound result. All experiments performed on Think Pad $X200$ personal computer, which is equipped with 5 GB RAM, 2.4 GHz dual-core CPU, and Windows 7 operating system.

In this paper, Some detailed definitions on these test instances can be referred to Table 1.

Table 1. Nine benchmark test functions

Test functions	Function expression	Range of variables	Dimension of decision space	Optimum
Sphere	$f_1(x) = \sum_{i=1}^{D} x_i^2$	[−100, 100]	30	0
Schwefel's	$f_2(x) = \sum_{i=1}^{D} |x_i| + \prod_{i=1}^{D} |x_i|$	[−10, 10]	30	0
Quadric	$f_3(x) = \sum_{i=1}^{D} (\sum_{j=1}^{i} x_j)^2$	[−100, 100]	30	0
Step	$f_4(x) = \sum_{i=1}^{D} (|x_i + 0.5|)^2$	[−100, 100]	30	0
Quadric noise	$f_5(x) = \sum_{i=1}^{D} i * x_i^4 + random[0, 1)$	[−1.28, 1.28]	30	0
Rastrgin	$f_6(x) = \sum_{i=1}^{D} [x_i^2 - 10cos(2\pi x_i) + 10]$	[−5.12, 5.12]	30	0
Non-rastrgin	$f_7(x) = \sum_{i=1}^{D} [y_i^2 - 10cos(2\pi y_i) + 10]$ $y_i = \begin{cases} x_i, |x_i| < 0.5 \\ round(2x_i)/2, else \end{cases}$	[−5.12, 5.12]	30	0
Ackley	$f_8(x) = -20\exp(-0.2\sqrt{1/D\sum_{i=1}^{D} x_i^2})$ $- \exp(1/D\sum_{i=1}^{D} cos(2\pi x_i)) + 20 + e$	[−32, 32]	30	0
Griewank	$f_9(x) = 1/4000\sum_{i=1}^{D} x_i^2 - \prod cos(x_i/\sqrt{i}) + 1$	[−600, 600]	30	0
Stretched V Sine Function	$f_{11}(x) = \sum_{i=1}^{D-1} (x_{i+1}^2 + x_i^2)^{0.25}$ $\{sin^2[50(x_{i+1}^2 + x_i^2)^{0.1}] + 1\}$	[−10, 10]	30	0
Easom's	$f_{18}(x) = -cos(x_1)cos(x_2)$ $\exp(-[(x_1 - \pi)^2 + (x_2 - \pi)^2])$	[−100, 100]	2	0
Six-hump	$f_{19}(x) = (4 - 2.1x_1^2 + x_1^{4/3})x_1^2 + x_1x_2$ $+ (-4 + 4x_2^2)x_2^2$	[−1, 1]	2	−1.0316

4 Conclusions

In view of the shortcomings of the basic fireworks algorithm, the opposition-based learning method and the approach of adaptive adjustment of explosion radius are integrated into the FA algorithm, and an adaptive fireworks explosion optimization algorithm with opposition-based learning (AFAOL) is proposed to solve some complex single objective optimization problems in the paper. The AFAOL use the OBL operator to expand the exploration range of the algorithm, which is beneficial for the algorithm to find the global optimum. In addition, the computational resource of the algorithm is utilized effectively by the use of adaptive fireworks explosion radius. The AFAOL is tested together with the other four algorithms, and the experimental results show that the proposed algorithm has significant advantages in the convergence and solution accuracy, and the AFAOL will be a promising optimizer.

Acknowledgement. This work was supported by Natural Science Foundation of China (51675 265), Jiangxi province humanities and social sciences key base project (JD17127), Key Research and Development Project of Jiangxi Province (20071BBE50049), and Jiangxi Province Education Science Planning Project (17YB276). The authors gratefully acknowledge these supports.

References

1. Colormi, A., Dorigo, M., Maniezzo V.: Distributed optimization by ant colonies. In: Proceedings of the 1st European Conference on Artificial Life, pp. 134–142. Elsevier, Amsterdam (1991)
2. Kennedy, J., Eberhart, R.C.: Particle swarm optimization. In: Proceedings of IEEE International Conference on Neural Networks, pp. 1942–1948. IEEE Press, Piscataway (1995)
3. Kirkpatrick, S., Gelartt, C.D., Vecchi, M.P.: Optimization by simulated annealing. Science **220**(11), 650–761 (1983)
4. Karaboga, D., Basturk, B.: A powerful and efficient algorithm for numerical function optimization: artificial bee colony (ABC) algorithm. J. Glob. Optim. **39**(3), 459–471 (2007)
5. Tan, Y., Zhu, Y.: Fireworks algorithms for optimization. In: Proceedings of International Conference on Swarm Intelligence, pp. 355–364. IEEE Press, Piscataway (2010)
6. Cao, J., Jia, H., Li, T.: A fireworks explosion optimization algorithm. Comput. Eng. Sci. **33** (1), 138–142 (2011)
7. Cao, J., Ji, Y.-F.: An improved fireworks explosion optimization algorithm and its convergence analysis. Comput. Eng. Sci. **34**(1), 90–93 (2012)
8. Cao, J., Li, T.T., Jia, H.: Fireworks explosion optimization algorithm with genetic operators. Comput. Eng. **36**(23), 149–151 (2010)
9. Zheng, Y.J., Xu, X.L., Ling, H.F., et al.: A hybrid fireworks optimization method with differential evolution operators. Neurocomputing **148**, 75–80 (2012)
10. Zheng, S., Janecek, A., Li, J., et al.: Dynamic search in fireworks algorithm. In: 2014 IEEE Congress on Evolutionary Computation, Beijing, China, pp. 3222–3229 (2014)
11. Zhang, B., Zhang, M.X., Zheng, Y.J.: A hybrid biogeography-based optimization and fireworks algorithm. In: 2014 IEEE Congress on Evolutionary Computation, pp. 3200–3206 (2014)

12. Tizhoosh, H.R.: Opposition-based learning: a new scheme for machine intelligence. In: Proceedings of International Conference on Computational Intelligence for Modeling Control and Automation, pp. 695–701. IEEE, Piscataway (2005)
13. Zhou, X., Wu, Z., Wang, H., et al.: Elite opposition-based particle swarm optimization. Acta Electronica Sinica **41**(8), 1647–1652 (2013)
14. Zhou, X.Y., Wu, Z.J., Wang, M.W.: Artificial bee colony algorithm based on orthogonal experimental design. J. Softw. **26**(9), 2167–2190 (2015)
15. Tang, K., Li, X.-D., Suganthan, P.N., et al.: Benchmark Functions for the CEC's 2010 Special Session and Competition on Large-Scale Global Optimization. Nature Inspired Computation and Applications Laboratory, USTC, Hefei (2009)

An Improved Multi-objective Fireworks Algorithm

Dongming Zhan[1] and Chengwang Xie[2(✉)]

[1] School of Software, East China Jiaotong University, Nanchang 330013, China
[2] Guangxi Colleges and Universities Key Laboratory of Scientific Computing
and Intelligent Information Processing, Guangxi Teachers Education University,
Nanning 530023, China
chengwangxie@163.com

Abstract. In the real world, there exist numerous multi-objective optimization problems (MOPs), and some representative multi-objective optimizers are proposed to meet the challenge. In this paper, an improved multi-objective fireworks algorithm (*i*MOFA) is presented to solve some complicated continuous MOPs. There are three significant features employed in *i*MOFA. Firstly, an initialization approach of homogenization with randomization is used to generate an even distributed initial population. Secondly, a method of fine-grained controlling radius is adopted to improve the search efficiency. In addition, a strategy of simplified *k*-nearest neighbor is utilized to maintain the diversity of external archive. Our *i*MOFA is compared with the other four multi-objective evolutionary algorithms (MOEAs) using IGD performance indicator on 12 benchmark MOP instances, and the numerical results show that the proposed *i*MOFA has the overall performance advantages in convergence and diversity.

Keywords: Fireworks algorithm · Multi-objective optimization problem
Multi-objective fireworks algorithm

1 Introduction

There exist many optimization problems required to be optimized simultaneously in the real world. In general, a multi-objective optimization problem (MOP) can be mathematically defined as follows.

$$\begin{cases} \min\ F(\mathbf{x}) = (f_1(\mathbf{x}), f_2(\mathbf{x}), \cdots, f_m(\mathbf{x}))^T, \\ s.t.\ \mathbf{x} \in \Omega \end{cases} \tag{1}$$

where $\mathbf{x} = (x_1, x_2, \cdots, x_n)$ is an n-dimensional decision vector bounded in the decision space Ω, the mapping function $F : \Omega \to R^m$ defines m objective functions, and R^m is called the objective space. In this paper, we only consider a MOP whose objectives are continuous and the feasible region is closed. In a multi-objective optimization problem, there are often conflicts among various objectives and it is impossible to find a single solution that is optimal in terms of all objectives. Therefore, it is preferable to search for a set of Pareto optimal or non-dominated solutions [1].

© Springer Nature Singapore Pte Ltd. 2018
K. Li et al. (Eds.): ISICA 2017, CCIS 873, pp. 204–218, 2018.
https://doi.org/10.1007/978-981-13-1648-7_18

During the last twenty years, evolutionary algorithms (EAs) have been widely used to solve MOPs, because of their generality (they require little specific domain information) and because of their population-based search nature, which allows them to produce multiple Pareto-optimal solutions in a single run [2, 3]. A variety of Pareto-based multi-objective evolutionary algorithms (MOEAs) have been proposed, including the non-dominated sorting genetic algorithm (NSGA) [4] and NSGA-II [2], the strength Pareto evolutionary algorithm (SPEA) [5] and SPEA2 [6], the Pareto archived evolution strategy (PAES) [7], the Pareto differential evolution algorithm (PDE) [8], etc. Besides these traditional MOEAs, Particle Swarm Optimization (PSO) has also been extended to solve MOPs. PSO is a bio-inspired metaheuristic mimicking the social behavior of bird flocking or fish schooling. Due to the high convergence speed and ease of implementation, PSO has become very popular to solve MOPs. Coello et al. [9] presents a multi-objective particle swarm algorithm (CMOPSO), in which Pareto dominance is incorporated into PSO, and a secondary repository of particles is used to guide their own flight. Sierra et al. [10] proposes an improved multi-objective particle swarm optimizer OMOPSO, which is based on Pareto dominance and the combination of crowding factor, different turbulence operators, and ε-dominance to balance the capacities between the exploration and exploitation. Nebro et al. [11] presents a Speed-constrained Multi-objective PSO (SMPSO), which is characterized by the use a strategy to limit the velocity of the particles, the use of polynomial mutation as a turbulence factor and an external archive storing the non-dominated solutions found during the search. The experimental results indicate that SMPSO obtains the remarkable results in terms of both optimization accuracy and convergence speed.

Moreover, Tan and Zhu [12] proposes a fireworks explosion algorithm (FA) to solve some complicated single objective optimization problems. The FA is a relatively new heuristic method inspired by the phenomenon of fireworks explosion. The algorithm selects in the search space a certain number of locations, each for exploding a firework to generate a set of sparks (new solutions); the fireworks and the sparks with good performance (fitness) are chosen as the locations for the next generation's fireworks, and the evolutionary process continues until a desired result is obtained. Numerical experiments on a number of benchmark functions show that the FA can converge to a global optimum with a smaller number of function evaluations than that of particle swarm optimizer.

In view of the advantages of FA, such as simple implementation, high precision, good convergence and robustness, some researchers try to modify the framework of FA to solve MOPs, then some multi-objective fireworks algorithms (MOFAs) have been proposed recently. Zheng et al. [13] develops a hybrid Multi-objective fireworks optimization algorithm (MOFOA) for variable-rate fertilization in oil crop production. The MOFOA evolves a set of solutions to the Pareto optimal front by mimicking the explosion of fireworks, uses the concept of Pareto dominance for individual evaluation and selection, and combines differential evolution (DE) operators to increase information sharing among the individuals. Milan et al. [14] describes fireworks algorithm adjusted for solving multi-objective radio frequency identification (RFID) network planning problem, and the hierarchical approach to objectives is used to implement the multi-objective optimizer. Liu et al. [15] proposes a multi-objective fireworks algorithm based on S-metric (S-MOFWA). In the S-MOFWA, the explosion amplitudes of fireworks and the number of sparks are calculated according to each firework's S-metric, the next generation of fireworks are selected by their S-metric and an external archive with fixed size is used to maintain non-dominated solutions found ever, by a novel updating strategy. Sarfi *et al.* [16] presents a multi-objective fireworks optimization (MOFWA) to find the most economic operating condition not only minimize the fuel cost, but also to find the best environmentally friendly solution without violating any constrains. The simulation results show the superiority of MOFWA to find the optimal set of solution with less computation time compared to NSGA-II. It is clear that the existing multi-objective fireworks algorithms have achieved some success on some complicated MOPs, but the performance of MOFAs still has room for improvement.

In this paper, an improved multi-objective fireworks algorithm (*i*MOFA) is proposed to improve the performance in solving some complex MOPs, the main contributions of the paper can be summarized as follows.

(1) An initialization way of homogenization combining with randomization is used to generate the diversified initial population.
(2) A strategy of fine-grained controlling the explosion radius of the firework is employed to improve the search efficiency.
(3) A simplified k-nearest neighbor method is adopted to maintain the diversity of external archive to save the computational cost.

The above three strategies are combined to improve the performance of *i*MOFA for continuous MOPs.

2 Backgrounds

2.1 Basic Concepts

Definition 1 (*Pareto Dominance*): A decision vector **x** is said to dominate another decision vector **y** (denoted as $\mathbf{x} \prec \mathbf{y}$) if and only if $(\forall i \in \{1, 2, \cdots, m\} : f_i(\mathbf{x}) \leq f_i(\mathbf{y}) \wedge (\exists j \in \{1, 2, \cdots, m\} : f_j(\mathbf{x}) < f_j(\mathbf{y}))$.

Definition 2 (*Pareto-Optimal*): A decision variable vector **x** is said to be *Pareto optimal* if and only if there exists no **y** in the decision space such that $\mathbf{y} \prec \mathbf{x}$.

Definition 3 (*Pareto-Optimal set*): The set **PS** includes All Pareto optimal vectors, defined by $\mathbf{PS} = \{ \mathbf{x} \mid \neg \exists \mathbf{y} \in \Omega : \mathbf{y} \prec \mathbf{x} \}$.

Definition 4 (*Pareto-Optimal Front*): The set **PF** includes the values of all the objective functions corresponding to the Pareto-optimal solutions in **PS**, such as $\mathbf{PF} = \{F(\mathbf{x}) = (f_1(\mathbf{x}), f_2(\mathbf{x}), \cdots, f_m(\mathbf{x}))^T \mid \mathbf{x} \in \mathbf{PS}\}$.

In this paper, the best solutions (possibly suboptimal) produced by an algorithm can be also be treated as a **PF**, when the true **PF** defined in Definition 4 is unavailable in practical cases. To distinguish these two types of **PF**, $\mathbf{PF}_{\text{true}}$ is used to refer to the true (or optimal) **PF** as defined in Definition 4, while $\mathbf{PF}_{\text{known}}$ is employed to represent the best solutions produced by an algorithm [3].

2.2   Basic Fireworks Algorithm

In this section, we simply introduce the basic fireworks algorithm (FA) for single objective problems. The FA is a heuristic algorithm inspired by the phenomenon of fireworks explosion. At each generation, the algorithm selects some quality points as fireworks, which generate a number of sparks to search the local space around them respectively. The evolutionary process continues until at least one spark reaches a desired optimum, or the stopping criterion is met.

In the algorithm presented in [12], the number of sparks and the amplitude of explosion for each firework x_i is respectively defined as follows.

$$s_i = s_m \cdot \frac{f_{max} - f(x_i) + \delta}{\sum\limits_{i=1}^{|Pop|} (f_{max} - f(x_i)) + \delta}, \qquad (2)$$

$$A_i = A \cdot \frac{f(x_i) - f_{min} + \delta}{\sum\limits_{i=1}^{|Pop|} (f(x_i) - f_{min}) + \delta}, \qquad (3)$$

where s_m and A are control parameters, $|Pop|$ is the size of the population, f_{max} and f_{min} are respectively the maximum and minimum objective values among the $|Pop|$ fireworks, and δ is a smallest constant to avoid zero-division-error.

To avoid overwhelming effects of splendid fireworks, the lower and upper bounds are defined for s_i as follows.

$$s_i = \begin{cases} round(a \cdot s_m) & if\ s_i < a \cdot s_m \\ round(b \cdot s_m) & if\ s_i > b \cdot s_m,\ a < b < 1, \\ round(s_i) & otherwise \end{cases} \qquad (4)$$

where a and b are constant parameters.

In explosion, sparks may undergo the effects of explosion from random z directions (dimensions). In the FA, the number of the affected directions randomly is obtained as follows.

$$z = round(d \cdot rand(0, 1)), \qquad (5)$$

where d is the dimensionality of the location x, and $rand(0,1)$ is an uniform distribution over $[0,1]$.

The location of a spark of the firework x_i is obtained using Algorithm 1. Mimicking the explosion process, a spark's location x_j is first generated. Then if the obtained location is found to fall out of the potential space, it is mapped to the potential space according the Algorithm 1.

To keep the diversity of sparks, another way of generating sparks: Gaussian explosion, which is shown in Algorithm 2. A function $Gaussian(1, 1)$, which denotes a Gaussian distribution with mean 1 and standard deviation 1, is used to define the coefficient of the explosion.

At each iteration of the algorithm, among all the current sparks and fireworks, the best location is always selected as a firework of the next generation. After that, $|Pop| - 1$ fireworks are selected with probabilities proportional to their distances to other locations.

Algorithm 1. Obtain the location of a spark

---

1. Initialize the location of the spark:$x_j = x_i$;

2. $z = round(d \cdot rand(0,1))$;

3. Randomly select z directions of x_j;

4. Calculate the displacement:$h = A_i \cdot rand(-1,1)$;

5. FOR each direction $x_k^j \in \{z$ pre-selected directions of $x_j\}$

6. $x_k^j = x_k^j + h$;

7.   IF $x_k^j < x_k^{min}$ or $x_k^j > x_k^{max}$ THEN

8.   map x_k^j to the potential space:$x_k^j = x_k^{min} + |x_k^j| \% (x_k^{max} - x_k^{min})$;

9.   END IF

10. END FOR

Algorithm 2. Obtain the location of a specific spark

---

1. Initialize the location of the spark:$x_j = x_i$;

2. $z = round(d \cdot rand(0,1))$;

3. Randomly select z directions of x_j;

4. Calculate the coefficient of Gaussian explosion:$g = Gaussian(1,1)$;

5. FOR each direction $x_k^j \in \{z$ pre-selected directions of $x_j\}$

6. $x_k^j = x_k^j \cdot g$;

7.   IF $x_k^j < x_k^{min}$ or $x_k^j > x_k^{max}$ THEN

8.   map x_k^j to the potential space:$x_k^j = x_k^{min} + |x_k^j| \% (x_k^{max} - x_k^{min})$;

9.   END IF

10. END FOR

---

3 Improved Multi-objective Fireworks Algorithm

3.1 Initialization Approach

A number of experiments and applications of MOEAs demonstrate that the well-distributed initial population is important for improving the performance of MOEAs.

However, the vast majority of the existing MOEAs only adopt the randomization way to generate initial population, which is not conducive to the performance of MOEAs due to the random error [17]. Therefore, a initialization method of homogenization combining with randomization is introduced to obtain an evenly distributed initial population, which endues iMOFA a good start of search solution space. The details of the initialization approach are as follows. Any decision variable x_i ($i = 1, 2, ..., n$) in the decision vector x is evenly divided into several equal subintervals, and the number of the subintervals is equal to the size of the evolutionary population. Then, given an individual, the value of any genetic locus is determined by two steps presented as follows. A subregion is randomly selected among these subintervals firstly, and the value of the genetic locus is generated randomly within the selected subregion. Every subinterval divided evenly has and only has one chance to produce the gene value, which can maximize the diversity of the gene expression of the initial population. Algorithm 3 shows the flow of the method of initialization population.

In Algorithm 3, N denotes the size of the population, and n denotes the dimensionality of the decision vector, at the same time, the range of the decision variable x_j ($j \in [1 : n]$) is defined as $[a_j, b_j]$. The proposed initialization method not only preserves the randomness of generation of the gene value within the randomly selected subregion, but also ensures the uniformity of the partition of the decision variable. Consequently, a set of initial solutions distributed evenly in the search space can be obtained if the Algorithm 3 is implemented.

Algorithm 3. Initialize the initial population

---

1. FOR $j = 1$ TO n

2. $\Delta_j = (b_j - a_j) / N$; //divide the decision variable x_j into

     N subregions

3. $\Lambda_j = \{[a_j, a_j + \Delta_j], [a_j + \Delta_j, a_j + 2\Delta_j], \text{L}, [a_j + (N-2)\Delta_j, a_j + (N-1)\Delta_j],$
     $[a_j + (N-1)\Delta_j, b_j]\}$;

4.   FOR $i = 1$ TO N

5.   select a subregion from the set Λ_j, and assigned a

          random value within the selected subinterval to x_j^i;

6.   Update the set Λ_j : delete the selected subregin

          derived from step5;

7. END FOR

8. END FOR

3.2 Fine-Grained Controlling Explosion Radius

The search manner based on the fireworks explosion optimization in iMOFA can be described as follows. First, a initial population with N fireworks is generated using Algorithm 3, and the N locations actually denote the initial N solutions to the MOP being solved. For example, the i-th explosion location in the n-dimensional search space can be denoted as $x_i = (x_{i,1}, x_{i,2}, ..., x_{i,n})$, we set the firework to perform the uniform explosion, and the maximum explosion radius is preset to r, which means the largest permeating bound of the sparks can reach. Furthermore, let the number of explosion layers be ω, then the explosion radius of each layer will be $j \cdot r/\omega$ (j = 1, 2, ..., ω). As far as we know, the existing fireworks explosion algorithms always set the explosion radius to perform a nonlinear decrease with the increase of the number of iterations, which can guarantee the global exploration at the early stage and the local accurate search at the end of the FAs.

Note that the existing FAs do not take into account the radius differences among the same generational fireworks, especially those with different Pareto domination strengths. However, in the complex environment of multi-objective optimization, for the same generational fireworks, the non-dominated fireworks should endue with the relatively good qualities due to their being closer to the Pareto front, so they should search accurately in the smaller neighborhood space; On the contrary, The qualities of those who are dominated are relatively poor, and they are generally far away from the Pareto front, so they should explore in the larger range. Inspired by the above, the strategy of fine-grained controlling the explosion radius is adopted to improve the efficiency of iMOFA.

Based on the existing research results, we decrease the explosion radius in the nonlinear manner, and in the same population, the different explosion radii of the fireworks are assigned according to the differences of the Pareto dominance strength among individuals. The manner of decreasing the explosion radius in the paper is shown in Eq. 6.

$$r(t) = (\frac{T_{\max} - t}{T_{\max}})^{\alpha} \cdot (r_{initial} - r_{end}) + r_{end} \tag{6}$$

where $r_{initial}$ denotes the initial generation's explosion radius, and r_{end} indicates the last generational explosion radius, and $T_{\max}$ means the maximum number of evolution, and t denotes the current number of the iteration.

In addition, the process of computing the radii of the fireworks in the same population is shown as follows.

(1) counting the Pareto strength of the firework in the population, which is defined as follows.

$$S(i) = |\{j|i,j \in Pop(t) : i \prec j\}| \tag{7}$$

where $|\cdot|$ denotes the cardinality of the set, and $S(i)$ indicates the number of individuals which the individual i dominates, and $Pop(t)$ represents the t-th generation population.

(2) computing the maximum individual's fitness of the t-th generation population S_{max}, which is shown in Eq. 8.

$$S_{\max} = \max_{i \in Pop(t)} \{S(i)\} \tag{8}$$

(3) calculating the explosion radius of the t-th generation population $r(t)$:

$$r(t) = k \cdot (D_{max} - D_{min}) \cdot (\frac{T_{max} - t}{T_{max}})^{\alpha} + r_{end} \tag{9}$$

where $r(t)$ controls the explosion radius of the firework to an appropriate range through the scaling parameter k, and D_{max} and D_{min} respectively denote the maximum and minimum values of all individual decision variables in the current population, and α indicates the attenuation index of the explosion radius.

(4) computing the explosion radius of firework i ($i \in Pop(t)$) as follows.

$$r_i(t) = \begin{cases} r(t) \cdot \frac{S_{max} - S(i)}{S_{max} + \delta} & \textit{if } S(i) \neq S_{max} \\ r_{end} & \textit{if } S(i) = S_{max} \end{cases} \tag{10}$$

where δ is a smallest constant to avoid zero-division-error, and r_{end} denotes the preset minimum explosion radius.

For the explosion location i, the spark generated by firework i can be defined as follows.

$$x_i^* = x_i + r_j \cdot b_k \tag{11}$$

where x_i denotes the current location of firework i.

Considering the limitations of the time and space resources of the algorithm and the need of enough sparks generated by the firework, we specify only four cases to express the distances between the firework and its sparks, i.e., $r/4$, $r/2$, $3r/4$, and r, where r ($=r(t)$) denotes the explosion radius of the t-th generation population. Based on the specification above, the explosion radius in Eq. 11 can be determined as $r_1 = r$, $r_2 = 3r/4$, $r_3 = r/2$, $r_4 = r/4$. In Eq. 11, b_k ($k = 1, 2,..., L$) denotes the direction vector of explosion, and L is the total direction number of explosion for firework i. For the low-dimensional search space (such as $n \leq 10$), the algorithm uses the intuitive standard coordinate axis as the explosion direction, that is, the number of the explosion directions of the n-dimensional search space is $2n$. However, for the high-dimensional search space (e.g. $n > 10$), if the $2n$ explosion directions are still utilized, the algorithm will consume the time and space resources too much, so we randomly select $n/3$ explosion directions of each layer from the $2n$ standard coordinate axis, plus the opposite direction of each direction, composed of the $2n/3$ explosion directions.

In addition, due to the probability of running over the bounds for the sparks generated by the firework exploding, a bounded method is applied to constrain the locations of the sparks. Let x_i ($i = 1, 2, ..., n$) denote any decision variable in n-dimensional search space, and set $x_i \in [a_i, b_i]$, if the spark x_j generated by firework x_i exceeds the bound $[a_i, b_i]$, it will become an infeasible solution. Under these circumstances, we reset the value x_{ji} of the i-th decision variable of the spark x_j, so as to improve the search efficiency of the algorithm, and the resetting way is shown in Eq. 12.

$$x_{ji} = rand(a_i, b_i) \text{ if } x_{ji} < a_i \text{ or } x_{ji} > b_i \qquad (12)$$

where $rand(\cdot)$ is a random number that is evenly distributed over the interval $[a_i, b_i]$.

3.3 Selection of Sparks

After initializing the population using Algorithm 3, iMOFA views each firework in the current population as a explosion location which will produce a number of sparks after the explosion process. Afterwards, these fireworks and their sparks will be evaluated together to find better individuals involving in the next generation. The flow of the update of the population is shown in Algorithm 4.

In Algorithm 4, N denotes the size of the evolutionary population. In addition, the selection of better individuals in step 3 and step 5 is based on the individuals' fitness.

Algorithm 4. Update the population

---

1. Construct the non-dominated solution set of the current population NDS using the method of normalized sort.
2. IF ($|NDS| < N$)
3.   Select $N-|NDS|$ better individuals from the set $\{i | i \in Pop(t)\}$ to merge with the set NDS, so as to yield the next generation of population $Pop(t+1)$.
4. ELSE
5.   Pick up N individuals from the set NDS to construct the next generation population $Pop(t+1)$.
6. END

3.4 Maintain the Diversity of External Archive

In general, the non-dominated solutions in the MOEAs increase rapidly with the evolutionary process, and maximum capacity of the external archive is limited by the applicable computational resources. It is necessary to trim the archive and maintain the diversity of the external file simultaneously. Among the existing strategies of maintaining the diversity of external archive, the k-nearest neighbor method in SPEA2 [6] is especially prominent, regrettably, the time complexity of the k-nearest neighbor is still high. So, a simplified k-nearest neighbor strategy is introduced to preserve the diversity of the iMOFA. To be more precise, for each individual i the distances (in objective space) to all individuals j in archive and population are calculated and stored in a list. After sorting the list in increasing order, the k-th element gives the distance sought, denoted as θ_i^k. What is different from the original k-nearest neighbor method in SPEA2, we use k equal to the logarithm of the sample size, thus, $k = log(N + N')$. Afterwards, the density $D(i)$ corresponding to i is defined by

$$D(i) = \frac{1}{\theta_i^k + 2} \tag{13}$$

In the denominator, two is added to ensure that its value is greater than zero and that $D(i) < 1$. Finally, adding $D(i)$ to the raw fitness value $S(i)$ of an individual i yields its fitness $F(i)$:

$$F(i) = D(i) + S(i) \tag{14}$$

In the simplified version, the value of k is less than the one in SPEA2, so the time complexity of the iMOFA can be improved to a certain extent.

3.5 Flow of iMOFA

Based on the Sects. 3.1 to 3.4 above, Algorithm 5 describes the flow of iMOFA as follows.

Algorithm 5. Flow of iMOFA

---
1. Generate the N initial fireworks using the Algorithm 1, and copy the non-
   dominated individuals to the external archive, and set the iterator $t = 0$;
2. WHILE ($t < T_{max}$)
3.   FOR ($i = 1; i \leq N; i ++$)
4.   Generate the sparks utilized Eq. 6 to Eq. 12;
5.   END
6.   Execute Algorithm 4 to construct the next generation;
7.   Maintain the diversity of the external archive using the simplified
       k-nearest neighbor;
8. $t = t + 1$;
9. END WHILE

---

4 Experimental Results

4.1 Test Problems

Two types of test problems are adopted to evaluate the performance of iMOFA. First, the popular ZDT problems are used [18]. Moreover, the three-objective DTLZ test problems are used to further examine the performance of iMOFA in handling MOPs with more than two objectives [19]. Thus, we used a total of 12 test problems (ZDT1-ZDT4, ZDT6, and DTLZ1-DTLZ7) in our experimental studies. It is noted that for ZDT1-ZDT3, the number of decision variables is 30, while the number of decision variables in ZDT4 and ZDT6 is 10. The details of the ZDT and DTLZ test problems are available in [18, 19], respectively.

4.2 Performance Measure

The goal of solving MOPs is to find a uniformly distributed set that is close the $\mathbf{PF}_{true}$ as possible. As the inverted generational distance (IGD) metric [20] can examine convergence and diversity simultaneously, it is used to assess the performance of all the compared algorithms in our experimental studies.

Let V be a set of solutions that are uniformly distributed along $\mathbf{PF}_{true}$ and let V' be the set of best solutions (i.e., $\mathbf{PF}_{known}$) that are found by an algorithm. The IGD value of V to V', i.e., $IGD(V, V')$ is defined as

$$IGD(V, V') = \frac{\sum_{i=1}^{|V|} d(V_i, V')}{|V|} \tag{15}$$

where $|V|$ returns the size of the set V and $d(V_i, V')$ denotes the minimum Euclidean distance in objective space between V_i and the individuals in V'. IGD requires to know $\mathbf{PF}_{true}$ in advance. In general, a lower value of $IGD(V, V')$ is preferred as it indicates that V' obtains a more even coverage of $\mathbf{PF}_{true}$ and is closer to $\mathbf{PF}_{true}$.

4.3 Experimental Settings

In our experiments, in order to assess the performance of iMOFA, we compare it with respect to several types of nature-inspired heuristic algorithms for solving MOPs, including NSGA-II [2], SPEA2 [6], MOEA/D [21], and SMPSO [11].

The parameters settings of the compared algorithms are established as recommended in [2, 6, 11, 21], as summarized in Table 1. It is worth noting that the parameters of the compared algorithms are properly tuned to solve most of the MOPs adopted in our experimental studies. To allow a fair comparison, the parameters of iMOFA are set according to those of the compared algorithms.

Table . Parameters settings of all algorithms compared

Algorithms	Parameters settings
NSGA-II	$N = 100$, $p_c = 0.9$, $p_m = 1/n$, $\beta_c = 20$, $\beta_m = 20$
SPEA2	$N = 100$, $p_c = 0.9$, $p_m = 1/n$, $\beta_c = 20$, $\beta_m = 20$
MOEA/D	$N = 100$, $CR = 1.0$, $F = 0.5$, $p_m = 1/n$, $\beta_m = 20$, $T = 20$, $\alpha = 0.9$, $n_r = 2$
SMPSO	$C_1 \in [1.5, 2.5]$, $C_2 \in [1.5, 2.5]$, $p_m = 1/n$, $\beta_m = 20$
iMOFA	$N = 100$, $N' = 100$, $\beta_c = 20$, $\beta_m = 20$, $\alpha = 3$, $r_{end} = 10^{-6}$

In Table 1, N is the population size and N' is the size of the external archive; p_c is the crossover probability and p_m is the mutation probability; and β_c and β_m are the distribution indexes of SBX and polynomial mutation, respectively. In MOEA/D, T defines the size of the neighborhood in the weight coefficients, α controls the probability that parent solutions are chosen from T neighbors and n_r is the maximum number of parent solutions that are replaced by each child solution. C_1 and C_2 are two

Table 2. The Results of IGD Obtained by Five MOEAs on 12 Test MOPs

Functions		*i*MOFA	NSGA-II	SPEA2	SMPSO	MOEA/D
ZDT1	Mean	1.45E−3	5.31E−3	5.94E−3	5.68E−3	1.43E−3
	Std	1.21E−7	7.63E−8	1.65E−7	2.77E−6	1.27E−7
	t-test		+	+	+	=
ZDT2	Mean	5.19E−3	5.26E−3	1.61E−2	5.29E−3	2.16E−2
	Std	4.71E−7	7.92E−8	8.52E−4	7.65E−7	6.31E−3
	t-test		=	+	+	+
ZDT3	Mean	4.16E−3	4.43E−3	2.59E−3	8.69E−3	6.18E−3
	Std	1.19E−8	4.40E−5	2.47E−5	1.11E−5	1.48E−5
	t-test		+	−	+	+
ZDT4	Mean	5.46E−3	1.41E−2	5.99E−2	7.89E−1	2.36E−3
	Std.	6.17E−6	8.10E−5	2.68E−3	1.75E−1	4.28E−4
	t-test		+	+	+	−
ZDT6	Mean	1.22E−2	1.95E−2	3.75E−2	3.83E−3	8.17E−3
	Std	4.20E−7	4.26E−6	2.70E−5	1.06E−7	2.69E−6
	t-test		+	+	−	−
DTLZ1	Mean	3.33E−2	4.22E−2	3.39E−2	3.16E+1	7.61E−2
	Std	1.48E-5	6.05E−5	1.82E-5	4.31E+1	1.20E−6
	t-test		+	+	+	+
DTLZ2	Mean	1.36E−2	4.77E−2	3.82E−2	5.14E−2	4.62E−2
	Std	8.92E−7	5.91E−6	9.96E−7	5.96E−6	5.22E−6
	t-test		+	+	+	+
DTLZ3	Mean	6.28E−2	4.33E−1	4.67E−1	1.30E−1	6.58E−2
	Std	1.19E−2	3.13E−1	3.28E−1	1.18E−1	1.16E−2
	t-test		+	+	+	=
DTLZ4	Mean	2.31E−2	3.14E−2	3.23E−2	1.08E−0	2.03E−2
	Std	3.82E−5	3.64E−5	3.80E−5	1.25E−1	3.11E−5
	t-test		+	+	+	−
DTLZ5	Mean	1.84E−3	1.11E−3	1.01E−3	6.06E−4	4.26E−3
	Std	9.58E−6	3.56E−6	2.69E−6	1.79E−6	5.98E−5
	t-test		−	−	−	+
DTLZ6	Mean	2.36E−4	1.21E−2	1.10E−3	5.39E−5	5.71E−4
	Std	1.14E−5	4.12E−4	4.84E−5	1.48E−7	4.36E−5
	t-test		+	+	−	+
DTLZ7	Mean	9.16E−4	7.52E−3	6.05E−3	8.32E−4	1.93E−2
	Std	1.24E−5	2.76E−5	2.87E−5	1.16E−5	7.63E−3
	t-test		+	+	−	+
Better(+)			10	10	8	7
Same(=)			1	0	0	2
Worse(−)			1	2	4	3
Score			9	8	4	4

control parameters randomly picked within the range [1.5, 2.5] in SMPSO. For iMOFA, N' is the size of external archive, and α is the attenuation index of explosion radius; and r_{end} is the minimum preset radius, and we set $r_{end} = 10^{-6}$ based on the research in [22]. It is noted that the settings of N and N' listed in Table 1 are only applied for the ZDT problems and that the maximum number of function evaluations is set to 25000. For solving the three-objective DTLZ test problems, the maximum number of function evaluations is set to 10^5.

All the experiments are run 30 times (using different random seeds), the mean IGD values (mean), and the corresponding standard deviations (std) of which are collected for comparison. The best results are identified in boldface in the comparison tables. Moreover, in order to have a statistically sound conclusion, the two-paired t-test is further conducted to assess the statistical significance of the difference between the results obtained by iMOFA and those obtained by the other algorithms with a significance level $\alpha = 0.05$. All the experiments of this paper are performed on Think Pad X200 PC, and the computer configuration involves 5G memory, and 2.4 GHz dual-core CPU, and Windows 7 X64 operating system.

The results of the IGD values obtained by the five MOEAs on 12 test MOPs are shown in Table 2. Note that the t-test value is the result of the t-test on the same MOP as iMOFA and other peering MOEAs, and the signs, i.e., "+", "=", and "−", respectively indicate the IGD value obtained by iMOFA is better than, equal to, and worse than that of the corresponding comparison MOEA on the same test instance in the t-test, which denotes the significance of the distinction between the results. And "Score" indicates the number of iMOFA being significantly better than the peering MOEA in 12 benchmark MOPs, that is, the difference between the number of obtaining "+" and the one gaining "−" for all MOEAs involved.

It can be seen from Table 2 that iMOFA and SMPSO both obtain four optimal IGD values on 12 test MOPs; and MOEA/D and SPEA2 achieve two and one, respectively; Unfortunately, NSGA-II fails to obtain any optimal IGD value. In addition, from the results of t-test shown in Table 2, our iMOFA obtains the scores which are all more than zero, i.e., compared with NSGA-II, SPEA2, SMPSO, and MOEA/D, the scores of iMOFA obtained are nine, eight, four, and four, respectively. The results of the t-test show that the IGD values of iMOFA on all test MOPs are generally superior to the other four peering MOEAs.

Since the IGD indicator can reflect the convergence and diversity of MOEA simultaneously, and the results of Table 2 show that iMOFA has the best IGD performance on the whole, which indicates that the proposed iMOFA combined the strategies of initialization population, fine-grained controlling radius, and simplified k-nearest neighbor method, can better balance the capabilities between the global exploration and local exploitation.

5   Conclusions

In reality, the complex MOPs are emerging, and it is necessary to develop some new type of multi-objective optimization algorithms to meet the challenge. In this paper, an improved multi-objective fireworks algorithm (iMOFA) is presented to solve some continuous MOPs. The *i*MOFA uses the method of homogenization and randomization to initialize the population; and utilizes the strategy of fine-grained controlling explosion radius to save search resources; and employs the simplified k-nearest neighbor method to maintain the diversity of the external archive. These three methods are combined and implemented at different stages of the algorithm to improve the performance on some complex MOPs. *i*MOFA companied with the other four MOEAs experiments with the IGD indicator on 12 benchmark MOPs. The experimental results show that the *i*MOFA has a significant advantage over the other four peering MOEAs on the performance of convergence and diversity. In the future, some more difficult MOPs and those in engineering practices will be utilized to evaluate the performance of *i*MOFA fully.

Acknowledgement. This work was supported by National Natural Science Foundation of China (NSFC), under Grant No. 61763010, and partially supported by Guangxi Colleges and Universities Key Laboratory of Scientific Computing and Intelligent Information Processing with Grant No. GXSCIIP201604.

References

1. Deb, K.: Multi-Objective Optimization Using Evolutionary Algorithms. Wiley, New York, NY, USA (2001)
2. Deb, K., Pratap, A., Agarwal, S., Meyaruvan, T.: A fast and elitist multiobjective genetic algorithm: NSGA-II. IEEE Trans. Evol. Comput. **6**(2), 182–197 (2002)
3. Lin, Q., Chen, J., Zhan, Z.-H., et al.: A hybrid evolutionary immune algorithm for multiobjective optimization problems. IEEE Trans. Evol. Comput. **20**(5), 711–729 (2016)
4. Srinivas, N., Deb, K.: Multiobjective optimization using nondominated sorting in genetic algorithms. Evol. Comput. **2**(3), 221–248 (1994)
5. Zitzler, E., Thiele, L.: Multiobjective evolutionary algorithms: a comparative case study and the strength Pareto approach. IEEE Trans. Evol. Comput. **3**, 257–271 (1999)
6. Zitzler, E., Laumanns, M., Thiele, L.: SPEA2: improving the strength Pareto evolutionary algorithm. Technical report 103, Computer Engineering and Networks, Swiss Federal Institute of Technology Zurich in Switzerland (2001)
7. Knowles, J.D., Corne, D.W.: The Pareto archived evolution strategy: a new baseline algorithm for Pareto multiobjective optimization. In: Congress on Evolutionary Computation (CEC 1999), vol. 1, pp. 98–105. IEEE Press, Piscataway (1999)
8. Abbass, H.A., Sarker, R., Newton, C.: PDE: a Pareto-frontier differential evolution approach for multi-objective optimization problems. In: Proceedings of the Congress on Evolutionary Computation, vol.2, 2001, pp. 971–978. IEEE Service Center, Piscataway (2001)
9. Coello Coello, C.A., Pulido, G.T., Lechuga, M.S.: Handling multiple objectives with particles swarm optimization. IEEE Trans. Evol. Comput. **8**(3), 256–279 (2004)
10. Sierra, M.R., Coello Coello, C.A.: Improving PSO-based multi-objective optimization using crowding, mutation and $\in$-dominance. In: Coello Coello, C.A., Hernández Aguirre, A., Zitzler, E. (eds.) EMO 2005. LNCS, vol. 3410, pp. 505–519. Springer, Heidelberg (2005). https://doi.org/10.1007/978-3-540-31880-4_35
11. Nebro, A.J., et al.: SMPSO: a new PSO-based metaheuristic for multi-objective optimization. In: Proceeding of IEEE Symposium on Computer Intelligence in Multicriteria Decision Making, Nashville, TN, USA, pp. 66–73 (2009)

12. Tan, Y., Zhu, Y.: Fireworks algorithm for optimization. In: Tan, Y., Shi, Y., Tan, K.C. (eds.) ICSI 2010. LNCS, vol. 6145, pp. 355–364. Springer, Heidelberg (2010). https://doi.org/10.1007/978-3-642-13495-1_44
13. Zheng, Y.-J., Song, Q., Chen, S.-Y.: Multiobjective fireworks optimization for variable-rate fertilization in oil crop production. Appl. Soft Comput. **13**, 4253–4263 (2013)
14. Tuba, M., Bacanin, N., Beko, M.: Fireworks algorithm for RFID network planning problem. In: The 25th International Conference Radioelektronika, pp. 440–444. IEEE Conference Publications (2015)
15. Liu, L., Zheng, S., Tan, Y.: S-metric based multi-objective fireworks algorithm. In: 2015 IEEE Congress on Evolutionary Computation (CEC), pp. 1257–1264. IEEE Conference Publications (2015)
16. Sarfi, V., Niazazari, I., Livani, H.: Multiobjective fireworks optimization framework for economic emission dispatch in microgrids. In: 2016 North American Power Symposium (NAPS), pp. 1–6. IEEE Conference Publications (2016)
17. Cheng-wang, X.I.E., Xiu-fen, Z.O.U., Xue-wen, X.I.A., et al.: A multi-objective particle swarm optimization algorithm integrating multiply strategies. Acta Electronica Sinica (in Chinese) **43**(8), 1538–1544 (2015)
18. Zitzler, E., Deb, K., Thiele, L.: Comparison of multiobjective evolutionary algorithms: empirical results. Evol. Comput. **8**(2), 173–195 (2000)
19. Deb, K., Thiele, L., Laumanns, M., Zitzler, E.: Scalable test problems for evolutionary multiobjective optimization. In: Abraham, A., Jain, L., Goldberg, R. (eds.) evolutionary multiobjective optimization, pp. 105–145. Springer, London (2005). https://doi.org/10.1007/1-84628-137-7_6
20. Li, H., Zhang, Q.F.: Multiobjective optimization problems with complicated Pareto sets, MOEA/D and NSGA-II. IEEE Trans. Evol. Comput. **13**(2), 284–302 (2009)
21. Zhang, Q., Li, H.: MOEA/D: a multiobjective evolutionary algorithm based on decomposition. IEEE Trans. Evol. Comput. **11**, 712–731 (2007)
22. Xie, C., Xu, L., Zhao, H., et al.: Multiobjective fireworks optimization algorithm using elite opposition-based learning. Acta Electronica **44**(5), 1180–1188 (2016)

Evolutionary Design of a Crooked-Wire Antenna

Lumin Ye[1], Bin Lan[1], Yi Yuan[2(✉)], Jianqing Sun[3(✉)], Yongzhi Sun[3], and Sanyou Zeng[1(✉)]

[1] School of Mechanical Engineering and Electronic Information, China University of Geosciences, Wuhan 430074, China
1459193124@qq.com, 874432599@qq.com, sanyouzeng@gmail.com
[2] Beijing Industry and Trade Technicians College, Beijing 100089, China
yuan0387@126.com
[3] No 8511 Research Institute of CASIC, Nanjing 210007, China
sunjianqing126@126.com, nanshen01@126.com

Abstract. This paper presents a new crooked-wire antenna. The antenna design is first modeled as a constrained optimization problem (COP). Then a differential evolution solves the COP. Consequently, an antenna is designed. The main difference of the antenna design in this paper from traditional evolutionary antenna is that the objective in the COP can potentially enhance the robustness of the antenna and widen the frequency band without additional computational cost. Actually, the objective is the sum of variance of the gain, axial ratio and VSWR over the frequency band. This paper chooses NASA LADEE satellite antenna design as an example to verify the method of this paper. The performance of the evolved antenna meets the requirements. Furthermore, the VSWR of the antenna is less than 1.5 over the required frequency range, much less than the requirement 2. The axis ratio is 2.06 dB at the beam direction which is much better than that of the antenna designed by the referred literature.

Keywords: Constrained optimization problem (COP)
Differential evolution algorithm · Small satellite antenna

1 Introduction

It has seen that an obvious increase of computational resources to solve science and engineering problems in the last thirty years; these progresses have led to the development of advanced numerical algorithms [1]. Recently, evolutionary methods have been widely applied in fields as diverse as Artificial Intelligence [2,3], engineering [4], bioinformatics [5] and economics [6]. Moreover, the further improvement in computational power for the near future will enhance the role of efficient numerical approaches in the solving complex problems.

Since the early 1990s, regarding the field of applied electromagnetics [7], the design of evolutionary antenna has been investigated by researchers [8].

© Springer Nature Singapore Pte Ltd. 2018
K. Li et al. (Eds.): ISICA 2017, CCIS 873, pp. 219–231, 2018.
https://doi.org/10.1007/978-981-13-1648-7_19

In recent years, the field has grown as computer speed has increased, electro-magnetics simulators have improved and evolutionary algorithms have improved. There are many researches developed in this field. Wire antenna in [8], quadri-filar helical antenna in [9] and X-Band antenna for NASA's Space Technology 5Mission [10] are all obtained through evolutionary computation and have great practical applicability. A systematic approach was introduced to the shape optimization of compact, single-aperture MIMO antennas based on character-istic modes and genetic algorithms [11]; Wu presented a method of controlling antenna radiation pattern based on 3-D printing of special dielectric materi-als [12]. Jiao and Zeng [13] integrated dynamic and multi-objective techniques into evolutionary algorithms to solve antenna array problems with many local optima.

We design a crooked wire antenna and choose NASA LADEE antenna [15] as a example to verify the method of this paper. The geometric structure of the antenna is mainly composed of a crooked wire for radiation, a coaxial feed line and a metal cup for reflection. Differential evolution [14] is used to optimize the crooked-wire antenna. The performance of the antenna is simulated and evaluated with Ansoft HFSS.

After Sect. 1, this paper is organized as follows: Sect. 2 gives the related work about the way to solving constrained optimization problem (COP) and the DE algorithm. Section 3 describes the formation of antenna design as COP, which includes antenna requirements, antenna structure, objective function and constrained function. Then antenna constrained optimization problem is solved by DE, and the antenna is fabricated Next. Simulated results are compared with the referred literature, and the difference between them is analyzed and discussion in Sect. 4. Then Sect. 5 is the summary of this paper. Finally, Sect. 6 gives the Acknowledgment about this paper.

2   Related Work

In this section, a constrained optimization problem and differential evolution will be introduced.

2.1   Constrained Optimization Problem

The minimization COP includes a objective function, some equality or inequality constraints and a suitable search space. The COP can be described as (1):

$$
\begin{aligned}
min \quad & y = f(\overrightarrow{x}) \\
st: \quad & \overrightarrow{g(\overrightarrow{x})} \leqslant \overrightarrow{0} \\
where \; & \overrightarrow{g(\overrightarrow{x})} = (g_1(\overrightarrow{x}), g_2(\overrightarrow{x}), ...g_n(\overrightarrow{x})) \\
& \overrightarrow{x} = (x_1, x_2, ...x_n) \in \mathbf{X} \\
& X = \{\overrightarrow{x} \,|\, \overrightarrow{l} \leq \overrightarrow{x} \leq \overrightarrow{u}\} \\
& \overrightarrow{l} = (l_1, l_2, ..., l_n) \\
& \overrightarrow{u} = (u_1, u_2, ..., u_n)
\end{aligned}
\tag{1}
$$

The $f(\overrightarrow{x})$ is objective function. $g(\overrightarrow{x}) \leqslant \overrightarrow{0}$ is the constraint; $\overrightarrow{0}$ is the constraint boundary, $\overrightarrow{x}$ is the solution variable and $X = [\overrightarrow{l}, \overrightarrow{u}]$ is the solution space; $\overrightarrow{l}$ is the lower boundary of the solution space and $\overrightarrow{u}$ is the upper boundary of the solution space.

The antenna design problem will be solved by DE, and we normalize the solution space, that is, $X = [\overrightarrow{0}, \overrightarrow{1}]$; and $X = [\overrightarrow{l}, \overrightarrow{u}]$ is normalized as follow (2):

$$\overrightarrow{x_l} = \frac{\overrightarrow{x_l} - \overrightarrow{l_l}}{\overrightarrow{u_l} - \overrightarrow{l_l}} \tag{2}$$

If there is equality constraints $h(\overrightarrow{x})$, the equality constraints convert the inequality constraints, that is, $|h(\overrightarrow{x})| - \varepsilon \leq 0$, ε is very small valve, such as 0.0001.

A feasible solution satisfies $\overrightarrow{x} = (x_1, x_2, ...x_n) \in X$ and $g(\overrightarrow{x}) \leqslant \overrightarrow{0}$, so the disaggregation of the feasible solution as follow (3):

$$S_F = \{\overrightarrow{x} : \overrightarrow{x} \in X, g(\overrightarrow{x}) \leqslant \overrightarrow{0}\} \tag{3}$$

The Constraint violation values of a solution $\overrightarrow{x}$ is (4):

$$G_i(\overrightarrow{x}) = max\{g_i(\overrightarrow{x}), 0\}, i = 1, 2, 3, ..., m. \tag{4}$$

The violation value of a solution $\overrightarrow{x}$ is (5):

$$\psi(\overrightarrow{x}) = \frac{1}{m} \sum_{l=1}^{m} \frac{G_i(\overrightarrow{x})}{max\{G_i(\overrightarrow{x})\}}, \overrightarrow{x} \in P(0). \tag{5}$$

where P(0) is initial population. If there is $max\{G_i(\overrightarrow{x})\} \leq 1, i = 1, 2, ..., m$, defining $max\{G_i(\overrightarrow{x})\} = 1$.

2.2   Differential Evolution

In this paper, we use the DE algorithm [14] to solve the above constraints optimization problem. The differential evolution has many different schemes [16], here the DE strategy DE/rand/1/bin is used, which is described as follows: Sect. 2.2.

DE/rand/1/bin
Generate the initial population $P = \{\overrightarrow{x_1}, \overrightarrow{x_2}, ..., \overrightarrow{x_{NP}}\}$, NP is the size.
Evaluate the objective value and constraint value of each individual in P
while The halting criterion is not satisfied **do**
 for $i = 1$ to NP **do**
  Select randomly $a \neq b \neq c \neq i$
  $j_{rand} = \text{rndint}(1, n)$
  for $j = 1$ to n **do**
   if $\text{rndreal}_j(0,1) < CR$ or $j_{rand} == j$ **then**
    $\overrightarrow{u_{ij}} = \overrightarrow{x_{aj}} + F \times (\overrightarrow{x_{bj}} - \overrightarrow{x_{cj}})$
   else
    $\overrightarrow{u_{ij}} = \overrightarrow{x_{ij}}$
   end if
  end for i
 end for j
 for $i = 1$ to NP **do**
  Evaluate the offspring $\overrightarrow{u_i}$
  if $\overrightarrow{u_i}$ is better than $\overrightarrow{x_i}$ **then**
   $\overrightarrow{x_i} = \overrightarrow{u_i}$
  end if
 end for
end while

A better solution is judged by the objective value and constraint value, and the following rules are as follows:

1. if two solutions are the same feasible solution, the one with smaller objective value is better.
2. A feasible solution is better than a infeasible solution.
3. if two solutions are both of the infeasible solution, the one with smaller violation objective value is better.

3   Formulating Antenna Design as COP

In the stage of modeling the antenna design problem as a COP, there are three steps, namely the determination of the solution space, the establishment of the objective function and the establishment of constraint function.

3.1   Antenna Requirements

The requirements of the LADEE antenna are shown in Table 1:

Table 1. Antenna requirements.

Parameter	Requirement(s)
Frequency	2200 MHz–2290 MHz
Polarization mode	Right-handed circular polarization
Input impedance	50Ω
VSWR	≤2
Gain pattern range	≥9 dB, $0° \leq \phi \leq 360°$, $-20° \leq \theta \leq 20°$
Size	$diameter \leq 229\,$mm, $height \leq 127\,$mm

According to the antenna design experience, each section of wire is neither too short (provisions not smaller than 1/10 of the wavelength) or too long (no longer than half a wavelength), angle of two adjacent wires can not be too narrow, otherwise it can lead to instability in electromagnetic computing software (angle of not less than 20°), combined with the requirements, evolutionary design requires to consider the constraints as follows:

Gain	$\geq 9dB$, $0° \leq \phi \leq 360°$; $-20° \leq \theta \leq 20°$
Length of each piece of wire(L)	$\lambda/10 \leq L \leq \lambda/2$
Adjacent angles	$\alpha \geq 20°$

3.2   Representation of the Antenna Structure

The overall structure of the LADEE wire antenna is shown in Fig. 1:

The antenna is mainly composed of seven crooked wire antenna, coaxial feed and a metal cup for reflection.

The structure of wire: The crooked-wire consists of 7 segments which stays in a cubic with size of 14.0 cm × 14.0 cm × 10.0 cm.

To sum up, the solution space of LADEE wire antenna structure variables is shown in Table 2.

As can be seen from Table 2, the 19 variables determine the geometric structure model of LADEE wire antenna. The solution space(x) is (6):

$$\overrightarrow{x} = (z_0, x_1, y_1, z_1, x_2, y_2, z_2, x_3, y_3, z_3, x_4, y_4, z_4, x_5, y_5, z_5, x_6, y_6, z_6) \quad (6)$$

The solution space as (7):

$$
\begin{aligned}
X &= \{\overrightarrow{x} | \overrightarrow{l} \leq \overrightarrow{x} \leq \overrightarrow{u}\} \\
\overrightarrow{x} &= (z_0, x_1, y_1, z_1, x_2, y_2, z_2, x_3, y_3, z_3, x_4, y_4, z_4, x_5, y_5, z_5, x_6, y_6, z_6) \\
\overrightarrow{l} &= (3, -70, -70, 0, -70, -70, 0, -70, -70, 0, -70, -70, 0, -70, -70, 0, -70, -70, 0) \\
\overrightarrow{u} &= (12, 70, 70, 100, 70, 70, 100, 70, 70, 100, 70, 70, 100, 70, 70, 100, 70, 70, 100)
\end{aligned}
\quad (7)
$$

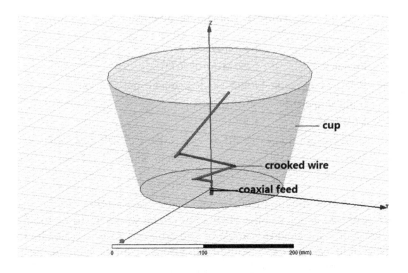

Fig. 1. LADEE wire antenna optimization model.

Table 2. The solution space of LADEE wire antenna structure variables.

Parameters	Range (mm)
z0	[3, 12]
x1	[−70, 70]
y1	[−70, 70]
z1	[0, 100]
x2	[−70, 70]
y2	[−70, 70]
z2	[0, 100]
x3	[−70, 70]
y3	[−70, 70]
z3	[0, 100]
x4	[−70, 70]
y4	[−70, 70]
z4	[0, 100]
x5	[−70, 70]
y5	[−70, 70]
z5	[−0, 100]
x6	[−70, 70]
y6	[−70, 70]
z6	[0, 100]

3.3   Objective Function

In this paper, the objective function is the sum of the variance of the gain, the axis ratio, and the VSWR (8), which can increase the robustness of the antenna design. The robustness of the antenna can be enhanced according to the performance characteristics of the antenna optimization design without additional computational cost.

$$f(\overrightarrow{x}) = \sum_{\phi} \sum_{\theta} (GVariance_{(\phi,\theta)} + ARVariance_{(\phi,\theta)}) + VSWRVariance$$

$$(8)$$

Among them, (ϕ,θ) is the point in the space spherical coordinate system, ϕ and θ are the azimuth and the elevation angles of the space spherical coordinate system, respectively, and the range of ϕ and θ is the design range of the antenna. Over the frequency band, $GVariance_{(\phi,\theta)}$ is the gain performance variance of the antenna at (ϕ,θ), $ARVariance_{(\phi,\theta)}$ is the axial ratio performance variance of the antenna at (ϕ,θ), and VSWRVariance is the VSWR performance variance of the antenna.

In this paper, the specific variance of antenna performance is calculated as follows (9):

$$GVariance_{(\phi,\theta)} = \sum_{freq} (Gain_{(\phi,\theta,freq)} - MeanG_{(\phi,\theta)})^2$$

$$MeanG_{(\phi,\theta)} = \sum_{(freq)} \frac{Gain_{(\phi,\theta,freq)}}{len_{(freq)}}$$

$$ARVariance_{(\phi,\theta)} = \sum_{freq} (Axial_{(\phi,\theta,freq)} - MeanAR_{(\phi,\theta)})^2$$

$$MeanAR_{(\phi,\theta)} = \sum_{(freq)} \frac{Axial_{(\phi,\theta,freq)}}{len_{freq}}$$

$$(9)$$

$$VSWRVariance = \sum_{freq} (VSWR_{freq} - MeanVSWR_{freq})^2$$

$$MeanVSWR_{freq} = \sum_{freq} \frac{VSWR_{freq}}{len_{freq}}$$

Among them, $GainVariance_{(\phi,\theta,freq)}$ and VSWRVariance are the variance of the gain and the VSWR; $Gain_{(\phi,\theta,freq)}$ is the gain of the antenna with a frequency of freq at (ϕ,θ), $Axial_{(\phi,\theta,freq)}$ is the axial ratio of the antenna with a frequency of freq at (ϕ,θ), $VSWR_{freq}$ is the VSWR of the antenna with a frequency of freq. Over the frequency band, Len_{freq} is the number of frequency points and freq is a frequency point.

3.4   Constraint Function

In the antenna optimization design of this paper, the constraint conditions of the antenna optimization design includes the gain constraint function, the axis ratio

constraint function and the VSWR constraint function. The constraint functions
are established as in Eq. (10).

$$gGain_{(\phi,\theta,freq)}(\overrightarrow{x}) = GainGoal - Gain_{(\phi,\theta,freq)} \leq 0$$
$$gVSWR_{freq}(\overrightarrow{x}) = VSWR_{freq} - VSWRGoal \leq 0$$
$$LinearlyPolarizationgAxial_{(\phi,\theta,freq)}(\overrightarrow{x}) = AxialGoal - Axial_{(\phi,\theta,freq)} \leq 0$$
$$CircularPolarizationgAxial_{(\phi,\theta,freq)}(\overrightarrow{x}) = Axial_{(\phi,\theta,freq)} - AxialGoal \leq 0$$
$$(10)$$

Among them, $\overrightarrow{x}$ is the solution variable, that is, the variable of the antenna
problem; ϕ and θ are the azimuth and the elevation; freq is the frequency point;
GainGoal is the objective value of the gain, $Gain_{(\phi,\theta,freq)}$ is the gain value with
a frequency of freq at (ϕ,θ). VSWRGoal is the objective value of the VSWR,
$VSWR_{freq}$ is the VSWR at the freq point; AxialGoal is the objective value of
the axial ratio, $Axial_{(\phi,\theta,freq)}$ is the axial ratio value with a frequency of freq at
(ϕ,θ).

3.5   Constrained Optimization Problem

Then a COP is established as in Eq. (11).

$$min: f(\overrightarrow{x}) = \sum_{\phi}\sum_{\theta}(GVariance_{(\phi,\theta)} + ARVariance_{(\phi,\theta)}) + VSWRVariance$$

$$gGain_{(\phi,\theta,freq)}(\overrightarrow{x}) = GainGoal - Gain_{(\phi,\theta,freq)} \leq 0$$
$$gVSWR_{freq}(\overrightarrow{x}) = VSWR_{freq} - VSWRGoal \leq 0$$
$$LinearlyPolarizationgAxial_{(\phi,\theta,freq)}(\overrightarrow{x}) = AxialGoal - Axial_{(\phi,\theta,freq)} \leq 0$$
$$CircularPolarizationgAxial_{(\phi,\theta,freq)}(\overrightarrow{x}) = Axial_{(\phi,\theta,freq)} - AxialGoal \leq 0$$
$$(11)$$

Combining the requirements of LADEE antenna in Table 1 to antenna design
COP in Eq. (11). The Eq. (11) changes into Eq. (12).

$$min: f(\overrightarrow{x}) = \sum_{\phi=0°}^{\phi=360°}\sum_{\theta=-20°}^{\theta=20°}(GVariance_{(\phi,\theta)}) + VSWRVariance$$
$$st: \quad gGain_{(\phi,\theta,freq)}(\overrightarrow{x}) = 9 - Gain_{(\phi,\theta,freq)} \leq 0$$
$$gVSWR_{freq}(\overrightarrow{x}) = VSWR_{freq} - 2 \leq 0 \qquad (12)$$

4   Solving Antenna Design by de

4.1   Setting DE Parameters

Using DE algorithm to solve the wire antenna, the parameters are set as follows:

1. Generations: T = 1000.
2. Population size: NP = 50.
3. Scale factor: F = 0.5.
4. Crossover rate: CR = 0.9.

4.2   Results and Discussion

The optimal antenna structure variables obtained by calculation is shown in Table 3, and the wire-structure is shown in Fig. 2; the Z axis don't include 3 mm-wire on the base.

Table 3. Optimal structure variable of LADEE wire antenna.

Parameters	Range (mm)
z0	4.44490259
x1	17.77588285
y1	10.21144474
z1	10.92184169
x2	−25.05202435
y2	−17.00814535
z2	17.0381978
x3	35.33219432
y3	−40.88013844
z3	36.15822884
x4	−4.59462421
y4	41.37324182
z4	30.43065683
x5	48.94386604
y5	20.61263286
z5	43.19187752
x6	−29.9285143
y6	−8.77754503
z6	98.23205633

The axial ratio of the evolved antenna at the central frequency point of the 2245 MHz is 2.06 dB. It is smaller than the 6.6 dB presented in the references [15], which meets the antenna design performance requirements. The axial ratio is shown in Fig. 3.

The VSWR of the evolved antenna in frequency range is shown in Fig. 4. As can be seen from the figure, the VSWR is less than 1.5 in the range of 2200 MHz–2290 MHz, much smaller than 2, which meets the performance requirements.

The RHCP gain of the evolved antenna at the central frequency point of the 2245 MHz is shown in Fig. 5. That can be seen in the antenna radiation range $-20° \leq \theta \leq 20°$, $0° \leq \phi \leq 360°$ in the gain is greater than 9 dB, maximum 12.4 dB, which meets the performance requirements. And the 3-D polar plot of the RHCP gain is shown in Fig. 6.

Simulated elevation cuts for the antenna are shown in Fig. 7.

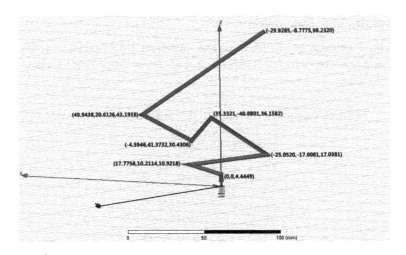

Fig. 2. Optimal structure of LADEE wire antenna.

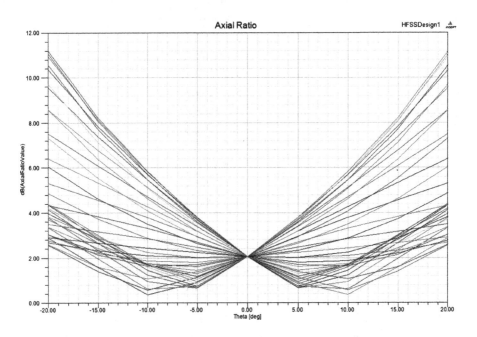

Fig. 3. The axial ratio of evolved antenna at the frequency point of 2245 MHz.

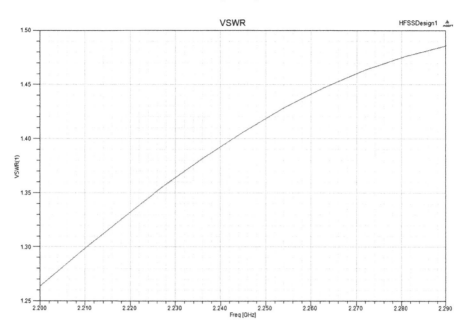

Fig. 4. The VSWR of the evolved antenna ranges of 2200 MHz–2290 MHz.

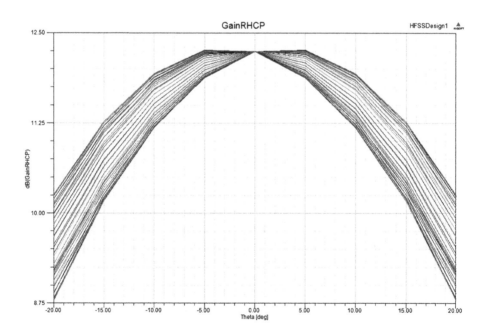

Fig. 5. The RHCP gain results of LADEE evolved antennas at 2245 MHz.

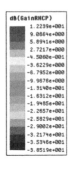

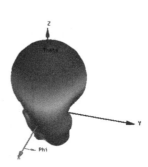

Fig. 6. The 3-D polar plot of the RHCP gain at 2245 MHz.

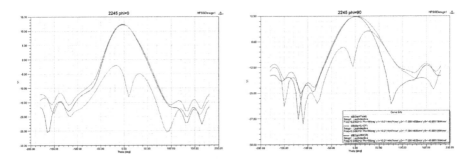

Fig. 7. Elevation plots for $\phi = 0°$ and $\phi = 90°$ at 2245 MHz for the evolved antenna.

5   Conclusion

The main difference of the antenna design in this paper from traditional evolutionary antenna is that the construction of the objective. The minimization of the objective can potentially enhance the robustness of the antenna and widen the frequency band without additional computational cost.

This paper chooses NASA LADEE satellite antenna design as an example to verify the method of this paper. The frequency of the LADEE evolved antenna is between 2200 MHz and 2290 MHz. The evolved antenna by the proposed method in this paper meets the requirements. Furthermore, the VSWR of the antenna is less than 1.5 over the required frequency range, much less than the requirement 2. The axis ratio is 2.06 dB at the beam direction which is much better than that of the antenna designed by the referred literature.

The future work is to manufacture the evolved antenna.

Acknowledgment. The authors are very grateful to the anonymous reviewers for their constructive comments to this paper. This work is supported by the National Science Foundation of China under Grant 61673355, 61271140 and 61203306.

References

1. Arianos, S., Araque Quijano, J.L., Vipiana, F., Dassano, G.: Application of evolutionary algorithms in the design of compact multi-band antennas. In: Joint 2012 IEEE International Symposium on Antennas and Propagation and USNC-URSI National Radio Science Meeting, APSURSI 2012, Chicago (2012)
2. Wang, Z., Qin, L., Yang, W.: A self-organising cooperative hunting by robotic swarm based on particle swarm optimisation localisation. Int. J. Bio Inspir. Comput. **7**(1), 68–73 (2015)
3. Xu, X., Guan, Y., Hong, B., Ke, W., Piao, S.: A humanoid robot path planning method based on virtual force-directed particle swarm optimisation. Int. J. Wirel. Mobile Comput. **9**(4), 325–331 (2015)
4. Bibao, M.N., Ser, J.D., Salcedo-Sanz, S., Casanova-Mateo, C.: On the application of multi-objective harmony search heuristics to the predictive deployment of fire-fighting aircrafts: a realistic case study. Int. J. Bio Inspir. Comput. **7**(5), 270–284 (2015)
5. Pei, Z., Zhou, Y., Chen, C., Liu, L., wang, Q.: Critical public opinion location and intelligence theme clustering strategy-based biological virus event detection and tracking model. Int. J. Wireless Mobile Comput. **9**(2), 192–198 (2015)
6. Kim, H., Cho, S.: Usability of spectrum overlay in the tv band in urban areas: focus on service availability range and spectrum utilisation efficiency. Int. J. Wirel. Mobile Comput. **9**(2), 115–124 (2015)
7. Weile, A.S., Michielssen, E.: Genetic algorithm optimization applied to electromagnetics: a review. IEEE Trans. Antennas Propag. **45**(3), 343–353 (1997)
8. Altshuler, E.E., Linden, D.S.: Wire-antenna designs using genetic algorithms. IEEE Antennas Propag. Mag. **39**(2), 33–43 (1997)
9. Lohn, J.D., Kraus, W.F., Linden, D.S.: Evolutionary optimization of a quadrifilar helical antenna. In: IEEE Antennas and Propagation Society Symposium 2004 Digest held in Conjunction with: USNC/URSI National Radio Science Meeting, Monterey, vol. 3, no. 34, pp. 2313–2316 (2004)
10. Hornby, G.S., Lohn, J.D., Linden, D.S.: Computer-automated evolution of an x-band antenna for nasas space technology 5 mission. Evol. Comput. **19**(1), 1–23 (2011)
11. Yang, B., Adams, J.J.: Systematic shape optimization of symmetric mimo antennas using characteristic modes. IEEE Trans. Antennas Propag. **64**(7), 2668–2678 (2016)
12. Wu, J., Abdelrahman, A., Liang, M., Yu, X., Xin, H.: Monopole antenna radiation pattern control via 3D-printed dielectrics. IEEE Trans. Antennas Propag. **65**(8), 3869–3876 (2017)
13. Jiao, R., Zeng, S., Alkasassbeh, J.S., Li, C.: Dynamic multi-objective evolutionary algorithms for single-objective optimization. Appl. Soft Comput. **61**, 793–805 (2017)
14. Storn, R., Price, K.: Differential evolution - a simple and efficient heuristic for global optimization over continuous spaces. J. Global Optim. **11**(4), 341–359 (1997)
15. Lohn, J.D., Linden, D.S., Blevins, B., Greenling, T., Allard, R.R.: Automated synthesis of a lunar satellite antenna system. IEEE Trans. Antennas Propag. **63**(4), 1436–1444 (2015)
16. Price, K., Storn, R.M., Lampinen, A.: Differential Evolution: A Practical Approach to Global Optimization. Natural Computing Series, pp. 1–24. Springer, Heidelberg (2005). https://doi.org/10.1007/3-540-31306-0

Typical Constrained Optimization Formulation in Evolutionary Computation Not Suitable for Expensive Optimization

Sanyou Zeng[1], Ruwang Jiao[1(✉)], Changhe Li[2], Bin Lan[1], Huanhuan Li[3], Jianqing Sun[4], and Yongzhi Sun[4]

[1] School of Mechanical Engineering and Electronic Information, China University of Geosciences, Wuhan 430074, China
sanyouzeng@gmail.com, ruwangjiao@gmail.com
[2] School of Automation, China University of Geosciences, Wuhan 430074, China
[3] School of Computer Science, China University of Geosciences, Wuhan 430074, China
[4] No 8511 Research Institute of CASIC, Nanjing 210007, China

Abstract. The typical formulation of a constrained optimization problem (COP), which has the constraint form of $\mathbf{g(x)} \leq \mathbf{0}$, has widely used in evolutionary computation field. However, it is not suitable for Gaussian processes (GPs) based expensive optimization. In this paper, a more general and suitable formulation of the COP, which has the constraint form of $\mathbf{l_g} \leq \mathbf{g(x)} \leq \mathbf{u_g}$, is recommended for the expensive optimization. Modeling a real world COP as a typical formulation will probably introduce additional constraints and dependencies among the objective and constraints while that as a suitable one introduces none additional. In the case of typical formulation, the additional constraints and dependencies have to be handled and the handling would cost additional computational resource, especially, the additional dependencies would lead to degenerating the performance of expensive optimization technologies since most of the technologies are based on the assumption of mutual independency among the objective and constraints. Experiments show that the performance of the expensive optimization technologies in the aspect of precision on solving the problems with suitable formulation is better than that with typical one. However, we could not verify the expense of additional computational resource since there are few expensive optimization technologies dealing with dependent objective and constraints.

Keywords: Evolutionary computation · Expensive optimization
Constrained optimization · Gaussian Process
Surrogate-assisted evolutionary algorithm

1 Introduction

Most of the science and engineering optimization problems in the real world are highly constrained. These constrained optimization problems present serious challenges to existing optimization methods. Evolutionary algorithms (EAs)

© Springer Nature Singapore Pte Ltd. 2018
K. Li et al. (Eds.): ISICA 2017, CCIS 873, pp. 232–245, 2018.
https://doi.org/10.1007/978-981-13-1648-7_20

have been successful in solving optimization problems, which are ubiquitous in constrained optimization problems (COPs). Constraint handling techniques based on EAs can be categorized into following approaches: penalty functions [1], stochastic ranking [2], feasibility rules [3,4], ε constrained methods [5], multi-objective methods [6,7], novel special operators [8,9] and ensemble of constraint-handling techniques [10,11].

Nevertheless, many of engineering optimization problems require expensive computer simulations for evaluating candidate solutions. Traditional EAs cannot directly solve them in that a large number of fitness evaluations are unaffordable. To overcome this barrier, surrogate-assisted EAs (SAEAs) have been developed, where part of the expensive function evaluations are replaced by computationally cheap approximate models, often known as surrogates or meta-models. Commonly used surrogates include Gaussian process (GP), radial basis function network (RBFN), support vector machine (SVM), and polynomial regression (PR) [12]. GP or Kriging model, has increasingly been employed as surrogates in evolutionary single and multi-objective optimization [13–19].

Maximizing the expected improvement (EI) [20] is one of the most popular infill sampling criteria used in selecting sample solutions for updating GP model. Using EI is advantageous since it is likely to be larger at under sampled areas near to the global optimum and offers solutions of both exploitation and exploration of the GP model. The original EI is only for unconstrained optimization problems. For the case of COPs, some attempts have been made to extend unconstrained EI into constrained EI. Schonlau [21] proposed a constrained EI by maximizing the multiplication of the EI and the probability feasibility (PF), which are both statistical measures determined from GP models of objective and constraints. In [22], expected violation was introduced where a threshold on violation was used instead of the actual violation, thereby allowing a larger sampling possibility around constraint bounds. This strategy was used to assess expected violations of promising points using a GP model. Durantin et al. [23] proposed a three-objective method which takes into account the EI, PF and prediction accuracy of the constraints.

However, the typical formulation of the COP adopted by the aforementioned technologies usually have the constraints with form $\mathbf{g}(\mathbf{x}) \leq \mathbf{0}$. Notably, in the case that there is an equality constraint $h(\mathbf{x}) = 0$, it has to be converted into two inequality constraints, i.e., $h(\mathbf{x}) - \delta \leq 0$ and $-h(\mathbf{x}) - \delta \leq 0$ where δ is a positive close-to-zero tolerance for numerical computation, i.e., $\delta = 0.0001$. It leads to two problems. One is that it increases a constraint. The other is that it introduces dependency since $h(\mathbf{x}) - \delta \leq 0$ and $-h(\mathbf{x}) - \delta \leq 0$ are strongly dependent between each other. Handling additional constraints and dependency will cost computational resources. In this paper, we introduce an suitable formulation of COP with another form of constraints $\mathbf{l_g} \leq \mathbf{g}(\mathbf{x}) \leq \mathbf{u_g}$. It will not cause the above two problems. No additional constraints and no additional dependency are introduced, which means that no additional resources will be consumed.

The remainder of the paper is organized as follows. The detailed description of related techniques is provided in Sect. 2. The typical formulation and the suitable one of a COP are presented and their interconversion is discussed in Sect. 3. Constrained expected improvements (CEIs) for the COP with the suitable formulation are discussed in Sect. 4. A GP surrogate-assisted EA with CEI as fitness function is presented in Sect. 5. Experimental results are presented in Sect. 6. Finally, conclusion and future work are discussed in Sect. 7.

2   Related Work

2.1   Gaussian Process

To model an unknown function $f(\mathbf{x})$, GP assumes that $f(\mathbf{x})$ at any point $\mathbf{x}$ is a Gaussian random variable $F(\mathbf{x}) \sim N(\mu, \sigma^2)$, where μ and σ are two constants independent of $\mathbf{x}$. For any $\mathbf{x}$, $f(\mathbf{x})$ is denoted as a sample of $F(\mathbf{x})$. The correlation between $F(\mathbf{x})$ and $F(\mathbf{x}')$ is only related to the distance between the $\mathbf{x}$ and $\mathbf{x}'$. The correlation between $\mathbf{x}$ and $\mathbf{x}'$ has the form

$$Corr(F(\mathbf{x}), F(\mathbf{x}')) = exp(- \sum_{i=1}^{n} \theta_i |x_i - x_i'|^{p_i}) \tag{1}$$

where $p_i \in [1, 2]$ and $\theta_i > 0$. In each coordinate direction, larger p_i can be interpreted as a parameter increasing the smoothness of the response surface, while larger θ_i indicates greater activity of nonlinearity.

1. Hyper Parameter Estimation: Given N solutions $\mathbf{x}^{(1)}, ..., \mathbf{x}^{(N)}$, we denote Gaussian random vector $F^N = (F(\mathbf{x}^{(1)}), ..., F(\mathbf{x}^{(N)}))^T$ and their tested function values $f^N = (f(\mathbf{x}^{(1)}), ..., f(\mathbf{x}^{(N)}))^T$ is regarded as a sample of F^N, then the hyper parameters $\mu, \sigma, \theta_1, ..., \theta_n$, and $p_1, ..., p_n$ can be estimated by maximizing the likelihood that the joint Gaussian probability density function of F^N at f^N

$$\frac{1}{(2\pi\sigma^2)^{N/2}\sqrt{det(\mathbf{C})}} exp[- \frac{(f^N - \mu\mathbf{1})^T \mathbf{C}^{-1}(f^N - \mu\mathbf{1})}{2\sigma^2}] \tag{2}$$

where $\mathbf{C}$ represents a $N \times N$ correlation matrix whose (i, j)-element has the form of $Corr(F(\mathbf{x}^{(i)}), F(\mathbf{x}^{(j)}))$, and $\mathbf{1}$ is a $N \times 1$ vector of ones.
To maximize (2), the values of μ and σ^2 must be

$$\hat{\mu} = \frac{\mathbf{1}^T \mathbf{C}^{-1} f^N}{\mathbf{1}^T \mathbf{C}^{-1} \mathbf{1}} \tag{3}$$

and

$$\hat{\sigma}^2 = \frac{(f^N - \mathbf{1}\hat{\mu})^T \mathbf{C}^{-1}(f^N - \mathbf{1}\hat{\mu})}{N}. \tag{4}$$

Replacing (3) and (4) with (2) eliminates the unknown parameters μ and σ from (2). Consequently, the likelihood function depends only on θ_i and p_i $(i = 1, ..., n)$. Then (2) can be maximized to calculate estimates of $\hat{\theta}_i$ and $\hat{p}_i$. Thus from (3) and (4), the estimates $\hat{\mu}$ and $\hat{\sigma}^2$ can be readily obtained.

2. Best Linear Unbiased Prediction: Given the hyper parameter estimates $\hat{\theta}_i, \hat{p}_i, \hat{\mu}$ and $\hat{\sigma}^2$, one can predict $f(\mathbf{x})$ at any untested point $\mathbf{x}$ based on the tested f^N at the points $\mathbf{x}^{(1)}, ..., \mathbf{x}^{(N)}$. The best linear unbiased predictor of $f(\mathbf{x})$ is

$$\hat{f}(\mathbf{x}) = \hat{\mu} + \mathbf{r}^T C^{-1}(f^N - \mathbf{1}\hat{\mu}) \tag{5}$$

and its mean squared error is

$$s^2(\mathbf{x}) = \hat{\sigma}^2[1 - \mathbf{r}^T C^{-1}\mathbf{r} + \frac{(1-\mathbf{1}^T C^{-1}\mathbf{r})^2}{\mathbf{1}^T C^{-1}\mathbf{1}}] \tag{6}$$

where $\mathbf{r}$ represents the $N \times 1$ vector of correlations $Corr(F(\mathbf{x}), F(\mathbf{x}^{(i)}))$ $(i = 1, ..., N)$. $N(\hat{f}(\mathbf{x}), s^2(\mathbf{x}))$ is regarded as a predictive distribution for $F(\mathbf{x})$ on the tested $F^N = f^N$ at the points $\mathbf{x}^{(1)}, ..., \mathbf{x}^{(N)}$.

2.2   Expected Improvement (EI) Without Constraints

After building a surrogate model, a criterion for measuring the quality of a new untested point should be defined. EI has been proposed in [20] and it achieves a trade-off between exploitation and exploration when deciding which point to evaluate.

Suppose $N(\hat{f}(\mathbf{x}), s^2(\mathbf{x}))$ is a GP model for a function $F(\mathbf{x})$ on the tested $f^N = (f(\mathbf{x}^{(1)}), ..., f(\mathbf{x}^{(N)}))^T$ at $\mathbf{x}^{(1)}, ..., \mathbf{x}^{(N)}$, and the best value of $F(\mathbf{x})$ over all the f^N is f_{min} (in an unconstrained problem). The improvement of $F(\mathbf{x})$ at an untested point $\mathbf{x}$ is

$$I(\mathbf{x}) = max\{f_{min} - F(\mathbf{x}), 0\}. \tag{7}$$

Thus, the **expected improvement (EI)** on $F^N = f^N$ is

$$E[I(\mathbf{x})|f^N] = E[max\{f_{min} - F(\mathbf{x}), 0\}|f^N]. \tag{8}$$

It can be expressed as [20]

$$E[I(\mathbf{x})|f^N] = (f_{min} - \hat{f}(\mathbf{x}))\Phi(\frac{f_{min}-\hat{f}(\mathbf{x})}{s(\mathbf{x})}) + s(\mathbf{x})\phi(\frac{f_{min}-\hat{f}(\mathbf{x})}{s(\mathbf{x})}) \tag{9}$$

where the predicted expectation $\hat{f}(\mathbf{x})$ and the square root of the predicted variance $s(\mathbf{x})$ are calculated according to Eqs. (5) and (6). $\Phi(.)$ is the standard normal cumulative distribution function and $\phi(.)$ is the standard normal probability density function, they can be expressed as follows

$$\begin{aligned} \Phi(x) &= \frac{1}{\sqrt{2\pi}} \int_{-\infty}^{x} exp(-\frac{t^2}{2})dt, \\ \phi(x) &= \frac{1}{\sqrt{2\pi}} exp(-\frac{x^2}{2}). \end{aligned} \tag{10}$$

In the expensive optimization in this paper, the solution $\mathbf{x}$ is optimized by maximizing its EI value $E[I(\mathbf{x})|f^N]$. In this sense, $E[I(\mathbf{x})|f^N]$ is the estimated fitness value of the solution $\mathbf{x}$. $E[I(\mathbf{x})|f^N]$ will enlarge with either $s(\mathbf{x})$ enlarge or $f_{min} - \hat{f}(\mathbf{x})$ enlarge, maximizing $E[I(\mathbf{x})|f^N]$ provides an elegant balance between global exploration and local exploitation.

3   Two Formulations of COPs and Their Interconversion

3.1   Two Formulations of COPs

One formulation of the COP is

$$
\begin{aligned}
\min \quad & f(\mathbf{x}) \\
\text{st}: \quad & \mathbf{g}(\mathbf{x}) = (g_1(\mathbf{x}), g_2(\mathbf{x}), ..., g_m(\mathbf{x})) \leq \mathbf{0} \\
\text{where } & \mathbf{x} = (x_1, x_2, ..., x_n) \in \mathbf{X} \\
& \mathbf{X} = \{\mathbf{x} | \mathbf{l_x} \leq \mathbf{x} \leq \mathbf{u_x}\} \\
& \mathbf{l_x} = (l_{x_1}, l_{x_2}, ..., l_{x_n}), \mathbf{u_x} = (u_{x_1}, u_{x_2}, ..., u_{x_n})
\end{aligned}
\tag{11}
$$

where $\mathbf{x}$ is the solution vector, $\mathbf{l_x}$ and $\mathbf{u_x}$ are the lower and upper bounds of the solution space, $f(\mathbf{x})$ denotes the objective function, $\mathbf{g}(\mathbf{x}) \leq \mathbf{0}$ represents the vector of constraints and $\mathbf{0}$ is the constraint boundary. If a solution $\mathbf{x} \in \mathbf{X}$ satisfies all constraints, it is feasible, otherwise it is infeasible. The formulation in Eq. (11) has widely used in the evolutionary computation field. We call it as **typical formulation**.

If an equality constraint $h(\mathbf{x}) = 0$ occurs, it can be changed into two inequality constraints, i.e., $h(\mathbf{x}) - \delta \leq 0$ and $-h(\mathbf{x}) - \delta \leq 0$, where δ is a tolerance, and usually set $\delta = 0.0001$. This change increases one constraint.

To avoid extra constraints, this paper introduces another formulation of COP:

$$
\begin{aligned}
\min \quad & y = f(\mathbf{x}) \\
\text{st}: \quad & \mathbf{l_g} \leq \mathbf{g}(\mathbf{x}) \leq \mathbf{u_g} \\
\text{where } & \mathbf{g}(\mathbf{x}) = (g_1(\mathbf{x}), g_2(\mathbf{x}), ..., g_m(\mathbf{x})) \\
& \mathbf{l_g} = (l_{g_1}, l_{g_2}, ..., l_{g_m}), \mathbf{u_g} = (u_{g_1}, u_{g_2}, ..., u_{g_m}) \\
& \mathbf{x} = (x_1, x_2, ..., x_n) \in \mathbf{X} \\
& \mathbf{X} = \{\mathbf{x} | \mathbf{l_x} \leq \mathbf{x} \leq \mathbf{u_x}\} \\
& \mathbf{l_x} = (l_{x_1}, l_{x_2}, ..., l_{x_n}), \mathbf{u_x} = (u_{x_1}, u_{x_2}, ..., u_{x_n})
\end{aligned}
\tag{12}
$$

where $\mathbf{l_x}$ and $\mathbf{u_x}$ denote the lower bound and upper bound of the solution space, $\mathbf{l_g} \leq \mathbf{g}(\mathbf{x}) \leq \mathbf{u_g}$ is the vector of constraints, and $\mathbf{l_g}$ and $\mathbf{u_g}$ are the lower and upper constraint bounds respectively. Actually, similar version of the formulation in Eq. (12) can be found in literature [24]. In the case of an equality constraint $h(\mathbf{x}) = 0$, it can be changed into one inequality constraint, i.e., $-\delta \leq h(\mathbf{x}) \leq \delta$. Notably, this change does not increase the number of constraints.

To contrast to the typical formulation in Eq. (11), we call Eq. (12) as the **suitable formulation**. The difference between the two formulations of the COPs is that the typical formulation in Eq. (11) has the constraint form $\mathbf{g}(\mathbf{x}) \leq \mathbf{0}$ while the suitable formulation in Eq. (12) has a more general form $\mathbf{l_g} \leq \mathbf{g}(\mathbf{x}) \leq \mathbf{u_g}$.

For a COP, either in Eq. (11) or in Eq. (12), the goal is to minimize $f(\mathbf{x})$ subject to $\mathbf{x}$ satisfying m constraints $g_i(\mathbf{x})(i = 1, ..., m)$. We assume that the objective function $F(\mathbf{x})$ and constraints functions $G_i(\mathbf{x})(i = 1, ..., m)$ are Gaussian processes in this paper. For easy description, some notations are introduced in the following.

Given a point $\mathbf{x}$, we denote Gaussian random vector:

$$\mathbf{G}(\mathbf{x}) = (G_1(\mathbf{x}), ..., G_m(\mathbf{x})).$$

Given N points $\mathbf{x}^{(1)}, ..., \mathbf{x}^{(N)}$, we denote Gaussian random vector:

$$F^N = (F(\mathbf{x}^{(1)}), ..., F(\mathbf{x}^{(N)}))^T,$$

$$G_1^N = (G_1(\mathbf{x}^{(1)}), ..., G_1(\mathbf{x}^{(N)}))^T, ..., G_m^N = (G_m(\mathbf{x}^{(1)}), ..., G_m(\mathbf{x}^{(N)}))^T,$$

$$and \ \ G^{mN} = (G_1(\mathbf{x}^{(1)}), ..., G_m(\mathbf{x}^{(1)}), ..., G_1(\mathbf{x}^{(N)}), ..., G_m(\mathbf{x}^{(N)}))^T.$$

And their tested function values:

$$f^N = (f(\mathbf{x}^{(1)}), ..., f(\mathbf{x}^{(N)}))^T,$$

$$g_1^N = (g_1(\mathbf{x}^{(1)}), ..., g_1(\mathbf{x}^{(N)}))^T, ..., g_m^N = (g_m(\mathbf{x}^{(1)}), ..., g_m(\mathbf{x}^{(N)}))^T$$

$$and \ \ g^{mN} = (g_1(\mathbf{x}^{(1)}), ..., g_m(\mathbf{x}^{(1)}), ..., g_1(\mathbf{x}^{(N)}), ..., g_m(\mathbf{x}^{(N)}))^T$$

are regarded as samples of F^N, G_1^N, ..., G_m^N and G^{mN}.

In this paper, $\mathbf{x}^{(1)}, ..., \mathbf{x}^{(N)}$ are regarded as tested points while the $\mathbf{x}$ is regarded as an untested point. In this way, F^N, G_1^N, ..., G_m^N and G^{mN} are known data while $F(\mathbf{x})$ and $\mathbf{G}(\mathbf{x})$ are required to be predicted. F^N, G_1^N, ..., G_m^N and G^{mN} can replace each other with f^N, g_1^N, ..., g_m^N and g^{mN} respectively, and $F(\mathbf{x})$ and $\mathbf{G}(\mathbf{x})$ can replace each other with $f(\mathbf{x})$ and $\mathbf{g}(\mathbf{x})$ respectively.

The same as Subsect. 2.1, one can predict $f(\mathbf{x})$, $g_i(\mathbf{x})(i = 1, 2, ..., m)$ at any untested point $\mathbf{x}$ based on the f^N, $g_i^N (i = 1, 2, ..., m)$ at the tested points $\mathbf{x}^{(1)}, ..., \mathbf{x}^{(N)}$.

3.2   Change Two Formulations into Each Other

A real world COP can be modeled either as a suitable formulation or a typical one, and the two formulations can be changed into each other.

Change Suitable Formulation into Typical One. The constraints of suitable formulation of COP in Eq. (12) $\mathbf{l_g} \leq \mathbf{g}(\mathbf{x}) \leq \mathbf{u_g}$ can be rewritten as two groups of constraints $\mathbf{g}(\mathbf{x}) - \mathbf{u_g} \leq \mathbf{0}$ and $\mathbf{l_g} - \mathbf{g}(\mathbf{x}) \leq \mathbf{0}$ which have the constraint form of typical COP in Eq. (11). Then the suitable formulation is changed into the following typical formulation

$$
\begin{aligned}
\min \quad & f(\mathbf{x}) \\
st: \quad & \mathbf{g}(\mathbf{x}) - \mathbf{u_g} \leq \mathbf{0} \\
& \mathbf{l_g} - \mathbf{g}(\mathbf{x}) \leq \mathbf{0} \\
where \ & \mathbf{g}(\mathbf{x}) = (g_1(\mathbf{x}), g_2(\mathbf{x}), ..., g_m(\mathbf{x})) \\
& \mathbf{l_g} = (l_{g_1}, l_{g_2}, ..., l_{g_m}), \mathbf{u_g} = (u_{g_1}, u_{g_2}, ..., u_{g_m}) \\
& \mathbf{x} = (x_1, x_2, ..., x_n) \in \mathbf{X} \\
& \mathbf{X} = \{\mathbf{x} | \mathbf{l_x} \leq \mathbf{x} \leq \mathbf{u_x}\} \\
& \mathbf{l_x} = (l_{x_1}, l_{x_2}, ..., l_{x_n}), \mathbf{u_x} = (u_{x_1}, u_{x_2}, ..., u_{x_n}).
\end{aligned}
\tag{13}
$$

Note, in the case of the original real-world problem having the suitable formulation of COP, the change of suitable formulation into typical one will double the number of constraints and introduce additional dependencies, which leads to two difficulties in the expensive optimization. Firstly, handling the doubled constraints and the additional dependencies increase the expense of computational resources since the expensive optimization method constructs a model for each constraint and the doubled constraints increase the dimension of the covariance matrix while the additional dependencies determine the dimension could not be reduced. Secondly, most technologies for expensive optimization would not work any more since these technologies are based on the assumption of mutual independency between the objective and constraints.

Change Typical Formulation into Suitable One. The constraints of typical formulation of COP in Eq. (11) $\mathbf{g}(\mathbf{x}) \leq \mathbf{0}$ can be regarded as having negative infinite lower bound and zero upper bound. That is $-\infty \leq \mathbf{g}(\mathbf{x}) \leq \mathbf{0}$. Then the typical formulation is changed into the following suitable formulation

$$
\begin{aligned}
&\min \quad y = f(\mathbf{x}) \\
&\text{st}: \quad \mathbf{l_g} \leq \mathbf{g}(\mathbf{x}) \leq \mathbf{u_g} \\
&\text{where } \mathbf{g}(\mathbf{x}) = (g_1(\mathbf{x}), g_2(\mathbf{x}), ..., g_m(\mathbf{x})) \\
&\qquad\quad \mathbf{l_g} = (-\infty, -\infty, ..., -\infty), \mathbf{u_g} = (0, 0, ..., 0) \\
&\qquad\quad \mathbf{x} = (x_1, x_2, ..., x_n) \in \mathbf{X} \\
&\qquad\quad \mathbf{X} = \{\mathbf{x} | \mathbf{l_x} \leq \mathbf{x} \leq \mathbf{u_x}\} \\
&\qquad\quad \mathbf{l_x} = (l_{x_1}, l_{x_2}, ..., l_{x_n}), \mathbf{u_x} = (u_{x_1}, u_{x_2}, ..., u_{x_n}).
\end{aligned}
\tag{14}
$$

The change of typical formulation into suitable one does not increase the number of constraints. In the case of the original real-world problem having the typical formulation of COP, solving the original real-world problem by optimizing the suitable formulation will spend the same computational resources as by optimizing the typical one.

In view of the suitable formulation in Eq. (12), the typical formulation in Eq. (11) is a simplified version of suitable formulation showing in Eq. (14).

Based on the above discussion, this paper suggests choosing the suitable formulation of COP for the expensive optimization based on Gaussian processes.

4   Constrained Expected Improvements for Suitable Formulation of COP

Since the suitable formulation of COP in Eq. (12) for the expensive optimization is chosen and the constrained expected improvement (CEI) of a point $\mathbf{x}$ is regarded as the fitness of the solution in the expensive evolutionary optimization, the CEI for the suitable formulation is discussed in this paper. We assume that the objective function $F(\mathbf{x})$ and constraint functions $G_i(\mathbf{x})(i = 1, ..., m)$ are mutually independent, and the CEI is discussed in two situations: infeasible situation and feasible one.

In the infeasible situation where all the tested solutions are infeasible, the CEI of a solution $\mathbf{x}$ is defined as the probability of feasibility

$$CEI(\mathbf{x}) = \prod_{i=1}^{m} P\{\{l_{g_i} \leq G_i(\mathbf{x}) \leq u_{g_i}\}|g_i^N\}$$
$$= \prod_{i=1}^{m} [\Phi(\frac{u_{g_i} - \hat{g}_i(\mathbf{x})}{s_{g_i}(\mathbf{x})}) - \Phi(\frac{l_{g_i} - \hat{g}_i(\mathbf{x})}{s_{g_i}(\mathbf{x})})]. \tag{15}$$

In the feasible situation where the algorithm has tested some feasible solutions, the improvement of a solution $\mathbf{x}$ is defined as the objective improvement under satisfying constraints which is

$$I_{c,N}(\mathbf{x}) = \begin{cases} f_{min}^N - F(\mathbf{x}), & F(\mathbf{x}) \leq f_{min}^N \text{ and } l_{g_i} \leq G_i(\mathbf{x}) \leq u_{g_i} \\ 0, & otherwise \end{cases} \tag{16}$$

where f_{min}^N is the best fitness value over the feasible points in the tested N points.

With the assumption of the mutual independence among $F(\mathbf{x})$ and $G_i(\mathbf{x})(i = 1, ..., m)$, the CEI of the solution $\mathbf{x}$ based on $F^N = f^N$ and $G^{mN} = g^{mN}$ is the multiplication of the expected improvement of the objective and the probability of feasibility of the solution [21], which is in the following

$$CEI(\mathbf{x}) = E[I_{c,N}(\mathbf{x})|f^N, g^{mN}]$$
$$= E[I(\mathbf{x})|f^N] \times \prod_{i=1}^{m} P\{\{l_{g_i} \leq G_i(\mathbf{x}) \leq u_{g_i}\}|g_i^N\}$$
$$= [(f_{min}^N - \hat{f}(\mathbf{x}))\Phi(\frac{f_{min}^N - \hat{f}(\mathbf{x})}{s_f(\mathbf{x})}) + s_f(\mathbf{x})\phi(\frac{f_{min}^N - \hat{f}(\mathbf{x})}{s_f(\mathbf{x})})] \tag{17}$$
$$\times \prod_{i=1}^{m} [\Phi(\frac{u_{g_i} - \hat{g}_i(\mathbf{x})}{s_{g_i}(\mathbf{x})}) - \Phi(\frac{l_{g_i} - \hat{g}_i(\mathbf{x})}{s_{g_i}(\mathbf{x})})]$$

where $E[I(\mathbf{x})|f^N]$ is calculated using Eq. (9), and $\prod_{i=1}^{m} P\{\{l_{g_i} \leq G_i(\mathbf{x}) \leq u_{g_i}\}|g_i^N\}$ is the probability of feasibility (PF) of the solution $\mathbf{x}$.

The CEI of a solution $\mathbf{x}$ is summarized as

Infeasible situation:
$$CEI(\mathbf{x}) = \prod_{i=1}^{m} [\Phi(\frac{u_{g_i} - \hat{g}_i(\mathbf{x})}{s_{g_i}(\mathbf{x})}) - \Phi(\frac{l_{g_i} - \hat{g}_i(\mathbf{x})}{s_{g_i}(\mathbf{x})})]$$
Feasible situation: $\qquad\qquad\qquad\qquad\qquad\qquad\qquad\qquad$ (18)
$$CEI(\mathbf{x}) = [(f_{min}^N - \hat{f}(\mathbf{x}))\Phi(\frac{f_{min}^N - \hat{f}(\mathbf{x})}{s_f(\mathbf{x})}) + s_f(\mathbf{x})\phi(\frac{f_{min}^N - \hat{f}(\mathbf{x})}{s_f(\mathbf{x})})]$$
$$\times \prod_{i=1}^{m} [\Phi(\frac{u_{g_i} - \hat{g}_i(\mathbf{x})}{s_{g_i}(\mathbf{x})}) - \Phi(\frac{l_{g_i} - \hat{g}_i(\mathbf{x})}{s_{g_i}(\mathbf{x})})]$$

Note, the CEI in Eq. (18) is for the suitable formulation of COP. In the case that the original real-world problem has the typical formulation of COP, we have

---

Algorithm 1. GP surrogate-assisted EA

1: Use LHD to obtain initial samples and test them using real expensive function;
2: Find the best solution $\mathbf{x}_{best}$ from the tested initial sample;
3: Cluster the tested initial solutions into several small clusters, see Algorithm 2;
4: For every cluster, build local predictive GP surrogate models for objective and each constraint respectively;
5: If the predefined stopping criterion is met, output $\mathbf{x}_{best}$; otherwise go to *step 6*;
6: Apply DE/rand/1/bin [25] to maximize the CEI, CEI in Eq. (18) for the suitable formulation of COP, CEI in Eq. (19) for the typical one;
7: Test the real objective and constraints values of the solution which has the maximum CEI;
8: Add the new tested solution to the cluster to which it belongs to. If the number of solutions in this cluster is not exceeded L, update the local GP model; else re-cluster all the tested points and re-build models for each cluster. Update $\mathbf{x}_{best}$ and go back to *step 5*.

$\mathbf{l_g} = (-\infty, -\infty, ..., -\infty)$ and $\mathbf{u_g} = (0, 0, ..., 0)$ in view of the suitable formulation in Eq. (12). The calculation of the CEI can be simplified as

Infeasible situation:
$$CEI(\mathbf{x}) = \prod_{i=1}^{m} \Phi(\frac{-\hat{g}_i(\mathbf{x})}{s_{g_i}(\mathbf{x})})$$

Feasible situation:
$$CEI(\mathbf{x}) = [(f_{min}^N - \hat{f}(\mathbf{x}))\Phi(\frac{f_{min}^N - \hat{f}(\mathbf{x})}{s_f(\mathbf{x})}) + s_f(\mathbf{x})\phi(\frac{f_{min}^N - \hat{f}(\mathbf{x})}{s_f(\mathbf{x})})] \times \prod_{i=1}^{m} \Phi(\frac{-\hat{g}_i(\mathbf{x})}{s_{g_i}(\mathbf{x})}).$$

$$(19)$$

The CEI in Eq. (19) is actually for the typical formulation of COP.

5   A GP Surrogate-Assisted EA with CEI Fitness

In this section, a GP surrogate-assisted EA is designed and the CEI is integrated into the EA as the fitness estimation. The algorithm framework is listed in Algorithm 1 and it is based on a famous algorithm: Efficient Global Optimization (EGO) [20].

In *step 1* of Algorithm 1, Latin hypercube design (LHD) [26] is chosen to get initial sampling data since it samples the design space very uniformly. A key issue in an SAEA is how to adopt a reasonable amount of computational effort to build a good model for selecting the most potential candidate points. When the number of tested solutions N is large, the computational overhead may be unbearable if all solutions are directly used to build surrogate models to estimate and predict a solution. Clustering all the tested solutions into several small clusters and then building several local predictive models based on these clusters is an effective method. The clustering adopted in Algorithm 2 is based on [27].

Building or updating a Gaussian model for a cluster need to maximize the likelihood function in Eq. (2) to determine the hyper parameters $\hat{\theta}$ and $\hat{p}$. In

---

Algorithm 2. Clustering Operator

Input: Tested points $\{\mathbf{x}^{(1)}, ..., \mathbf{x}^{(N)}\}$, maximum number of points in each cluster L.
Output: Cluster set $\mathbb{G}$.

1: Create a temporary cluster set $\mathbb{G}$, where each individual $\mathbf{x}$ in the sampling data constitutes a distinct cluster;
2: If $|\mathbb{G}| = 1$, go to *step 6*, else go to *step 3*;
3: Calculate the distance of all pairs of clusters. The distance d_c of two cluster $\mathbf{C}_1$, $\mathbf{C}_2 \in \mathbb{G}$ is regarded as the average distance between pairs of points across the two clusters:

$d_c = \frac{1}{|\mathbf{C}_1|.|\mathbf{C}_2|} \sum\limits_{\mathbf{x}_1 \in \mathbf{C}_1, \mathbf{x}_2 \in \mathbf{C}_2} d(\mathbf{x}_1, \mathbf{x}_2)$, where $d(\mathbf{x}_1, \mathbf{x}_2)$ is the Euclidean distance

between two individuals $\mathbf{x}_1$ and $\mathbf{x}_2$;
4: Chosen two clusters $\mathbf{C}_1$ and $\mathbf{C}_2$ with minimal distance d_c;
5: If the chosen two clusters $|\mathbf{C}_1| + |\mathbf{C}_2| < L$, merge clusters $\mathbf{C}_1$ and $\mathbf{C}_2$ into $\mathbf{C}_1$ and delete the cluster $\mathbf{C}_2$ from $\mathbb{G}$, go to *step 2*. Else go to *step 6*;
6: Return the cluster set $\mathbb{G}$.

---

this paper, DE/rand/1/bin [25] is adopted to maximize the likelihood due to its good global search ability.

6   Numerical Experiments and Discussions

In this section, we validate the suitability of the suggested formulation by verifying the success of Algorithm 1 on solving the COP with the suitable formulation and the degeneration with the typical one.

6.1   Test Problems

We provide two problems P_1 and P_2. Their feasibility rates are 0.0000% and 17.0753%, respectively. Their suitable formulations are as

$$\textbf{P1 – Suitable}$$
$$\min f(\mathbf{x}) = x_1^2 + (x_2 - 1)^2 + x_3^2$$
$$st: \; -0.0001 \le g(\mathbf{x}) \le 0.0001 \tag{20}$$
$$g(\mathbf{x}) = x_2 - x_1^2 - x_3^2$$
$$-1 \le x_1, x_2 \le 1, \frac{1}{\sqrt{2}} \le x_3 \le 10$$

and

$$\textbf{P2 – Suitable}$$
$$\min f(\mathbf{x}) = \sum_{i=1}^{4} x_i^2 + sin^2(\sum_{i=1}^{4} ix_i)$$
$$st: \; 0 \le g(\mathbf{x}) = \sum_{i=1}^{4} x_i \le 100 \tag{21}$$
$$-10 \le x_1, x_2, x_3, x_4 \le 100.$$

The typical formulations are as

P1 – Typical

$$\min f(\mathbf{x}) = x_1^2 + (x_2 - 1)^2 + x_3^2$$

$$\text{st}: \; g_1(\mathbf{x}) = x_2 - x_1^2 - x_3^2 - 0.0001 \le 0 \tag{22}$$
$$g_2(\mathbf{x}) = -(x_2 - x_1^2 - x_3^2) - 0.0001 \le 0$$
$$-1 \le x_1, x_2 \le 1, \; \frac{1}{\sqrt{2}} \le x_3 \le 10$$

and

P2 – Typical

$$\min f(\mathbf{x}) = \sum_{i=1}^{4} x_i^2 + sin^2(\sum_{i=1}^{4} ix_i)$$

$$\text{st}: \; g_1(\mathbf{x}) = \sum_{i=1}^{4} x_i - 100 \le 0 \tag{23}$$
$$g_2(\mathbf{x}) = -\sum_{i=1}^{4} x_i \le 0$$
$$-10 \le x_1, x_2, x_3, x_4 \le 100.$$

Compared with constraints in the typical formulation, each of the constraint in the suitable formulation has doubled into two constraints and the two constraints are strongly dependent, actually, the correlation coefficient is -1.

6.2   Parameter Settings

Algorithm parameter settings in Algorithm 1:

- The number of initial samples generated using LHD: $11n - 1$, where n is the number of variables of the problem.

Clustering operator settings in Algorithm 2:

- The maximum number of points in each cluster $L = 50$.

DE/rand/1/bin [25] settings:

- Population size: 30.
- Number of generations: 500.
- Crossover probability: $CR = 0.9$.
- Scaling factor F is randomly given in the range of $[0.0, 1.0]$.

Other settings:

- The maximum test number: 100.
- The number of independent runs: 25.

6.3   Results and Discussions

We use Algorithm 1 to solve two problems with suitable formulation in Eqs. (20) and (21) and typical ones in Eqs. (22) and (23) respectively.

$f(\mathbf{x})$ and $g(\mathbf{x})$ in suitable formulations Eqs. (20) and (21) are approximate independent, and can be regarded as satisfying the assumption of independencies. Then the CEI for suitable formulation in Eq. (18) holds. The CEI in Algorithm 1 will work.

However, $g_1(\mathbf{x})$ and $g_2(\mathbf{x})$ in typical formulations Eqs. (22) and (23) are strongly dependent. The assumption of independencies are violated. The CEI for typical formulation in Eq. (19) may not be true. Equation (19) has to be used to calculate the CEI in Algorithm 1 since we have no other CEI could handle dependency.

The obtained results of the two problems for these two formulations are listed in the form of 'best, mean, worst and standard deviation' in Table 1.

From Table 1, the performance of Algorithm 1 in the aspect of precision on solving the problems with suitable formulations is better (but not much better) than that with typical ones.

Table 1. Results for 2 problems with typical and suitable formulations respectively.

	P1-Typical	P1-Suitable		P2-Typical	P2-Suitable
Best	0.75020	**0.75010**	Best	0.04406	**0.04106**
Mean	0.75201	**0.75149**	Mean	0.23243	**0.21964**
Worst	0.75632	**0.75416**	Worst	**0.38691**	0.56455
Std	2.08e-03	**1.45e-03**	Std	**1.12e-01**	1.57e-01

7   Conclusion and Future Research

A suitable formulation of COP is recommended for the GP based expensive optimization since the widely used typical formulation is not suitable. Modeling a real world COP as the typical formulation will probably introduce additional constraints and dependencies among the objective and constraints while that as the suitable one introduces none additional. Handling the additional constraints and dependencies would spend additional computational resource, especially, the additional dependencies would lead to degenerating the performance of expensive optimization technologies since most of the technologies are base on the assumption of mutual independency between the objective and constraints. We did not verify the expense of additional computational resource since there are few expensive optimization technologies dealing with dependent objective and constraints. However, experiments verified that precision performance of GP surrogate-assisted EA in Algorithm 1 on solving a COP with the suitable formulation is better than that with the typical one, where the CEI is based on the assumption of mutual independent among the objective and constraints.

The future work is to study expensive optimization technology based on dependent objective and constraint and verify the increase of computational resource where an expensive optimization technology solves a COP with the suitable formulation changed into the typical one.

Acknowledgments. This work was supported by the National Science Foundation of China (No.s: 61673355, 61271140 and 61203306).

References

1. Saha, C., Das, S., Pal, K., Mukherjee, S.: A fuzzy rule-based penalty function approach for constrained evolutionary optimization. IEEE Trans. Cybern. **46**(12), 2953–2965 (2016)
2. Runarsson, T.P., Yao, X.: Stochastic ranking for constrained evolutionary optimization. IEEE Trans. Evol. Comput. **4**(3), 284–294 (2000)
3. Maesani, A., Iacca, G., Floreano, D.: Memetic viability evolution for constrained optimization. IEEE Trans. Evol. Comput. **20**(1), 125–144 (2015)
4. Wang, Y., Wang, B., Li, H., Yen, G.G.: Incorporating objective function information into the feasibility rule for constrained evolutionary optimization. IEEE Trans. Cybern. **46**(12), 2938–2952 (2016)
5. Takahama, T., Sakai, S.: Constrained optimization by the constrained differential evolution with an archive and gradient-based mutation. In: Proceedings of IEEE Congress on Evolutionary Computation, Barcelona, Spain, pp. 1–9 (2010)
6. Zeng, S., Jiao, R., Li, C., Wang, R.: Constrained optimisation by solving equivalent dynamic loosely-constrained multiobjective optimisation problem. Int. J. Bio-Inspired Comput. (2018, to be published)
7. Zeng, S., Jiao, R., Li, C., Li, X., Alkasassbeh, J.S.: A general framework of dynamic constrained multiobjective evolutionary algorithms for constrained optimization. IEEE Trans. Cybern. **47**(9), 2678–2688 (2017)
8. Sarker, R.A., Elsayed, S.M., Ray, T.: Differential evolution with dynamic parameters selection for optimization problems. IEEE Trans. Evol. Comput. **18**(5), 689–707 (2014)
9. Hamza, N.M., Essam, D.L., Sarker, R.A.: Constraint consensus mutation-based differential evolution for constrained optimization. IEEE Trans. Evol. Comput. **20**(3), 447–459 (2015)
10. Mallipeddi, R., Suganthan, P.N.: Ensemble of constraint handling techniques. IEEE Trans. Evol. Comput. **14**(4), 561–579 (2010)
11. Bu, C., Luo, W., Yue, L.: Continuous dynamic constrained optimization with ensemble of locating and tracking feasible regions strategies. IEEE Trans. Evol. Comput. **21**(1), 14–33 (2016)
12. Guo, D., Jin, Y., Ding, J., Chai, T.: Heterogeneous ensemble based infill criterion for evolutionary multi-objective optimization of expensive problems. IEEE Trans. Cybern. **PP**(99), 1–14 (2018)
13. Zhang, Q., Liu, W., Tsang, E., Virginas, B.: Expensive multiobjective optimization by MOEA/D with gaussian process model. IEEE Trans. Evol. Comput. **14**(3), 456–474 (2010)
14. Liu, B., Zhang, Q., Gielen, G.G.: A gaussian process surrogate model assisted evolutionary algorithm for medium scale expensive optimization problems. IEEE Trans. Evol. Comput. **18**(2), 180–192 (2014)

15. Cheng, R., Jin, Y., Narukawa, K., Sendhoff, B.: A multiobjective evolutionary algorithm using gaussian process-based inverse modeling. IEEE Trans. Evol. Comput. **19**(6), 838–856 (2015)
16. Svenson, J., Santner, T.: Multiobjective optimization of expensive-to-evaluate deterministic computer simulator models. Comput. Stat. Data Anal. **94**, 250–264 (2016)
17. Zhan, D., Cheng, Y., Liu, J.: Expected improvement matrix-based infill criteria for expensive multiobjective optimization. IEEE Trans. Evol. Comput. **21**(6), 956–975 (2017)
18. Ohno, H.: Empirical studies of Gaussian process based Bayesian optimization using evolutionary computation for materials informatics. Expert. Sys. Appl. **96**, 25–48 (2018)
19. Liu, H., Ong, Y.S., Cai, J., Wang, Y.: Cope with diverse data structures in multi-fidelity modeling: a Gaussian process method. Eng. Appl. Artif. Intel. **67**, 211–225 (2018)
20. Jones, D.R., Schonlau, M., Welch, W.J.: Efficient global optimization of expensive black-box functions. J. Global Optim. **13**(4), 455–492 (1998)
21. Schonlau, M., Welch, W.J., Jones, D.R.: Global versus local search in constrained optimization of computer models. Lect. Notes Monogr. Ser. **34**, 11–25 (1998)
22. Audet, C., Dennis Jr., J., Moore, D.W., Booker, A., Frank, P.D.: A surrogate model-based method for constrained optimization. In: Proceedings of the 8th Symposium Multidisciplinary Analysis and Optimization, Long Beach, CA, USA, pp. 1–8 (2000)
23. Durantin, C., Marzat, J., Balesdent, M.: Analysis of multi-objective Kriging-based methods for constrained global optimization. Comput. Optim. Appl. **63**(3), 903–926 (2016)
24. Santner, T.J., Williams, B.J., Notz, W.I.: The Design and Analysis of Computer Experiments. Springer Series in Statistics. Springer, New York (2003). https://doi.org/10.1007/978-1-4757-3799-8
25. Storn, R., Price, K.V.: Differential evolution a simple and efficient heuristic for global optimization over continuous spaces. J. Global Optim. **11**(4), 341–359 (1997)
26. Stein, M.: Large sample properties of simulations using latin hypercube sampling. Technometrics **29**(2), 143–151 (1987)
27. Zitzler, E., Laumanns, M., Thiele, L.: SPEA 2: Improving the strength pareto evolutionary algorithm for multiobjective optimization. In: Proceedings of the Evolutionary Methods for Design, Optimization and Control with Applications to to Industrial Problems, pp. 95–100 (2001)

Data Mining – Association Rule Learning

Maize Gene Regulatory Relationship Mining Using Association Rule

Jianxiao Liu[1,2(✉)], Chaoyang Wang[1], Haijun Liu[2], Yingjie Xiao[2],
Songlin Hao[1], Xiaolong Zhang[1], Jianchao Sun[1], and Huan Yu[1]

[1] College of Informatics, Huazhong Agricultural University,
Wuhan 430070, China
liujianxiao321@163.com
[2] National Key Laboratory of Crop Genetic Improvement,
Huazhong Agricultural University, Wuhan 430070, China

Abstract. How to mine the gene regulatory relationship, and thus to construct gene regulatory network (GRN) is of utmost interest and has become a challenging computational problem for understanding the complex regulatory mechanisms in cellular systems. In this work, we use the association rule mining method to infer the gene regulatory relationship through the steps of mining frequent set, generating rules and rule merging. This method can not only get different types of gene regulatory relationships, but also get regulatory direction among genes. Experiment results show the effectiveness of this method. In all, the association rule mining method can effectively mine gene regulatory relationships of our maize gene expression dataset.

Keywords: Association rule · Gene regulatory · Gene expression
Maize

1 Introduction

Gene regulatory network construction refers to analyze expression data, use the methods of bioinformatics to construct the gene regulatory network. It is helpful to study the interaction between genes from the system level, explain the new function of genes and the mechanism of life process.

A variety of approaches have been proposed to get gene regulatory relationship and infer GRNs from gene expression data, such as Boolean model [1], differential equation method [2], regression method [3], etc. The information-theoretic approaches are increasingly used for reconstructing GRNs recently. Several mutual information (MI)-based methods have been successfully used to infer GRNs, such as ARACNE [4], PCA-CMI [5], CMI2NI [6], etc. However, the MI-based method cannot determine the direction of gene regulatory relationship. This method often calculates the relationship between two genes, but it could not calculate the regulatory relationship among multiple genes.

Due to the advantages of dealing with continuous and nonlinear expression data, processing data with noise, flexible acquisition and representation the relationship between multiple genes of Bayesian network, this method is more and more used in the research of gene network construction. At present, there are some research work using

© Springer Nature Singapore Pte Ltd. 2018
K. Li et al. (Eds.): ISICA 2017, CCIS 873, pp. 249–258, 2018.
https://doi.org/10.1007/978-981-13-1648-7_21

the Bayesian network structure learning algorithm, such as *K2* [7], Hill Climbing (*HC*), Greedy Search (*GS*), Recently, some research work use the prior knowledge to improve the efficiency and accuracy of Bayesian network method, such as using the cocitation in PubMed and schematic similarity in gene ontology annotation, using the information of transcription factor.

Association rule mining method has been widely used in data mining area, and it can discover the relationship among data items in massive data. It has the characteristics of supporting indirect relationship mining, processing the boolean and numerical data, and getting the hidden rules of specific data. At present, there are some research work of using association rule mining method to study the relationship of biological and medical data, such as promoter sequences prediction [8], hearing impairment [9], clinical parameters identification [10], identifying complications of cerebral infarction [11], autism susceptibility genes prediction [12], microarray data classification [13], gene expression data clustering [14], etc. In this work, we use the association rule method to mine the gene regulatory relationships in flexible using the steps of mining frequent set, generating rules and rule merging. Several types of gene regulatory relationship (including 1:1, 1:*n*, *n*:1 and *m*:*n*) and regulatory direction can be got using this approach. In addition, we use the gene expression dataset of maize global germplasm collection with 368 elite inbred lines to validate the proposed method.

2 Methods

2.1 Association Rule

Association rule mining is one of the most active research methods in data mining It can be used to find the correlation between things, and firstly used in finding the relationship between different commodities in the supermarket transaction database. Supposing $I = \{i_1, i_2, ..., i_m\}$ represents the item set, and i_k ($k = 1, 2, ..., m$) represents each item in I. $D = \{T_1, T_2, ..., T_n\}$ represents the transaction set. Each transaction T_i is a subset of the item set I, $T_i \subseteq I$. The transaction number which support the item set A in D is called the support number of A, denoted as *Count*(A).

Definition 1. *Support*. It refers to the number of transaction that support A divided by the total number of transaction in D, denoted as *Supp*(A), as shown in Eq. (1). In the equation, $|D|$ represents the total number of transaction in D.

$$Supp(A) = Count(A) / |D| \tag{1}$$

Supposing A and B are subset of the item set I, $A \subseteq I$, $B \subseteq I$, $A \cap B = \emptyset$, we use the following logical implication to denote the association rule: $R: A \Rightarrow B$. A is called as rule of the left-hand side, also known as the head of rule. B is called as rule of the right-hand side, also known as the tail of rule.

Definition 2. *Support* of rule R: $A \Rightarrow B$. It refers to number of transaction that support $A \cup B$ divided by the total number of transaction in D, denoted as $Supp(R)$ or $Supp$ $(A \cup B)$, as shown in Eq. (2).

$$Supp(R) = Count(A \cup B) / |D| \tag{2}$$

Definition 3. *Confidence* of R: $A \Rightarrow B$. It refers to the number of transaction which includes A and B simultaneously divided by the number of transaction which includes A only in D, denoted as $Conf(R)$ or $Conf(A \Rightarrow B)$, as shown in Eq. (3).

$$Conf(A \Rightarrow B) = p(B \subseteq D | A \subseteq D) = \frac{p(A \subseteq D \wedge B \subseteq D)}{p(A \subseteq D)} = \frac{Supp(A \cup B)}{Supp(A)} \tag{3}$$

If the support of item set is larger than the minimum support that user has been set, we call this item set as frequent item set. The *Apriori* algorithm is one of the most influential association rules mining methods. Its core idea is the recursive algorithm based on the frequent item set. The process of *Apriori* algorithm includes the following two stages:

(1) Deducing frequent item set by the candidate item set. It firstly calculates the frequency of all the items in I, and the support will be calculated using Eq. (1). The items whose support is greater or equal to the minimum support will be selected, resulting in one dimensional frequent item set.
(2) Getting association rules based on the frequent item set. It generates rules based on the obtained frequent item set, and then calculates the confidence according to Eq. (3). The association rules will be generated according to the principle of confidence is larger than the minimum confidence. In accordance with this method, until it fails to generate a higher dimension of frequent item set.

2.2 Gene Regulatory Relationship Mining Using Association Rule

(1) Data preprocessing

Due to some genes are not expressed in some materials, it is inevitable that the transcriptome data is missing in some materials. We denote different materials in the transcriptome data as samples, and use the mean or qqnorm method to fill the missing data. The transcriptome expression data will be processed in the scope of $[-3, 3]$. Then we discrete the expression data in order to facilitate using the association rule mining method. The following three data discretization methods are mainly used: *Interval*, *Quantile* and *clustering*. Supposing gene set $G = \{g_1, g_2, ...g_i, ..., g_n\}$, material set $M = \{m_1, m_2, ... m_j, ..., m_m\}$, we can get the expression matrix $GM = \{gm_{ij}, 1 \leq i \leq n, 1 \leq j \leq m\}$. When to do N-value discretization, for example, we denote the sequence of multi-value discretization results about gene g_i as "g_i-0, g_i-1, ..., g_i-N". The expression of different materials is seen as the transaction T in D. The sequence of m_j

can be denoted as "g_1-0, g_2-1,..., g_i-N", and the length of each sequence is n. For example, supposing $i = 4$, $j = 6$, $N = 3$, the gene expression matrix after discretization is shown in Table 1. In the table, we can see the sequence of m_2 is "g_1-2, g_2-0, g_3-0, g_4-1" and the sequence of m_5 is "g_1-0, g_2-2, g_3-1, g_4-2".

Table 1. Gene expression matrix

Genes	Materials					
	m_1	m_2	m_3	m_4	m_5	m_6
g_1	g_1-0	g_1-2	g_1-1	g_1-2	g_1-0	g_1-2
g_2	g_2-1	g_2-0	g_2-2	g_2-1	g_2-2	g_2-1
g_3	g_3-0	g_3-0	g_3-1	g_3-2	g_3-1	g_3-0
g_4	g_4-2	g_4-1	g_4-0	g_4-0	g_4-2	g_4-1

(2) Gene regulatory relationship mining

Firstly, we will calculate the frequency of all the items in I and deduce the frequent item set. All the item set can be denoted as $I = \{g_i-k, 1 \leq i \leq n, 1 \leq k \leq N\}$, such as g_1-0, g_2-1, g_3-1, etc. The *Support* of item set is shown in Table 2.

Table 2. The Support of item set

Item	Support
g_1	$Supp(g_1-0) = 2/6$, $Supp(g_1-1) = 1/6$, $Supp(g_1-2) = 3/6$
g_2	$Supp(g_2-0) = 1/6$, $Supp(g_2-1) = 3/6$, $Supp(g_2-2) = 2/6$
g_3	$Supp(g_3-0) = 3/6$, $Supp(g_3-1) = 2/6$, $Supp(g_3-2) = 1/6$
g_4	$Supp(g_4-0) = 2/6$, $Supp(g_4-1) = 2/6$, $Supp(g_4-2) = 2/6$

Then we will deduce the frequent item set based on the calculated frequency of all the items in I. The item whose support is larger than the minimum support will be selected to form frequent item set.

Secondly, we calculate confidence based on the obtained frequent item set and get the association rules whose confidence is greater than the minimum confidence.

(3) Rule merging

Through above two stages, we can get the strong association rules whose support and confidence is larger than the minimum support and confidence. The form of the association rule will be denoted as $\{g_i-k \Rightarrow g_a-c, 1 \leq i \leq n, 1 \leq a \leq n, 1 \leq k \leq N, 1 \leq c \leq N\}$. It can learn the association rules among multiple genes and get several types (including 1:1, 1:n, n:1 and m:n) of gene regulatory relationships, such as $g_1-2 \Rightarrow g_3-0$ and g_1-2, $g_3-0 \Rightarrow g_4-1$, etc. It can be seen that the association rule method divides gene g_i into N parts. We will merge the rules and change the regulatory relationship of multiple parts of a gene to the regulatory relationship among multiple genes.

Firstly, we merge the initial association rule according to the left-hand side rule merging method about *Support*. For example, for gene g_1, g_2, and the association rule $g_1-i \Rightarrow g_2-j$ ($0 \le i \le 2$, $0 \le j \le 2$). For the rule of $g_1-i \Rightarrow g_2-0$, we can get $s_{i,0}$ using Eq. (4).

$$s_{i,0} = \frac{Num(g_1 - i, g_2 - 0)}{Num(Total)} \tag{4}$$

Then we can get s^0 using $s^0 = s_{0,0} + s_{1,0} + s_{2,0}$ for the rules of $g_1 \Rightarrow g_2-0$. Similarly, we can get $s^1 = s_{0,1} + s_{1,1} + s_{2,1}$ using the rules of $g_1 \Rightarrow g_2-1$, and get $s^2 = s_{0,2} + s_{1,2} + s_{2,2}$ using the rules of $g_1 \Rightarrow g_2-2$.

Secondly, we process the association rules according to the left-hand side rule merging method about *Confidence*. For the rule of $g_1-i \Rightarrow g_2-0$ ($0 \le i \le 2$), we can get $c_{i,0}$ using Eq. (5).

$$c_{i,0} = \frac{Num(g_1 - i, g_2 - 0)}{Num(g_1 - i)} \tag{5}$$

Then we can get c^0 using $c^0 = c_{0,0} + c_{1,0} + c_{2,0}$ for the rules of $g_1 \Rightarrow g_2-0$. Similarly, we can get $c^1 = c_{0,1} + c_{1,1} + c_{2,1}$ using the rules of $g_1 \Rightarrow g_2-1$, and get $c^2 = c_{0,2} + c_{1,2} + c_{2,2}$ using the rules of $g_1 \Rightarrow g_2-2$. At the same time, we process the gene regulatory rules according to the right-hand side rule merging approach of *Support* and *Confidence*. The order of the above merging steps has no influence to the final results, so we can adjust the merging steps randomly, as long as they are completed.

According to Eqs. (1) and (3) we can see that the part of numerator is identical and the part of denominator is different when to calculate the total support and confidence. But when to consider the relationship between two objects only, the number of total items is as many as the number of item of an object. Therefore, the final value of the support and confidence is equal. We use a unified name of *Reference* to represent their value, and we can get the *Reference* calculation approach of gene regulatory relationship, as shown in Eq. (6). In the equation, n denotes the discretization value. The larger of the *Reference* indicates the stronger of the relationship between two objects.

$$Reference = \sum_{i=0}^{n} \sum_{j=0}^{m} S_{i,j} \tag{6}$$

According to the above calculation method, each rule has a corresponding *Reference* value. We will sort the rules in order from the large to small according to the value of *Reference*. Then we select some rules to do the experiment in the sorted rule, the detailed content will be introduced in Sect. 3.2.

3 Experiment Results

3.1 Dataset

We have used the transcriptome quantification dataset about the germplasm collection with 527 elite inbred lines (association mapping panel, AMP). This population is released from the major temperate and tropical/subtropical breeding programs. China, CIMMYT and the Germplasm Enhancement of Maize (GEM) project in the US have participated in the program. The AMP population represents maize genetic diversity in maize improvement [15]. We have assayed all of the lines by the 50K Maize SNP array (commercially available from Illumina). In addition, we have used the deep RNA sequencing technology to get the expression data about 368 of the 527 lines of kernels harvested 15 days after pollination (DAP) [16]. The dataset of germplasm and transcriptome is shown in Table 3. All the dataset can be got through http://www.maizego.org/.

Table 3. Dataset

Index	Data
Germplasm resources	The association mapping panel (AMP) includes the following different 527 populations: 143 lines for NSS, non-stiff-stock, 33 for SS, Stiff-stock, 232 for TST, tropical and Semi-tropical, 119 for MIXED
Transcriptome quantification	The quantitative expression of maize whole kernel about 15 days after pollination of 28769 genes

3.2 Experiment

As has been described in Sect. 2.2, we sort the gene regulatory relationship according to the value of *Reference*, and then select some rules in the sorted rules to do the experiment. The selection ratio in the following experiment refers to the selection criteria on the basis of the sorted gene regulatory relationships. It also means the selected rules whose *Reference* is larger than the specific value. We use the form of (A, B) to express the experiment comparison result in the following experiment. A refers to the common number of gene regulatory relationship obtained by the association rule mining method with *gene_links* divided by the number of gene regulatory relationships that are selected in the scope of specific selection ratio. B refers to the common number of gene regulatory relationship obtained by the association rule method with *gene_links* divided by the total number of gene regulatory relationship in *gene_links*.

(1) Comparison of N-value discretization

Different discretization values will have great influence on the result of association rule method. In the case of setting the minimum *Confidence* to 0.6, setting the minimum *Support* to 0.3, using the mean missing data processing method, we compare the learning result of different discretization values of *Interval*, *Quantile* and *Kmeans*. The result is shown in Table 4.

Table 4. Result of the different discretization methods

Methods and N-value		Selection ratio			
		10%	30%	50%	80%
Interval method	2-value	(0.45, 0.17)	(0.39, 0.46)	(0.34, 0.66)	(0.25, 0.78)
	3-value	(0.49, 0.19)	(0.37, 0.44)	(0.32, 0.63)	(0.28, 0.87)
	4-value	(0.51, 0.2)	(0.25, 0.29)	(0.24, 0.45)	(0.28, 0.86)
	5-value	(0.51, 0.19)	(0.26, 0.29)	(0.25, 0.46)	(0.3, 0.88)
Quantile method	2-value	(0.84, 0.08)	(0.63, 0.2)	(0.58, 0.3)	(0.54, 0.45)
	3-value	(0.0, 0.0)	(0.0, 0.0)	(0.0, 0.0)	(0.0, 0.0)
	4-value	(0.0, 0.0)	(0.0, 0.0)	(0.0, 0.0)	(0.0, 0.0)
	5-value	(0.0, 0.0)	(0.0, 0.0)	(0.0, 0.0)	(0.0, 0.0)
Kmeans method	2-value	(0.3, 0.12)	(0.3, 0.35)	(0.28, 0.56)	(0.28, 0.89)
	3-value	(0.25, 0.09)	(0.31, 0.34)	(0.32, 0.58)	(0.32, 0.92)
	4-value	(0.47, 0.11)	(0.47, 0.34)	(0.46, 0.56)	(0.39, 0.76)
	5-value	(0.75, 0.08)	(0.6, 0.19)	(0.56, 0.3)	(0.56, 0.48)

We can see the learning effect of association rule mining method is different in the case of setting different discretization values. It has better learning effect when to use the 3-value discretization method. In addition, we can see the *Quantile* discretization method could not learn any gene regulatory relationships when to do the discretization of 3-value, 4-value and 5-value. This method is not suitable to our association rule mining method.

(2) Comparison of different discretization methods

Different discretization methods will have influence on the result of association rule method. In the case of setting the minimum *Confidence* to 0.4, setting the minimum *Support* to 0.3, using the mean missing data processing, we compare the learning result of three different methods (*Interval*, *Quantile*, *Kmeans*) of N-value discretization. The result is shown in Table 5.

It can be seen that the learning effect is different in the case of using different discretization methods. Combination with the result in Table 8, we can see the learning effect of the *Interval* method is best when the selection ratio is set to 10%–50%, and the effect of *Kmeans* method becomes better when the selection ratio is set to 80%. On the whole, the learning effect of the *Interval* method is better than the *Kmeans* method.

(3) Comparison of different values of the minimum confidence

In the case of using the mean missing data processing and 3-value *Interval* discretization method, we compare the learning result of different values of the minimum *Confidence* (0.4, 0.6, 0.8) of specific value of the minimum *Support* (0.1, 0.3, 0.5). The result is shown in Table 6.

It can be seen that the learning effect is different in the case of setting different values of the minimum *Confidence*. The learning effect of association rule method is about same when to set the minimum *Confidence* to 0.4 or 0.6. But when to set the minimum *Confidence* to 0.8, the value of *B* is less than the value when setting

Table 5. Result of 2-value discretization of different methods

N-values and methods		Selection ratio			
		10%	30%	50%	80%
2-value	Interval	(0.45, 0.17)	(0.39, 0.46)	(0.34, 0.66)	(0.25, 0.78)
	Quantile	(0.84, 0.08)	(0.63, 0.2)	(0.58, 0.3)	(0.54, 0.45)
	Kmeans	(0.3, 0.12)	(0.3, 0.35)	(0.28, 0.56)	(0.28, 0.89)
3-value	Interval	(0.49, 0.19)	(0.37, 0.44)	(0.32, 0.63)	(0.28, 0.87)
	Quantile	(0.0, 0.0)	(0.0, 0.0)	(0.0, 0.0)	(0.0, 0.0)
	Kmeans	(0.25, 0.09)	(0.31, 0.34)	(0.32, 0.58)	(0.32, 0.92)
4-value	Interval	(0.51, 0.2)	(0.25, 0.29)	(0.23, 0.45)	(0.29, 0.89)
	Quantile	(0.0, 0.0)	(0.0, 0.0)	(0.0, 0.0)	(0.0, 0.0)
	Kmeans	(0.47, 0.11)	(0.47, 0.34)	(0.46, 0.56)	(0.39, 0.76)
5-value	Interval	(0.51, 0.19)	(0.26, 0.29)	(0.25, 0.47)	(0.3, 0.88)
	Quantile	(0.0, 0.0)	(0.0, 0.0)	(0.0, 0.0)	(0.0, 0.0)
	Kmeans	(0.75, 0.08)	(0.6, 0.19)	(0.56, 0.3)	(0.56, 0.48)

Table 6. Result of different values of the minimum confidence

Support and minimum confidence		Selection ratio			
		10%	30%	50%	80%
Support = 0.1	0.4	(0.51, 0.2)	(0.35, 0.41)	(0.33, 0.65)	(0.28, 0.87)
	0.6	(0.51, 0.2)	(0.35, 0.41)	(0.33, 0.65)	(0.28, 0.88)
	0.8	(0.52, 0.2)	(0.36, 0.41)	(0.34, 0.64)	(0.29, 0.87)
Support = 0.3	0.4	(0.49, 0.19)	(0.37, 0.44)	(0.32, 0.63)	(0.28, 0.87)
	0.6	(0.49, 0.19)	(0.37, 0.44)	(0.32, 0.63)	(0.28, 0.87)
	0.8	(0.49, 0.19)	(0.37, 0.42)	(0.33, 0.62)	(0.28, 0.85)
Support = 0.5	0.4	(0.55, 0.2)	(0.42, 0.44)	(0.37, 0.65)	(0.31, 0.87)
	0.6	(0.55, 0.2)	(0.42, 0.44)	(0.37, 0.65)	(0.31, 0.87)
	0.8	(0.55, 0.19)	(0.43, 0.44)	(0.37, 0.63)	(0.31, 0.86)

Confidence to 0.4 or 0.6 in the case of different selection ratios. For example, we can see B is 0.65 when the minimum *Support* is set to 0.5, the minimum *Confidence* is set to 0.6 and selection ratio is set to 50%. But it is 0.63 when minimum *Confidence* is set to 0.8.

(4) Comparison of different values of the minimum support

In the case of using the mean missing data processing and 4-value *Interval* discretization method, we compare the learning result of different values of the minimum *Support* (0.1, 0.3, 0.5) of specific value of the minimum *Confidence* (0.4, 0.6, 0.8). The result is shown in Table 7.

It can be seen that the learning effect of association rule method is different in the case of setting different values of the minimum *Support*. In combination with the above

Table 7. Result of different values of the minimum support

Confidence and minimum support		Selection ratio			
		10%	30%	50%	80%
Confidence = 0.4	0.1	(0.53, 0.21)	(0.5, 0.59)	(0.39, 0.77)	(0.29, 0.9)
	0.3	(0.51, 0.2)	(0.25, 0.29)	(0.23, 0.45)	(0.29, 0.89)
	0.5	(0.73, 0.22)	(0.35, 0.31)	(0.33, 0.5)	(0.36, 0.87)
Confidence = 0.6	0.1	(0.53, 0.21)	(0.51, 0.6)	(0.4, 0.77)	(0.29, 0.89)
	0.3	(0.51, 0.2)	(0.25, 0.29)	(0.24, 0.45)	(0.28, 0.86)
	0.5	(0.73, 0.22)	(0.35, 0.31)	(0.33, 0.5)	(0.36, 0.87)
Confidence = 0.8	0.1	(0.55, 0.19)	(0.51, 0.54)	(0.39, 0.69)	(0.28, 0.8)
	0.3	(0.55, 0.19)	(0.26, 0.27)	(0.24, 0.42)	(0.28, 0.76)
	0.5	(0.73, 0.22)	(0.35, 0.31)	(0.33, 0.49)	(0.36, 0.86)

experiment result, we can see the learning effect of association rule method is the best when the minimum *Support* is set to 0.1. In addition, the learning effect becomes better with the increasing of the selection ratio.

Through the above experiment results, we can see the association rule mining method can have better learning effect using the following parameters: setting the minimum *Confidence* to 0.6, setting the minimum *Support* to 0.1, using the mean missing data processing method, using *Interval* and 3-value discretization method.

4 Conclusion

In this work, we infer gene regulatory relationships from gene expression data using the *Apriori* association rule mining approach. In this method, we firstly do the missing value imputation and discretization to the expression data, and use the form of $gm_{ij}-k$ to denote the item set I. Then we get the gene regulatory relationships in flexible through the steps of mining frequent set and generating rules. This approach uses the recursive algorithm to do the judgement about the frequent items until it fails to generate a higher dimension of frequent item set. Finally, it uses the left-hand side and right-hand side rules merging method about *Support* and *Confidence* to merge the rules, and thus to get the gene regulatory relationships comprehensively. Experiment results on the expression data of maize global germplasm show this method can get more number of gene regulatory relationships.

Acknowledgements. This research is supported by the National Natural Science Foundation of China under Grant No. 31601078, the Natural Science Foundation of Hubei Province under Grant No. 2016CFB231, the Fundamental Research Funds for the Central Universities under grant No. 2662018JC030, No. 2015BC017.

References

1. Shmulevich, I., Dougherty, E., Zhang, W.: From Boolean to probabilistic Boolean networks as models of genetic regulatory networks. Proc. IEEE **90**, 1778–1792 (2002)
2. Sakamoto, E., Iba, H.: Inferring a system of differential equations for a gene regulatory network by using genetic programming. In: Proceedings of the 2001 Congress on Evolutionary Computation, vol. 1, pp. 720–726 (2001)
3. Gardner, T., Di Bernardo, D., Lorenz, D.: Inferring genetic networks and identifying compound mode of action via expression profiling. Science **301**, 102–105 (2003)
4. Margolin, A., Nemenman, I., Basso, K.: ARACNE: an algorithm for the reconstruction of gene regulatory networks in a mammalian cellular context. BMC Bioinform. **7**, S7 (2006)
5. Zhang, X., Zhao, X., He, K.: Inferring gene regulatory networks from gene expression data by path consistency algorithm based on conditional mutual information. Bioinformatics **28**, 98–104 (2012)
6. Zhang, X., Zhao, J., Hao, J.: Conditional mutual inclusive information enables accurate quantification of associations in gene regulatory networks. Nucleic Acids Res. **43**, e31 (2015)
7. Chen, X., Anantha, G., Wang, X.: An effective structure learning method for constructing gene networks. Bioinformatics **22**, 1367–1374 (2006)
8. Czibula, G., Bocicor, M., Czibula, I.: Promoter sequences prediction using relational association rule mining. Evol. Bioinform. Online **8**, 181 (2012)
9. Iltanen, K., Kiviharju, S., Ao, L.: Clustering and summarising association rules mined from phenotype, genotype and environmental data concerning age-related hearing impairment. In: MedInfo, pp. 452–456 (2013)
10. Sengupta, D., Sood, M., Vijayvargia, P.: Association rule mining based study for identification of clinical parameters akin to occurrence of brain tumor. Bioinformation **9**, 555 (2013)
11. Jung, S., Son, C., Kim, M.: Association rules to identify complications of cerebral infarction in patients with atrial fibrillation. Healthc. Inform. Res. **19**, 25–32 (2013)
12. Gong, L., Yan, Y., Xie, J.: Prediction of autism susceptibility genes based on association rules. J. Neurosci. Res. **90**, 1119–1125 (2012)
13. Giugno, R., Pulvirenti, A., Cascione, L.: MIDClass: microarray data classification by association rules and gene expression intervals. PLoS ONE **8**, e69873 (2013)
14. Sethi, P., Alagiriswamy, S.: Association rule based similarity measures for the clustering of gene expression data. Open Med. Inform. J. **4**, 63 (2010)
15. Fu, J., Cheng, Y., Linghu, J.: RNA sequencing reveals the complex regulatory network in the maize kernel. Nat. Commun. **4**, 2832 (2013)
16. Yang, X., Gao, S., Xu, S., Zhang, Z., Prasanna, B., Li, L., Li, J., Yan, J.: Characterization of a global germplasm collection and its potential utilization for analysis of complex quantitative traits in maize. Mol. Breed. **28**, 511–526 (2011)

Database Reengineering Scheme from Object-Oriented Model to Flattened XML Data Model

Yue Liu[1(✉)] and Xukun Wu[2]

[1] College of Information Engineering, Jiangxi College of Applied Technology,
Ganzhou 341000, Jiangxi, China
liuyue1229@qq.com
[2] UX Department, Guangzhou Boguan Info Tech Ltd., Guangzhou 510000,
Guangdong, China

Abstract. To deal with issues of data model transformation and to make combined use of different data models, this paper proposed a database reengineering scheme which includes schema translation and data conversion from object-oriented data model to flattened XML data model with is-a and cardinality data semantics preservation. In this paper, conceptual schema of object-oriented data model and flattened XML data model are elaborated by UML class diagram and XML Schema Definition(XSD) graph respectively. Logical schema is described by UNISQL class definition and XSD respectively. This paper firstly analyzes class definitions from OODB and sorts them from most independent classes to most dependent classes. Secondly schema translation is processed by mapping class to complexType and then creating element definitions in XSD file. Thirdly data conversion is processed automatically. Performance tests have shown that the proposed database reengineering scheme is reliable and efficient.

Keywords: Data reengineering · Objected-oriented data model
Flattened XML data model

1 Introduction

In the recent decades, with the big bang of data today, not only the amount of data increases sharply, also the kinds and complexity of data augment greatly. Different data models emerged to meet different requirements. Object-oriented databases (OODB) emerged in the mid-80's in response to the feeling that relational databases were inadequate for certain classes of applications [1] and also in response to the popularity of object-oriented programming. Object-oriented data model is powerful because each object represents a real-world entity with the ability to relate to itself and to other objects. It is superior in representing and storing multimedia, intensive and redundant foreign key data (e.g. GPS data, graphic, voice, video data and so on). However, we seldom see OODB in industry. OODBs are more common in labs, in research area, like artificial Intelligence (AI), multimedia management and aerospace area and so on. XML database appeared in 90s with the development of Internet, and it has been the standard format for data exchange between different partners [2]. XML data model

© Springer Nature Singapore Pte Ltd. 2018
K. Li et al. (Eds.): ISICA 2017, CCIS 873, pp. 259–268, 2018.
https://doi.org/10.1007/978-981-13-1648-7_22

inherits the strengths of XML, preponderating in data exchanging, integration and interoperability, especially in ecommerce and multi-platform information system.

Both XML and object-oriented databases play significant role nowadays, and data exchange happens frequently on Internet. We choose flattened XML as OO data's target data model because of its relational table structure. Relational table structure is most user friendly and has a strong mathematical foundation of relational algebra to implement the constraints of major data semantics such as is-a and cardinality and so on to meet users' data requirements. This paper elaborates a database reengineering scheme which largely improves the efficiency and conveniences of schema translation and data conversion from object-oriented data model to flattened XML data model.

It already has been many attempts translating relational data model to other models, e.g. [3, 4]. In industries, there are several tools that concentrate on the composition of XML documents from relational data and relational-object data, such as IBM XML Extender and XPERANTO. XPERANTO [5] is a middleware solution to publish object-relational data as XML documents. However, most of such tools relies heavily on manual work and human experts. According to [6, 7], it is able to preserve most of the model features in the translation from XML to object-oriented data model. In [8–10], object-oriented schema firstly is reversed reengineered into object graph, which includes nesting links (cardinality). Then, the object graph is transformed into XML schema and the object-oriented data is mapped into corresponding XML document(s). However, it only introduces the methodology to preserve one-to-one cardinality data semantic.

2 Database Reengineering Scheme

The database reengineering scheme in this paper is divided into two parts: firstly, schema translation from classes to flattened XSD file(s); secondly, data conversion from instance of classes to flattened XML document(s), where one XSD file relates to one corresponding XML document.

2.1 Schema Translation

Object-oriented databases schema is a collection of classes and flattened XML schema is one or several XSD file(s), thus in this part of our scheme, the input is classes in OODB and the output is flattened XSD files. The key of the translation is mapping: mapping classes into XSD "complexType", mapping unique "OID" via "ID" type in XSD, mapping class attribute names and data types into XSD sub-element names and primitive type or "complexType", mapping the association of classes via "ID" and "IDREF" type in XSD and mapping inheritance relationship between classes via "extensions" element and add an attribute to present parent "complexType" name in XSD.

In is-a data semantic preservation, superclass and subclass in OODB will be translated into "complexType" with a sub "extension" basing on a super one, which is used for inheritance the attributes from superclass. Moreover, in subclass's "complexType", "oid" attribute is ignored and a "base" attribute with "IDREF" type is added. This translation is shown in Fig. 1.

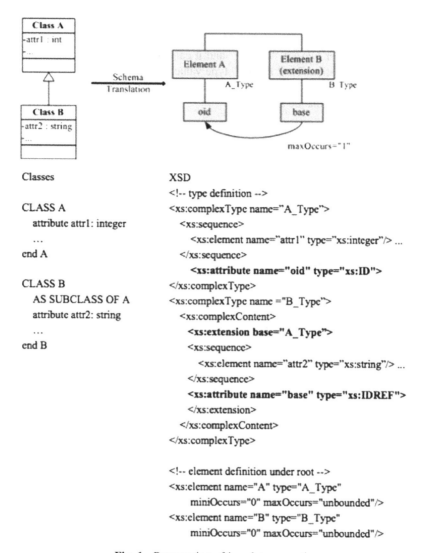

Fig. 1. Preservation of is-a data semantic

In cardinality data semantic preservation, a class has nested class(es) in OODB. They are associated via "OID" and "stored OID". "OID" identifies the instance of class and "stored OID" denotes the reference to an instance. In one-to-one data semantic, association attribute in OODB schema will be translated into an sub-element with "IDREF" type and "maxOccurs = "1"" in XSD. In one-to-many data semantic, it is similar to one-to-one case, but we replace the "maxOccurs = "1"" with "maxOccurs = "unbunded"" in XSD. In many-to-many data semantic, there are two possibilities in OODB. One possibility is that two classes associated with each other, and the other possibility is that one class associated with other two (or more) classes. Classes in both two possibilities can be translated into three "complexType" in XSD and one "complexType" has multiple sub-elements with "IDREF" type and "maxOccurs = "unbunded"". This "complexType" associates to the other two via "IDREF" and it can be origin from source database or created during translation. The "IDREF" type value in XSD can be regarded as "stored OID" in OODB and "ID" type value can be regarded as "OID" in OODB. We show a case of one-to-many data semantic as in Fig. 2.

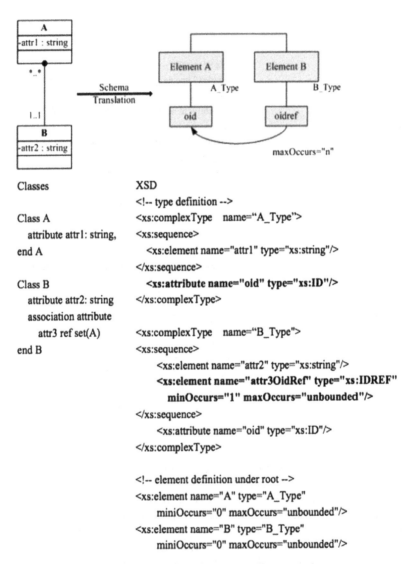

Fig. 2. Preservation of one-many data semantic

We discussed class-to-complexType mapping in above. Flattened XML element definition in XSD is base on "complexType". A root element is on top level, sub-elements are instances of classes. The instances of the same class have the same element name but differ with "oid" attribute.

2.2 Data Conversion

In this part, the input of data conversion is object data and the output is flattened XML document(s). The whole conversion is processed by program and under the support of class definition from OODB and XSD file(s) from schema translation. Commonly, root element is on level one, entities are on level two and there are no more levels. But in our scheme, attributes of the entity construct level three, which can be regarded as attribute elements. As attributes of entity can not be separated with entity, entity on level two and its attribute elements on level three can be consolidated as level two.

Data semantics preservation of data conversion can be verified by data dependencies as follows. Inclusion dependency (ID: A $\subseteq$ B): The set of values appearing in attribute A must be a subset of the set of values appearing in attribute B. Functional dependency (FD: R.X -> R.Y): Given a relation R, attribute Y of R is functionally dependent on attribute X of R, iff each X value in R has associated with it precisely one Y value. Multi-valued dependency (MVD: R.A ->-> R.B): Given a relation R with attributes A, B, C. The B is multi-dependent on A iff in every legal value of R, the set of B values matching a given AC pair value depends on the A value, and is independent of the C value.

Data semantics preservation in data conversion follows the schema translation. In is-a data semantic preservation, one instance of subclass will be converted into two object element under root, converting all the attributes value (including the attributes inheriting from superclass and new attributes in subclass) to one object element with sub one's "complexType" as defined in XSD and converting attributes value that inherited from super class into another object element with super one's "complexType" under root. This two element is is-a related by "oid" attribute and "base" attribute in flattened XML document.

Cardinality data semantic preservation in data conversion is easily realized via "ID" type and "IDREF" type in XML. All instances on the many side will be converted into an sub-element with "IDREF" type and "maxOccurs" choice under entity element. On the one side, each entity element has an "oid" value (or "base" value of subclass entity). The "oid" attribute in XML is "ID" type, which is unique in an XML document. Thus, "IDREF" value on the many side points to the "ID" value in the document and forms the association relationship. It is similar in one-to-one, one-to-many and many-to-many cases. The difference is the "maxOccurs" choice value defined in XSD during schema translation, which specifies the occurrence times of "IDREF" subelement in XML data document. In many-to-many data semantic, there is one element associating two "many" side via multiple "IDREF" subelement. It can be newly created if there are no such entity in source database.

2.3   Implementation

Pre-process. Analyzing and ordering classes from OODB.
A very important process. Pre-process helps increasing efficiency of data conversion by executing in parallel. We do not elaborate this process here.

Algorithm 0. Analyzing and layering classes.

```
Input: Classes definition from OODB
Output: A sorted list of entity (Seq", ClassName, Level,
Parent, Association)

Begin
  Step 0. Create an empty list.
  Step 1. Find the most independent classes first, and
add them into the list with level 1.
  Step 2. Then find the classes only dependent on level
1, and add them into the list with level 2, their    su-
perclass name and association class name.
  Step 3. Then find the classes only dependent on level 1
and level 2, and add them into the list with    level 3,
their superclass name and association class name.
  ...
  Step n. Then find the classes only dependent on level
1, level 2, ..., level n-1, and add them into the list
with level n, their superclass name and association class
name.
End
```

Process 1. Schema translation.
Basing on the list from Pre-process, schema translation can be done automatically by program. Figure 3 shows a general flow of schema translation.

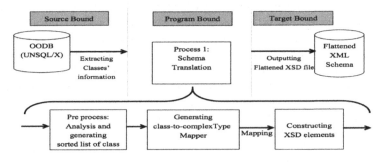

Fig. 3. General flow of schema translation

Algorithm 1. Schema translation from classes to flattened XSD files.

```
Input: Classes definition from OODB and the sorted class
list from Pre-process
Output: flattened XSD file

Begin

  Create a XSD file with corresponding head.
  For each class c in the list from top to bottom
    do
    Read definition of class c from OODB
    Create a "xs:complexType" with name "c_Type" in XSD
    If c has parent, it is a subclass,
       then do
         Create a "xs:extension" element and "base" on the
super "xs:complexType" name.
         Create a "xs:attribute" element with name "base"
and type "xs:IDREF".
      Else do
        Create an "xs:attribute" element with name "oid"
and type "xs:ID".
      End if.
  For each attribute a in class c
    do
    If a is a association attribute
       then do
         Create a "xs:element" element with corresponding
name indicating the reference associate and type
"xs:IDREF".
      If a refer to set of another class
         then do
         Add "maxOccurs="unbounded"" into the element.
      End if.
      Else
        Create a "xs:element" element with corresponding
name and type.
      End if.
  End for.

End.
```

Process 2. Data conversion.

Following the sequence in class list, executing data conversion of same level in parallel. According to the parent and association field in the list, program can easily find the corresponding data in OODB and guarantee the "IDREF" referring to data already converted in flattened XML document. Figure 4 shows a general view of data conversion.

Algorithm 2. Data conversion from instance data of classes to flattened XML documents.

```
Input: Instances data and classes definition from OODB,
flattened XSD file from Process 1 and the sorted list
from Pre-process
Output: flattened XML document

Begin
  Create a XML document with root element.
  For all class c of same level in the list from level 1
to level n do in parallel
    Execute query ("select oid, * from c") in UNISQL/X.
  For each tuple in the result do
    Create an element with name "c" and type "c_Type".
    If c has parent p, c is a subclass of p, then do
      Add "base" attribute into the element with oid val-
ue.
      Create an element with name "p", type "p_Type" and
same "oid" attribute value.
      Convert attributes name and value inherited from p
class into subelements of p element.
    End if.
    Convert all the attributes name and value into
subelements of c element.
      If encountering association attribute then do
      Finding the corresponding "oid" value(s) already
transformed into XML element.
      Assign the "oid" value to the subelement value with
"xs:IDREF" type.
      End if.
    End for.
End.
```

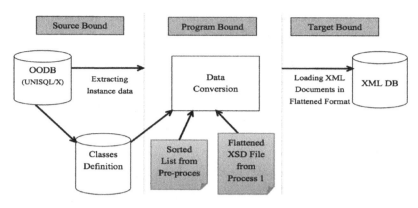

Fig. 4. General flow of data conversion

3 Performance Analysis

Following the implementation described above, we have a preliminary performance analysis. The test environment is shown in Table 1.

Table 1. Test environment

	Operating system	Specification
Source	Windows server 2008	UNISQL/X
Program	Mac OS X 10.12	Java 8, JDBC, XQuery API
Target	Mac OS X 10.12	eXist-db 2.1

Firstly, we q and load data into UNISQL/X and measure the time consumption. Then, we extract from and load file into eXist-db via JDBC and measure the time consumption. The result is shown in Table 2.

Table 2. Time consumption

	UNISQL/X	eXist-db (flattened XML)	Dataset
Query	0.5 s	0.006 s	400
Load	0.5 s	0.8 s	400

4 Conclusion

In this paper we proposed a feasible and efficient data reengineering scheme to transform OO data model to flattened XML data model while preserving is-a and cardinality data semantics. In our proposal, there are two phases in data reengineering: schema translation and data conversion, we also give detailed implementation method which includes three processes and three algorithms. At last, we show preliminary experiment results considering the efficiency of our scheme.

Acknowledgement. This work was jointly supported by Natural Science Foundation of China (61773296), the Education Department of Jiangxi Province of China Science and Technology research projects with the Grant No. GJJ151433, GJJ161687, GJJ161688 and GJJ161691.

References

1. Stajano, F.: A Gentle Introduction to Relational and Object Oriented Databases. ORL Technical Report TR-98-2 (1998)
2. Bray, T., et al.: Extensible Markup Language (XML) 1.0 (Fourth Edition), pp. 115–146 (2006)
3. Maatuk, A.M.: Migrating relational databases into object-based and XML databases. J. Object Technol. **9**(9), 145–161 (2009)
4. Fong, J., Pang, F., Bloor, C.: Converting relational database into XML document. In: Proceedings of 12th International Workshop on Database and Expert Systems Applications 2001. IEEE (2001)
5. Carey, M., Florescu, D., Ives, Z., Lu, Y., Shanmugasundaram, J., Shekita, E., Subramanian, S.: XPERANTO: publishing object-relational data as XML. In: Workshop on the Web and Databases, Informal Proceedings (2000)
6. Ahmad, U., et al.: An integrated approach for extraction of objects from XML and transformation to heterogeneous object oriented databases. arXiv preprint cs/0402007 (2004)
7. Chung, T.-S., et al.: Extracting object-oriented database schemas from XML DTDs using inheritance. In: Bauknecht, K., Madria, S.K., Pernul, G. (eds.) EC-Web 2001. LNCS, vol 2115, pp. 49–59. Springer, Heidelberg. https://doi.org/10.1007/3-540-44700-8_5
8. Naser, T., et al.: Transforming object-oriented databases into XML. In: IEEE International Conference on Information Reuse and Integration, IRI 2007. IEEE (2007)
9. Naser, T., AlHajj, R., Ridley, M.J.: Reengineering XML into object-oriented database. In: IEEE International Conference on Information Reuse and Integration, IRI 2008. IEEE (2008)
10. Naser, T., Alhajj, R., Ridley, M.J.: Two-way mapping between object-oriented databases and XML. Informatica (Slovenia) **33**(3), 297–308 (2009)

A Modified Shuffled Frog Leaping Algorithm
for Constructing DNA Codes

Zhenghui Liu[1], Bin Wang[1], Changjun Zhou[1], Xiaopeng Wei[2], and Qiang Zhang[2(✉)]

[1] Key Laboratory of Advanced Design and Intelligent Computing, Dalian University,
Dalian 116622, China
[2] Department of Computer Science and Technology, Dalian University of Technology,
Dalian 116024, China
zhangq@dlut.edu.cn

Abstract. High-quality DNA codes satisfying combinatorial constraints are important in the efficiency of DNA computing and other applications (e.g., DNA probes, DNA tagging). The key of constructing DNA codes is to find maximum sets of DNA codes that satisfy combinatorial constraints. In this paper, a modified Shuffled Frog Leaping Algorithm (MSFL) is proposed to construct DNA codes. 50 instances with Constant GC-content are carried out and when the length of codes is smaller than 13 the MSFL is able to improve some lower bounds, several of which are best-known. Comparing with previous works, the proposed algorithm is more efficient for constructing DNA codes.

Keywords: DNA codes · MSFL · GC-content · Lower bounds

1 Introduction

Generally, DNA codes are described as sets of codes of fixed length over the alphabet {A, C, G, T}. The problems of constructing DNA codes are to get maximum DNA codes satisfying combinatorial constraints and to apply these codes to molecular computation [1], DNA nanostructures [2], molecular tags [3, 4], gene chip [5–7], information storage [8] and other areas. For the efficiency of DNA computing and other applications involving synthetic DNA codes, it is very vital to obtain a lot of high-quality sets of DNA codes. Meanwhile, these DNA codes in the high-quality sets can greatly reduce the erroneous hybridizations between DNA codes and make DNA codes hybridize with their complements effectively and accurately according to the Watson-Crick base pairing.

In the past, a lot of previous lower bounds were able to be improved by algorithmic constructions and coding theory constructions. In 2001, Marathe et al. [9] made use of the method of coding theory to improve many previous lower bounds. In 2002, Ming et al. [10] introduced a construction of temple-map method for constructing DNA codes and succeeded in applying codes to DNA chips. In the same year, Tulpan et al. [11] successfully proposed a stochastic local search algorithm (SLS); their SLS initialized a random set of DNA codes, replaced the codes violating one of the constraints with probability and finally lead to the best set of codes. King [12] employed lexicographic

© Springer Nature Singapore Pte Ltd. 2018
K. Li et al. (Eds.): ISICA 2017, CCIS 873, pp. 269–278, 2018.
https://doi.org/10.1007/978-981-13-1648-7_23

algorithm to discuss the best lower and upper bounds and improved some previous lower bounds in 2003. With the method of various neighborhoods, Tulpan and Hoos [13] improved SLS, examined the method of generating various neighborhoods, analyzed the efficiency of algorithm and finally obtained some new best codes. Gaborit and King [14] used the coding theory for constructing DNA codes and managed to get many new better codes in many instances. Montemanni and Smith [15] introduced four new local search algorithms (i.e. Seeding Building (SB), Clique Search (CS), Hybrid Search (HS) and Iterated Greedy Search (IGS)) carefully for constructing DNA codes and combined them into the variable neighborhood search (VNS). Meanwhile, Roberto and Derek compared the efficiency of five algorithms in constructing sets of DNA codes in many instances. More details about VNS, the reader can obtain systematic knowledge from [16]. Chee and Ling [17] introduced clique search and a new stochastic local search that was similar to Simulated Annealing method removing the cooling process to lead to some new best sets of short codes. Niema [18] employed a computer algebra system for constructing DNA codes and a lot of new best DNA codes was obtained. Montemanni et al. [19] provided some details on their three different local search algorithms, which were Variable Neighborhood Search (VNS), Simulated Annealing (SA) and Evolutionary Algorithm (EA), and obtained many best sets of DNA codes. Tulpan et al. [20] didn't think hybridization strength between DNA codes was an efficient constraint and they reported many good lower bounds obtained by four different methods. Varbanov [21] proposed additive self-dual codes based on GF(4) to construct several new sets of DNA codes. Limbachiya et al. [22] surveyed many algorithmic constructions and coding theory constructions, and provided some tables of good lower bounds.

In 2003, Shuffled Frog Leaping Algorithm (SFLA) was proposed by Eusuff and Lansey [23] to solve the optimal of water distribution network design, showing its ability to tackle discrete optimization problems. Memetic evolution was the vital idea of SFLA and it was conducted in the form of infection of information between the best frog and other frogs in the population. Because of the exchange of information, SFLA was able to search the solution fast. Meanwhile, since SFLA was proposed, solving some complicated problems by the modified SFLA [24–28] has recently attracted the considerable interest of many researchers. Hence, SFLA is introduced and used to solve the problem of constructing DNA codes.

In this paper, the work focuses on constructing DNA codes with combinatorial constraint, and in order to improve the previous lower bounds, a MSFL that is different from other algorithms is proposed. The MSFL adopts an improved acceptance probability function based on [17] and a random way based on [23] of updating the frogs of the individual. The results with combinatorial constraint, reported in the Section Results, are compared with previously best-known lower bounds obtained by VNS and other methods. And some lower bounds that match and exceed the previous results obtained by the algorithm are marked. This indicates that the MSFL is more efficient in constructing DNA codes.

The paper is organized as follows: the Sect. 2 is related constraints and explanation; the Sect. 3 is modified shuffled frog leaping algorithm; the Sect. 4 is results; the Sect. 5 is conclusions and prospect.

2   Related Constraints and Explanation

In the past, researchers [9, 11, 12, 15, 17–19] focused on constructing sets of DNA codes satisfying one or more of several constraints, such as Hamming distance constraint (HD), Reverse Complementary Hamming distance constraint (RC), Constant GC-content constraint (GC) and Reverse Hamming distance (RC) constraint. Marathe et al. [9] accounted for four different constraints carefully in their paper. They suggested that HD limited false conflict between DNA codes and HD was regarded as a basic constraint on the DNA codes. The paper focuses on three constraints of them, which are HD, RC and GC.

Henceforth, λ denotes a DNA code of fixed length n over the alphabet $\{A, C, G, T\}$. λ^r denotes the inverse code of λ. λ^c denotes the Watson-Crick complementary code of λ. The Watson-Crick base pairing can be expressed as the following forms: $A^c = T$, $T^c = A$, $C^c = G$, $G^c = C$. Relevant knowledge and explanation about these constraints is as the following:

- **Hamming distance Constraint (HD):** Hamming distance constraint can always reduce these nonspecific hybridizations between DNA codes. HD can be explained: for two random and distinct codes λ_1 and λ_2 of a set of DNA codes, $H(\lambda_1, \lambda_2) \geq d$, where $H(\lambda_1, \lambda_2)$ denotes the number of the whole positions from which λ_1 and λ_2 differ, that is [17]

$$H(\lambda_1, \lambda_2) = n - \left| \left\{ 1 \leq i \leq n : \lambda_{1_i} = \lambda_{2_i} \right\} \right| \tag{1}$$

  where the subscript i denotes a random position of code. It is necessary to be noted that λ_1 and λ_2 are two distinct codes and the parameter d ranges from 3 to n ($n \leq 12$) in this paper. Related details about HD can be also found in [9, 12].
  Example 1. Let $\lambda_1 = $ TCGGAT and $\lambda_2 = $ TTCGGA, then $H(\lambda_1, \lambda_2) = 4$.
- **Reverse Complementary Hamming Distance Constraint (RC):** as with HD, RC is important in reducing nonspecific hybridizations between DNA codes. RC can be explained: for two random codes λ_1 and λ_2 of a set of DNA codes, $H(\lambda_1, (\lambda_2^r)^c) \geq d$, where λ_1 and λ_2 may be the same code and $H(\lambda_1, (\lambda_2^r)^c)$ denotes the number of the whole positions from which λ_1 and $(\lambda_2^r)^c$ differ, that is [17]

$$H(\lambda_1, (\lambda_2^r)^c) = n - \left| \left\{ 1 \leq i \leq n : \lambda_{1_i} = (\lambda_2^r)_i^c \right\} \right| \tag{2}$$

  The reader is referred to [9, 15, 17–19] for a more detailed description of Reverse Complementary Hamming distance constraint.
  Example 2. Let $\lambda_1 = $ TCGGAT and $\lambda_2 = $ TTCGGA, then $\lambda_2^r = $ AGGCTT, $(\lambda_2^r)^c = $ TCCGAA, and $H(\lambda_1, (\lambda_2^r)^c) = 2$.
- **Constant GC-content Constraint (GC):** C and G are held together by three hydrogen bonds. A and T are held together by two hydrogen bonds. The number of hydrogen bonds of DNA reflects thermal stability (i.e. high GC-content usually

means high thermal stability [17]), so Constant GC-content constraint is described as a significant factor of DNA codes with great thermal stability. Generally, the GC-content of DNA codes is 50%. Concrete explanation about Constant GC-content constraint is as the following: for a random code δ of a codes set, total number of character G and C in δ is n/2 (4 ≤ n ≤ 12), where GC(δ) denotes the GC-content of the code δ, that is [17]

$$GC (\delta) = \left| \{1 \leq i \leq n : \delta_i \in \{C, G\} \} \right| \qquad (3)$$

where the subscript i denotes a random position of code δ_i.
Example 3. Let δ = AGTCAGAG, then GC (δ) = 4.

Henceforth, we let HD + GC + RC denote combinatorial constraints consisting of HD, GC and RC. And we let $A_4^{GC,RC}$(n,d,w) denote the size of the optimal set of DNA codes satisfying HD + GC + RC.

3   Modified Shuffled Frog Leaping Algorithm

Since SFLA was proposed in 2003, it has not been used for constructing DNA codes construction. Also, memetic evolution of SFLA is able to provide a good search eighborhood. We now introduce and improve it to obtain more high-quality DNA codes. And Stochastic local search algorithm (SLS) in [11] was usually fast and efficient and showed its excellent performance in searching high-quality DNA codes sets. A new improved SLS in [17] played a great role in the research process of DNA codes construction and it suggested that the efficiency of designing SLS lied in the specification of initializing the sets of DNA codes, selecting from suitable search neighborhood and measuring efficient sets of DNA codes, acceptance probability and terminating condition. These ideas above are also the essential perspectives that lead to the MSFL.

3.1   An Improved Acceptance Probability Function

As with [17], the acceptance probability is considered as the key of the MSFL. And the improved accept function that we use is the following definition

$$f(\Delta W) \begin{cases} 1 & (\Delta W \geq 0) \\ (-1) \cdot (\Delta W/c) & (-3 \leq \Delta W < 0) \\ 0 & (\Delta W < -3) \end{cases} \qquad (4)$$

where ΔW denotes the change of working set workSet of DNA codes after the individual (i.e. a DNA code) is added to the working set workSet and c is a variable constant influence factor ($10^4 < c < 10^7$). ΔW is treated as the evaluation criteria measuring the individual fitness in the population Γ. More details of the accept function can also be consulted in [17].

Note that in the function (4), we adopt an improved and variable acceptance probability for constructing DNA codes. And we observe that $10^4 < c < 10^5 (\Delta W = -1)$, $10^5 < c < 10^6 (\Delta W = -2)$ and $10^6 < c < 10^7 (\Delta W = -3)$ always work well. In the following section, we report a more detailed description of the improved acceptance function that is used in updating Γ and modifying the SFLA.

3.2 The Formula of Updating the Population

In addition, the formula that the algorithm uses to update the worst individual of the population Γ is the equation

$$P'_w = P_w + \mu \cdot (P_b - P_w) \tag{5}$$

where μ is a random number $(0 < \mu < 1)$; P_b is the best individual of the population Γ; P_w is the worst individual of the population Γ; P'_w is the new individual produced by the worst individual leaps according to the equation in the search space. Importantly, the formula adopted to update the best individual P_b of the population Γ is the equation

$$P'_b = \tau \cdot P_b \tag{6}$$

where τ is a random number $(0 < \tau < 1)$. Note that when ΔW of the best individual P_b of the population Γ is smaller than -3, the population is randomly generated.

3.3 Constructing DNA Codes Based on the MSFL

Generally, the population Γ is randomly generated. There are 20 individuals in the population Γ. And it is necessary to be noted that the search space that we use consists of specific DNA codes, each of which satisfies GC and RC (i.e. the GC-content of each individual is n/2 and the RC distance of each individual is not smaller than d). Meanwhile, the best individual P_b of the population Γ is considered as a solution of the working codes set workSet. Then, the population Γ is divided into 10 memeplexs. Thus, there are a better adaptable individual and a worse adaptable individual in every memeplex. The worse frog P_w in every memeplex changes its position according to the Eq. (5) and the best frog P_b changes its position according to the Eq. (6). When ΔW of the best individual P_b of the population Γ is smaller than -3, the population, each individual of which itself satisfies GC and RC, is randomly generated. Thus, the evolution of Γ is finished. Then, it does a descending sort according to ΔW when Γ completes the evolution. The best frog P_b in Γ is saved and added into the set TempSet according to the function (4). With the procedures of evolving the population, searching P_b and adding P_b into TempSet according to the probability, when the size of TempSet equals 4, the algorithm uses exhaustive search algorithm and obtain the maximum compatible (i.e. satisfy all the constraints) set addSet in the set TempSet. Then, the algorithm removes all those codes from the set workSet that are incompatible with the set addSet and get the union of sets addSet and workSet. The optimal DNA codes set bestcode is saved and the algorithm is terminated when the size of bestcode is not smaller than the value A

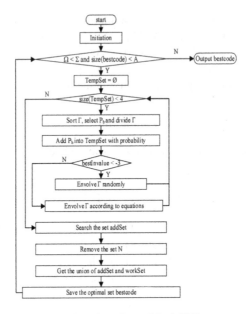

Fig. 1. Flow chart of the MSFL

and the computation time Ω of the MSFL is greater than the ending time Σ, otherwise the algorithm is continued.

```
MSFL(Σ, n,d,A, Γ, workSet,bestcode)
    While(Ω < Σ and size(bestcode)<A)
        TempSet := Φ;
        While(size(TempSet) < 4)
            bestInValue := the ΔW of P_b;
            Sort and select P_b from Γ according to ΔW;
            P_b := the best frog of Γ;
            Divide Γ into 10 memeplexs;
            TempSet: = TempSet ∪P_b with probability f(ΔW) where:
                P_b is compatible with workSet or a part of workSet and Γ is
                always in low incompatible value.
            If bestInValue < -3
                    Evolve Γ randomly;
            Else
                    Evovle Γ according to equations (5) and (6);
            EndIf;
        EndWhile;
        Search the maximum compatible set in the TempSet;
        addSet := the maximum compatible set in the TempSet;
        N := { υ ∈ workSet:H(υ,ξ) < d or H(υ,(ξ^τ)^c) < d,ξ ∈ addSet};
        workSet := workSet ∪ addSet - N;
        If (size(workSet) > size(bestcode))
            bestcode := workSet;
        EndIf;
    EndWhile;
```

Fig. 2. Pseudo Code of the MSFL

Flow chart and Pseudo Code of the MSFL are presented in Figs. 1 and 2. The sets bestcode and workSet represent respectively the optimum set obtained and the working set. Parameter TempSet represents a random set made up of 4 codes that are saved with the probability. The set addSet obtained by using exhaustive search algorithm is an optimum compatible set. Parameter N consists of all those codes from the set workSet, each of which is incompatible with the set addSet. Importantly, the algorithm merges the set addSet and the set workSet in each iteration. Thus, parameter workSet is renewed. Also, when the size of the set workSet is larger than the size of the set bestcode, the algorithm renews the parameter bestcode.

4 Results

All the instances obtained by the MSFL are carried out on Intel(R) Xeon(R) CPU E5-2603 v3 @ 1.60 GHz/64 GB RAM machine with the following parameter settings: $\Gamma = 30 * 24\,h$ for instances with $4 \leq n \leq 9$, $\Gamma = 60 * 24\,h$ for instances with $10 \leq n \leq 12$. Note that only one run was considered for each instance. Also, experimental environment and the parameter settings of the MSFL are different from the ones of [13, 14, 17–19] and found experimentally during some instances with relatively short codes.

The paper uses the lower bounds in table of [19] as benchmarks for the performance of the MSFL. In Tables 1 and 2, the entry V represents the results cited in [19], and the entry M represents the results obtained by the MSFL.

Table 1. Lower bounds for $A_4^{GC,RC}(n, d, w)$

n\d		3	4	5	6	7	8	9	10	11	12
4	V	6	2	-	-	-	-	-	-	-	-
	M	**6**	**2**	-	-	-	-	-	-	-	-
5	V	15	3	1	-	-	-	-	-	-	-
	M	**15**	**3**	**1**	-	-	-	-	-	-	-
6	V	42	16	4	2	-	-	-	-	-	-
	M	**43**	**16**	**4**	**2**	-	-	-	-	-	-
7	V	126	35	11	2	1	-	-	-	-	-
	M	**129**	34	**11**	**2**	**1**	-	-	-	-	-
8	V	417	101	27	12	2	2	-	-	-	-
	M	**431**	**104**	26	11	**2**	**2**	-	-	-	-
9	V	1199	269	67	21	-	2	1	-	-	-
	M	**1257**	**277**	65	19	-	**2**	**1**	-	-	-
10	V	3927	821	176	49	17	-	2	2	-	-
	M	**4106**	**862**	**183**	**50**	15	-	**2**	**2**	-	-
11	V	10993	2404	471	117	37	14	-	2	1	-
	M	**11518**	**2468**	**476**	**118**	36	13	-	**2**	**1**	-
12	V	-	6186	1381	311	87	29	12	4	2	2
	M	-	**7051**	1272	300	85	**29**	10	**4**	**2**	**2**

Table 2. The best-known lower bounds for $A_4^{GC,RC}(n, d, w)$

n\d		3	4	5	6	7	8	9
6	B	44^C	16^C	-	-	-	-	-
	M	43	**16**	-	-	-	-	-
7	B	135^C	36^C	11^C	2^C	-	-	-
	M	129	34	**11**	**2**	-	-	-
8	B	528^G	128^R	28^C	12^T	-	-	-
	M	431	104	26	11	-	-	-
9	B	1354^G	275^C	67^C	21^R	-	-	-
	M	1257	**_277_**	65	19	-	-	-
10	B	4542^G	860^N	210^N	54^G	17^R	-	-
	M	4106	**862**	183	50	15	-	-
11	B	14405^G	2457^G	477^C	117^C	37^R	14^R	-
	M	11518	**_2468_**	476	**_118_**	36	13	-
12	B	-	14784^N	1848^N	924^R	87^R	29^R	12^R
	M	-	7051	1272	300	85	**29**	10

In Table 1, the entries in bold and underscore are 15 lower bounds that surpass the lower bounds reported in [19] and the 23 entries only in bold match the lower bounds. Table 1 suggests that the MSFL is more efficient with state-of-the-art VNS on most of the instances with n ≤ 12 considered. Particularly, the algorithm is capable of improving and matching 38 lower bounds over 50 instances considered, showing its ability to improve upon results achieved by VNS of Montemanni and other researchers. It is interesting to observe that the algorithm works well when $7 \leq n \leq 11$ and $3 \leq d \leq 5$.

In Table 2, the good lower bounds obtained by coding theory constructions and algorithmic constructions are reported and the meanings of the superscripts are as follows: 'C' means previously good lower bounds from [17]; 'R' means previously good lower bounds reported in [19]; 'G' means previously good lower bounds obtained from [14]; 'N' means previously good lower bounds cited in [18]; 'T' means previously good lower bounds of [13]. Table 2 suggests that the algorithm is able to obtain 4 new best-known lower bounds and 4 lower bounds that match the previously good results. But for other instances, the algorithm can't lead to best lower bounds.

5   Conclusions and Prospect

In this paper, we propose a MSFL based on SFLA for solving the problems of constructing sets of DNA codes with HD + GC + RC. The MSFL is carried out in 50 instances. Meanwhile, it is able to lead to 15 lower bounds which are better compared with results obtained by VNS and improve the previous good lower bounds in 4 instances (i.e. $A_4^{GC,RC}(9, 4, 4) = 277$, $A_4^{GC,RC}(10, 4, 5) = 862$, $A_4^{GC,RC}(11, 4, 5) = 2468$, $A_4^{GC,RC}(11, 6, 5) = 118$). This shows its potential application for improving lower bounds. When the length n of DNA code is larger than 11 and the d of HD and RC is not smaller than 6, the algorithm doesn't obtain some better lower bounds than those in [19]. By comparing Table 1 with Table 2, it is important to observe that the lower bounds obtained by the MSFL remain far from the best-known results in some instances. This suggests that there's some room for further improvements in the study of the algorithm.

Acknowledgment. This work is supported by the National Natural Science Foundation of China (Nos. 61425002, 61751203, 61772100, 61402066, 61672121, 61672051, 61572093, 61402067, 61370005, 31370778), Program for Changjiang Scholars and Innovative Research Team in University (No. IRT_15R07), the Program for Liaoning Innovative Research Team in University (No. LT2015002), the Basic Research Program of the Key Lab in Liaoning Province Educational Department (Nos. LZ2014049, LZ2015004), Scientific Research Fund of Liaoning Provincial Education (Nos. L2015015, L2014499), and the Program for Liaoning Key Lab of Intelligent Information Processing and Network Technology in University.

References

1. Frutos, A.G., Liu, Q., Thiel, A.J., Sanner, A.M., Condon, A.E., Smith, L.M., Corn, R.M.: Demonstration of a word design strategy for DNA computing on surfaces. Nucleic Acids Res. **25**(23), 4748–4757 (1997)
2. Reif, J.H., LaBean, T.H., Seeman, N.C.: Challenges and applications for self-assembled DNA nanostructures? In: Condon, A., Rozenberg, G. (eds.) DNA 2000. LNCS, vol. 2054, pp. 173–198. Springer, Heidelberg (2001). https://doi.org/10.1007/3-540-44992-2_12
3. Brenner, S., Lerner, R.A.: Encoded combinatorial chemistry. Proc. Nat. Acad. Sci. **89**(12), 5381–5383 (1992)
4. Braich, R.S., et al.: Solution of a satisfiability problem on a gel-based DNA computer. In: Condon, A., Rozenberg, G. (eds.) DNA 2000. LNCS, vol. 2054, pp. 27–42. Springer, Heidelberg (2001). https://doi.org/10.1007/3-540-44992-2_3
5. Schena, M., Shalon, D., Davis, R.W., Brown, P.O.: Quantitative monitoring of gene expression patterns with a complementary DNA microarray. Science **270**(5235), 467–470 (1995)
6. Tulpan, D., Andronescu, M., Chang, S.B., Shortreed, M.R., Condon, A., Hoos, H.H., Smith, L.M.: Thermodynamically based DNA strand design. Nucleic Acids Res. **33**(15), 4951–4964 (2005)
7. Condon, A.: Designed DNA molecules: principles and applications of molecular nanotechnology. Nat. Rev. Genet. **7**(7), 565–575 (2006)
8. Yazdi, S.M.H.T., Yuan, Y., Ma, J., Zhao, H., Milenkovic, O.: A rewritable, random-access DNA-based storage system. arXiv preprint arXiv:1505.02199 (2015)
9. Marathe, A., Condon, A., Corn, R.M.: On combinatorial DNA word design. J. Comput. Biol. **8**(3), 201–219 (2001)
10. Ming, L., Lee, H.J., Condon, E.A., Corn, R.M.: DNA word design strategy for creating sets of non-interacting oligonucleotides for DNA microarrays. Langmuir **18**(3), 805–812 (2002)
11. Tulpan, D.C., Hoos, H.H., Condon, A.E.: Stochastic local search algorithms for DNA word design. In: Hagiya, M., Ohuchi, A. (eds.) DNA 2002. LNCS, vol. 2568, pp. 229–241. Springer, Heidelberg (2003). https://doi.org/10.1007/3-540-36440-4_20

12. King, O.D.: Bounds for DNA codes with constant GC-content. Electron. J. Comb. **10**(1), 1–13 (2003)
13. Tulpan, D.C., Hoos, H.H.: Hybrid randomised neighbourhoods improve stochastic local search for DNA code design. In: Xiang, Y., Chaib-draa, B. (eds.) AI 2003. LNCS, vol. 2671, pp. 418–433. Springer, Heidelberg (2003). https://doi.org/10.1007/3-540-44886-1_31
14. Gaborit, P., King, O.D.: Linear constructions for DNA codes. Theor. Comput. Sci. **334**(1), 99–113 (2005)
15. Montemanni, R., Smith, D.H.: Construction of constant GC-content DNA codes via a variable neighbourhood search algorithm. J. Math. Modell. Algorithms **7**(3), 311–326 (2008)
16. Hansen, P., Mladenović, N., Pérez, J.A.M.: Variable neighbourhood search: methods and applications. Ann. Oper. Res. **175**(1), 367–407 (2010)
17. Chee, Y.M., Ling, S.: Improved lower bounds for constant GC-content DNA codes. IEEE Trans. Inf. Theor. **54**(1), 391–394 (2008)
18. Niema, A.: The construction of DNA codes using a computer algebra system. Ph.D thesis. University of Glamorgan (2011)
19. Montemanni, R., Smith, D.H., Koul, N.: Three metaheuristics for the construction of constant GC-content DNA codes. Lect. Notes Manage. Sci. **6**, 167–175 (2014)
20. Tulpan, D., Smith, D.H., Montemanni, R.: Thermodynamic post-processing versus GC-content pre-processing for DNA codes satisfying the hamming distance and reverse-complement constraints. IEEE/ACM Trans. Comput. Biol. Bioinf. **11**(2), 441–452 (2014)
21. Varbanov, Z., Todorov, T., Hristova, M.: A method for constructing DNA codes from additive self-dual codes over GF (4). ROMAI J. **10**(2), 203–211 (2014)
22. Limbachiya, D., Rao, B., Gupta, M.K.: The art of DNA strings: sixteen years of DNA coding theory. arXiv preprint arXiv:1607.00266 (2016)
23. Eusuff, M.M., Lansey, K.E.: Optimization of water distribution network design using the shuffled frog leaping algorithm. J. Water Resour. Plan. Manage. **129**(3), 210–225 (2003)
24. Bhattacharjee, K.K., Sarmah, S.P.: Shuffled frog leaping algorithm and its application to 0/1 knapsack problem. Appl. Soft Comput. **19**, 252–263 (2014)
25. Darvishi, A., Alimardani, A., Vahidi, B., Hosseinian, S.: Shuffled frog-leaping algorithm for control of selective and total harmonic distortion. J. Appl. Res. Technol. **12**(1), 111–121 (2014)
26. Jadidoleslam, M., Ebrahimi, A.: Reliability constrained generation expansion planning by a modified shuffled frog leaping algorithm. Int. J. Electr. Power Energy Syst. **64**, 743–751 (2015)
27. Moramelia, D., Iglesiasrey, P.L., Martinezsolano, F., Munozvelasco, P.: The efficiency of setting parameters in a modified shuffled frog leaping algorithm applied to optimizing water distribution networks. Water **8**(5), 182 (2016)
28. Orouji, H., Mahmoudi, N., Fallah-Mehdipour, E., Pazoki, M., Biswas, A.: Shuffled frog-leaping algorithm for optimal design of open channels. J. Irrig. Drain. Eng. **142**(10), 06016008 (2016)

Clustering Based Prediction of Financial Data by ARMA Model

Duobiao Ning, Siyu Zhang, Wenfei Chen$^{(\boxtimes)}$, and Xinqiao Yu

Computer Science and Technology Department, Chengdu Neusoft University,
Chengdu 611844, Sichuan, China
wenfei_chen@163.com

Abstract. It is an important factor to predict the financial data for the decision making by a number of institutions. First, operating recorders of customers are clustered into different classes' sets in the process of financial accounts management. Second, customers are organized as time series in our model, and the ARMA model is used to predict the result. Our data comes from YuEBao which is used to verify the above method of preprocessing and k-means clustering which is used to predict the time series for the result here. Compared with the actual data, our method outperforms the other method in accuracy.

Keywords: Time series · ARMA model · Financial data · K-means

1 Introduction

The essence of the financial theory research is the way to construct a prediction model and reality in order to reduce the maximum error. With financial phenomenon, the market is not always efficient or weak form efficient market [1]. However, it is possible using historical data to predict the future market conduct by researchers [2]. With the development of computer science and the Internet, volume data are used in the area of Business Management, Scientific Research, and Engineering Development. It is a challenge for governments and corporations in data processing and application. Especially, it is a problem that should be solved as soon as possible forecasting the several of data from financial institutions. When forecasting the monetary data, almost all methods are trying to deal directly instead of researching the behaviors of clienteles divided by different classes and stratum. The sketchy methods above do not guarantee the precision results correct.

Time series occupies a large proportion in the data analysis, and here the prediction of the financial phenomenon is equivalent to the time series problem to some extent. The methods of time series include traditional methods like the Auto-Regressive and Moving Average model (ARMA) [3], as well as the nonlinearity predicted method [4] such as Embedded Space method and Neural Network. Though the self-adaptive of the

W. Chen—A Project Supported by Scientific Research Fund of Sichuan Provincial Education Department under Grant No. 17ZB0002.

© Springer Nature Singapore Pte Ltd. 2018
K. Li et al. (Eds.): ISICA 2017, CCIS 873, pp. 279–287, 2018.
https://doi.org/10.1007/978-981-13-1648-7_24

nonlinearity method is very strong, it is too difficult to apply because of the large parameters and serious model quality [5]. First, the paper proposes a method which clusters the customer data as the features and demands. Second, the paper makes model identification by the ARMA model and gets the forecast result.

2 Overview

Table 1 presents the parameters of financial data. The *report_date* is the date when customers operate their accounts. The *user_id*, *sex* and *city* are used to mark customers' identity and features. The *picnic* means the operating time in one day. The *yBanlance*, *purchase_amt* and *redeem_amt* are all parameters about the situations of customers' assets – the antianaphylaxis date makes sure that balance of one customer today equals *yBalance + purchase_amt − redeem_amt*.

Table 1. Parameters of data from YuEBao

Parameter	Data type	Parameters meaning
report_date	string	Data
user_id	bigint	User ID
sex	string	Sexy
city	string	Country
picnic	bigint	Operating frequency
yBalance	bigint	Balance of yesterday
purchase_amt	bigint	Purchase of today
redeem_amt	bigint	Redemption of today

In order to predict the total balance every day in August 2014, we use the YuEBao data from 1st July 2013 to 31st July 2014 in this prediction model. We preprocess the data including aggregate and handling missing values firstly and then use k-means algorithm to structure the data classification model as the indexes of Table 1. Next, we get the result of the ARMA model. Finally, we discuss and evaluate the result data by the above method.

3 Financial Data Prediction Model

In this section, we detail our method used to predict financial data. The process is shown in Fig. 1.

3.1 K-Means Algorithm [6]

It is our objective to predict the change of user's financial data. Operating behaviors and cash flow are almost distinct between different users, so it can result in an error when unifying and summarizing customers' data directly. Therefore it is necessary to cluster customers' data.

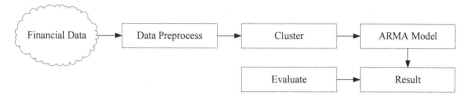

Fig. 1. Framework of our method

The clustering algorithm is used to sort based on no-guidance. It purposes to minimize the distance between the same objects, and maximize the distance between different objects on the contrary. The current study divides clustering into classification method, hierarchy method, and density method. To classification method, a given set $D = \{x1, x2,..., xn\}$ with N observations put them into $K(\leq N)$ partitions of $C = \{c1, c2,..., ck\}$ which make sure that every observation is in a single partition and minimize the sum of distance functions of each point in the cluster to the K center. According to the data features, the indexes shown in Table 1 are described by numeric data including Boolean data. In this case, we try to use the k-means algorithm to classify the users.

As Eq. (1) represents below [7], Euclidean Distance is used to calculate the distance sum of points to their set center μ_k,

$$S(c_k) = \sum_{x_i \in D_k} \|\mu_k - x_i\|^2 \tag{1}$$

The aim is to get the minimal sum of all sets according to Eq. (2),

$$S(C) = \sum_{k=1}^{K} S(c_k) = \sum_{k=1}^{K} \sum_{x_i \in D_k} \|\mu_k - x_i\|^2 = \sum_{k=1}^{K} \sum_{i=1}^{N} d_{ki}\|\mu_k - x_i\|^2 \tag{2}$$

And here,

$$d_{ki} = \begin{cases} 1, & x_i \in c_k \\ 0, & x_i \notin c_k \end{cases} \tag{3}$$

Obviously, as Least Square method and Lagrange principle are shown, μ_k should be the average of the points c_k set.

3.2 ARMA Model

Extract a data unit into a two-tuple (T, O). Here T is the time variables, and O is the data variables. They are corresponding to each other. We use them to build *Auto-Regressive and Moving Average model* (ARMA) which is created by Box etc. It describes the correlation information of a smooth sequence and establishes the model to predict the time series data. ARMA includes *Auto-Regressive model* (AR), *Moving Average method* (MA) and ARMA. The first two are a special situation of the last model. Time series data y_t is the linear function of rand error of current and past, also former status.

It is expressed as Eq. (4):

$$y_t = \delta_1 y_{t-1} + \delta_2 y_{t-2} + \ldots + \delta_p y_{t-p} - \theta_1 \varepsilon_{t-1} - \theta_2 \varepsilon_{t-2} - \theta_q \varepsilon_{t-q} + \varepsilon_t + C \qquad (4)$$

Here C is a constant, ε_t is the error term, δ_p is the regression coefficient and θ_q is the smoothing coefficient of the model. The Eq. (4) is denoted by ARMA(p, q).

The correlation measurement of time series is evaluated by autocorrelation function (ACF) and partial autocorrelation function (PACF), and the universal model is the Least Square method.

3.3   The Steps of Prediction Modal

The algorithm is shown below:
Input: Users' financial data matrix M, quantity of clusters K ;
Output: Predicted data sequence;
1. Preprocess M;
2. Build the classification model;
3. Do
4. $(p, q) \leftarrow$ ACF and PACF;
5. While residual sequence is smooth;
6. $K \leftarrow$ ARMA(p, q);
7. Return K.

4   Experiment

All data used in this model are from YuEBao [8]. It is organized by 567 YuEBao customers' data and operated situation from 2013/07/01 to 2014/08/31. The structure is shown in Table 1, and the behavior records count 99451.

4.1   Preprocess Data and Build the Classification Model

We collected the total balance of every user at every day. Another import factor is users' operating frequencies which describe the behavior habits. Higher frequencies there were, the more transformation had happened, so we need to calculate the operating frequencies of everyone. At last, if there were a few missing values in the set, we will delete them directly.

Before we use the k-means algorithm to process data, we must normalize the data to avoid the effect resulted from the skimble-scamble dimension of different indexes.

The results of Table 2 tell us there are big gaps in balance, roll-in, roll-out and picnic between three kinds of users, especially between the first, the second and the third. For example, the first class of people is men, the second class is women and the third one includes men and women with a ration of 57.5%.

4.2 Build ARMA Model

In Table 2 we have got three kinds of sets. Because of the same data format and similar variation, we choose the first data to predict as a sample. And the other two sets can be processed in the same way.

Table 2. Clustering results

Classes	First	Second	Third
user_count	262	265	40
user_proportion	46.21%	46.74%	7.05%
yBalance_mean	495428.2	641324.5	11269500
capital_roll-in	2328440	2037581	70853336
capital_roll-out	1823084	1448886	58762538
pinci_mean	173.7366	156.2151	285.825
sex	1	0	0.575

ARMA model is mixed by the AR model and MA model, and we need to find the two parameters because that they are the key points of the model-regression coefficient and smoothing coefficient.

We can find the first set expresses an obvious growth trend and unstable growth rate in Fig. 2. And then we use PACF to get the value of regression coefficient, drawn the PACF picture and find the value by truncation. In Fig. 3, we found the ACF at the fourth truncation, the regression coefficient is 4.

With the same method, we drew the PACF picture of MA model presented in Fig. 4 and found the ACF at the second truncation. So the value of smoothing coefficient is 2.

We have to verify the residual series of H-P Filter data by regression coefficient and smoothing coefficient. As Fig. 5 shown, the series leveled off. So it is effective.

4.3 Result

We have gotten the p and q, and used ARMA (p, q) to predict the data for one month (2014/08) by the data last year. And we got the result shown in Table 3.

We made a comparison of the result between predicted data and the real data, and the outcome is presented in Fig. 6. And the trend fitting is shown in Fig. 7.

The result of a one-day delaying described the data variable trend well.

At last, we used the method above to predict the whole data and compared with the result on direct ARMA algorithm. As is shown in Fig. 8, the error sum of squares of our method is $3.968 * 10^{12}$ relative to the $2.378 * 10^{16}$ from the later one. So our method is better.

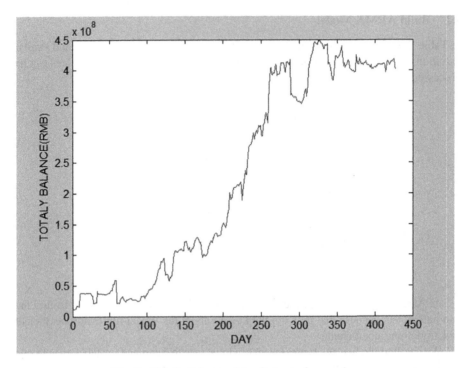

Fig. 2. Totally balance of the first set of every day

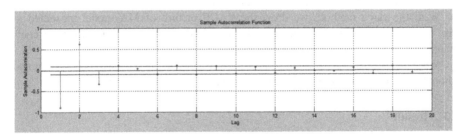

Fig. 3. PACF result of AR

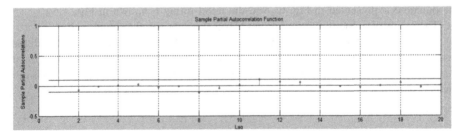

Fig. 4. PACF result of MA

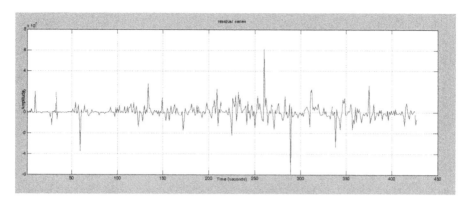

Fig. 5. Residual series of data

Table 3. Predict result

Date	20140801	20140802	20140803	20140804
Totally balance	396467367.5	397640091.9	396706926.3	399093285.2
Date	20140805	20140806	20140807	20140808
Totally balance	399891123.3	399166009.7	402112422.5	400568765.2
Date	20140809	20140810	20140811	20140812
Totally balance	404454908.4	404290750.4	404025453.3	403688735.9
Date	20140813	20140814	20140815	20140816
Totally balance	403433581.5	403770821.5	402928748.6	405094695.1

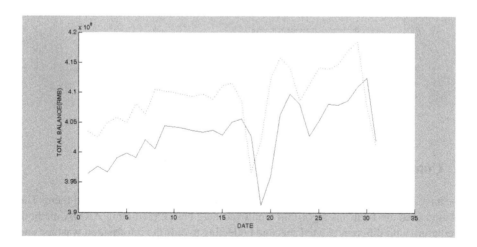

Fig. 6. Predicted and original data

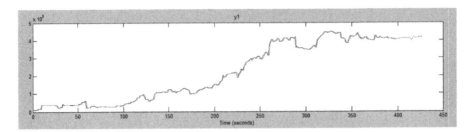

Fig. 7. Trend comparing

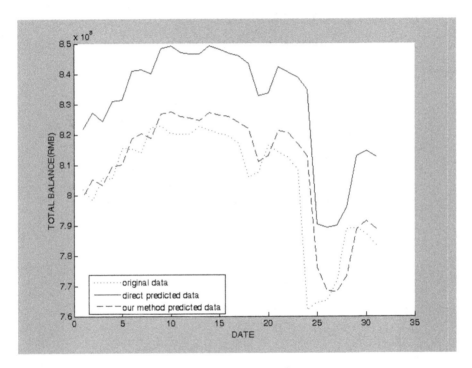

Fig. 8. Method comparing

5   Conclusion

We have researched the YuEBao customers' balance variation and predicted them. The method uses the K-means algorithm employed here before predicting ARMA method. Predicting with different classed sets and collecting avoids the information wastage and keeps the data correct as far as possible. The experiment verifies the method which describes the financial data well compared with the direct method. Clustering model is appropriate for financial time-series data and provides references for prediction method. It supports the theory for enterprise decision.

References

1. Hamid, K., Suleman, M.T., Ali Shah, S.Z., Akash, R.S.I.: Testing the weak form of efficient market hypothesis empirical evidence from Asia-Pacific markets. Int. Res. J. Financ. Econ. (58) 121–133 (2010)
2. Zemke, S.: On developing financial prediction system: pitfalls and possibilities. In: Proceedings of DMLL Workshop at ICML 2002 (2002)
3. Kalpakls, K., Gada, D., Puttagunta, V.: Distance measure for effective clustering of ARIMA time series. In: Proceedings of the IEEE ICDM, pp. 273–280 (2001)
4. Takens, F.: Detecting strange attractors in turbulence. In: Rand, D., Young, L.-S. (eds.) Dynamical Systems and Turbulence, Warwick 1980. LNM, vol. 898, pp. 366–381. Springer, Heidelberg (1981). https://doi.org/10.1007/BFb0091924
5. Maguire, L.P., Roche, B., Mcginnity, T.M.: Predicting a chaotic time series using a fuzzy neural network. Inf. Sci. **112**(1–4), 125–136 (1998)
6. Anil, K.J.: Data clustering: 50 years beyond K-means. Pattern Recognit. Lett. **31**(8), 651–666 (2010)
7. Wang, Q., Wang, C., Feng, Z., Ye, J.: Review of K-means clustering algorithm. Electron. Des. Eng. **20**(7), 21–24 (2012)
8. Aliyun homepage. https://tianchi.shuju.aliyun.com/competition/introduction.htm?spm=5176.100066.333.10.BcErHh&raceId=3

Research on Data Mining Algorithm
of Association Rules Based on Hadoop

Linrun Qiu[⊠]

Department of Computer Science,
Guangdong University of Science and Technology, Dongguan, China
79713159@qq.com

Abstract. Big data mining based on cloud computing is the hot topic of the industry research, this paper proposed an improved distributed Apriori algorithm. More importantly, In view of the poor performance of running Apriori algorithm in large data, the algorithm of association rule data mining based on Apriori algorithm is put forward, and the improved distributed Apriori algorithm based on Hadoop platform is proposed. The algorithm focuses on the application of association rules algorithm based on Hadoop in mass data mining. This paper describes the idea of improved Apriori algorithm on Hadoop platform, and presents the experimental test. The experimental results show that the improved algorithm of association rules based on Hadoop can effectively improve the Apriori algorithm for association rules of operation efficiency, and reduce the redundant association rules, and has the efficient advantage in dealing with massive data.

Keywords: Hadoop · Association rules · Apriori algorithm · Data mining

1 Introduction

Data mining is the extraction of implicit, previously unknown and potentially valuable knowledge and rules from large databases or data warehouses. It is the combination of artificial intelligence and database development, in which information decision-making system is in the forefront of research. The main algorithms of data mining are classification pattern, association rules, decision tree, sequence pattern, clustering model analysis, neural network algorithm [1]. The association rules are a very important research topic in the field of data mining. It is widely used in various fields, which can test the long-term knowledge pattern in the industry and discover the new laws. It is an important method to accomplish the data mining task effectively, Therefore, it is significant to study the association rules.

With the rising of cloud computing technology, how to extract knowledge from the massive data is currently an urgent problem to be solved. At the same time, people pay more and more attention to data mining technology in big data environment applications, many scholars made some improved research in the field, such as data classification, clustering and so on. As an important part of data mining, association rule mining plays an important role in data mining. The status of Aprioi, a classic algorithm in association rules, is more important [2]. However, the traditional serial algorithm

© Springer Nature Singapore Pte Ltd. 2018
K. Li et al. (Eds.): ISICA 2017, CCIS 873, pp. 288–297, 2018.
https://doi.org/10.1007/978-981-13-1648-7_25

can't meet the requirements of big data era. In recent years, cloud computing platform distributed system infrastructure to make full use of large clusters for high-speed computing and storage for big data mining provides a parallel computing framework. Based on the Apriori association rule mining algorithm, this paper proposes an improved algorithm based on Hadoop framework for distributed parallel mining algorithm, which can be applied to the mining of large data association rules, and improve the efficiency of association rule mining in Apriori algorithm under large data environment.

2   The Technology of Hadoop Platform

Hadoop is an open source parallel computing programming tool and a distributed file system developed by the Apache Software Foundation that allows users to develop a highly scalable distributed batch processing system with its MapReduce programming model without knowing the underlying details [3]. Hadoop's framework is the core design of HDFS and MapReduce. HDFS provides storage for massive amounts of data, and MapReduce provides calculations for massive amounts of data. Hadoop mainly by the HDFS (Hadoop Distributed File System) and MapReduce engine composed of two parts. The bottom is HDFS which stores the files on all storage nodes in the Hadoop cluster. The upper layer of HDFS is the MapReduce engine. The engine consists of JobTrackers and TaskTrackers. Hadoop is an open source distributed parallel programming framework that implements the MapReduce computing model. Programmers can use Hadoop to write programs that run programs on a computer cluster to handle massive amounts of data [4]. With the help of Hadoop framework and cloud computing core technology MapReduce to achieve data calculation and storage, and HDFS distributed file system and HBase distributed database is well integrated into the cloud computing framework, so as to achieve cloud computing distributed, parallel computing and Storage, and the ability to achieve good processing of large-scale data. The Hadoop storage architecture is shown in Fig. 1. Particle swarm algorithm.

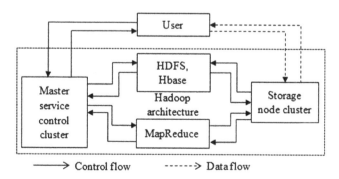

Fig. 1. The architeture of Hadoop storage

Hadoop uses a master/slave architecture in both distributed and distributed storage. Distributed storage system that Hadoop file system, or referred to as HDFS. NameNode is located on the HDFS master and directs the underlying DataNode to perform underlying I/O tasks. Secondary Name Node (SNN) monitors HDFS cluster status. JobTracker connects applications and Hadoop clusters [5]. TaskTracker manages the operation of each node. MapReduce is Google's proposed software architecture, but also Hadoop programming model. The data is matched in the MapReduce as a key/value pair:

Map: (K1, V1)→list(K2, V2)
Combiner: (K2, list(V2))→list(K2, V2)
Reduce: (K2, list(V2))→list(K3, V3).

Combiner can be used as part of the Map function, the Map output of the same key results in the results of the merger, similar to the local Reduce. The purpose of doing so can reduce the data transmission of the network and improve the performance of data processing.

The master service control cluster is equivalent to the controller part, which is responsible for receiving the application request and answering according to the request type. Storage node cluster is equivalent to the memory part, is a huge disk array system or a large amount of data storage capacity of the cluster system, the main function is to deal with data access [6]. Hadoop has a master server (called JobTracker) for scheduling and managing other computers (called TaskTracker), and JobTracker can run on any computer in the cluster. Users do not directly read through the Hadoop architecture and HDFS and HBase access to data, thus avoiding a large number of read operations may cause system congestion. After the user passes the information from the Hadoop architecture to the master service control cluster, the user performs the read operation directly with the storage node. The Hadoop cluster structure is shown in Fig. 2.

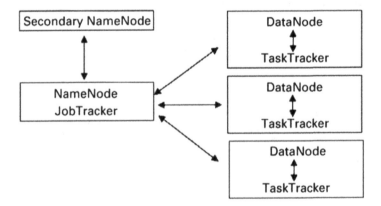

Fig. 2. Structure of Hadoop cluster

Spark is a highly fault-tolerant, memory-based distributed cluster computing framework. Relative to the Hadoop cluster computing framework, Spark cluster computing is stored in memory, and Hadoop is stored on the distributed file storage system HDFS (Hadoop Distributed File System). Hadoop computing process will produce a lot of input and output files, so the processing time is longer, you also need to rely on storage resources. Spark's computations share memory in the form of Resilient Distributed Data sets (RDDs), which are more efficient in iterative or interactive computing. Spark supports more data processing interfaces than Hadoop's MR (Map and Reduce) interfaces that process data. Spark as a distributed computing framework relies on resource containers, early support for Mesos and Yarn as resource containers, and later also supports stand-alone cluster standalone mode as well as Local mode. Which Yarn mode compatible with Hadoop, Spark is conducive to the deployment of Hadoop based on the deployment of the framework to improve performance.

3   The Data Mining Algorithm of Association Rules

3.1   Association Rules

Association Rules is a logical implication of $X = \{x_1, x_2, \ldots, x_m\} \rightarrow Y = \{y_1, y_2, \ldots, y_n\}$, where X and Y are called the precedence and succession of association rules, Association rules suggest that there may be strong relationships between the pilot and the followers [7]. Let $i = \{i_1, i_2, \ldots, i_m\}$ be a set of m different items, given a transaction database D, where each transaction T is a set of items in I, which is TI, T There is a unique identifier TID. If the set X I and XT, then the transaction set T contains the item set X. An association rule is the implication of the form X Y, where XI, YI, $X \cap Y = \emptyset$. (1) It has the support degree s, that is, the transaction with at least s% in the transaction database D contains $X \cup Y$; (2) It has the confidence c, that is, the transaction database D contains X's business has at least c% and also contains Y. The association rule mining problem is to find the association rule with the minimum support minsup and the minimum confidence minconf in the transaction database D [8]. The association rule mining process consists of two phases:

(1)   to find out that there is support for all strong itemsets X in the transaction database support (X) is not less than the minimum support minsup given by the user, then X is the large item set.

(2)   the use of strong items to generate association rules. For each strong set A, if B A, $B \neq$, and support (A)/support (B) $\geq$ minconf, then the associated rule B (A–B).

Definition 1. Support: Suppose rule $x_1 \rightarrow y_1$, support that x1 as a pilot and y1 as a successor in the data set at the same time the probability.

Definition 2. Confidence: Suppose rule $x_1 \rightarrow y_1$, trust degree represents the number of occurrences of x1 as lead and y1 as subsequent successions divided by x1 as the number of occurrences of the pilot.

Definition 3. Frequent set: the support of the rule is not less than the support threshold, the rule is called the frequent set; otherwise, the non-frequent set.

3.2 Apriori Algorithm

In practice, there are many kinds of association rules mining algorithms, among them, the most classic is Apriori algorithm which is proposed by RAgrawal and RSrikant in 1994 and it is the basic association rules of data mining algorithms. Apriori algorithm for many areas, such as recommended systems, social networks, e-commerce platform. Apriori algorithm uses "in a given transaction database D, any subset of the strong set of items are strong items set; any weak sets of superset are weak items" This principle of the transaction database for multiple scans, the first scan (Ie, the large k-1 term set L k-1) and the function Apriori-gen are used to generate the candidate for the k-1-second scan before the k (k > 1) And then determine the number of support for each element in Ck during the scanning process. Finally, the large k - item set L_k is calculated at the end of each scan. When the candidate k - item set C_k is empty When the end [9]. The frequent item sets of the algorithm are generated as follows:

Input: data set, minimum support threshold minSup, minimum confidence threshold minConf;

Output: association rules;

(1) The data of the data set by line scan, combined with minSup, produce frequent 1-item set, and the initial k = 1;

(2) Under the action of L_k cut and connected to produce a collection of Ck + 1. Combined with minSup, resulting in frequent (k + 1) - item sets;

(3) If L_k + 1 is not empty, then k is incremented by 1, and step (2) is performed again. Otherwise,

(4) Get all the frequent itemsets from step (3). Combined with minConf, generate strong association rules, the algorithm ends.

4 The Improved Apriori Algorithm

The main problems in the practical application of Apriori algorithm are as follows: The more times the database is scanned; the longer the operation time is. each stage of the Ck is too large; can't be updated, can't directly deal with numerical data; database The association rule mining can't be applied directly; performance and efficiency are low. Therefore, based on the Apriori algorithm, this paper proposes an improved Hadoop platform based distributed association rule mining algorithm.

Spark is a general-purpose, large-scale data processing engine that primarily provides an abstract object based on memory computing RDD, allowing users to load data into memory and use it repeatedly [10]. Spark programming model reference MapReduce, the difference is that Spark memory-based computing features in some applications on the experimental performance more than MapReduce100 times. Spark platform written entirely by the Scala language, Scala is a fusion of object-oriented and functional programming language, Which is specifically designed for distributed, streamlined and concurrency.

The basic operations of RDD include transformation and action. You can construct a new RDD from a Scala collection or a Hadoop data set or generate new RDDs from existing RDDs, such as map, filter, groupBy, and reduceBy [11]. Action is calculated by RDD to get one or a group of values, such as count, collect and save. Spark all the conversion is inert, not directly calculate the results, just remember to apply to the basic data set (such as a file) on These conversion actions. Only when a request to return to the results of the Driver action, these conversions will really run, this design allows Spark to run more effectively.

4.1 The Apriori Algorithm Based on Hadoop

The idea of matrix based Apriori algorithm is implemented on Hadoop. The concept of matrix is introduced. It only needs 2 times to scan data set D, combined with Spark technology framework, and improves the efficiency of association rule mining by using local pruning property and global pruning property to improve the process of frequent itemsets generation [12]. Transaction data set and frequent item sets Hadoop-based HDFS file system. The matrix of the behavior of the transaction set, the matrix as a collection of items, vector matrix storage variables of 0 and 1, can reduce the data storage space, reduce the number of scanning, According to the operation rules of the vector, the support of the item set can be generated quickly by using the AND operation in the matrix. According to the Spark internal mechanism, the entire Spark programming framework is based on the operation of RDD [13], the specific algorithm description as follows:

(1) Scan the transaction database, seeking a set of 1 sets of frequent.

(2) The transaction database stored on HDFS is an RDD, RDD is split into n data blocks, and these data blocks are allocated to m work nodes for processing.

(3) Construct a local matrix. Let Di be a data block in the transaction database ($1 \leq i \leq n$). Assume that the number of L1s is H and the number of transactions in Di is J, The H $\times$ (J + 2) matrix Gi is constructed by L1 and Di, where the first column is the term in L1, The last column is the support count for the corresponding item in L1, The remaining columns in Gi are the transactions T in Di, If there is an item corresponding to L1 in T, the corresponding position is set to 1, otherwise it is set to 0.

(4) Use Gi to generate local support for candidate item sets, The candidate set containing k items is the set of k items in the first column of Gi, It is only necessary to perform the AND operation on the corresponding k rows in Gi to calculate the local support for the set of candidates containing k items. In the case of a k candidate set, you only need to "OR" with the row of the last column greater than k, Can greatly reduce the number of candidate items. Using local pruning properties, delete the local support degree is less than the threshold of the item.

(5) Use the ReduceByKey operation to get the global support for the candidate set. If the global support is greater than the support threshold, the item is added directly to the frequent item set L; If the global support is less than the support threshold, the global pruning property is used to scan the database again, and finally the set of frequent item sets L.

4.2   The Implementation of Apriori in Hadoop Platform

Transplant the Apriori algorithm to the Hadoop platform based on the Spark programming framework [14], using the Scala language program, the distributed Apriori algorithm to achieve the pseudo-code is as follows:

Input: Data set D (stored as a data block in Hadoop's distributed file system), minimum support threshold min_sup.
Output: The set of frequent itemsets in D.
(1) Solving L_1
instans=sc.textfile(D);
L1=instans.map(_,1).reduceByKey(_ + _).filter(_ > min_sup);
(2) Construct a local matrix
Matrix G = Ø; // initialize the H × (J + 2) matrix
foreach(1 in L1){}
foreach(t in Di){}
if(1 in t)
    G.add (1); // If the item 1 in l is in transaction t, the corresponding position is 1
else
    G.add (0); // otherwise, the corresponding position 0
(3)To find the local candidate set
for (1 < k< max L) {//maxL is the number of rows of matrix G
    for (0 = <m <maxL) {
        Count = 0;
        for (m <n <maxL) {
            while (count <k) {// count count, make sure k lines do "with" operations
                // The last column value is less than k, no "AND" operation
                if (G [m] [maxL-1] <k)
                    break;
                else
                    Count + = 1; }
            Local_sup_count = [use "AND" operation on 'k column items'of G];
            Ck.add (<k column items, local_sup_count>);
}}}
(4) calculate the global support, get frequent item sets L
/ * Apply global pruning policies to items with global support less than minimum support, traverse transaction databases for pruning */
    GCk = Ck.reduceByKey (_ + _). Filter (_. 2 <min_sup);
// items with global support greater than minimum support are added directly to frequent itemsets
    L = instans.map (_, gCk) .reduceBykey (_ + _). Filter (_ > min_sup) ;
    L + = Ck.reduceByKey (_ + _). Filter (_. 2> min_sup) .add (k items, sup_count);
    return L;

5   Experimental Results and Analysis

In order to verify the performance of the algorithm, Six nodes in the cluster environment are deployed in the laboratory LAN. the software list as follows: operating system centos 7.2, Java version jre-8.0, Hadoop version hadoop-2.7.2, using Eclipse neon configured MapReduce distribution Programming environment. The test data is based on scrapy crawling the day cat electric business platform in the first quarter of 2017 clothing category user shopping dataset. According to the distributed Apriori algorithm, the Map function, Combiner function and Reduce function are written, and the distributed Apriori algorithm is compared with the traditional Apriori algorithm. The comparison results are shown in Figs. 3 and 4, where the abscissa indicates the amount of data and the ordinate represents the calculation time.

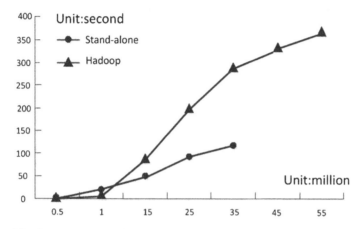

Fig. 3.   Apriori algorithm and distributed Apriori algorithm comparison

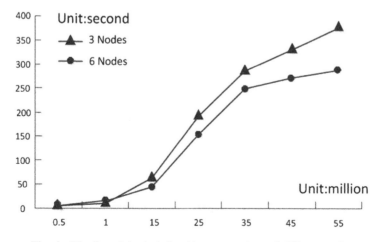

Fig. 4.   Distributed Apriori algorithm comparison of different nodes

Compared with the experimental data, the distributed Apriori algorithm has no obvious advantage, but with the increase of the data, the time of the two algorithms is linearly increasing. When the amount of data grows to 350,000, the advantages of the distributed Apriori algorithm begin to appear. At this time, the Apriori algorithm can't continue to run because of insufficient memory, and the distributed Apriori algorithm can continue to run and the slope of the time change becomes slow. So As the amount of data increases, the operation time slows down.

Figure 4 Comparison of experimental data Note: In the case of the same amount of data, DataNode nodes increase, you can reduce the algorithm's computing time. In theory, when the number of nodes doubled, the operation time should be halved, but the actual situation is not the case. The reason why the operation log analysis is combined with the system may be: in the reduce stage of the algorithm, each node needs to calculate the support degree of each candidate set, different calculation amount of different candidate sets and different number of frequent item sets, Each node has a different running time. But on the whole, increase the number of nodes can effectively reduce the system running time.

6 Conclusion

In this paper, an improved mining algorithm of association rules is proposed based on Apriori algorithm. And the implementation of Apriori algorithm is described in Hadoop platform, which proves the advantages of the distributed association rule algorithm in dealing with big data. With the rapid growth of cloud data, the traditional data classification algorithm based on statistics and machine learning method in the handling of big, heterogeneous and complex Web data, the use of ordinary association rules algorithm on the server test is very serious, based on The Hadoop platform's distributed association rule algorithm can be used to effectively allocate these tests on a cluster machine. Experimental results show that the algorithm is simple and easy to implement, and can effectively improve the efficiency of association rules mining.

Acknowledgements. The work is supported in part by Department of Education of Guangdong Province under Grant 2015KQNCX188.

References

1. Rai, N., Jain, S., Jain, A.: Mining interesting positive and negative association rule based on improved genetic algorithm. Int. J. Adv. Comput. Sci. Appl. 5(1), 160–165 (2014)
2. Gupta, M.K., Sikka, G.: Association rules extraction using multi-objective feature of genetic algorithm. In: Proceedings of the World Congress on Engineering and Computer Science, pp. 23–25 (2013)
3. Zhao, L., Jin, X., Sun, L., et al.: Association rule mining method based on niching genetic algorithm. Comput. Eng. 34(10), 163–165 (2013)
4. Lammel, R.: Google's MapReduce programming model - revisited. Sci. Comput. Prog. (S0167-6423) 70(1), 1–30 (2012)

5. Mccreadie, R.M.C., Macdonald, C., Ounis, I.: On single-pass indexing with MapReduce. In: Association Computing Machinery, New York, USA (2009)
6. He, B., Yang, K., Fang, R., Lu, M., Govindaraju, N.K., Luo, Q., Sander, P.V.: Relational joins on graphics processors. In: ACM SIGMOD 2014 (2014)
7. Mei, S., Kun, Z.: Big-data analytics: challenges, key technologi and prospects. ZTE Commun. **11**(2), 11–17 (2013)
8. Yang, X., Liu, Z., Fu, Y.: MapReduce as a programming model for association rules algorithm on Hadoop. In: The 3rd International Conference on Information Sciences and Interaction Sciences (ICIS), pp. 99–102 (2010)
9. Dean, J., Ghemawat, S.: MapReduce: a flexible data processing tool. Commun. ACM **53**(1), 72–77 (2013)
10. Lin, J., Dyer, C.: Data-Intensive Text Processing with MapReduce (2010)
11. Yadav, C., Wang, S., Jkumar, M.: An approach to improve Apriori algorithm based on association rule mining. In: 2013 Fourth International Conference on Computing, Communications and Networking Technologies (ICCCNT), pp. 1–9. IEEE, USA (2013)
12. Abadeh, M.S., Hamid, M., Jafar, H.: Design and analysis of genetic fuzzy systems for intrusion detection in computer networks. Expert Syst. Appl. **38**, 7067–7075 (2014)
13. Wu, K., Hao, J., Wang, C.: Application of fuzzy association rules in intrusion detection. In: International Conference on Internet Computing and Information Services, pp. 269–272 (2011)
14. Wu, Y., Qin, Y., Song, J.: Research overview of intrusion detection algorithms based on association rules. Comput. Eng. Des. **32**(3), 834–838 (2013)

Data Mining – Data Management Platforms

Data-Driven Phone Selection for Language Identification via Bidirectional Long Short-Term Memory Modeling

Xiao Song[1,2], Qiang Cheng[3], Jingping Xing[4], and Yuexian Zou[1(✉)]

[1] ADSPLAB/Intelligent Lab, SECE, Peking University, Shenzhen, China
zouyx@pkusz.edu.cn
[2] PKU Shenzhen Institute, Shenzhen, China
[3] Shenzhen Press Group, Shenzhen, China
[4] Shenzhen Securities Information Co., Ltd, Shenzhen, China

Abstract. In this paper, we propose a new phone selection method to select more suitable phones with higher score for language identification (LID), which is more similar to target language. A data-driven approach is developed for the phone selection to avoid using complex semantic knowledge which benefits from significant reduction in the manual cost of learning different languages. Recently, bidirectional long short-term memory (BLSTM) can provides more accurate content frame alignments with sequence information from longer duration, which has improved automatic speech recognition (ASR) performance. In principle, the output of BLSTM based ASR contains more candidates in form of phone lattice, which can reduces adverse effect of many practical factors, such as variations of channels, noises and accents. Therefore, initial phones sequences are extracted from phone lattice firstly which are generated by speech recognition results of BLSTM based ASR system. Second, asymmetrical distance between each phone and target language is proposed and then applied to weight the initial phones sequences. Accordingly, language-related phones are selected from the weighted phones. Finally, the selected phones are used to re-score input sentences for the LID system. Intensive experiments have been conducted on AP16-OLR Challenge to validate the effectiveness of our proposed method. It can be seen from results, these selected phones are more effective to LID than the rest phones. Our method gives improvement up to 39.96% in terms of C_{avg} compared with method without using phone selection.

Keywords: Phone selection · Data-driven
Bidirectional long short-term memory · Language identification

1 Introduction

There are many state-of-the-art language identification (LID) systems still include acoustic modeling [1, 2], even though several high level approaches based on phonotactic and prosody were widely used as meaningful complementary information sources [3, 4]. Recently, method of using i-vector front-end features followed by a classification stage was proposed to compensate speaker and session variability [5, 6].

© Springer Nature Singapore Pte Ltd. 2018
K. Li et al. (Eds.): ISICA 2017, CCIS 873, pp. 301–312, 2018.
https://doi.org/10.1007/978-981-13-1648-7_26

In addition, deep neural networks (DNNs) for LID [7] was shown high preference than that of i-vector based approach. Moreover, long short-term memory (LSTM) with the ability to model sequential data makes it a suitable candidate for acoustic LID systems [8]. The methods discussed previous were only implemented with proprietary DNNs and LSTMs, which are not available for research teams. Besides, the effects of well-known practical factors, such as variations of channel, noise and accent, were not considered in LID systems, which may degrade the performance of the LID systems.

In this paper, we propose an approach of data-driven phone selection method for improving LID results via a non-proprietary bidirectional LSTM (BLSTM). The details include:

(1) The BLSTM is used to build a non-proprietary neural network for modeling acoustic features. While an automatic speech recognition (ASR) system is generated based on both N-gram language model (LM) and the BLSTM acoustic model (AM). The BLSTM provides more accurate content frame alignments with sequence information from longer duration. Obviously, the output of BLSTM based ASR contains more candidates in form of phone lattice [9]. Accordingly, the recognition result can reduce the adverse effects of channel, noises and accents. It is well-known that phone lattice keeping both the scores and time alignment information is an essential part of many state-of-the-art LID system.

(2) A new data-driven phone selection method is proposed to improve the LID results. First, we extract initial phones sequences (speech recognition results) from the phone lattice. Next, different from [10, 11], where [10] used a training data selection method to develop compact models in their hierarchical LID framework, [11] proposed a phonetic unit selection method to represent speech information spoken in a different language, we use asymmetrical distance to determine the phone selection to obtain more "important" phones from the initial phones sequences. When the asymmetrical distance is generated, Gaussian mixture model (GMM) is employed to replace all phones in target language. The GMM can better model and represent the phone set than directly using them in the target language. Besides, the asymmetrical distance is used to weight the initial phones sequences. Finally, selecting some important language-related phones from these weighted phones re-score the input sentences, which produces better LID results. Two other considerations about the data-driven method lie in avoiding to use complex semantic knowledge and to reduce computational cost of learning different languages.

(3) The performance of our method is evaluated on AP16-OLR7 dataset which is provided by the oriental language recognition challenge on APSIPA 2016 (AP16-OLR challenge). It is the benchmark corpus for evaluation of LID [12]. We implement the data-driven phone selection method for LID system (denoted as DDPS-LID) and compare it with the conventional LID system without using phone selection.

The paper is organized as follows. In Sect. 2, we briefly introduce the principle of the LID system with data-driven phone selection (DDPS-LID). In Sect. 3, we present our proposed data-driven phone selection method (DDPS). In Sect. 4, we describe the dataset and evaluation metrics. Experiments and results are showed in Sect. 5. Finally, the conclusions are presented in Sect. 6.

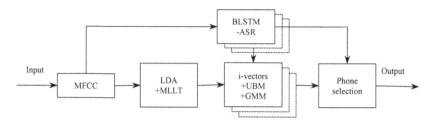

Fig. 1. LID system with phone selection.

2 Proposed LID System

Figure 1 shows the proposed LID system integrated with phone selection module (DDPS-LID). It's noted that acoustic features are needed to compute first. The well-known Mel-frequency cepstral coefficients (MFCC) features are widely used for LID system. In the LID task, the target languages are known in prior, the scores from LID system are comparable across languages and then the highest score refers to the final result of the LID system. Therefore, the score for each language hypothesis will be computed by LID system.

Shown in the top of Fig. 1, BLSTM is used for acoustic modeling and N-gram is used for language modeling. The acoustic model (AM) and language model (LM) are language independent. Considering the limitation of training data, we use phoneme as the modeling units in the AM and LM. The output of the BLSTM-ASR is taken as one of the input to phone selection module.

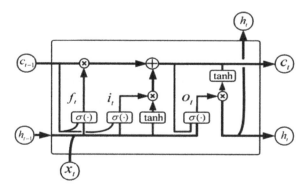

Fig. 2. Bidirectional Long short-term memory (BLSTM) block.

In BLSTM, recurrent connections and special network units called memory blocks contain memory cells with self-connections storing the temporal state of the network which changes with the input to the network at each time step. Figure 2 illustrates a single BLSTM block. In addition, they have multiplicative units called gates to control the flow of information into the memory cell and out of the cell to the rest of the network.

Mathematically, a set of cells can be described by the following forward operations iteratively over time $t = 1, 2, \ldots, T$:

$$i_t = \sigma(W^{(xi)}x_t + W^{(hi)}h_{t-1} + W^{(ci)}c_{t-1} + b^{(i)}) \tag{1}$$

$$f_t = \sigma(W^{(xf)}x_t + W^{(hf)}h_{t-1} + W^{(cf)}c_{t-1} + b^{(f)}) \tag{2}$$

$$c_t = f_t c_{t-1} + i_t \tanh(W^{(xc)}x_t + W^{(hc)}h_{t-1} + b^{(c)}) \tag{3}$$

$$o_t = \sigma(W^{(xo)}x_t + W^{(ho)}h_{t-1} + W^{(co)}c_t + b^{(o)}) \tag{4}$$

$$h_t = o_t \tanh(c_t) \tag{5}$$

Where i_t, o_t, f_t, c_t, and h_t are vectors, with the same dimensionality, which represent 5 different types of information at time t of the input gate, output gate, forget gate, cell activation, and hidden layer, respectively [13]. $\sigma(\cdot)$ is the logistic sigmoid function, W's are the weight matrices connecting different gates, and b's are the corresponding bias vectors. All the weight matrices are full except the weight matrix $W^{(ci)}$ is diagonal.

Shown in the bottom of Fig. 1, after the Linear Discriminant Analysis (LDA) and Maximum Likelihood Linear Transform (MLLT) module, new features can be obtained. Then the joint Universal Background Model (UBM), i-vectors and GMM are determined accordingly, which are taken as another input to phone selection module.

For each language, those previous steps as the input to the phone selection module, key module, are completed separately. It is noted that, with unknown input utterances, the BLSTM-ASR module generates the phones sequences across each language hypothesis in the form of phone lattice, which are termed as the initial phones sequences. Therefore, initial frame-by-frame scores can be determined from the initial phones sequences. Based on language hypothesis, the distance between each initial phone and the language can be calculated. Then, the phone selection module utilizes the distances calculated above to weight the initial frame-by-frame scores of the initial phones sequences. Finally, select some phones with higher scores to re-score the input sentence, which is the final output score of the input sentence.

3   Proposed Data-Driven Phone Selection Method

As shown in Fig. 1, the phone selection module is a key for DDPS-LID system. In this section, the proposed DDPS method is introduced in details.

3.1 Phone Lattice Generation

In order to describe the principle of phone selection module, we firstly need to understand the phone lattice generation in a LID system. Actually, there are various definitions of phone lattice in literature. Researches [14, 15] show that there is a common point that the alignment and scores information are correct in the phone lattice and the completeness of such information (i.e. no high scores phones sequences are missing). Followed by the method proposed, in [9], the phone lattice is obtained by Kaldi toolkit.

3.2 Asymmetrical Distance Between Phones and Target Languages

For selecting the phones from the initial phones sequences, we need to determine a language-related distance. A distance metric is proposed to measure the distance between the input sentences and the target languages.

First, the distance $D(ph_a, ph_b)$ between two phones is defined as:

$$D(ph_a, ph_b) = -\log P(ph_a \mid ph_b) \tag{6}$$

where ph_a and ph_b are phone a and phone b, respectively. $P(ph_a \mid ph_b)$ is a conditional probability.

Second, the asymmetrical distance $D(ph_a, L_j)$ between a phone and a target language is defined as:

$$D(ph_a, L_j) = -\frac{1}{N_{ph(L_j)}} \sum_{k=0}^{N_{ph(L_j)}} \log P(ph_a \mid ph_k) \tag{7}$$

where L_j is target language j, $N_{ph(L_j)}$ is the number of phones in target language j.

Third, we can obtain the asymmetrical distance $D(utt_i, L_j)$ between an input sentence and a target language as follows:

$$D(utt_i, L_j) = -\frac{1}{N_{ph(utt_i)} \times N_{ph(L_j)}} \sum_{m=0}^{N_{ph(utt_i)}} \sum_{k=0}^{N_{ph(L_j)}} \log P(ph_m \mid ph_k) \tag{8}$$

where utt_i is i-th sentence, $N_{ph(utt_i)}$ is the number of phones in i-th sentence.

Finally, we use a GMM of target language to replace all of the phones in this target language. This way can avoid learning all phones of every target language. Instead, the GMM can better model and represent the phone set. So Eq. (8) is simplified as follows:

$$D(utt_i, L_j) = -\frac{1}{N_{ph(utt_i)}} \sum_{m=0}^{N_{ph(utt_i)}} \log P(ph_m \mid GMM(L_j)) \tag{9}$$

where $GMM(L_j)$ represents the GMM of target language j.

3.3 Proposed Data-Driven Phone Selection Method

The proposed DDPS method consists of four steps:

(1) using Eq. (9) to calculate the asymmetrical distances between an input sentence and a target language;
(2) a *weight* factor is proposed to weight the initial phones sequences, thus, *weighted phones* can be obtained for the input sentence;
(3) the important language-related phones are selected from the weighted phones to obtain *selected phones*, which are used to represent the input sentence;
(4) the input sentence is re-scored by the selected phones to obtain *final score* of the input sentence.

Specifically, the *weight* factor and the *final score* of input sentence are computed in Eqs. (10) and (11), respectively.

$$
\begin{aligned}
W(utt_i, L_j) &= \frac{1}{D(utt_i, L_j)} \\
&= \{w_1(utt_{(i,1)}, L_j), w_2(utt_{(i,2)}, L_j), \ldots, w_k(utt_{(i,k)}, L_j), \ldots, w_{N_{ph(utt_i)}}(utt_{(i,N_{ph(utt_i)})}, L_j)\} \quad (10) \\
&= \{\frac{1}{d_1(utt_{(i,1)}, L_j)}, \frac{1}{d_2(utt_{(i,2)}, L_j)}, \ldots, \frac{1}{d_k(utt_{(i,k)}, L_j)}, \ldots, \frac{1}{d_{N_{ph(utt_i)}}(utt_{(i,N_{ph(utt_i)})}, L_j)}\}
\end{aligned}
$$

$$
Score(utt_i, L_j) = \frac{\sum_{k=0}^{N_{ph(utt_i)}} \alpha w_i(utt_{(i,k)}, L_j) * Ph\_Score(utt_{(i,k)}, L_j)}{\sum_{k=0}^{N_{ph(utt_i)}} \alpha w_i(utt_{(i,k)}, L_j)} \quad (11)
$$

where $W(utt_i, L_j)$ is the weight factor of i-th sentence and language j. $Score(utt_i, L_j)$ is the final score of i-th sentence and language j. $utt_{(i,k)}$ is the k-th phone of i-th sentence. $w_i(utt_{(i,k)}, L_j)$ is the *weight* factor of k-th phone of i-th sentence and language j, which is the derivative of $d_i(utt_{(i,k)}, L_j)$. $d_i(utt_{(i,k)}, L_j)$ is the distance of k-th phone of i-th sentence and language j. $Ph\_Score(utt_{(i,k)}, L_j)$ is the initial score of k-th phone of i-th sentence and language j, which is extracted from the phone lattice generated by the BLSTM-ASR system of language j. The parameter α is given as:

$$
\alpha = \begin{cases} 1, & \text{if } utt_{(i,k)} \in U_{\widetilde{(i)}} \\ 0, & \text{otherwise.} \end{cases} \quad (12)
$$

where $U_{\widetilde{i}}$ is the set of the selected phones in i-th sentence. $\alpha = 1$ represents the $utt_{(i,k)}$ is selected, otherwise the $utt_{(i,k)}$ is not selected.

3.4 Proposed DDPS-LID System

As discussed above, the initial phones with higher score associated with language L_j will lead to that the i-th input sentence is classified as language L_j greatly. Therefore, the selection operation is necessary.

Our proposed DDPS-LID system includes following steps:

(1) the initial phones sequence is generated by the BLSTM-ASR of language L_j in the form of phone lattice for the i-th input sentence;
(2) the initial score, $Ph\_Score(utt_{(i,k)}, L_j)$, is extracted from the phone lattice;
(3) the scores of the i-th input sentence, $Score(utt_i, L_j)$, are obtained across each language hypothesis by the four steps of the proposed DDPS method described as in Sect. 3.3;
(4) the scores of the i-th input sentence across each language hypothesis are comparable and the highest score is the final LID result for the i-th input sentence.

4 Datasets and Evaluation Metrics

4.1 Datasets

In this study, AP16-OL7 database is used to evaluate our proposed DDPS-LID system, which is provided by SpeechOcean[1] in the AP16-OLR challenge [12]. The multilingual database includes seven oriental languages, 71 h of speech signals in total: Cantonese in China mainland and Hong Kong (ct-cn), Indonesian in Indonesia (id-id), Japanese in Japan (ja-jp), Korean in Korea (ko-kr), Russian in Russia (ru-ru), Vietnamese in Vietnam (vi-vn), Mandarin in China (zh-cn), where original scripts and lexicon are available. 24 speakers are included in each language, 18 of them are selected as training set ($\sim$ 8 h) and the other 6 are in test set ($\sim$ 2 h), where male and female are balanced. The details are given in Table 1.

Table 1. AP16-OL7 data profile.

Datasets		Training set	Testing set
Code	Description	Hours	Hours
ct-cn	Cantonese in China Mainland and Hong Kong	7.71	2.48
id-id	Indonesian in Indonesia	7.48	3.16
ja-jp	Japanese in Japan	5.82	2.16
ko-kr	Korean in Korea	5.99	1.92
ru-ru	Russian in Russia	9.92	2.99
vi-vn	Vietnamese in Vietnam	8.46	2.94
zh-cn	Mandarin in China	7.66	2.63

---

[1] www.speechocean.com.

4.2   Evaluation Metrics

In order to assess the performance of LID systems, three different metrics are used. As the main error measure to evaluate the capabilities of one to all language detection, C_{avg} (average cost) is used as defined in the AP16-OLR challenge evaluation plan (as in LRE15). C_{avg} is a measure of the cost of taking bad decisions, and thus, it considers not only discrimination, but also the ability of setting optimal thresholds.

Further, metric Equal Error Rate (EER) is used to evaluate the performance, which considers only scores of each individual language. The EER metrics are not related to the decision result, but the quality of the scoring. Therefore it evaluates the verification system from different angle. Detailed information can be found in LRE'09 evaluation.

In the AP16-OLR challenge, the target languages are known in prior, and the scores are comparable across languages, which means that OLR can be treated as a language identification task, for which the language obtaining the highest score in a trail is regarded as the identification result. For such an identification task, identification recognition results (IDR) is a widely used metric, which treats errors on all languages equally serious.

5   Experimental Settings and Results

5.1   Experimental Settings

We used BLSTM layers with recurrent and non-recurrent projections for ASR as suggested in kaldi. The neural networks were created using alignments from GMM system trained using Kaldi. Following this, a GMM was trained firstly with standard 13-dimensional MFCC features, using Hann windows of 25-ms frames shifted by 10 ms each time for each target language $(GMM(L_i),\ i \in \{1,2,3,...,7\})$. The forced alignment given by a GMM system was used to create frame-level acoustic targets. At the same time, the GMM is adopted to replace the set of phones in the target language for calculating the distance between an unknown input sentence with a known target language (described in Sects. 3.2, 3.3 and 3.4). Recognition is performed by combining the acoustic probabilities yielded by the network with the state transition probabilities from the HMM and the word transition probabilities from the LM, which can be done efficiently for speech using weighted finite state transducers.

The BLSTM had three hidden layers, with 640 cells in each layer. BLSTM was trained by using stochastic gradient descent (SGD) with one weight update per utterance, and the truncated back propagation through time (BPTT) learning algorithm. Mealwhile, learning rate ranging from 5×10^{-4} to 5×10^{-5} that are exponentially decayed during training was used. At each layer, a constant delay of -3 (or $+3$) along both directions was adopted.

In phonotactics modeling, each different phonotactics N-gram language model was trained using the phonetic sequences of the challenge training data for each tokenizer. For that purpose, the SRILM toolkit had been used and 4-gram back-off models smoothened using Kneser-Ney discounting are obtained. Finally, for an input speech, each BLSTM model was evaluated and log-likelihood scores were obtained.

5.2 Experimental Results

In our proposed DDPS method, the key points are that how to select phones and how many phones selected from all of the weighted phones in an unknown input sentence. We used an example to explain these two key points, showed in Tables 2 and 3, respectively.

Table 2. An example of the scores of some weighted phones in one input sentence under different language *hypothesis*.

	ik1	eon2	d	ing3	il	n
ct-cn	25.7	45.3	13.7	16.4	13.5	32.2
	"e	"oj	dZ	"i	n	"i
id-id	4.1	6.5	4.3	7.5	2.9	7.5
	ie3	ong4	d	shi4	n	an4
zh-cn	27.9	27.6	0.3	46.3	13.4	27.4

As an example shown in Table 2, the ground-truth of the input sentence is language "ct-cn", and the scores of initial phones sequences were weighted under different language hypothesis (shows scores of some weighted phones). In fact, the scores between each phone and each language are different. For each language hypothesis, we selected the weighted phones with relatively higher scores. If just selecting one phone, the phone *eon2* could be selected under language "ct-cn" hypothesis. Because that the phone *eon2* has the highest score compared with other weighted phones in the input sentence under language "ct-cn" hypothesis. If selecting two phones, the top two phones *eon2* and *n* could be selected under language "ct-cn" hypothesis.

Table 3. The different scores in sentences level for the input sentence and each language hypothesis by using the data-driven phone selection method

Lan.	Scores of one utterance by choosing various numbers of weighted phones					
	Top 2	Top 4	Top 6	Top 8	Top 10	All
ct-cn	38.8	31.5	26.0	**22.7**	18.2	15.3
id-id	7.1	6.2	5.5	4.9	4.38	4.1
ja-jp	12.0	11.2	10.1	9.12	7.92	6.9
ko-kr	10.7	8.9	8.0	7.1	6.43	5.6
ru-ru	**55.2**	**36.9**	**26.4**	20.8	17.2	14.9
vi-vn	12.7	10.6	9.25	8.4	7.42	6.4
zh-cn	37.2	30.3	26.0	22.6	**18.7**	**15.6**

Table 3 shows the different scores in sentences level for the input sentence and each language hypothesis by using the DDPS method. From Table 3, when selecting the weighted phones with top two weighted scores in all phones, we can see that the highest score of the whole sentence is 55.2 compared with under each language hypothesis. And the recognition result for this utterance is language "ru-ru". However, with increasing the number of the selected weighted phones, we can find the result is changed. Selecting top eight weighted phones obtained "ct-cn" result, where the highest score of the sentence is 22.7 under language "ct-cn" hypothesis, which is the correct result. But selecting top ten weighted phones obtain "zh-cn" result, where the highest score of the sentence is 18.7 under language "zh-cn" hypothesis. Therefore, the selected "degree" of the weighted phones was explored on average performance, showed in Fig. 3.

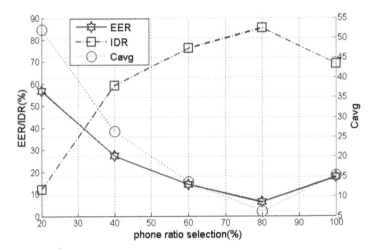

Fig. 3. Performance (EER, IDR and C_{avg} on average) by using the proposed data-driven phone selection method, where selecting various proportion of the weighted phones shows the different results.

Figure 3 shows the average performance on three evaluation metrics EER, IDR and C_{avg} (described in Sect. 4.2). From Fig. 3, we can discover that when choosing the top 20% of the weighted phones in the input sentences, the results are very bad, no matter what the metric is. With the increasing of the proportion of selecting weighted phones, the results are improved. When selecting 70% $\sim$ 85% of the phones, the performance become better. However, selecting more than 85% of phones, especially when using all phones, the result is worse than the use of partial selected weighted phones.

Table 4 summarizes the results of some teams participating in the AP16-OLR challenge and our results obtained in terms of C_{avg}, EER and IDR (on average). For each target language, by using the proposed method, we obtained better result than our submitted in the challenge (our submitted result is obtained by using the initial scores). We can also find that the best result is Haizhou Li team from Singapore. We didn't

Table 4. Methods performance on AP16-OL7

Methods	C_{avg}	EER%	IDR%
Haizhou Li,Singapore	1.13	1.09	97.56
NTUT,Taiwan,China	5.86	5.88	87.02
MMCL_RUC,China	6.06	6.16	86.21
NTU,Singapore	14.72	17.44	71.44
Our-submitted	36.99	40.26	31.91
TLO,China	50.00	53.34	12.37
No phone selection	15.29	17.63	69.06
Phone selection 60%	13.65	14.28	76.22
Phone selection 80%	6.11	6.31	85.39

exceed the state-of-art team, even break through the third team (C_{avg} = 6.06). But, their methods were proprietary and complex [16]. Compared with them, we propose a non-proprietary, simpler solution in LID task.

6 Conclusions

In this paper, we proposed a data-driven phone selection approach for language identification (LID). The bidirectional long short-term memory (BLSTM) in automatic speech recognition (ASR) was designed to obtain the phone lattice for its better speech recognition capability. Next, a data-driven phone selection method was proposed where asymmetrical distance between each phones and target languages was used to weight the phones from lattice. These weighted phones were used to re-score the input sentence which was the final score of the input sentence. Finally, intensive experiments had been conducted to evaluate the proposed method on the AP16-OLR challenge. Our method gave an improvement of 39.96% in terms of C_{avg} with respect to the LID without using phone selection method. In the future, we will further evaluate the quality of the selected phones for improving the performance of LID systems.

Acknowledgements. This work is partially supported by Key Technologies Research & Development Program of Shenzhen (No: JSGG20150512160434776) and Key Technologies Research & Development of Data Retrieval and Monitoring via Multi-layer Network (No: JSG G20160229121006579).

References

1. Torres-Carrasquillo, P.A., Singer, E., Gleason, T., McCree, A., Reynolds, D.A., Richardson, F., Sturim, D.E.: The MITLL NIST LRE 2009 language recognition system. In: Acoustics Speech and Signal Processing (ICASSP) IEEE International Conference on 2010, pp. 4994–4997 (2010)
2. Gonzalez-Dominguez, J., Lopez-Moreno, I., Franco-Pedroso, J., Ramos, D., Toledano, D.T., Gonzalez-Rodriguez, J.: Multilevel and session variability compensated language recognition: ATVS-UAM systems at NIST LRE 2009. IEEE J. Sel. Top. Sig. Proc. **4**(6), 1084–1093 (2010)

3. Ferrer, L., Scheffer, N., Shriberg, E.: A comparison of approaches for modeling prosodic features in speaker recognition. In: International Conference on Acoustics, Speech, and Signal Processing, pp. 4414–4417 (2010)

4. Martinez, D., Lleida, E., Ortega, A., Miguel, A.: Prosodic features and formant modeling for an ivectorbased language recognition system. In: Acoustics, Speech and Signal Processing (ICASSP) IEEE International Conference on 2013, pp. 6847–6851 (2013)

5. Dehak, N., Torres-Carrasquillo, P.A., Reynolds, D.A., Dehak, R.: Language recognition via i-vectors and dimensionality reduction. In: Interspeech ISCA, pp. 857–860 (2011)

6. Martinez, D., Plchot, O., Burget, L., Glembek, O., Matejka, P.: Language recognition in ivectors space. In: Interspeech ISCA, pp. 861–864 (2011)

7. Lopez-Moreno, I., Gonzalez-Dominguez, J., Plchot, O., Martinez, D., Gonzalez-Rodriguez, J., Moreno, P.: Automatic language identification using deep neural networks. In: Acoustics, Speech and Signal Processing (ICASSP) IEEE International Conference on 2014, pp. 5337–5341 (2014)

8. Gonzalez-Dominguez, J., Lopez-Moreno, I., Sak, H., Gonzalez-Rodriguez, J., Moreno, P.J.: Automatic language identification using long short-term memory recurrent neural networks. In: Interspeech, pp. 2155–2159 (2014)

9. Povey, D., Hannemann, M., Boulianne, G., Burget, L., Ghoshal, A., Janda, M., Karafiat, M., Kombrink, S., Motlicek, P., Qian, Y., et al.: Generating exact lattices in the WFST framework. In: Proceedings of ICASSP, pp. 4213–4216 (2012)

10. Irtza, S., Sethu, V., Fernando, S., Ambikairajah, E., Li, H.: Out of set language modelling in hierarchical language identification. In: Interspeech 2016, pp. 3270–3274 (2016)

11. Lopez-Otero, P., Docio-Fernandez, L., Garcia-Mateo, C.: Phonetic unit selection for cross-lingual query-by-example spoken term detection. In: Automatic Speech Recognition and Understanding (ASRU) IEEE Workshop on 2015, pp. 223–229 (2015)

12. Wang, D., Li, L., Tang, D., Chen, Q.: AP16-OL7: a multilingual database for oriental languages and a language recognition baseline, submitted to APSIPA 2016.pdf

13. Graves, A., Mohamed, A., Hinton, G.E.: Speech recognition with deep recurrent neural networks. In: International Conference on Acoustics, Speech, and Signal Processing (2013)

14. Sak, H., Saraclar, M., Güngör, T: On-the-fly lattice rescoring for real-time automatic speech recognition. In: Interspeech, pp. 2450–2453 (2010)

15. Ortmanns, S., Ney, H., Aubert, X.: A word graph algorithm for large vocabulary continuous speech recognition. Comput. Speech Lang. 11, 43–72 (1997)

16. Irtza, S., Sethu, V., Fernando, S., Ambikairajah, E., Li,H.: Out of set language modelling in hierarchical language identification. In: Interspeech 2016, pp. 3270–3274 (2016)

Multi-document Summarization via LDA and Density Peaks Based Sentence-Level Clustering

Baoyan Wang[1,2], Yuexian Zou[1(✉)], Jian Zhang[3], Jun Jiang[4], and Yi Liu[2]

[1] ADSPLAB/Intelligent Lab, School of ECE, Peking University, Beijing, China
zouyx@pkusz.edu.cn
[2] PKU Shenzhen Institute, Shenzhen, China
[3] Dongguan University of Technology, Dongguan, China
[4] Shenzhen Press Group, Shenzhen, China

Abstract. In this paper, we present a novel unsupervised extractive multi-document summarization method by ranking sentences based on the integrated sentence scoring method. The cluster-based methods tend to ignore informativeness of words and Latent Dirichlet Allocation (LDA) based methods are inclined to extract the longish sentences and cannot remove redundancy directly. Those methods select sentences with higher score to generate summaries but not necessarily to the optimal summaries. Our method takes four key issues of sentences into account concurrently by applying LDA to calculate term weighting of words and evaluate the informativeness of sentences and then applying Density Peaks Clustering (DPC) to assess relevance and diversity of sentences simultaneously. Our method achieves the best property on the DUC2004 dataset, which outperforms the state-of-the-art methods, such as DUC2004 Best, R2N2_ILP [3], and WCS [13].

Keywords: Multi-document summarization
The integrated sentence scoring method · Latent Dirichlet Allocation
Density Peaks Clustering

1 Introduction

Given the exponential rate of the volume of information data overload on the WWW, consumers are flooded with a variety of electronic documents i.e. news, blogs, e-books. Therefore, there are urgent requirements for multi-document summarization (MDS) now more than ever, as it focuses on creating a condensed and informative summary for the large set of original documents that facilitates readers quickly grasp the general information of them. Most existing researches are extractive methods, which aim at extracting sentences firsthand from the given documents and assembling sentences together to generate the final summary. In this work, we address the task of generic extractive summarization from multiple documents. An effective summarization method always considers the four key issues properly: [1–3] Relevance, Diversity, Informativeness and Length Constraint.

© Springer Nature Singapore Pte Ltd. 2018
K. Li et al. (Eds.): ISICA 2017, CCIS 873, pp. 313–323, 2018.
https://doi.org/10.1007/978-981-13-1648-7_27

The methods of extractive-based summarization can be classified into two categories: supervised methods that depend on the provided document-summary pairs, while unsupervised ones based upon the properties derived from document sets. On one hand, the supervised methods generally tend to regard the summarization task as a classification or regression issue [1, 4]. For those supervised methods, a fair amount of annotated data is required, which are costly and time-consuming. On the other hand, the unsupervised methods are continued to be researched. They tend to score and then rank sentences based on semantic, linguistic or statistic grouping extracted from the original documents. Typical existing methods include graph-based methods [5], matrix factorization based methods [6], submodular functions based methods [7], topic model based methods [8, 9], etc. Some of them might just consider one or more key issues, which can be improved farther. Some papers consider reducing the redundancy to hold the diversity of summary [10, 11], i.e. MMR.

It is also appropriate and natural to process MDS with clustering. The cluster based methods [12, 13] tend to divide sentences into groups through clustering method and then rank the sentences on the basis of their saliency scores. [14] ranked sentences with Density Peaks Clustering (DPC) [17] ignoring the informativeness of words and selected sentenced based greedy algorithm, which cannot guarantee the optimal summary. [12] presented their studies about document summarization using the notion of Latent Dirichlet Allocation (LDA) as the representation of documents and mixture models to capture the topics and pick up the sentences, which tend to select longish sentences and cannot remove redundancy directly. Inspired by the applications of cluster-based methods for MDS and LDA for topic model, we propose an integrated sentence scoring method based on LDA combined with DPC to extract sentences with more informativeness, higher relevance, and better diversity under the limitation of length for sentences ordering. Different from their work, our main contributions are:

(1) LDA combined with DPC are firstly adopt for unsupervised multi-document summarization, which improve the performance mutually.
(2) We put forward the integrated sentence scoring method with better scalability to rank sentences, which can be interpreted more intuitively.
(3) We leverage the information of three levels: term level, sentence level and cluster level, which are more comprehensive.

The structure of this paper is as follows. In Sect. 2 we present the integrated sentence scoring method based on LDA and DPC for MDS in detail. Section 3 gives the evaluation of our method on the open benchmark datasets DUC2003 and DUC2004. In Sect. 4, we conclude the paper with directions for the future study.

2   Proposed MDS Method

We propose a new MDS method termed as the integrated sentence scoring method, which use LDA and DPC to take relevance, diversity, informativeness and length constraint into account simultaneously. Sentences are scored in the four aspects, and then the scores are log linearly combined. Finally, the sentences are extracted to generate optimal summary based on dynamic programming algorithm.

2.1 Pre-processing

In order to represent sentences rationally and reprocess them expediently, it is indispensable to carry out t preprocessing step. After the given collection of English documents, $C_{corpus} = \{d_1, d_2,..., d_i,..., d_{cor}\}$, in which d_i denotes the i-th document in C_{corpus}, splitting apart into individual sentences, $S = \{s_1, s_2, ..., s_i, ..., s_{sen}\}$ where s_i means the i-th sentence in C_{corpus}, all words stemming is performed by Porter's stemming algorithm.

2.2 The Integrated Sentence Scoring Method

(1) Informativeness Score
A good summary sentence should contain enough information. LDA is employed to represent sentences and calculate the informativeness. LDA model is viewed as breaking down the set of documents into themes by representing the document as a mixture of themes with a probability distribution representing the significance of the theme for that document. The themes in turn are represented as a mixture of words with a probability representing the importance of the word for that theme [9]. We utilize Gibbs sampling [8] for the LDA parameters inference including the probability distribution $P(T_k|s_i)$ and $P(W_n|T_k)$. Those variables are always sampled once for each word of the given document. $P(W_n|T_k)$ denotes the distribution matrix of the topics over the words, while $P(T_k|s_i)$ presents the distribution matrix of sentences over the topic. The informativeness of a sentence $SC_I(i)$ is calculated by the sum of the weightings of non-stop words in the sentence.

$$P(T_k|C_{Corpus}) = \sum_{i=1}^{sen} P(T_k|s_i) / \sum_{k=1}^{K} \sum_{j=1}^{sen} P(T_k|s_j) \tag{1}$$

$$w_n = \sum_{k=1}^{K} P(W_n|T_k)P(T_k|C_{Corpus}) \tag{2}$$

$$SC_I(i) = \sum_{j=1}^{num_w} w_{ij}, \quad w_{ij} = \begin{cases} w_j & W_j \in s_i \\ 0 & else \end{cases} \tag{3}$$

where w_n is the n-th word's weighting, w_{ij} is the weighting of j-th word in i-th sentence, num_w is the non-stop word's number of the corpus and $P(T_k|C_{corpus})$ presents the probability distribution of topics over the whole corpus.

(2) Relevance Score
We employ the relevance score to quantify the degree concerning how much the relevance is between one with the other sentences of the document set. The DPC proposed

based on two potential hypotheses. One of the potential hypotheses is that cluster centers should be characterized by a higher density than their neighbors. Proceeding from it we consider that a sentence will be deemed to be more relevant and more representational when it possesses higher density, namely of more similar sentences. In our method, the text representation is the vector space model and the representation vector of each sentence generate from (3). The similarity between sentences is calculated by cosine similarity method. Thus we define the function as follows, which calculate the Relevance Score $SC_R(i)$ in sentences level for every sentence s_i:

$$SC_R(i) = \sum_{j=1}^{K} f(Sim_{ij} - \omega), \quad f(x) = \begin{cases} 1 & x \geq 0 \\ 0 & else \end{cases} \tag{4}$$

where K denotes the number of sentences in the set of documents, and Sim_{ij} is the similarity value between the i-th and j-th sentence. ω represents the predefined threshold of density in DPC.

(3) Diversity Score
We show the diversity score to ensure that the sentences of the optimal summary should not be analogical. The set of documents always includes one central theme and some subthemes. The summary should contain the most evident theme beyond doubt. In order to better comprehend the whole document set, it's also necessary to make the sub-themes displayed in the final summary. Put another way, sentences of the summary ought to be less reduplicated with one another in order to eliminate redundancy. Another underlying assumption of DPC is that cluster centers also are characterized by a relatively large distance from points with higher densities. According to that, the scores of the similar sentences can be ensured to obtain larger gap. Furthermore, the sentences with higher diversity scores could be selected in comparison with all the other sentence of the document set. Therefore the diversity of the summary can be guaranteed globally. In our method, the diversity score $SC_D(i)$ is calculated in clusters level as follows.

$$SC_D(i) = \min_{j:SC_R(j)>SC_R(i)} (1 - Sim_{ij}) \tag{5}$$

We set the diversity value of the sentence maximum of $(1 - Sim_{ij})$ conventionally [15], which possesses the highest relevance score.

(4) Length Constraint
The sentence with the longer length usually possesses the more information. Moreover, the human summarizers always are apt to generate masses of shorter sentence for summary. When only consider the informativeness of sentences, the longish sentences tend to be selected, which is incongruent to the human habit. In the real summary system, it is usually restrained of the amount of words. When we select the longish sentences, the number of selected sentences is fewer. As a consequence, it is in sore need of providing the length constraint score. The length of sentences l_i is distributed in a large scale. In this situation, we bring in the taking logarithm smoothing method to solve this issue. Therefore, the length constraint score SC_L is calculated as follows.

$$SC_L = \log(\max_j l_j/l_i + 1) \tag{6}$$

(5) The Integrated Sentence Scoring Method

In order to extract the sentences with more information, higher relevance, and better diversity under the limitation of length, we proposed an integrated sentence scoring method all sidedly considering the four above objectives. For adapting to the integrated sentence score method, $SC_I(i)$, $SC_R(i)$, $SC_D(i)$ and $SC_L(i)$ should be normalized by divided their own highest values firstly.

$$SC(i) = \alpha \log SC_R(i) + \beta \log SC_D(i) + \gamma \log SC_I(i) + \log SC_L(i) \tag{7}$$

where the parameters α, β, and γ of the integrated scoring method are applied to adjust weightings of the four scores. We conduct a series of experiments on the standard datasets to tune and obtain the optimal the parameters settings.

We should generate a summary by extracting sentences under the restriction of the demanded length L. As every sentence is measure by an integrated score, the score sum of extracted sentences in summary should be as high as possible. Therefore the summary generation is considered as the 0–1 knapsack problem.

$$\arg\max \sum \left(SC(i) \times x_i \right)$$
$$Subject\ to\ \sum_i l_i x_i \le L, x_i = \{0, 1\} \tag{8}$$

To alleviate the NP-hard problem, we introduce the dynamic programming (DP) algorithm [23] to extract the sentences from the document set until the required length of ultimate summary is reached.

3   Experimental Setup

3.1   Dataset and Evaluation Metrics

The open benchmark datasets DUC2003 and DUC2004, from Document Understanding Conference, are employed in our experiments. DUC2004 consists of 50 news document sets and 10 documents related to each set. Length Limit of summary is 665 bytes. DUC2003 consists of 60 news document sets and about 10 documents for each set. The structures of both datasets are similar. Therefore, we choose DUC2003 as the development dataset for parameters tuning and DUC2004 for evaluation. There are four human generated summaries provided as ground truth for each news set. We observe that the sentences of summaries are not strictly selected in their entirety, but changed considerably.

We apply widely used ROUGE version 1.5.5 toolkit [16] to evaluate the performance of the summary system in our experiments. We select three Rouge evaluation metrics, Rouge-1, 2, SU, in all of embodied evaluation metrics. Rouge-1 is used to measure the co-occurrence of the identical words between the summary of our method and the anno-tated summary. Rouge-2 is used to measure the co-occurrence of the 2-g while Rouge-SU is applied to measure the co-occurrence of skip-grams. Rouge-2 and Rouge-SU concern more over the readability of the ultimate summary. We show the average value of the F-measure scores of the used metrics in the execute phase. Note that the higher ROUGE scores, the more similar between generated summary and annotated one.

3.2   Parameter Settings

We investigate how parameters α, β and γ, the topic number (T) of LDA and the density threshold ω relate to our method by a series of experiments on DUC2003. The three parameters α, β and γ are set ranging from 0 to 1.5 respectively at the step size of 0.1. The best values of the parameters are selected by comparing all of the results. As shown in Fig. 1, one parameter is tuned while set the others on their best values. The results of tuning parameters are shown in Fig. 1. We find that $\alpha = 0.6$, $\beta = 0.5$ and $\gamma = 0.9$ produce a better performance than $\alpha = 1$, $\beta = 1$ and $\gamma = 1$, which indicates effective degree of the

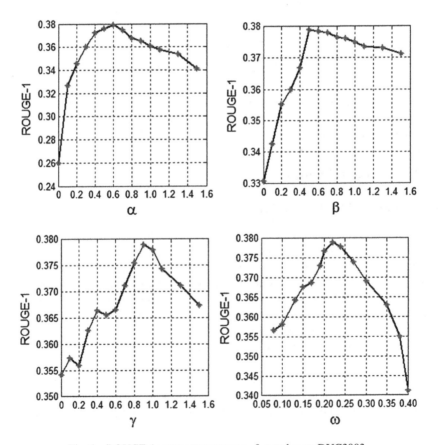

Fig. 1.   ROUGE-1 versus parameter α, β, γ and ω on DUC2003.

four scores are different for the integrated sentence scoring method. When α, β, and γ equal zero successively, we observe that the property of our proposed method become worse and dropped at most with α, followed by others. In other words, the relevance score plays a most important role compared with others in our method. The information score enhances the result by term level information while the diversity score eliminates redundancy in cluster level of sentences. Our method works best when density threshold$\omega$is about equal to 0.22. We adopt T as 20, 50 and 100, and choose the best value 50 for the summarization task.

3.3 Experimental Results

We compare different term weighting schemes with ours firstly: (1) BOOL (presence or absence); (2) TF (term frequency); (3) ISF (inverse sentence frequency); (4) TF-ISF (combine TF with ISF). The results of these experiments are listed in Table 1. It can be seen that BOOL term weighting achieves better results compared with that of TF, ISF and TF-ISF. The cause may lie in the frequency of term repetition occur less in sentences. Our method (OURS) gets better results than other rivals. It is probably because LDA weights terms by using its mixture model, which describes the structure of the documents more fully.

Table 1. Validity of different term weighting schemes

Methods	ROUGE-1	ROUGE-2	ROUGE-SU
TF+DPC	0.38756	0.09278	0.13729
ISF+DPC	0.37461	0.08755	0.12863
TF-ISF+DPC	0.38109	0.08934	0.13243
BOOL+DPC	0.39047	0.09559	0.13916
OURS	**0.39893**	**0.09910**	**0.14530**

We choose the following typical or recent published approaches for genic multi-document summarization in comparison with our method. The results of these methods are listed in Table 2. We divided the baseline methods into four categories:

(1) The best human summarizer's performance;
(2) DUC best: The best participating team in DUC 2004;
(3) Cluster based method: RTC (Rank Through Clustering) [12]; WFS-NMF (Weighted Feature Subset Nonnegative Matrix Factorization) [6]; ClusterHITS (Cluster-based HITS) [13]; KM (Kmeans);
(4) Others: BSTM (Topic Model) [9]; MSSF (Submodular Functions) [7]; MCKP (Maximum Coverage Problem) [10]; WCS (Weighted Consensus Scheme) [13]; R2N2_ILP (Recursive Neural Networks) [1].

Table 2. Overall the property comparison of our method and baselines on DUC2004. Remark: "–" indicates that the method does not officially report the results.

Methods	ROUGE-1	ROUGE-2	ROUGE-SU
Best Human	0.41820	0.10500	–
DUC best	0.38224	0.09216	0.13233
KM	0.34872	0.06937	0.12115
RTC	0.37475	0.08973	–
ClusterHITS	0.36463	0.07632	–
WFS-NMF	0.39330	0.11210	0.13540
MCKP	0.38640	0.09240	0.13330
BSTM	0.39065	0.09010	0.13218
MSFF	–	0.09897	0.13951
R2N2_ILP	0.38780	0.09860	–
OURS	**0.39893**	**0.09910**	**0.14530**

For better demonstrating the results, we visually illustrate the comparison between our method with the start-of-art methods in Fig. 2. From Table 3 and Fig. 2, we can have the following observations: the experimental results of all methods are adequate for a 95% confidence interval. Our method achieves a result close to the best human summarizer's performance. What's more, our method distinctly outperforms the DUC04 best participating work. Besides, it also can be seen that our method outperforms other competitors significantly on the ROUGE-1 and ROUGE-SU metric. It can be attributed to the integrated sentence scoring method to combine LDA with DPC, which promotes

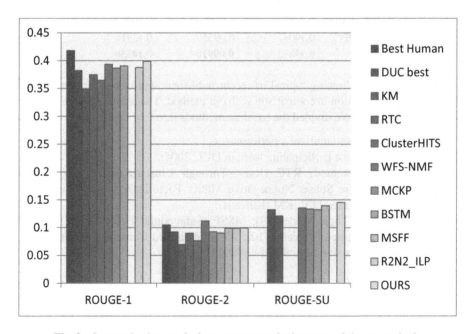

Fig. 2. Summarization results between ours and other state-of-the-art methods

the results mutually and ensure higher quality of the summaries. Compared with other cluster based method, our method removes redundancy when clustering and considers the informativeness of sentences. Our method performs slightly worse than WFS-NMF, MSSF and R2N2_ILP on ROUGE-2 score. It may due to our unigram-centric approach upon which text representation is built. Besides, those methods are complex and even need multiple features and postprocessor. Also, it is extensible for our integrated scoring method to introduce more features like position features.

Figure 3 shows the example of a document set and its result. We pick up one from the document sets randomly, D30028t, to show the human summary and the extractive summary of our method. The D30028t document set talks about the rising tension between Syria and Turkey. Looking at the results by our method in Table 8, each of the sentences represents one cluster respectively and summarizes well specific topics of each cluster. Note that the sentences of the extractive summary for each topic are not just discriminative but they also present the essence of the topic. The contents of the two summaries are very similar. Firstly, all of them talk the current emergency situation between Syria and Turkey. Then, the related countries, such as Egypt, Lebanon, and Greece, expatiate on their standpoints respectively. Besides, the selected summary sentence is completely dissimilar to the summaries of other topics and at the same time it is very relevant to the core event.

Multi-Documents

Tensions between Syria and Turkey increased as Turkey sent 10,000 troops to its border with Syria.
The dispute comes amid accusations by Turkey that Syria helping Kurdish rebels based in Syria.
Kurdish rebels have been conducting cross border raids into Turkey in an effort to gain Kurdish autonomy in the region.
Egyptian President Mubarek has been involved in shuttle diplomacy to the two states in an effort to defuse the situation and Iraq also has offered to mediate the dispute between the two countries.
Although Israel has tried to demonstrate its neutrality, Lebanon has charged That Israel is the cause of the tensions between Syria and Turkey.

Human Summary

- The talks in Damascus came as Turkey has massed forces near the border with Syria after threatening to eradicate Kurdish rebel bases in the neighboring country.
- Egypt already has launched a mediation effort to try to prevent a military confrontation over Turkey's allegations that Syria is harboring Turkish Kurdish rebels.
- Turkey accuses Syria of harboring Turkish Kurdish rebels fighting for autonomy in Turkey's southeast; it says rebel leader Abdullah Ocalan lives in Damascus.
- Lebanon on Monday denied it is harboring Kurdish rebels and blamed Israel for the rising tension between Syria and Turkey.
- Greece accused Turkey of undermining the whole region's stability through its stand-off with Syria over the alleged harboring of Kurdish rebels.

Extractive Summary

Fig. 3. Example of summary produced by our summarizer and the reference summary

4   Conclusion

In this paper, we proposed an unsupervised method to deal with the issue of multi-document summarization. We take informativeness, relevance, diversity and length constraint into account concurrently by employing LDA and DPC technique and the integrated sentence scoring method to generate the optimal summary. In our method, LDA was applied to acquire the information of sentences in terms level, while DPC was applied to survey the relevance among sentences in sentences level and diversity of sentences in clusters level in the meantime. Considering the length problem, we propose a length constraint score. By combining the four score of sentences, we finally extract sentences based dynamic programming algorithm. A series of experiments on DUC2003 and DUC2004 datasets demonstrate the excellent effectiveness of our method. The performance of our proposed method achieves a significant progress over a set of typical or recent published approaches. In our future work, we will delve into the text representation methods to acquire the semantic of sentence and combine with our integrated sentence scoring method effectively.

Acknowledgements. This work is partially supported by Key Technologies Research & Development Program of Shenzhen (No: JSGG20150512160434776) and Shenzhen Science & Research projects (No: JCYJ20160331104524983).

References

1. Cao, Z., Wei, F., Dong, L., Li, S., Zhou, M.: Ranking with recursive neural networks and its application to multi-document summarization. In: AAAI, pp. 2153–2159 (2015)
2. Li, L., Zhou, K., Xue, G.-R., Zha, H., Yu, Y.: Enhancing diversity, coverage and balance for summarization through structure learning. In: Proceedings of the 18th International Conference on World Wide Web, pp. 71–80 (2009)
3. Ma, T., Wan, X.: Multi-document summarization using minimum distortion. In: 2010 IEEE International Conference on Data Mining. IEEE (2010)
4. Liu, H., Yu, H., Deng, Z.-H.: Multi-document summarization based on two-level sparse representation model. In: AAAI, pp. 196–202 (2015)
5. Mei, Q., Guo, J., Radev, D.: DivRank: the interplay of prestige and diversity in information networks. In: Proceedings of the 16th ACM SIGKDD International Conference on Knowledge Discovery and Data Mining, pp. 1009–1018 (2010)
6. Wang, D., Li, T., Ding, C.: Weighted feature subset non-negative matrix factorization and its applications to document understanding. In: 2010 IEEE International Conference on Data Mining, pp. 541–550 (2010)
7. Li, J., Li, L., Li, T.: Multi-document summarization via submodularity. Appl. Intell. **37**, 420–430 (2012)
8. Arora, R., Ravindran, B.: Latent Dirichlet allocation and singular value decomposition based multi-document summarization. In: 2008 Eighth IEEE International Conference on Data Mining, pp. 713–718 (2008)
9. Wang, D., et al.: Multi-document summarization using sentence-based topic models. In: Proceedings of the ACL-IJCNLP 2009 Conference Short Papers. Association for Computational Linguistics (2009)

10. Takamura, H., Okumura, M.: Text summarization model based on maximum coverage problem and its variant. In: Proceedings of the 12th Conference of the European Chapter of the Association for Computational Linguistics, pp. 781–789 (2009)
11. Goldstein, J., Mittal, V., Carbonell, J., Kantrowitz, M.: Multi-document summarization by sentence extraction. In: Proceedings of the 2000 NAACL-ANLP Workshop on Automatic summarization, vol. 4, pp. 40–48 (2000)
12. Cai, X., Li, W.: Ranking through clustering: an integrated approach to multi-document summarization. IEEE Trans. Audio Speech Lang. Process. **21**, 1424–1433 (2013)
13. Wan, X., Yang, J.: Multi-document summarization using cluster-based link analysis. In: Proceedings of the 31st Annual International ACM SIGIR Conference on Research and Development in Information Retrieval, pp. 299–306. ACM (2008)
14. Zhang, Y., et al.: Clustering sentences with density peaks for multi-document summarization. In: Proceedings of Human Language Technologies: The 2015 Annual Conference of the North American Chapter of the ACL (2015)
15. Wang, B., Zhang, J., Liu, Y., Zou, Y.: Density peaks clustering based integrate framework for multi-document summarization. CAAI Trans. Intell. Technol. **2**(1), 26–30 (2017)
16. Lin, C.-Y: Rouge: a package for automatic evaluation of summaries. In: Text Summarization Branches Out: Proceedings of the ACL-04 Workshop (2004)
17. Rodriguez, A., Laio, A.: Clustering by fast search and find of density peaks. Science **306**, 1910–1913 (2014)

The Dynamic Relationship Between Bank Credit and Real Estate Price in China

Xiaofan Wang[(✉)] and Li Zhou[(✉)]

Beijing Wuzi University, Beijing, China
sunnyday1010@163.com, zhouli@bwu.edu.cn

Abstract. Real estate prices is an important indicator to measure the quality of macroeconomic, playing a vital role in national economy. The real estate market fluctuations, affects the whole macroeconomic operation, in turn, macroeconomic fluctuations, also will inevitably have a profound impact on the real estate market. The real estate market depends on financial support and, at the same time, bank credit is the main means of financial support. Bank credit not only play a crucial role to the prosperity of the real estate market, but also is a great force to promote the rapid development of the real estate market. Through theoretical analysis and empirical research, the statistical model of Chinese bank credit and the real estate price mechanism and effect is established in this paper. We select the relevant index to the studies of bank credit and real estate prices as a variable, collecting the quarterly data from 2006 to 2015, using the ADF test, Johansen co-integration test and granger test to study the dynamic relationship between price volatility and bank credit. Furthermore, we contrive to put forward relevant suggestions, making active contribution to the better advancement of real estate finance.

Keywords: Real estate prices · Bank credit · Dynamic relationship

1 Introduction

1.1 Research Context

In recent years, China's house prices rose at a staggering rate, real estate nearly became the most popular assets that every wealth class generally keen to buy and hold. And the real estate price strong volatility after continuous brewing fermentation might even lead to financial crisis. A few years ago, the subprime crisis in the United States is closely related to some enterprises like Freddie Mac and Fannie Mae. In the 20th century, Japan's real estate economy bubble and southeast Asian financial crisis also are closely related to the wild fluctuation of real estate price.

Bank credit to the change of the supply of real estate industry, generally spreading to the surrounding of the real economy, the small amount change of bank credit will have a great impact on the real estate market, as well as slowly conduct to the global economy industry. Therefore, the research on the relationship between the bank credit and real estate prices are necessary.

© Springer Nature Singapore Pte Ltd. 2018
K. Li et al. (Eds.): ISICA 2017, CCIS 873, pp. 324–338, 2018.
https://doi.org/10.1007/978-981-13-1648-7_28

1.2 Research Significance

It is well-known that the real estate industry is a new type of urbanization construction and the pillar industry of national economy, which plays an important role in making active contribution to employment, pulling economic increase and improving the living standard of people. However, with the advancement of the real estate industry, the appearance of over-speculation, over-rapid development and prices bubbles bring negative effects on the industry itself and relevant industry. At the same time, because of highly intensive properties of the real estate capital, there is a natural linkage between commercial bank credit and the real estate market, from the development of new buildings to real estate sales, this process requires a lot of commercial bank credit funds to support, and the cycle of this process is relatively long. Therefore, commercial bank credit makes active contribution to this process.

From the perspective of supply of real estate, for a long time, the loan of real estate development of commercial banks is the most important source of capital for the investment of real estate development, which keeps the capital chain of real estate industry running smoothly.

From the perspective of the demand of real estate, currently, because of the large value of real estate, when people buy real estate, they generally adopt the method of the one-time pay, rather than the most common way as well as the main way in a one-time pay, commercial bank mortgage loan. Banks provide loans for home buyers, while mortgage purchases are the main way for commercial housing consumption. Based on this, this study is based on the fact that the relationship between housing price and bank credit is of great realistic significance and can provide some theoretical basis and suggestions for government policy. At the beginning of the market operation, the real estate industry belongs to high risk and high return industry of sudden huge profits, commercial banks can not only perform higher loan interest rate, such a large company for deposit precipitation, but also bring a lot of personal mortgage customers, personal savings deposits and a series of related resources. Due to the profit-driven motivation, the higher comprehensive income leads to the tendency of commercial banks to increase the credit support of the development loans and personal mortgage loans of the real estate industry. Not only commercial banks' credit support is vital to the real estate industry, but the credit of the real estate industry is also a significant part of the business bank.

1.3 Research Purposes

This study has two main objectives: first, whether the credit support of commercial Banks to the real estate industry is related to the price of real estate, and whether the relationship is causal. Whether it is a single causal relationship or reciprocal causality. If their relationship is interactive, how the two interact with each other; Second, if the relationship exists, how the supply and demand of the real estate market are affected by the real estate credit of commercial banks. Exploring these two issues is helpful for maintaining the stability of the real estate market and financial market, and providing a useful reference for monetary policy, new regulation or adjustment of real estate macro-control policies.

2 Analysis on the Current Situation of China's Real Estate Credit Market

Whether real estate or financial, is an important industry in the economic development of our country, so to adjust industry of both state and the relationship between the two is one of the important conditions to ensure the steady and rapid economic development of our country. By studying the development situation of these two industries to regulate the policy, the property price and credit are the key indicators. In order to study the impact of bank credit on the fluctuation of property price, it is necessary to understand the specific content of the real estate financial system to analyze the main influencing factors.

2.1 China's Real Estate Finance System

China's real estate financial system is centered on credit. From the perspective of real estate financial market, consumer financing is dominated by commercial banks, because only commercial banks can handle personal housing mortgage loans according to the regulations of the state. And due to the increased number of housing accumulation fund, covering small, and asset securities in the secondary market blank, so both developers and individual are mainly dependent on commercial bank credit. China's capital market is under development, is not yet mature, so most of the private room cannot get money by capital development tools such as stocks and bonds, and financing has become a very difficult thing. This indicates that China's real estate financial institutions are single and have fewer tools.

2.2 The Effect of Bank Credit on Real Estate Price Fluctuation

Real estate industry relative to other industry has long construction cycle, the characteristics of capital occupying large proportion, but developers invest owned fund that occupy small proportion into the real estate, the vast majority of borrowing money from the financial system; In China, the financial system is not sound enough and the monetary sector is not perfect. Most of the development funds of real estate enterprises depend on the credit of commercial banks. The credit scale and credit direction of commercial banks will definitely influence the investment decisions of both sides of the real estate market, and the real estate prices will also produce violent fluctuations. For suppliers, through commercial banks, the real estate developers have a lot of money, for commercial housing construction funds will increase, the housing supply will increase, in the case of demand remain unchanged, falling house prices; For the demand side, the expansion of bank credit, bring needs more money, with more investment and speculative demand in the market need, in the case of supply remains unchanged, soaring house prices. Comprehensive consideration, the construction period of the real estate industry is long, the supply of commercial housing will not change greatly in the short term, and the expansion of bank credit will only deepen the price rise. Once the housing price rises, the value of the bank's holdings of real estate is higher, and on the other hand, bank profits will increase and the bank will expand the credit in the face of increasing profits. The expansion of bank credit, which prompted a new round of housing price rises, was so reciprocated that the real estate bubble came into being.

2.3   Real Estate Prices Fluctuate in Response to Bank Credit

The role of house prices to bank credit also needs to be discussed from the perspective of demand and supply. Rising house prices, on the one hand, allow investors and speculators to keep up with expectations that house prices will continue to rise, prompting them to borrow money from banks to buy property. On the other hand, banks will also relax their credit conditions and expand the credit scale in the face of rising house prices, thus further driving demand and keeping prices rising. For the supply side, the housing price rises, the real estate developers are more willing to undertake real estate development in order to get more profit from it, the demand for funds also correspondingly increases. Commercial banks' profit comes from the spread of borrowing funds. In pursuit of profit maximization, commercial banks will relax the credit conditions for developers and expand credit. If house prices fall, investment demand and speculative demand in commodity markets will fall, and individuals and developers will reduce credit. Falling house prices could lead to real estate collateral prices, real estate collateral to determine the ability of individuals and developers to obtain loans, once the real estate collateral prices, personal and developers access to loans will be less, credit scale natural decrease in the financial markets. The fall in mortgage prices in the banks led to a reduction in net capital and a further reduction in credit in the market as banks tighten credit policy.

3   Research Ideas and Methods

3.1   Research Ideas of the Thesis

On the relationship between bank credit and real estate prices research have a lot of related theory can use for reference, in this article, through the introduction of real estate sales price index and our country actual credit balance, according to a study in China in recent years, the interaction mechanism between the two micro levels is analyzed, and the equation of interaction between variables is obtained based on the data. Drawing the technical roadmap (Fig. 1).

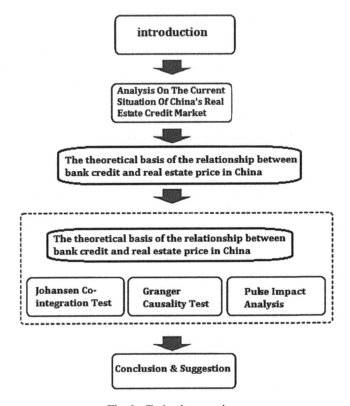

Fig. 1. Technology roadmap

3.2 Research Method

This paper will firstly analyze the stability of real credit and real house prices. Then the co-integration test is used to determine whether there is a long-term equilibrium between the two changes. The VAR model of multivariable is established after the data difference processing. Based on the VAR model and the causal relationship between the two changes using Granger test. Finally, the impulse response function and variance decomposition method are used to analyze the dynamic effect of a change of error term on the system.

4 Empirical Study on the Relationship Between Bank Credit and Real Estate Price in China

4.1 Data and Variable Selection

About prices, this paper collected the national commercial housing sales area and quarterly sales data from 2003 to 2015, and the original data conversion into national residential quarter average selling prices, then its standardization for the first quarter of 2003 as the base price index (HP).

About credit, this paper mainly studies the relationship between total bank credit and housing prices, instead of the relationship between mortgage and the housing price, so this article uses medium and long-term credit quarter balance of financial institutions in our country as sample data of bank credit.

As for the choice of control variables, credit and real estate cycles are driven by common economic factors. On the one hand, the credit cycle is mainly determined by macro economy and expectations (especially GDP, interest rate, etc.). On the other hand, the state of economic activity exerts important influence on the real estate market. The change of macro economy will cause the demand and supply adjustment of real estate, which will affect the change of real estate investment and price. These external shocks can be either from the demand side, such as the change of the income, interest rates, and population factors, also can be from the supply side, such as labor and construction costs, as well as change of the limitation of land development and so on. In order to more accurately reflect the relationship between credit and real estate prices, based on China's real GDP and the actual loan interest rate (R) as control variable to control the external shocks from the demand side, at the same time to join China's CPI variables in the model to control the impact from the supply side.

In this paper, the standard indexation of all data (except interest rate) is carried out, which is 100 in the first quarter of 2003. In order to eliminate the influence of the possible variance of variance, the logarithm of all variables other than interest rate is taken (Table 1).

Table 1. Variables, symbols, and meanings in the model

Variable	Symbol	Connotation
Real estate price	HP	China real estate sales price index (logarithm)
Actual credit	LOAN	The actual loan balance of our financial institutions (logarithm)
Gross domestic product	GDP	China's real GDP (logarithm)
Consumer price index	CPI	China's consumer price index (logarithm)
Interest rate	R	China's loan real interest rate

4.2 Data Source and Description

In the study, the quarterly data of Chinese Banks' mid - and long-term credit data and the average price of real estate sales in China were selected from the first quarter of 2003 to the fourth quarter of 2015. The average price of credit, real estate sales, GDP and CPI came from the *China economic sentiment monthly report*, which came from the people's bank of China website. Then, standardizing the raw data.

4.3   Unit Root Test

Analyze the time series is the premise of assurance of stationary sequence, rather than an unsteady time series participate in regression modeling analysis, leading to spurious regression problem, thus generally analyzed before, it is necessary to use unit root test method to test the original variable sequences, determine the stability of the sequence. The method includes graphical intuitive judgment and unit root test, while the latter is more accurate and important. In the meantime, it can determine the integration order.

The null hypothesis of ADF test is H_0: Time series has unit roots. If the sequence is a stationary sequence, you can continue to modeling; If the sequence is a non-stationary sequence, the method of differential treatment or co-integration is required. In the face of China's real estate price HP sequence and the bank credit LOAN sequence in our country, the trend chart is presented in this paper, then we do the unit root test (Figs. 2 and 3).

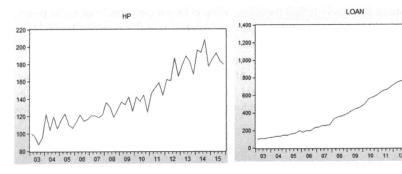

Fig. 2. HP sequence trend chart   **Fig. 3.** LOAN sequence trend chart

As shown in Figs. 2 and 3, the HP sequence and LOAN sequence are seen as non-stationary sequences. ADF unit root test is further carried out. The test results are shown in Tables 2 and 3.

Table 2. Unit root test results of HP sequence

Null hypothesis: HP has a unit root		Exogenous: constant, linear trend	
		t-Statistic	Probability
Augmented Dickey-Fuller test statistic		−1.428876	0.8396
Test critical values	1% level	−4.161144	
	5% level	−3.506374	
	10% level	−3.183002	

Table 3. Unit root test results of LOAN sequence

Null hypothesis: LOAN has a unit root		Exogenous: constant, linear trend	
		t-Statistic	Probability
Augmented Dickey-Fuller test statistic		−0.897098	0.9479
Test critical values	1% level	−4.161144	
	5% level	−3.506374	
	10% level	−3.183002	

By unit root test results of HP sequence in Table 2, we can find that the probability of ADF value is 0.8396, significantly greater than 0.05, significance level of the original sequence can't through the inspection, so acceptable condition of HP sequence is null hypothesis with a unit root, thus HP series is nonstationary, then as for HP sequence, difference treatment and cointegration analysis could be carried out. By unit root test results of LOAN sequence, we can find that the probability of ADF value is 0.9479, significantly greater than 0.05, significance level of the original sequence can't through the inspection, so acceptable condition of LOAN sequence is null hypothesis with a unit root, thus LOAN series is nonstationary, then as for LOAN sequence, difference treatment and cointegration analysis could be carried out.

The unit root test of sequence Ln LOAN and sequence Ln HP was performed according to the same procedure, List the test results (Table 4).

Table 4. Sequence Ln LOAN and Sequence Ln HP unit root test results

Variable	Test type (c, t, k)	ADF value	P value	Conclusion
Ln HP	(c, t, 3)	−2.0360	0.5672	Unstable
Ln LOAN	(c, t, 3)	0.3598	0.9984	Unstable

4.4   Co-integration Test

The co-integration theory is a statistical analysis method, which is mainly used to study the long-term equilibrium relationship between non-stationary sequences. When variable of time series analyzed are non-stationary sequence, except with difference processing for the original sequence variable, cointegration analysis could be conducted, inspect whether linear combination of the non-stationary variables is stationary sequence or not, if the linear combination of the nonstationary sequence are stationary variables, there exists a long-term equilibrium relationship between these variables. In the analysis of the co-integration of variables, the most commonly used is the Johansen co-integration test method. We can see that the Johansen co-integration test results (Table 5) and the Johansen co-integration test expression (Table 6).

Table 5. Johansen co-integration test results

Null hypothesis	Eigenvalue	Trace statistic	Critical value (5%)	Probability	Conclusion
None	0.3079	19.347	15.4947	0.0125	Reject null hypothesis
At most 1	0.0343	1.679	3.8414	0.1951	Accept null hypothesis

Table 5 shows the test results of trace statistics and maximal characteristic root statistics. These two statistics are used to determine the number of co-integration relationships between variables in the Johansen co-integration test. The Johansen co-integration test is conducted in order of the number of co-integration relationships from 0 to k − 1, until the corresponding original hypothesis is rejected.

In Table 5, through the test for the trace statistic, we can determine that the original hypothesis *None* said there was no co-integration relationship. Under the assumption, we can calculate the probability of trace statistics, 0.0125, you can reject the null hypothesis, thinking there are at least a co-integration relationship; Another null hypothesis *at most 1* said there is at most a co-integration relationship. Under the assumption, we can calculate the probability of trace statistics, 0.1951, is greater than 0.05, accepting the null hypothesis, thinking there is no co-integration relationship. Therefore, through the trace statistics of these two sequences respectively, it can be estimated the co-integration relationship between Ln HP and Ln LOAN.

Table 6. Johansen co-integration test expression

Unrestricted Adjustment Coefficients(alpha):		
D (Ln HP)	0.027992	-0.007694
D (Ln LOAN)	0.015158	0.004641
1 Cointegrating Equation(s): Log likelihood 162.2417		

Normalized cointegrating coefficients (standard error in parentheses)		Adjustment coefficients (standard error in parentheses)	
Ln HP 1.000000	Ln LOAN -0.168349 (0.02759)	D (Ln HP) -0.353093 (0.11771)	D (Ln LOAN) -0.191204 (0.06743)

Similarly, the rule of judgment of the maximum eigenvalue is the same as the trace statistic. From Table 5, it can be seen that the probability of the maximum eigenvalue under this hypothesis is 0.0139, which can reject the original hypothesis and think that there is at least one co-integration relationship. Another null hypothesis *at most 1* said there is at most a co-integration relationship. The probability of the maximum eigenvalue under this hypothesis is 0.1951, is greater than 0.05, which can accept the original hypothesis and think that there is no co-integration relationship. Therefore, it can be said that the test results of the maximum eigenvalues of the study are consistent with the test results of the trace statistics. Both of them believe that there is a co-integration relationship between Ln HP and Ln LOAN.

Table 6 shows the co-integration relationship with the largest logarithmic likelihood value, which is also the co-integration relationship of regression in VEC. The standardized co-integration relationship value refers to the co-integration relationship expressions after the coefficients before the first variable are normalized to one, which can be easily written into the final co-integration equation. The co-integration equation in this study is:

$$\text{Ln HP} = 0.1683 * \text{Ln LOAN} \tag{1}$$

Through this co-integration relationship expressions, both Ln HP and Ln LOAN are positively correlated long-term equilibrium relations: every 1% increase in credit, the price of housing increases 0.16%. The significance T value of the estimated parameter value of the cointegration relation can be calculated by the standard deviation in the lower brackets and the value of the parameter estimation. In addition, its T value is also reported in the results of VEC estimation.

The adjustment coefficient value refers to the adjustment speed after the dynamic relationship between variables in the VEC model deviates from the co-integration relationship. The adjustment coefficient of the equation of D (Ln HP) in this study is −0.35309, and the adjustment coefficient of the equation of D (Ln LOAN) is −0.1912. If the adjustment coefficient is negative, the deviation from the non-equilibrium error will be corrected. If the value of the adjustment coefficient is positive, the non-equilibrium error is not only not corrected, but the error will be greater. The adjusted coefficients of the calculation have at least one negative value, if all positive, indicating that the cointegration relationship is invalid. In this study, the adjustment coefficient of the equation of D (Ln HP) is negative, indicating that the co-integration relationship is valid and the operation of the house price in the short term is bound by the long-term equilibrium of credit variables.

4.5 Vector Error Correction Model

Sequence EC_t is the core part of vector error correction model, but the co-integration relationship can only explain the long-term equilibrium relationship and trend of each explanatory variable. The positive correlation between the actual credit balance and real house prices may be because credit growth has contributed to the rise in house prices, or it may be that rising house prices have contributed to the increase in credit balances. In order to define the actual relationship between house prices and credit balance, the following will be based on the co-integration relationship between variables, further establish the short-term volatility associated with long-term equilibrium of vector error correction model (VECM), the actual credit balances and real house prices conduct Granger causality test, thus further determine between them is the effect of forward, reverse, or both ways. The VECM model results can be obtained as shown in Table 7.

Table 7. VECM model results

Vector Error Correction Estimates				
Cointegrating Eq:	D (Ln HP(-1))	D (Ln LOAN(-1))	R (-1)	C
CointEq1	1.000000	-1.170027 (0.46826) [-2.49868]	0.606234 (0.28080) [2.15893]	0.018137
Error Correction:	D (Ln HP, 2)		D (Ln LOAN, 2)	D
CointEq1	-1.0819 -0.52817 [-2.04841]		0.777854 -0.31024 [2.05730]	-0.051892 -0.07707 [-0.67334]
D (Ln HP(-1), 2)	-0.587348 (0.47593) [-1.23410]		-0.744145 (0.27955) [-2.66190]	0.018878 (0.06944) [0.27184]

Table 7 shows the co-integration relationship in VEC, but in the VEC model, the co-integration relationship express into the form of error correction item is:

$$\text{Coint EQ1} = \ln \text{HP} - 1.17 * \ln \text{LOAN} + 0.018 \tag{2}$$

The error correction term expression and Johansen cointegration test of cointegration relationship is consistent, only in the Johansen cointegration test relation with a constant term, causing the coefficient estimate is slightly different. The error correction term calculated by this formula is the CointEQ1 variable in the error correction model. The middle and lower half part of Table 7 is the specific estimation coefficient vector of the VEC error correction model. The explanatory variables of the three equations in the VEC model are not the original sequence but the difference sequence of the original sequence, because the original sequence is non-stationary sequence, and the stability of the VEC model is guaranteed after the difference. VEC model explanation variable is the dependent variable lags behind, this is the same as the structure of VAR completely, the only difference is that: the VEC add the error correction term CointEQ1 calculated in the first part to VAR model with differential variable. The VEC model of this study can be written as:

$$\Delta Y_{t-1} = \begin{bmatrix} -1.08 \\ 0.78 \end{bmatrix} \text{Coint } EQ_{t-1} + \Delta Y_{t-1} + \ldots + \Delta Y_{t-4} + \theta + \varepsilon_t \tag{3}$$

4.6   Granger Causality Test

The above-mentioned co-integration test can only show that there is some kind of long-term stable relationship between bank credit and real estate price, but it is not possible to determine the causality between the two variables. Therefore, this research will continue to conduct granger causality test between the two variables, in order to better understand the intrinsic law of interaction between China's bank credit and real estate prices.

The interdependence in the data variables is called the granger causality test. On the one hand, the original dependent variable determines the outcome variable in the causality relationship, on the other hand, the change resulting from the result variable is caused by the cause variable. This method can be used to test the variable relationship between causality and causality in the real estate credit market. Granger test define causality from time series: if use the past value of Y and the past value of the X to conduct the regression of Y, examining whether adding the lag value of X can significantly strengthen the explanation ability of the regression equation, if the lag value of X helps to improve the interpretation of the Y, arguing that sequence X is the Granger reason of sequence Y.

In this study, if bank credit LOAN is X and real estate price HP is Y, the test results are shown in Table 8.

Table 8. Granger test results

Null hypothesis	Lags	Obs	F-Statistic	Probability	Conclusion
LOAN does not granger cause HP	2	50	4.06704	0.0238	Reject
HP does not granger cause LOAN	2	50	0.21122	0.8104	Accept

It can be seen from Table 8 that as for sequence LOAN is not the original hypothesis of the granger cause of the sequence HP, the P value of F test is 0.0238, which was less than 0.05, which passed the hypothetical test and rejected the original hypothesis. Therefore, arguing that LOAN is the granger cause of HP, the credit balance of our bank is the granger cause of the real estate price. However, as for sequence HP is not the original hypothesis of the granger cause of the sequence LOAN, the P value of F test is 0.8104, which was significantly greater than 0.05, which did not pass the hypothetical test and accepted the original hypothesis. Therefore, it is believed that HP is not the granger cause of LOAN, it means that China's real estate price is not the granger cause of bank credit balance.

Through the granger test results above, it can be analyzed that, on the one hand, this paper considers that the bank credit in China is more of a policy variable, which has a certain degree of exogeneity. On the other hand, because of many factors affecting the bank credit growth, including the various industries of national economy, therefore, based on the control of the gross domestic product, the consumer price index, and long-term interest rates and other variables, it is difficult to find that real estate price changes can significantly affect the actual bank credit balance changes.

4.7   Impulse Response Function

Impulse response function is used to measure one standard deviation from a random disturbance impact, effect on current and future values of trajectory of endogenous variable, more intuitive to depict the dynamic interaction among variables and their effects. Based on the VAR model established in this paper, the impulse response function (IRF) between credit balance and real estate price will be portrayed to further analyze the short-term dynamic relationship between the two. The results of the impulse response function are shown in Fig. 4.

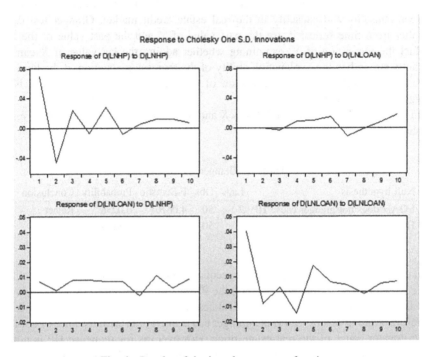

Fig. 4. Results of the impulse response function

From Fig. 4 in the top right corner of the impulse response of track can be seen that the growth rate of housing prices is the positive impact of credit growth and, in the first quarter is positive, but then started declining volatility negative, and falling into the third quarter, the lowest and then started to rise as positive, to the fourth phase of the maximum, then began to fluctuate around zero value, in the end, this effect tends to disappear.

Through this path diagram you can see, by the positive shock of growth rates of credit, in the first three quarters, growth rates of real house prices being falling status and negative, looking at the cause of the change has the following two points:

Influence of Policy Variables Under Macro-Control
In this study, bank credit in China is more of a macroeconomic policy variable with a certain exogeneity. When the economy booms, it is often accompanied by a rapid rise in house prices. Central banks tend to tighten credit, tighten liquidity and prevent overheating, which can lead to severe inflation. When the economy goes down, it is often accompanied by falling house prices, and central banks influence commercial banks in a variety of ways, boosting credit and stimulating the economy. As a result, growth rate of credit is rising in the short term and real house prices do not rise immediately.

The Impact of Popular Expectations

When the economic downturn began, the central bank to expand credit, to stimulate the economy, the public believes the central bank has admitted that the economic downturn, due to economic policy has the lag, the public would be expected that economic further deterioration in the short term, house prices will fall further; Vice versa. This will also lead to a rise in credit growth in the short term but lower house prices. A positive shock of growth rate of credit, from the third quarter, the growth rate of real house prices began to rise and changed to positive, illustrating that the expansion of bank credit promotes real estate market going into the prosperity and the growth rate of real house prices rising up, this process will take at least two to three quarters of the time. Credit growth will take a long time to make investment, speculation and consumption of real estate have a sufficient source of funds, it means that increase of income will promote the increase in the price of housing. In addition, the expansion of credit makes money supply increased in the economic entity, promoting raw materials and artificial price rising, increasing the cost of real estate development, promoting the real estate prices, this process also need a period of time.

From Fig. 4, the first part of the trajectory of the impulse response of values, it can be seen that the actual growth rate of credit was positively shocked by growth rate of real house prices, changes in the actual credit growth is not obvious, from the first period, beginning to rise until the third period up to the maximum, and then began to decline, but basically around zero value maximum fluctuated. The result is consistent with the real rate of real estate growth shown in the Granger causality test in the previous article.

5   Conclusion of Empirical Study

Compared with the developed countries abroad, as an emerging and transition of China's credit and real estate market, the relationship between the bank credit and the real estate price has particularity, it means that these two sides do not have a constant relationship for a long time, the influence of bank credit dominate in the change of house prices in the short term, which in turn is not established. Based on the empirical results, this paper has obtained the following conclusions:

In the sample period, this paper argues that there is no long-term co-integration relationship between actual bank credit and real estate price. Bank credit in China has been more of variables of a macro policy, and set up relatively late, the development of the real estate market in China is not enough mature, and with dynamic and turbulent government policy on the real estate market, it is difficult to find there is a long-term constant cointegration relationship between bank credit and real estate prices.

In the short term, the change of actual credit growth is dominant and in a certain period of time has a statistic stability, it means that the causal relationship and explanatory ability between the actual bank credit and the real estate prices is from the growth rate of the actual credit to the growth rate of real house prices, and obviously the opposite direction cannot be established. On the one hand, due to the policy of bank credit, it has some degree of externality; On the other hand, there are many factors that can affect the bank credit growth, including the industries of the national economy.

Therefore, in this paper, through controlling the gross national product, interest rates and other variables, it is difficult to find that change of growth rate of real house prices have significantly impact on change of growth rate of real bank credit.

Acknowledgement. This study is supported by the science training program of Information School of the Beijing Wuzi University in 2016.

References

1. Akin, O.: Prices and trading volume in the housing market: a model with down-payment effects. In: SERIES, vol. 5, no. 2–3, pp. 223–243 (2014)
2. Fitzpatrick, T., Mc Quin, K.: House prices and mortgage credit: empirical evidence from Ireland, vol. 75, no 1. The Manchester School (2015)
3. Goetzmann, W.: The subprime crisis and house price appreciation. In: Yale ICF Working Paper No. 1340577, February 2011
4. Borio, C., Lowe, P.: Assessing the risk of banking crises. BIS Q. Rev. **3**, 43–54 (2012)
5. Gerlach, S., Peng, W.: Bank lending and property prices in Hong Kong. J. Bank. Financ. **29** (2), 461–481 (2015). Author, F.: Article title. Journal **2**(5), 99–110 (2016)
6. Xiao, B.: Credit expansion and real estate prices in China. J. Shanxi Univ. Financ. Econ. **1**, 27–31 (2014)
7. Zhu, H.: The impact of commercial bank credit on real estate prices. Financ. Econ. **3**, 70–71 (2011)
8. Wen, F.: Real estate price fluctuations and financial fragility – based on empirical research in China. China Manag. Sci. **20**, 1–10 (2012)
9. Li, J., Shi, C.: The impact of bank credit on real estate price fluctuations. J. Shanghai Univ. Financ. Econ. **4**, 26–32 (2015)
10. Wu, K., Pi, S., Lu, G.: General equilibrium analysis of Chinese real estate market and financial market. Quant. Econ. Technol. Econ. Res. **10**, 24–32 (2014)

Big-Data Cloud Services Platform for Growth Enterprises with Adaptive Exception Handling and Parallelized Data Mining

Yazhi Wen[1], Hu Bo[2], and Bin Wen[3(✉)]

[1] School of Computer Science and Technology,
Huazhong University of Science and Technology, Wuhan 430074, China
wyazhi@hust.edu.cn
[2] Kingdee Research, Kingdee International Software Group Company Limited,
Shenzhen 518057, China
[3] School of Information Science and Technology,
Hainan Normal University, Haikou 571158, China
951714238@qq.com

Abstract. Large-scale growth companies need big-data analysis platform, at the same time it is difficult to self-build. Based on cloud computing technology, big-data analysis can moved to the cloud and utilize cheap SaaS model with the advantages of the "cloud + end" pattern for enterprise applications to help enterprises collect and store data to provide massive data analysis and visualization services. In the design of the platform, intelligent computing plays an important role. In order to deal with the exception handling of the requirements and scene changes for services composition, the adaptive optimization mechanism in runtime adopts the spatial search optimization algorithm SSOA. At the calculation management layer for the platform architecture, we have made a parallel transformation for the traditional data mining methods. K-Means parallelization method has been proposed. Tag-based domain expert finding has also been designed and implemented. Combination of theoretical research and empirical validation, the paper tries to provide a technical operational and cost-effective solution for big-data cloud services development.

Keywords: Big-data analysis · Spatial search optimization
Data mining · Parallelization

1 Introduction

With the gradual maturity of social networks, mobile bandwidth is rapidly increasing. More IoT devices, mobile terminals have access to the Internet, the resulting data and its growth rate has been more than any time in history [1].

© Springer Nature Singapore Pte Ltd. 2018
K. Li et al. (Eds.): ISICA 2017, CCIS 873, pp. 339–350, 2018.
https://doi.org/10.1007/978-981-13-1648-7_29

Big-data age has arrived. Big data and Cloud computing are the most important topics for organizations across the globe amongst the plethora of softwares [2]. Big-data era should encourage enterprises to take the initiative to adapt the changes through massive data analysis. Growth enterprises refer to the firms with the ability to utilize the unused resources continuously and show the overall expansion of the situation, and the future development is expected to good business for a long period of time (such as more than 3 years). Growth-oriented enterprises, they do not have the talent and technology, equipment to build big-data analysis platform. Service computing has become very popular nowadays in providing big data analytics [3]. To this end, we decided to use the form of cloud services for growth-oriented enterprises to design big-data analysis platform. Based on cloud computing technology, big-data analysis can moved to the cloud and utilize cheap SaaS model and SOA with the advantages of the "cloud + end" style for enterprise applications to help enterprises collect and store data to provide big-data analysis and visualization services.

The architecture of big-data analysis cloud service platform shown in Fig. 1. The main purpose of this platform is to realize the enterprise big-data analysis and related services application research, which involves more key technology. The platform consists of three vertical and five horizontal modules. The shadow part of the figure is the focus of this article. It mainly related to the following three aspects.

1. exception handling with intelligent algorithm
2. parallel data mining algorithm
3. parallel social network analysis algorithm

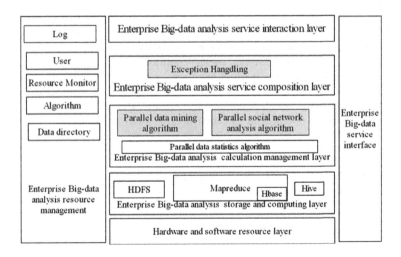

Fig. 1. Architecture of big-data analysis cloud service platform

Software and resource utilization have entered into the cloud for consumer use in the form of services. Services has been become the basic fashion to access and enlarge the capability of infrastructure [4]. Now the general process for service-based software production is as follow: producing services resource from service providers publishing services selecting services from service consumers services aggregation. But the process lacks of runtime exception handling and iterative services re-aggregation considering without starting from the requirements for services system adaptive mechanism to overcome the exceptions, services resource insufficiency and context changes.

Enterprise Big-data Analysis Calculation Management, which is the core layer of enterprise big-data analysis platform, including computing task scheduling, parallel data mining algorithm for enterprise application, data conversion module supporting multi-source data integration.

Currently, due to the dynamic instability of service composition on runtime, a variety of service exceptions or anomalies often appear [5,6]. Exception refers to the services failure (fault), network error or abnormal events caused by resource or requirements changes. Lack of exception handling mechanism, it will lead to these problems such as poor performance, resource waste, poor optimized services and even failure. So services resource provisioning must be able to actively produce. They should have adaptive runtime exception capability.

Typical data mining methods include classification and regression algorithms, typically Bayesian, hierarchical C45 decision trees. Clustering algorithms, such as K-Means, BIRCH algorithms, and so on. Association rule class algorithm, typically such as Apriori, timing correlation. However, the above traditional data mining algorithm is stand-alone support [7], does not meet the MapReduce parallel computing framework needs for big-data analysis. Therefore, the above data mining algorithms need to be modified to meet the needs of Mapreduce computing.

This paper aims at study the runtime exception handling self-adaptive method and parallelization transformation of data mining for big-data cloud services. We will mainly focus on adaptive SOA runtime exception handling mechanism and parallel data mining. This paper is structured as follows. Section 2 includes the adaptive runtime exception handling. Section 3 presents parallelization considerations for data mining to meet Mapreduce computing framework. Section 4 shows the experiment and empirical analysis with the approach. Conclusions with main contributions of proposed approach and further work plans are also touched upon in Sect. 5.

2   Adaptive Runtime Exception Handling

Service virtualization is designed to shield the heterogeneous properties of IT resource. Software abstraction expression will be decoupled with concrete IT resource to realize semantic equivalence mapping between web service of IT level and business functional requirements [8].

2.1 Runtime Adjustment Mechanism to Deal with Context Changes

We have systematical explored acquirement technology and services aggregation under the guidance of requirements model with stakeholders groups to participate. Here in this section, adaptive runtime exception handling proposal intends to carry out based on personalized customized services.

Considering technical feasibility and simplicity, data syndication with ATOM can be selected to encapsulate the requirements semantics of services. The requirements of each service are described by the ATOM protocol and are achieved in a subscription-notification pattern. Because the service description in Atom comes from the Requirements Symbol Ontology (RSO), custom services are also proactively delivered to satisfy related stakeholders.

The core of Scenario Self-Tuning Change Framework is semantic-driven custom management. It mainly includes three parts: custom manager, evaluator and aggregator. Aggregator performs related actions including search, match, and bind services based on RSO's business processes. For unsuccessful matching of service resources, Custom Managers are responsible for their custom production and management. The evaluator is responsible for evaluating and verifying the overall quality of matching services. They work together to cope with runtime scenario changes.

2.2 Self-Adaptive Custom Service Resources Optimization

The main core algorithms includes the services customer preferred selection, services aggregation with customization, self-adaptive optimization, services evaluation algorithm, customization management process, customization information feedback and services re-aggregation with joined custom services and application effectiveness analysis for services resource customized production.

Self-adaptive optimization algorithm partly adopts the early results of ours effort, namely SSOA (Space Search Optimization Algorithm) (see Algorithm 1). In custom resource provisioning, selection and search efficiency will be improved.

---

1 INPUT:solution set (population)
2 OUTPUT:optimization space

```
1:  Begin
2:    Initialization:
3:      1) Initialize a solution set (population) at random.
4:      2) Opposition-based space search.
5:    While (the termination conditions are not met)
6:      IF (rand(0, 1) < C_r) // C_r is a fixed given number
7:        Local space search:
8:          1) Generate a new space: Generate a new space based on three given solutions.
9:          2) Search the new space: Reflection, Expansion, and Contraction.
10:       Global space search: Cauchy search (Cauchy mutation).
11:     Else
12:       Opposition-based space search.
13:   End While
14: End
```

Algorithm 1. Space search optimization

Space search algorithm uses search operation to realize optimization object. Algorithm starts from the known solution, and also produces new subspace and searches this subspace.

The algorithm consists of three types of space search operations.

1. Local space search:
   The operation has been improved based on the Simplex algorithm (plus the search with constraint conditions)
2. Global space search:
   Essentially, the operation is Cauchy mutation.
3. Reverse operation:
   In order to speed up the convergence of the algorithm, the operation referred the "reverse number". The result has proved better than pure random search.
4. Algorithm characteristics:
   SSOA has stronger local search ability. For example, compared with most of the DE algorithm, SSOA algorithm has relatively stronger global search ability, this is because it includes Cauchy mutation. SSOA algorithm has faster convergence speed.
5. Algorithm advantages:
   The experimental results have showed that SSOA has faster convergence speed, and has more possibility to obtain the approximate solution or more precise value compared some famous DE algorithm. Especially in the high dimensional optimization problem, these advantages perform more outstanding.

2.3   Custom Service Resources Management Monitoring Mechanism

On-demand customization of services resource needs runtime abnormal monitoring to trigger the customization process. So services quality evaluation metrics should be explored. Monitoring mechanism needs to define the boundary conditions of abnormal action, and it is also a real-time system that must meet the demand of real-time triggering, releasing and feedback.

When an exception occurs, the process of services aggregation will be interrupted. Services aggregation should be restarted and worked again with the recovery of unavailable services. Implementation method is drawn lessons from the scientific workflow's transaction in that the process either completed or terminated. At the same time, related process operation data should be retained. For the terminated process aggregation again, resumed process does not affect the services resource providers. So related data tables of system database are required to design. These data records will support the aggregation restart again.

The main operation points of services resource custom management method are as follows.

1. We propose a personalized active custom approach driven by requirements fragments for services requester-centric SOA with adaptive mechanism. It will reinforce current status for customized services resource without runtime on-demand customization.

2. We have designed a full set of architecture and implementation including self-adaptive custom, services aggregation restart and exception handling monitor for abnormal case of user services. Also, feasibility and simplicity will be focused.
3. Through some mathematical methods such as custom optimization algorithm of services resource, exception handling ability with adaptability will be built to optimize and quantity services resource production in runtime.

2.4 Runtime Adaptive Adjustment Mechanism with the Changes

Now runtime adaptive adjustment is still a difficult problem for Internet-intensive software system. It needs to solve how to map requirements changes into architecture units at runtime. Driven by predictive control for SaaS components to induce requirements evolution, runtime architecture change will be realized, and the validity of predictive control can be proved in the aspect of requirements/architecture evolution [9]. But this method is not extended to SOA level with requirements change-driven architecture evolution.

We propose the runtime adaptive adjustment scheme combined with effective predictive control method [9] and MAPE-K control circuit model [10]. The proposed adjustment system divided into real-time monitor, analysis engine, software architecture adjustment manager, requirements evolution manager and Aspect execution engine. The core points of the scheme are as follows.

(1) How to select the predictive QoS value for runtime services resource? Wavelet transformation is the best choice.
(2) Software architecture adjustment mechanism based on QoS changes The runtime adaptive adjustment algorithm is designed (Algorithm 2) for solving the above core points.

---

1 INPUT:QoS value of service resources for SOA software in run time t and t+1; QoS value of the expected output.
2 OUTPUT:control operation vector in t+1 moment

  1: Begin
  2:   Initialization: training classification prediction model; the points of tags to improve training requirements model;
  3:   IF Classification prediction (QoS value of services resource in run time t and t+1, expected output of QoS value) = requirements
  4:     THEN
  5:       control operation vector in t+1 moment=tag improvement point (QoS value of services resource in t and t+1 moment at runtime, the expected output of QoS value)
  6:     ELSE
  7:       control operation vector in t+1 moment = architecture evolution (QoS value service resources in t and t+1 moment at runtime, the expected output of QoS value)
  8:   END IF
  9: RETURN control operation vector in t + 1 moment
10: End

Algorithm 2. Architecture adjustment approach

Due to the limitation of space, the details of the method described in this section, and the experimental validation will be described in subsequent papers.

3   Parallelization Considerations for Data Mining

For this platform, there are many sources of data. The first part is derived from enterprise management cloud services produced data, such as online accounting, online sales. The second part is derived from corporate social networking data, such as Cloud-Hub (Kingdee). The third part is derived from generic service data, such as Express 100, invoice query data. The fourth part is derived from the standard external data source, this part mainly to meet the business needs of their own business data. Additional data, such as system log data.

This section is mainly for the traditional mining algorithm modification, and the parallel processing should be investigated.

3.1   Research on Parallelization of K-Means Approach

K-means method is mainly divided n data objects into k clusters, that is, divide and rule. Based on large-scale data operations, it is clear that K-means on a single computer has been far from satisfying the processing of data clustering, and constant iterations have influenced the computing time [11,12].

In the literature [13], K-means is improved, and the distributed clustering algorithm DK-means is proposed. The algorithm can be implemented as a parallel implementation of K-means clustering algorithm. The algorithm has a distributed system with p sites, from which a site selected S_m is the main site, and the remaining p-1 sites are slave sites. Firstly, k clustering centers $C_1, C_2, ..., C_k$ are randomly generated at the primary site as global initial cluster centers and broadcast to all slave sites. Each site confirms the cluster to which the data object belongs, and obtains the local cluster center by the following formula 1. At the same time, from the site will be the site of the local clustering center point and the corresponding cluster data object sum $\{(c_{i1}, n_{i1}), ..., (c_{ik}, n_{ik})\}$ $(1 \leqslant i \leqslant p)$ to the main site. The master site calculates the global cluster center based on these clustering information.

$$C_j = \frac{n_{1j} \times c_{1j} + n_{2j} \times c_{2j} + \cdots + n_{pj} \times c_{pj}}{n_{1j} + n_{2j} + \cdots + n_{pj}}, (1 \leq j \leq k) \tag{1}$$

For the iteration of this process, E value can be stable until the global discriminant function, that is, the global clustering center is stable. Algorithm DK-means does not need to transmit massive data between nodes, only transfer the cluster center point and the totally data objects of the cluster. The traffic is very small, so the algorithm DK-means is very efficient. The clustering result of distributed clustering algorithm DK-means is equivalent to the clustering of distributed data using K-means algorithm.

Based on above parallelization considerations, this project will realize the K-Means process in combination with the characteristics of Mapreduce calculation framework. The basic principle of the process is as follows. According to the K-means clustering algorithm, the parallel principle analysis distributes all the data to different nodes. Each node only calculates its own data. Each node can read the cluster generated by the last iteration and calculate its own data points to determine which cluster the data point should belong to. Each node calculates a new cluster centers in one iteration based on its own data points. Consolidate the cluster centers calculated by each node to calculate the final actual cluster centers.

3.2 Industry-Oriented Mining Algorithm: Tag-Based Domain Expert Finding

For each enterprise social network users, the ability to quickly find the information they need is the most important. If a user wants to learn about a company's product, he needs to find a stakeholder who understands and is familiar with the product. So our problem is translated into the search for "authoritative users". For any one domain, we want to find the "authoritative user" problem is in fact an expert search problem in this area. If we can quickly rank all the experts in this field in the corporate Weibo network, then our problems can be solved (Fig. 2).

For any member v in the social network, there exists a label set

$$T_v = \{t_1, t_2, \cdots, t_{n-1}, t_n\}$$

where t_i is the label of member v, which can be the user's own label, or other user's evaluation tag for user v. If the member v has the tag t_i, then we can consider that the user v is good at this domain t_i. That is, the user v is an expert in the domain t_i. For the entire corporate Weibo network, all user tags will form a label set $\Gamma = \bigcup T_v$. If there is a label "data model" in Γ, we want to make a ranking among the experts in the "data model" of the whole enterprise microblogging network, which is beneficial to the users' search, study and the use of enterprise talents. Then we can calculate on this label "data model" for an expert ranking.

The size of the expert influence determines the rank of the experts. The following conditions determine the influence of experts.

1. For any user v, the size of his expert influence in the field is influenced by the number of users in the social network that give him the labeled t_i tag. In general, if the user can give the user v labeled tag t_i more, it shows that the greater influence of field t_i in the enterprise microblogging. The expert ranking is even more forward.
2. The greater the number of users who have a greater impact on the user's tag t_i, the greater the user's influence.

This understanding is very similar to the PageRank algorithm. The definition of PageRank comes from Larry Page as follows.

$$R_u = c \sum\nolimits_{v \in N^+(u)} \frac{R_v}{|N^-(v)|} = c \sum\nolimits_{v \in N^+(u)} \frac{R_v}{d_G^-(v)}$$

where u, v respectively represent two different pages. R_u and R_v denote the PageRank values for u and v, respectively. $N^+(u)$ represents the collection of pages that refer to the page u. $d_G^-(v)$ indicates the number of web links referenced by page v. In the linked network graph, $N^+(u)$ represents the set of vertices in the vertex u. $d_G^-(v)$ represents the output degree of the vertex v.

For this purpose, this project is based on PageRank implementation domain expert discovery algorithm. At the same time, a user-tag data network model is formed by abstracting on the basis of the original social network. Since the process of tagging another user is a one-way process, the user-tagged data network model is not exactly the same as the previous SN model. Here it is a directed graph.

We use ER to express the influence of the size of the experts. From the previous introduction we can see that for any user v, his influence in the field of t_i from all the field t_i experts to evaluate the user's influence. That is to say.

$$ER_v = c \sum_{u \in d_G^+(v)} N^+(u) \omega_{u \to v} * ER_u$$

Here, ER_v and ER_u denote the expert influence of user v and user u in t_i domain, respectively. C is a normalized constant. From the above formula can be seen, for the user v the expert influence depends entirely on the size of $d_G^+(v)$ and ER_u. In other words, the more tags the user v given experts for t_i domain, the greater the influence of his experts. The greater the influence of the user who gives him t_i domain expert tags, the greater his influence. This is a testament to the factors that we are discussing influencing the ranking of experts.

Here the presence of $\omega_{u \to v}$ is to illustrate the user u to the user v in the domain t_i of the tags generated by expertise. Here, we can believe that the user u own expert influence in line with the size of $d_G^-(u)$, evenly distributed to their own field set $N^-(u)$. That is, if the user gives a tag of the t_i domain expert to the t users, then each user will receive the expert influence of the value $ER_u/d_G^-[u]$ of the user u. So the above formula can be optimized for the following formula.

$$ER_v = c \sum_{u \in N_G^+(v)} N^+(u) ER_u/d_G^-[u]$$

4   Experiment and Empirical Analysis

Kingdee Cloud-Hub should be chosen as experimental carrier which integrated the Internet of things, cloud and big data technology to build a service-oriented Internet based application software system. Kingdee Cloud-Hub is a

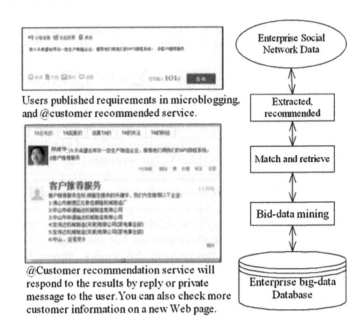

Fig. 2. Tag-based domain expert finding for recommend services

service-oriented platform faced with richness stakeholders, diversity personalized requirements, thus leading to a variety of personalized customization requirements. In this experimental carrier, we have developed a prototype of big-data cloud service system (see Fig. 3). Meanwhile, the platform adopts SOA development style and a large number of service resources (including Microsoft Asmx or Java Axis) have developed. The above characteristics conform to exploring

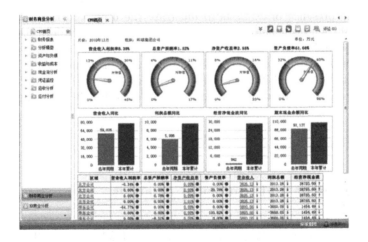

Fig. 3. User interface of prototype for big-data cloud service system

empirical collaborative SOA carrier requirements. Through continuous iteration, we can get a comprehensive CASE tool needs to support big-data analysis cloud services.

A large number of services have been developed for the platform. For example, only for expert traceability information service, there are more than a hundred WCF services.

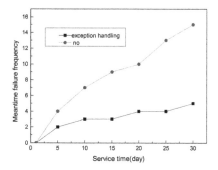

Fig. 4. Effect comparison for cloud servises platform using exception handling

In addition to platform-developed services, system also has called a large number of external services, such as map services, weather services and other physical services etc., which are typical SBS applications. In the early stage, platform operation is extremely unstable. No corresponding runtime exception handling mechanism is the main reason. Figure 4 is the comparison result for the two sets of same system platform running at the same time. One of platforms is running under the part of the designed exception handling mechanism in this paper. Comparison results show that the embedded runtime exception handling mechanism obviously improved the platform's ability to deal with all kinds of uncertainty.

5   Conclusions

The main contributions of this paper are briefly summarized as follows. (1) Self-adaptive exception handling architecture for active services resource provisioning has been built. (2) Runtime adaptive adjustment mechanism has been designed to deal with requirements and context changes. (3) For the traditional data mining algorithms, parallel transformation was modified.

This paper aims at study the adaptive runtime exception handling and parallel data mining for massive data processing. We will mainly focus on active services resource provisioning and adaptive SOA runtime exception handling mechanism. We also systematically studied the application of parallel data mining algorithms in cloud data. The solution will provide systematical support to build intelligent cloud services for big-data analysis.

Acknowledgment. This research has been partially supported by the Natural Science Foundation of China (No.61562024) and Huazhong University of Science and Technology undergraduate students to study abroad exchange project.

References

1. Dai, W., Qiu, L., Wu, A., Qiu, M.: Cloud infrastructure resource allocation for big data applications. IEEE Trans. Big Data **PP**(99), 1 (2017). (IEEE Early Access Articles)
2. Thingom, C., Yeon, G.: An integration of big data and cloud computing. In: Satapathy, S., Bhateja, V., Joshi, A. (eds.) Advances in Intelligent Systems and Computing, vol. 469, pp. 729–737. Springer, Singapore (2017). https://doi.org/10.1007/978-981-10-1678-3_70
3. Yu, W.D., Gottumukkala, A., Senthailselvi, D.A., Maniraj, P., Khonde, T.: Distributed big data analytics in service computing, pp. 55 – 60, Bangkok, Thailand (2017)
4. Liu, L.: Editorial: service computing in 2015. IEEE Trans. Serv. Comput. **8**(1), 1 (2015)
5. Allier, S., et al.: Multitier diversification in web-based software applications. IEEE Softw. **32**(1), 83–90 (2015)
6. Lemos, A.L., Daniel, F., Benatallah, B.: Web service composition: a survey of techniques and tools. ACM Comput. Surv. **48**(3), 1–41 (2015)
7. Zaki, M.J.: Parallel and distributed association mining: a survey. IEEE Concurrency **7**(4), 14–25 (1999)
8. Chouiref, Z., Belkhir, A., Benouaret, K., Hadjali, A.: A fuzzy framework for efficient user-centric web service selection. Appl. Soft Comput. **41**, 51–65 (2016)
9. Xiong, W., et al.: A self-adaptation approach based on predictive control for SaaS. Chin. J. Comput. **39**(2), 364–376 (2016)
10. Kephart, J.O., Chess, D.M.: The vision of autonomic computing. Computer **36**(1), 41–45 (2003)
11. Sultana, M., Paul, P.P., Gavrilova, M.L.: Social behavioral information fusion in multimodal biometrics. IEEE Trans. Syst. Man Cybern. Syst **PP**(99), 1–12 (2017). (IEEE Early Access Articles)
12. Besimi, N., Cico, B., Besimi, A.: Overview of data mining classification techniques: traditional vs. parallel/distributed programming models. In: 2017 6th Mediterranean Conference on Embedded Computing (MECO), pp. 1–4, June 2017
13. Ji, G., Ling, X., Yang, M.: Distributed clustering algorithm based on ensemble learning. Dongnan Daxue Xuebao (Ziran Kexue Ban)/J. Southeast Univ. (Natural Science Edition) **37**(4), 585–588 (2007)

Research on Automatic Generation of Test Cases Based on Genetic Algorithm

Lu Xiong$^{(\boxtimes)}$ and Kangshun Li

Software Department, Guangdong University of Science and Technology,
Dongguan, Guangdong, China
317771184@qq.com, likangshun@sina.com

Abstract. Automatic generation of test cases is a key problem in software testing, and also a hot issue in software testing research. After introducing the existing methods of automatic generation of software test cases, this paper focuses on the automatic generation method of test cases based on genetic algorithm. Summarizes the basic idea of the method and basic procedure, method of automatic generation of test cases according to the different classified as random method, static method and dynamic method of three categories, and combined with the literature analysis of the three kinds of methods and techniques to their characteristics, compares their ad-vantages and disadvantages. Finally, the shortcomings are pointed out, and the development direction is discussed.

Keywords: Software testing · Genetic algorithm · Test cases generation

1 Overview of Software Testing Technology

With the development of computer software and hardware technology, the application field of computer is more and more extensive, the application of software in various aspects is more and more powerful, and the complexity of software is also higher and higher. But how can we ensure the quality of the software and ensure the high reliability of the software? Undoubtedly, the necessary testing of software products is a very important link. Software testing is also the final audit of software requirements, design and coding before the software is put into operation [1]. Software testing is a means to ensure software quality, through testing can find the defects of the software, the measure of software quality, is the final review of software specification, design and encoding, verify whether the software implements all functional requirements in the design, so it is one of the most important aspects of software quality assurance [2].

Software testing is a process that can be carried out in 4 steps, namely, unit testing, integration testing, acceptance testing and system testing. Software testing can be a very good solution to the problems in software, in a test case is correct, the test by software testing can find software errors, can also enable developers to reduce error in the future development process through the analysis of the testing results, the testing personnel can also reduce the useless test to improve the case test experience. But generating test cases is a huge amount of work, and if you rely on manual input to complete, there is likely to be errors. Therefore, it is a difficult problem to realize the automatic generation of test cases.

© Springer Nature Singapore Pte Ltd. 2018
K. Li et al. (Eds.): ISICA 2017, CCIS 873, pp. 351–360, 2018.
https://doi.org/10.1007/978-981-13-1648-7_30

At present, the test cases automation generation methods mainly include 3 categories: random generation method, static generation method and dynamic generation method [3]. The random method refers to the random selection of test cases within the scope of the feasible domain, until all test requirements are met or the maximum test cases limit is reached [4]. The static structured test cases generation method does not need to execute the tested program. By transforming all the sentences on the path into the constraint system, the test cases set is solved by using good mathematical properties. The solution methods include interval arithmetic [5], symbolic execution method, [6, 7], etc. The dynamic test cases generation method is widely adopted by academic circles in recent years, by introducing the heuristic in generating data in exploration, using the feedback information of the measured program; determine which data can meet the test requirements. According to this feedback mechanism, the test cases is adjusted gradually until the test requirements are met. This method was put forward in the literature [7], and then a lot of work was carried out on this basis for [3, 9–21].

2 The Method of Automatic Generation of Test Cases Based on Genetic Algorithm

In recent years, meta heuristic algorithm has been widely used in this field, in which genetic algorithm is the most widely used technology [24]. GAs have been considered with increasing interest in a wide variety of applications [22]. These algorithms are used to search the solution space through simulated evolution of "survival of the fittest." These are used to solve linear and nonlinear problems by exploring all regions of state space and exploiting potential areas through mutation, crossover, and selection operations applied to individuals in the population [23].

(1) According to the problems demand, it will produce the first generation of population.
(2) Calculate the fitness value of the individual in the population (the fitness function is generated according to the problem demand).
(3) According to the problem demand, the number of iterations and the fitness threshold should be set. If the number of iterations is lower than the set value or the maximum fitness value in the population is lower than the set value, the cycle operation will be entered:
    a. Generation of offspring, then genetic operators (select operators, crossover operators, and mutation operators) are performed.
    b. Calculate the fitness value of the new population.
(4) When the new population chromosome fitness value or number of iterations reaches the set value, stop the algorithm.

In 1992, the paper [25] proposed the application of genetic algorithm to the automatic generation of structured test cases(GA) for the first time. The basic principles of GA are shown in Fig. 1.

In 2001, Michael et al. used two genetic algorithms to solve the local search problem in hill climbing method in paper [3]. In 2004, Khor et al. combined the genetic algorithm with formal concept analysis in the paper [26] to track the test cases and the

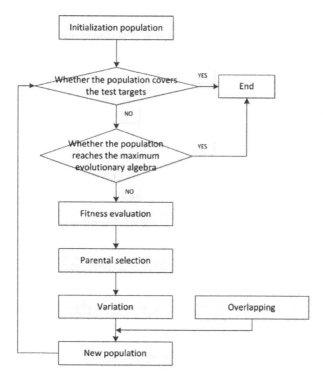

Fig. 1. Basic schematic diagram of GA algorithm

corresponding test execution relationship. In 2006, Xue et al. proposed a generative method based on the messy genetic algorithm in paper [27]. The test coverage rate was used as the evaluation function of the test input set to guide the generation process. In 2008, Alba et al. conducted a comparative analysis of parallel evolutionary algorithms and serial evolutionary algorithms for test cases generation problems [28]. In 2009, Awedikian et al. proposed an extended branch distance calculation method in paper [29], which guided the generation of test cases using the dependency between the genetic algorithm framework and variables.

Data generation method based on path test is one of the main methods of test case generation in software testing. At present, there are many ways to generate test case generation methods; especially the use of genetic algorithm to generate test cases is made great progress. The paper [30] proposes a method based on genetic algorithm and function minimization to generate test cases. The paper [32] proposes an adaptive SAGA algorithm to generate test cases automatically. The paper [33] proposed an improved genetic algorithm combining adaptive genetic algorithm with hill climbing algorithm to generate test cases automatically by HCGA. The paper [35] proposed differential evolution algorithm based on the basic genetic algorithm. The algorithm is a random heuristic search algorithm, simple and easy to use, the stability of the algorithm is good, the global search ability is strong, has been successfully applied in many fields easy to use, good stability, strong global search ability, have been successfully applied in many fields. In paper [36], the directional mutation operator is introduced into the genetic algorithm, and then the fitness function is designed according to the similarity and distance between the paths, which can effectively distinguish the degree of the individual.

3   The Method of Automatic Generation of Test Cases Based on Ant Colony Algorithm

Automatically generate test cases based on ant colony algorithm is referred to as ant colony optimization algorithm (ACO algorithm). McMinn et al. [25] introduces the bionic search and heuristic algorithm proposed by Dorigo et al. [26] to the new algorithm which is formed in the automatic generation of test cases.

The basic idea of ACO algorithm is to express the structure information of the application to some knowledge, then apply ant colony algorithm to complete the knowledge reasoning and generate the required test cases in the process of reasoning.

The basic principle is to put the ant colony intelligent body on the path to the graph G (ant refers to the data element that forms a test case code). The optimal path P is searched through the action of ant movement and release pheromones; so as to find the input variables that can cover the test day standard O. (cover the marking of edges in figure G). The ant's moving path on the graph G is decoded to form a test case.

The core process of ACO algorithm includes:

(1) Convert test questions to direct graph G;
(2) Construct an evaluation function F based on the graph G, which can be used to quantify the path that is covered by the pheromone;
(3) Possible solution generation algorithm and stop ratio criterion;
(4) update strategy information element;
(5) The ant colony is moved from node A of figure G to the strategy of node B.

The basic principles of ACO are shown in Fig. 2.

In 2003, McMinn et al. [25] introduced the ACO algorithm into the test case automatic generation, which was used to find the state migration sequence of the waiting program and optimize the evolutionary algorithm. In the same year, Doerner et al. [28] used the ACO algorithm to extract the available test paths for optimizing test coverage and testing costs. In 2005, Li et al. [29] used the ACO algorithm to perform state coverage tests based on UML. In 2007, Bo Fu [27] introduced the path mutation operator and the adaptive volatility coefficient in ACO algorithm, which improved the diversity of the ant path and outperformed GA in the number of iterations. In 2009, Li et al. [30] input fields of the application under test is divided into different subdo mains, based on the child domain to construct the search path diagram, and based on the path graph using the ACO algorithm to generate test cases, on the number of iterations of the algorithm is superior to GA. In 2009, Mingshi Chen et al. [31] improved the standard ACO calculation and proposed the polymorphic ant colony algorithm (PACA). PACA divides ant colony into reconnaissance ant, search ant and worker ant, thus narrow the search space of algorithm, enhance the collaboration between ant colony and improve search efficiency. The experiment of Mingshi Chen et al. proved that PACA is superior

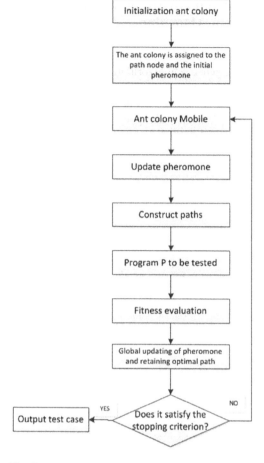

Fig. 2. Basic schematic diagram of ACO algorithm

to ACO algorithm in the execution time of test case generation. In 2010, Xu et al. [32] introduced the GAACO algorithm by introducing genetic algorithm in the ACO algorithm. The algorithm overcomes the poor local search ability of genetic algorithm and medium drought phenomenon, at the same time make up for the ant colony algorithm is time-consuming, easy to stagnation phenomenon, enhance the randomness generated test cases, quickness and global convergence. Xu et al. demonstrated that GAACO is superior to GA, ACO and RT in iteration times. The paper [31] proposes an automatic test case generation method based on ant colony algorithm. ACO algorithm is an enhanced learning system with distributed computing, easy to integrate with other algorithms, robust robustness, etc. However, due to the relative shortage of early information element in search, the search efficiency of the algorithm is reduced, and the positive feedback mechanism is prone to stagnation and drought.

4 The Method of Automatic Generation of Test Cases Based on Particle Swarm Algorithm

Automatic generation algorithm for test cases based on particle swarm (PSO), which is a new test case automatic generation algorithm [44] based on Kennedy and Eherhart's research on animal social behavior [45]. PSO has the characteristics of simple algorithm and good robustness. The interaction between the individual cognition of the test cases and the interaction of swarm intelligence and the dynamic use case data have a good orientation and convergence.

The basic principle of PSO algorithm is shown in Fig. 3.

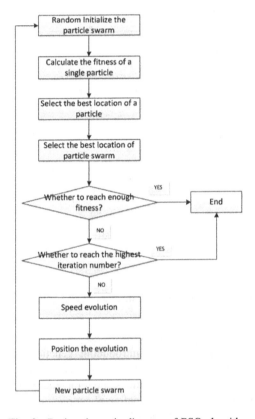

Fig. 3. Basic schematic diagram of PSO algorithm

In 2007, Windisch et al. [44] the PSO algorithm is introduced into the automatic test case generation algorithm and through the branch coverage test and GA are analyzed and compared, points out that the PSO is better than that of GA in the test of complex objects. In 2008, Li Aiguo et al. [46] branch function superposition method is used to construct the PSO adaptive value function, the measuring unit of execution path coverage testing, and work with GA is analyzed and compared, show that PSO algorithm in the number of iterations and the execution time is better than that of GA on the two indicators. That same year, Hla et al. [47] propose a fitness measurement method based on the comprehensive coverage, and compared with the random algorithm performance, analyzed the PSO algorithm is better than that of GA in complexity. In 2010, Cui et al. [48] constructed the EFAU path measurement function in the PSO algorithm. The PSO adaptive value is the similarity between the target path and the EFAU metric value. Cui et al. experimentally compared the iterative performance of GA, PSO and EFAU-PSO, proving that EFAU-PSO was superior to the other two algorithms in iteration times. Hong Zhou et al. [49] introduced GA in PSO algorithm, to enhance the diversity of the population of test cases in PSO, reduce the precocious characteristics of PSO algorithm. The number of iterations and execution time is better than GA. Nayak et al. [50] built PSO test case generation algorithm based on data flow diagram, the complexity of PSO and GA algorithms is analyzed. In the coverage test experiment, it verifies that the PSO algorithm has less iteration than GA. The paper [34] proposes an automatic test data generation method for software structure based on PSO algorithm. However, the above methods are difficult to code in different degrees, the parameter setting is difficult, the algorithm is tedious and the efficiency is not high.

PSO is superior to traditional algorithms in terms of its quick calculation and ease of use. But PSO is as precocious as any other intelligent algorithm. It is generally believed that PSO precocious is caused by the loss of diversity of particle swarm, and that leads to the majority of individuals in the group gathering in a very small search area, when the range of the search does not contain the optimal solution, the algorithm will be trapped in the local optimal.

4.1 Conclusions

Genetic algorithm is an effective method for automatic generation of structural test cases, which has great advantages and good prospects for development. However, the genetic algorithm also has the following shortcomings:

(1) Although genetic algorithm has strong global optimization ability, it always exist the possibility of falling into local optimum.
(2) How to choose the appropriate genetic algorithm? It is necessary to test and analyze how to select the appropriate parameters and so on.

In order to give full play to the role of genetic algorithm in software testing, the genetic algorithm will be further studied for the above problems in the future.

Acknowledgements. The work is supported in part by Department of Education of Guangdong Province under Grant 2015KQNCX193.

References

1. Fu, B.: Research on software intelligent test cases generation. Comput. Eng. **32**(14), 117–207 (2006)
2. Roy, C.K., Cordy, J.R., Koschke, R.: Comparison and evaluation of code clone detection techniques and tools: a qualitative approach. Sci. Comput. Program. **74**(7), 470–495 (2009)
3. Michael, C.C., et al.: Generating software test cases by evolution. IEEE Trans. Softw. Eng. **27**(12), 1085–1110 (2001)
4. Bird, D.L., Munoz, C.U.: Automatic generation of random self-checking test case. IBM J. Res. Dev. **22**(3), 229–245 (1983)
5. Wang, Z., Liu, C.: Interval arithmetic application in software testing. J. Softw. **9**(6), 438–443 (1998)
6. Clarke, L.A.: A system to generate test cases and symbolically execute programs. IEEE Trans. Softw. Eng. **2**(3), 215–222 (1976)
7. Ramamoorthy, C.V., et al.: Automated generation of program test cases. IEEE Trans. Softw. Eng. **2**(4), 293–300 (1976)
8. Miller, W., Spooner, D.L.: Automatic generation if floating-point test cases. IEEE Trans. Softw. Eng. **2**(3), 223–226 (1976)
9. Wegener, J., et al.: Testing real-time systems using genetic algorithms. Softw. Qual. J. **6**(2), 127–135 (1977)
10. Korel, B.: Automated software test cases generation. IEEE Trans. Softw. Eng. **16**(8), 870–879 (1990)
11. Jones, B.F., et al.: Automatic structural testing using genetic algorithms. Softw. Eng. J. **11**(5), 299–306 (1996)
12. Zheng, C., Li, J., Jing, X., He, Y.: A path-oriented automatic test cases generation method. J. Comput. Appl. Softw. **32**(11), 11–13 (2015)
13. Li, Z.: Automatic testing-case generation based on adaptive genetic algorithm. J. Comput. Syst. Appl. **25**(1), 192–196 (2016)
14. Jiang, Y., Dong, Q.: Approach to generate test cases based on genetic algorithm and branch coverage. J. Comput. Eng. Des. **36**(1), 112–117 (2016)
15. Wu, H., Li, H., Wan, J.: Improved genetic algorithm used in test cases. J. Comput. Syst. Appl. **25**(8), 200–205 (2016)
16. Liu, X., Yang, Y., Zhang, C., Wang, W.: A novel method for fuzzing test case generating based on genetic algorithm. J. Acta Electronica Sinica **45**(3), 552–556 (2017)
17. Song, Q.: Test Case Generation Based on Genetic Algorithm. J. Comput. Syst. Appl. **23**(11), 264–267 (2014)
18. Liu, D., Jin, B., Que, X.: Automatic generation of minimal test set based on a genetic algorithm. J. Comput. Technol. Dev. **26**(4), 86–89 (2014)
19. Ding, R., Dong, H., Zhang, Y., Feng, X.: Fast automatic generation method for software testing data based on key-point path. J. Softw. **27**(4), 814–827 (2016)
20. He, J., Zheng, L.: Automatic generation for software testing data based on a improved genetic algorithm. J. Softw. Guide **16**(9), 10–12 (2016)
21. Huang, X., Wang, X., Chang, D., He, G.: Application of modified differential evolution in test cases generation. J. Comput. Appl. **29**(6), 1722–1724 (2009)
22. Chen, Y., et al.: A genetic algorithm based on improved selection operator. Comput. Eng. Appl. **44**(2), 44–45 (2008)
23. Li, M., Kou, J., et al.: The Basic Principle and Application of Genetic Algorithm. Science Press, Beijing (2002)

24. Diaz, E., et al.: A tabu search algorithm for structural software testing. Comput. Oper. Res. **35**(10), 3052–3072 (2008)
25. Xanthakis, S., et al.: Application of genetic algorithm to software testing. In: Proceedings of the 5th International Conference on Software Engineering, pp. 625–636. IEEE Computer Society, Los Alamitos (1992)
26. Khor, S., Grogono, P.: Using a genetic algorithm and formal concept analysis to generate branch coverage test cases automatically. In: Proceedings of the 19th IEEE International Conference on Automated Software Engineering, pp. 346–349. IEEE Computer Society, Los Alamitos (2004)
27. Xue, Y., et al.: An automated approach for structural test cases generation based on messy GA. J. Softw. **17**(8), 1688–1697 (2006)
28. Alba, E., Chicano, F.: Observations in using parallel and sequential evolutionary algorithms for automatic software testing. Comput. Oper. Res. **35**(10), 3161–3183 (2008)
29. Awedikian, Z., et al.: Mc/Dc automatic test input data generation. In: Proceedings of the 11th Annual Conference on Genetic and Evolutionary Computation, pp. 1657–1664. ACM, New York (2009)
30. Kaixing, Lun, L.: Automatic test cases generation method. Comput. Technol. Dev. **9**(16), 53–55 (2006)
31. Fu, B., et al.: Automatic generation of software test cases based on ant colony algorithm. Comput. Eng. Appl. **43**(12), 97–99 (2007)
32. Guobin, et al.: Automatic generation of test cases based on adaptive SAGA. Microelectron. Comput. **8**(23), 10–13 (2006)
33. Wang, J., Ding, G., Song, H.: Automatic test cases generation based on improved adaptive genetic algorithm HCGA. Trans. Beijing Inst. Technol. **10**(27), 883–885 (2007)
34. Li, A., Zhang, Y.: Automatic test cases generation method for software structure based on PSO. Comput. Eng. **34**(6), 93–97 (2008)
35. Storn, R., Price, K.: Differential Evolution a Simple an Deficient Adaptive Scheme for Global Optimization over Continuous Space, TR 95-012. International Computer Science Institute, Berkley (1995)
36. Feng, J., Yu, L.: Genetic algorithm improvement in test cases generation. J. Comput. Aided Des. Comput. Graph. **27**(10), 2008–2014 (2015)
37. McMinn, P., Holcombe, M.: The state problem for evolutionary testing. In: Cantú-Paz, E., et al. (eds.) GECCO 2003. LNCS, vol. 2724, pp. 2488–2498. Springer, Heidelberg (2003). https://doi.org/10.1007/3-540-45110-2_152
38. Dorigo, M., Gambardella, L.M.: Ant colony system: a cooperative learning approach to the traveling salesman problem. IEEE Trans. Evol. Comput. **1**(1), 53–66 (1977)
39. Doerner, K., Gutjahr, W.J.: Extracting test sequences from a markov software usage model by ACO. In: Cantú-Paz, E., et al. (eds.) GECCO 2003. LNCS, vol. 2724, pp. 2465–2476. Springer, Heidelberg (2003). https://doi.org/10.1007/3-540-45110-2_150
40. Li, H., Lam, C.P.: Software test cases generation using and colony optimization. In: Proceedings of World Academy of Science, Engineering and Technology, pp. 1–4 (2005)
41. Li, K., Zhang, Z., Liu, W.: Automatic test data generation based on ant colony optlmization. In: Proceedings of the 5th International Conference on Natural Computation, pp. 216–220 (2009)
42. Chen, M., Liu, X., Liu, T.: Automatic generation of test cases based on polymorphic ant colony algorithm. Comput. Appl. Res. **26**(6), 2347–2348 (2009)
43. Xu, D., Li Z.: Study on test case automated generation technology based on genetic algorithm and ant colony optimization algorithm. In: Proceedings of International Conference on Electrical and Control Engineering, pp. 5655–5658. 11,1,1, Computer Society, Washington DC (2010)

44. Windisch, A., Wappler, S., Weglner, J.: Applying particle swarm optimization to software testing. In: Proceedings of the 9th Genetic and Evolutionary Computation Conference, pp. 1121–1128. ACM Press, New York (2007)

45. Kennedy, J., Eberhart, R.: Particle swarm optimization. In: Proceedings of the IEEE International Conference on Neutral Networks, pp. 1942–1948. IEEE Press, Piscataway (1995)

46. Li, A., Zhang, Y.: Automatic generation method of software structure test data based on PSO. Comput. Eng. **34**(6), 93–97 (2008)

47. Hla, K.H.S., Choi, Y.S., Park, J.S.: Applying particle swaTxn optimization to prioritizing test cases for embedded real time software retesting. In: Proceedings of the 8th IEEE International Conference on Computer and Information Technology Workshops, pp. 527–532. IEEE Computer Society, Washington DC (2008)

48. Cui, H.-H., Chen, L., Zhu, B., et al.: An efficient automated test data generation method. In: Proceedings of International Conference on Measuring Technology and Mechatronics Automation, pp. 453–456. IEEE, Computer Society, Washington DC (2010)

49. Zhou, H., Zhang, S., Liu, L.: The path test data based on ga-pso algorithm is automatically generated. Comput. Appl. Res. **27**(4), 1366–1369 (2010)

50. Nayak, N., Mohapatra, D.P.: Automatic test cases generation for data flow testing using particle swarm optimization. Contemp. Comput. **95**(2), 1–12 (2010)

A Routing Acceleration Strategy via Named Data Networking in Space-Terrestrial Integrated Networks

Feng Yang[1(✉)] and Di Liu[2]

[1] School of Physics and Mechanical and Electrical Engineering,
Hechi University, Yizhou, Guangxi, China
oyfo@163.com
[2] Computer School, Wuhan University, Wuhan, Hubei, China

Abstract. Content producer mobility support is remaining a challenging problem. In this paper, we consider it in a Future Space-Terrestrial Integrated Networks (FSTINs) scenario without handover management infrastructure. Specifically, we propose a routing acceleration strategy via Named Data Networking (NDN) in Space-Terrestrial Integrated Networks, as well as the acceleration algorithm. Performance evaluation results demonstrate that the proposed strategy can obtain the acceleration ratios from 4.8% to 9.4% at different nodes, and the average acceleration ratio is 7%.

Keywords: Routing acceleration strategy
Space-Terrestrial Integrated Network · NDN forwarding

1 Introduction

The change of philosophy in wireless networking design privileged the local "cellular" connectivity rather than the global interconnection. This model works well for terrestrial nodes especially desktop PCs talking to servers over wires, but not for our increasingly wireless mobile world, especially for Future Space-Terrestrial Integrated Networks (FSTINs) [1]. As in [1], FSTINs must be globally "anywhere-anytime", must be capable of assisting society in emergencies, and must be trustworthy. Terrestrial networks cannot effectively fulfil these basic requirements due to their intrinsic "local" nature. In the upcoming "Space 2.0" era, there are much more space nodes, especially MEO/LEO satellite and Near-Earth aircraft, they can equip powerful links and serve as either data producers/consumers or relay nodes. For this reason, FSTINs can be thought of as a "building" supported by two basic pillars: the "local" pillar made by terrestrial networking, and the "global" pillar made by space networking. In such a futuristic vision, Multi-layered Satellite Networks (MLSNs) [2] can serve as the space backbone of FSTINs. One possible scenario is show in Fig. 1.

In such a scenario, mobility is the norm rather than the exception, the current TCP/IP architecture is not competent and we need location-independent communication model [3]. A common criticism is the so-called location-identity conflation problem. Despite the enormous body of work on dealing with device mobility so as to

© Springer Nature Singapore Pte Ltd. 2018
K. Li et al. (Eds.): ISICA 2017, CCIS 873, pp. 361–370, 2018.
https://doi.org/10.1007/978-981-13-1648-7_31

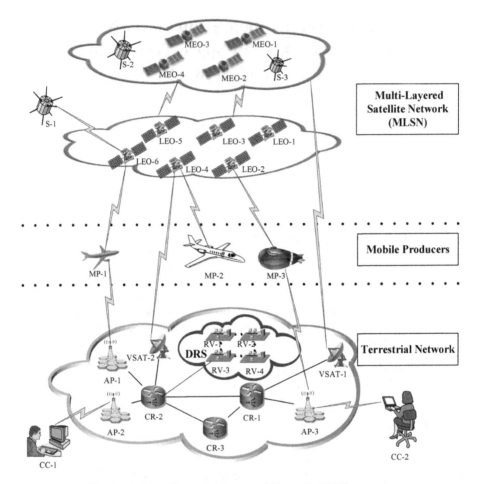

Fig. 1. A Space-Terrestrial Integrated Network (STIN) scenario

achieving location-independent communication, most known approaches fundamentally take one of just three different approaches [3]: (1) indirection routing; (2) name resolution; (3) name-based routing. Between them, name-based routing trying to solve this problem from a new perspective by changes the mobility problem from "delivering packets to a mobile node (MN)" to "retrieving data produced by an MN".

As a promising Future Internet architecture, Named Data Networking (NDN) [10] changes the network communication model from point-to-point packet delivery to named data retrieval without concerning the exact locations where data reside in. As a newly proposed network architecture, NDN has drawn much attention from networking researchers and has been applied in several fields, such as vehicular networking [11], video streaming [12] etc.

The NDN architecture naturally supports consumer mobility through its stateful forwarding plane and the receiver-driven paradigm. But the content producer mobility problem remains an active research topic [5–8]. Producer mobility leads to frequent routing update and low routing aggregation due to the locator/identifier binding properties. The quantitative comparison results [5] suggest that name-based routing may be augmented with addressing-assisted approaches to handle content producer mobility for highly dynamic scenarios, such as FSTINs.

In tracing-based mobility solutions [6, 8], whenever a Mobile Producer (MP) changes its Attachment Point (AP), it needs to inform its *Rendezvous* (RV) to create a "breadcrumb trail" that can be followed by *Interest* to reach it. This process can be referred as Attachment Update (AU) and it does not need to perform any location registration or update operation [13]. Existing solutions just consider situations that AUs occur between non-moving APs that reside in terrestrial network, e.g. MP-1 moves from AP-2 to AP-1, as illustrate in Fig. 2 (Green dashed line).

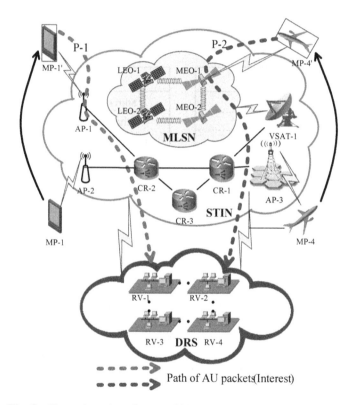

Fig. 2. Illustration of producer mobility in FSTIN (Color figure online)

In such an AU, the forwarding path (P-1) on which the AU packet will be routed could be considered as a known condition, just as most existing solutions. But such an assumption will not hold, if the current AP is also a space node. As in the AU that after MP-4' AP was changed from AP-3 to MEO-1, P-2 would across the Multi-Layered Satellite Networks (MLSNs) where the MEO-1 resides in. In MLSNs, all links are switch frequently, how such special *Interest* packets should be routed is remaining a challenging problem.

2 The Basic Design of the Mobility Management Scheme

This section presents a tracing-based producer mobility management scheme via NDN for highly dynamic scenarios. In particular, we focus on Space-Terrestrial Integrated scenarios, e.g., as shown in Fig. 1. We partly borrow the ideas of [8, 9]. RVs are used to guarantee the global reachability of MPs and all RVs form a Distributed Rendezvous System (DRS). Each object has a unique prefix and with a hierarchical name following it. The data's order of magnitude may be too high to processing them separately in Forwarding Information Base (FIB). We just store a soft-state in tFIB [9]. For long-term relocation of data producers, they need to re-publish data by using the prefix of its current RV, so as to avoid long-term path stretching.

We define a special traceable *Interest* packet–AU packet (AUP), the specific structure is shown in Fig. 3.

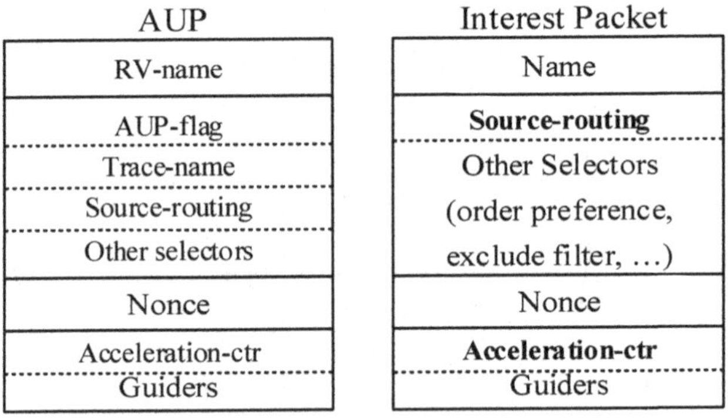

	AUP
RV-name
AUP-flag
Trace-name
Source-routing
Other selectors
Nonce
Acceleration-ctr
Guiders

Interest Packet
Name
Source-routing
Other Selectors
(order preference,
exclude filter, ...)
Nonce
Acceleration-ctr
Guiders

Fig. 3. The AUP and modified interest packet structure

"RV-name" is used to lead AUP Advancing toward its corresponding RV. "AUP-flag" is used to indicate whether the packet is an AUP or not. The specific processes are described in Sect. 3. "Trace-name" indicates which *Interest* will be traced, and it should be exactly same with the *Data* name. It is worth noting that here "Data name" is not restrict to a piece of content, it could be an aggregation one. "Source-routing" field is used to carry the route which the AUP will follow if exists. "Acceleration-ctr" field is optional. A typical MP mobility management process is shown in Fig. 4.

The steps are as follows:

Step 1. MP-4 sends one or more AUPs to its RV after its AP has been changed from AP-1 to LEO-1 (satellite).

Step 2. The following *Interest* packet will still be forwarded towards AP-1. Until this moment, Content Consumer (CC-1) has not aware of this mobility event.

Step 3. The *Interest* packets will trace the AUP at one Content Router (CR), e.g., CR-1.

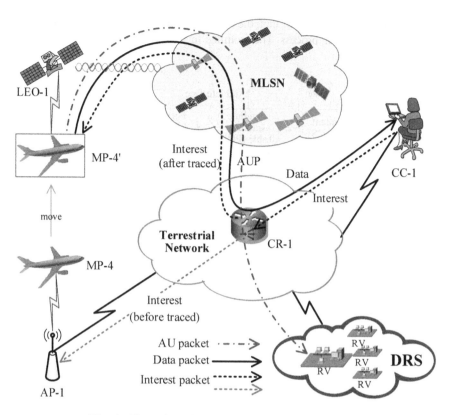

Fig. 4. Illustration of MP' mobility management process

Step 4. The following *Interests* will be forwarded towards MP-4 along the traced path, as well as the unsatisfied *Interests* which have been sent out before traced.
Step 5. MP-4 sends *Data* packet (s) back to CC-1 along the reverse path of *Interests*.

It is worth noting that RVs do not need to participate in producer/consumer communications, and they just act as the rendezvous of AUPs and the unsatisfied *Interests*.

Then, we need addressing the problem that how AUP could be forwarded in the space segment, especially in MLSNs.

3   Space-Terrestrial Integrated Forwarding

This section presents the addressing-assisted forwarding method in NDN. In MLSNs, all links are time-related and there has no stable infrastructure. The CRs will need to update their FIBs frequently, if we directly use pure name-based forwarding. We allocate each satellite a globally reachable name prefix and then calculate the source route of the space segment. After that, the source route will be attached in selector field, so as to assist forwarding process. The modified forwarding process is shown in Fig. 5.

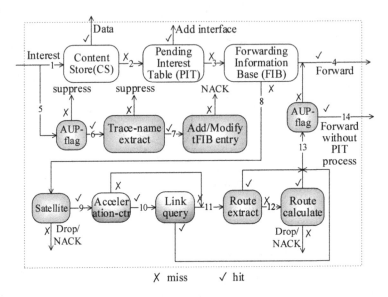

Fig. 5. The modified forwarding process

Process 1–4 just keep the same with the process that in original NDN architecture. Process 5 is used to judging whether the packet is an *Interest* or an AUP. If miss, then suppress, else process 6 extract the "Trace name" that has been attached in the selector fields which will be used in the next process. Process 7 generates the correspondence between "Trace-name" and "Interface", the former is the globally reachable prefix (addressing-assisted) and the latter is the interface through which the *Data* packet(s) can be sent back. After that, the added or modified tFIB entry could be used to guide the following forwarding process and the corresponding *Interests* will toward to the current AP and finally reach the MP.

It is worth noting that after process 7 has been done, the AUP will be suppressed due to the fact that the AUP has been forwarded and toward to its RV through standard process, e.g., process 1 to 4. Our method will not increase the communication burden of CR in terrestrial networks and the additional computation cost is very slight. In most cases, these *Interest* packets are regular ones and the additional cost is one logical judgment, as in process 5.

In MLSNs, the return value of process 3 is "miss" for most packets due to the above analyses. We propose the addressing-assisted forwarding processes. In process 8, if the node is not a satellite, then returns a "NACK" message, else goes to process 11 to extract attached source route, if process 11 return "hit", then go to process 13, else go to process 12 and calculate the source route. If process 13 returns "miss", that is, the packet is an *Interest*, then goes to process 4, else goes to process 14 and forwards the AUP without adding an entry in Pending Interest Table (PIT) so as to reduce entries.

4 Acceleration Strategy

The process 7 that in Fig. 5 can be used for acceleration control. Our motivation is that each space node can be autonomous decision routing and the S2G (Space to Ground) links has stronger transmission capability than S2S (Space to Space) links. Therefore, we can assign higher priority for S2G links to speed up the entire forwarding process.

Algorithm: Acceleration algorithm

Input: The remaining window period of the S2G link—t_r; The Source routing set--R_s; S2G
       link rate—r_2; Packet size—s_p;
Output: acceleration-ctr.
Begin
    Step1: Link assessment.
    For each packet
    if $s_p / r_2 > t_r$ then
    exit
    else goto Step2
    Step2: Acceleration control.
    Extract the remaining hops from $R_s \rightarrow h_t$
    Extract the remaining hops from $R_s \rightarrow h_r$
    if $h_r / h_t > 1/2$ **then**
        acceleration-ctr = 1
    else
        acceleration-ctr = 0
End

The basic idea of the algorithm is: we first evaluate the current S2G link, if the remaining window period of the S2G link can transmit the current packet then we make the decision whether or not to accelerate. The judgment is based on whether the number of remaining hops is greater than half the total number of hops or not.

5 Performance Evaluation

5.1 Scenario Description

We construct the terrestrial segment as illustrated in Fig. 1 and set one RV. We set the MLSN according to the "Iridium NEXT" project which starts in 2016 [15]. The link rate is set to 10 Mbps. The specific constellation parameters are shown in Table 1.

Table 1. Constellation parameters

Satellite type	Satellite number	Orbit period	Orbit altitude	Orbit inclination	RAAN	True anomaly
MEO [16]	2 × 5	360 min	10390 km	45°/135°	Interval 72°	–
LEO [15]	6 × 11	100 min	780 km	86.4°	Interval 45°	Interval 60°

5.2   Forwarding Performance

We store 1000 different Data objects in each MP and allocate each MP with 10 CCs, CCs are need to fetch back these objects from their' corresponding MP. These objects have an average size 1 Mbyte. Then we evaluate the forwarding performance, the results are shown in Fig. 6.

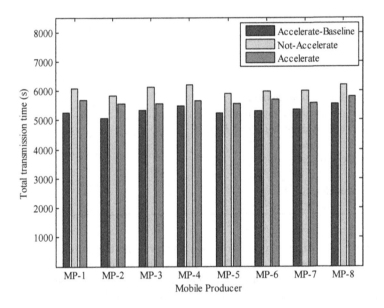

Fig. 6. Total transmission time

The total transmission time of different MPs have slightly difference. The accelerate-Baseline strategy is that: It will use SG link whenever a node has such one, and this strategy has the best acceleration performance, the acceleration ratios varying from 9% to 13.4% and the average acceleration ratio of all the 10 MPs is 11.5%. When we use the accelerate strategy that stated in Sect. 4, the acceleration ratios varying from 4.8% to 9.4% and the average acceleration ratio of all the 10 MPs is 7%.

6   Conclusion

In this paper, we propose a space-terrestrial integrated acceleration strategy based on NDN architecture for assisting space-terrestrial integrated cooperative communication in FSTIN scenarios. We first describe the FSTIN scenario and analyze the producer mobility problem in it, and then design a tracing-based producer mobility management scheme to reduce the FIB update operations in CRs, as well as an addressing-assisted forwarding method that can leverage NDN's stateful forwarding plane. Moreover, describe an efficient routing accelerate algorithm. Finally, we perform evaluations to verify the handover latency and forwarding efficiency of the acceleration strategy.

Acknowledgment. This work was supported by the University research projects of the education department of Guangxi, China (YB2014330).

References

1. Sacchi, C., Bhasin, K., Kadowaki, N., Vong, F.: Toward the "space 2.0" Era [Guest Editorial]. IEEE Commun. Mag. **53**(3), 16–17 (2015)
2. Guta, M., Ververidis, C., Drougas, A., Andrikopoulos, I., Siris, V., Polyzos, G., Baudoin, C.: Satellite-terrestrial integration scenarios for future information-centric networks. In: 30th AIAA International Communications Satellite System Conference. AIAA, 15043 (2012)
3. Li, D., Shen, X., Gong, J., Zhang, J., Lu, J.: On construction of China's space information network. Geomatics Inf. Sci. Wuhan Univ. **40**(06), 711–715 (2015)
4. Min, S.: An idea of China's space-based integrated information network. Spacecraft Eng. **22** (05), 1–14 (2013)
5. Gao, Z.Y., Venkataramani, A., Kurose, J., Heimlicher, S.: Towards a quantitative comparison of location-independent network architectures. ACM SIGCOMM Comput. Commun. Rev. **44**(04), 259–270 (2014)
6. Zhang, Y., Afanasyev, A., Burke, J., Zhang, L.: A survey of mobility support in named data networking. In: Workshop on Name-Oriented Mobility: Architecture, Algorithms and Applications, pp. 83–88. ACM (2016)
7. Kim, D.H., Kim, J.H., Kim, Y.S., Yoon, H.S., Yeom, I.: Mobility support in content centric networks. In: Proceedings of the Second Edition of the ICN Workshop on Information-Centric Networking, pp. 13–18. ACM (2012)
8. Zhang, Y., Zhang, H., Zhang, L.: Kite: a mobility support scheme for NDN. In: Proceedings of the 1st International Conference on Information-Centric Networking, pp. 179–180. ACM (2014)
9. Augé, J., Carofiglio, G., Grassi, G., Muscariello, L., Pau, G.: Anchorless producer mobility in ICN. In: Proceedings of the 2nd International Conference on Information-Centric Networking, 189–190. ACM (2015)
10. Zhang, L., Afanasyev, A., Burke, J., Jacobson, V., Crowley, P., Papadopoulos, C.: Named data networking. ACM SIGCOMM Comput. Commun. Rev. **44**(04), 66–73 (2014)
11. Grassi, G., Pesavento, D., Pau, G., Lixia, Z., Fdida, S.: Navigo: interest forwarding by geolocations in vehicular named data networking. In: IEEE 16th International Symposium on a World of Wireless, Mobile and Multimedia Networks, pp. 1–10. IEEE (2015)

12. Gusev, P., Burke, J.: NDN-RTC: real-time videoconferencing over named data networking. In: Proceedings of the 2nd International Conference on Information-Centric Networking, pp. 117–126 (2015)
13. Soliman, H., Castelluccia, C., Elmalki, K., Bellier, L.: Hierarchical mobile IPv6 (HMIPv6) mobility management. IETF RFC 5380 (2008)
14. Fraire, J.A., Finochietto, J.M.: Design challenges in contact plans for disruption-tolerant satellite networks. IEEE Commun. Mag. 53(5), 163–169 (2015)
15. GunterI: ridium-NEXT. http://space.skyrocket.de/doc_sdat/iridium-next.htm
16. Liu, H.Y., Sun, F.C.: Routing for predictable multi-layered satellite networks. Sci. China Inf. Sci. 56(11), 1–18 (2013)

Cloud Computing and Multiagent Systems – Service Models

Exploring Migration Issue Based on Multi-agent Modeling

Pengfei Liu[1(✉)], Xiaxu He[2], Weifeng Zhang[1], and Enkai Chen[1]

[1] South China Agriculture University, Guangzhou, China
pfl@outlook.com
[2] South China University of Technology, Guangzhou, China

Abstract. The purpose of the paper is to find out the behavior models of individual and group, also the path, rules and the relationship under the national environment in the process of population migration. Thus, the multi-Agent based modeling technology is used, to model the evolution process of population migration, including conceptual model, interaction rules and the rules of the game. Combined with typical immigration cases, the empirical research is carried out to verify the scientific nature of the model. The conclusions of this paper include: Agent based modeling method can integrate the actual situation and the behavior of individuals and groups; carrying out the modeling and simulation of long-term evolution of migration events can reveal the law of migration evolution. The study solves the problem of low visibility in current research of the population migration, and realizes a new method of intuitive and controllable concept of population migration, which provides experience and inspiration from a practical point of view.

Keywords: Migration behavior · Computer simulation · NetLogo
Multi-agent

1 Introduction

The essence of science is the process of modeling while the scientific learning is the process of student modeling [1]. Mainly using computer models, computer simulation simulates real situations so as to study characteristics that are difficult to be observed directly and predict the future trend. Visual controllable characteristics of computer simulation model make the expressionism of abstract theory enhanced, which makes it easier to be understood, and makes it possible to predict conclusions that can only be reached after a long-time evolution [2]. However, in the implementation of most domestic social researches, the research activities based on modeling and model application are rare among frontline researches, yet visualization and operability are the two fundamental elements in studies.

To study the sociological models, researchers are required to solve the specific problems by utilizing the thinking of "macro-micro-macro", summarizing the essential elements of the micro level through social phenomenon of the macro to construct models; through operating models, the cognitive subject observes the outputs of the model's key information, explains macroscopic phenomena or predicts.

© Springer Nature Singapore Pte Ltd. 2018
K. Li et al. (Eds.): ISICA 2017, CCIS 873, pp. 373–382, 2018.
https://doi.org/10.1007/978-981-13-1648-7_32

The current popular simulation systems mainly include Swarm, Repast and NetLogo. The application examples exhibited in other domestic subjects [3–6] and foreign cases [7–9] have provided useful inspirations for domestic sociological research workers in aspects of conducting research and teaching activities with Netlogo, and supported users to test theories repeatedly, conduct empirical researches and reproduce realistic problems with dynamic model.

However, the sociological models in Netlogo model base are few, the majority of them are relatively simple and out of line with national conditions, thus they did not play their proper exemplary roles. Although Netlogo supports lots of sociology teaching models to be written, owing to the limited software development levels of teachers in sociology, and the ordinary programmer's lack of concepts in sociology, the researches that apply Netlogo to sociology are still rare. In this paper, the hot issue "movement of population" will be used as a case, based on the current social situations and population migration theory, the effects of various factors on the migration behaviors in China will be showed dynamically through Netlogo simulation model. It not only embodies the construction processes and characteristics of multi-agent modeling, but also reflects its research effects and advantages in sociological population migration problems.

Study on movement of population is of special social significance to China. Firstly, movement of population may remarkably affect the population distribution pattern in China. Compared to any other nations in the world, Chinese nation attaches much more importance of hometown, and the so-called "It is difficult to move from the homeland" reflects the Chinese culture containing family and homeland standard, "Homesickness", "Go back home in embroidered clothes" and "The closer I approach my hometown, the more timid I am" explicitly or implicitly express strong attachments to hometown and reluctance of leaving hometown; but meanwhile, the living environments of the Chinese nation have been relatively tough and harsh since ancient times, there is a common saying, "Man struggles upwards; water flows downwards", which caused large-scale population migration trends in history, such as "Brave the journey to Northeast", "Leave for the West Col", "Hu-Guang fills Sichuan", and "The peacock flies to Southeast" since the reform and opening up. It follows that the economic factor is a very important variable in Chinese population migration; but on the contrary, many people living in economic less-developed regions are willing to suffer poverty and keep their faith and that they rarely migrate. Secondly, movement of population has both advantages and disadvantages. On one hand, it makes people develop their own strength more profitably and the economic development will be boosted; on the other hand, it may bring a lot of social problems, such as the typical "left-behind children" problem and the of rural hollowing; however, international migration can promote cultural exchange and integration, but it will lead to the loss of the social elite and wealth, even conflicts between different civilizations.

Movement of population and immigration is a thing that fortune and misfortune depend on each other, so it is necessary to analyze specific factors of its occurrence. In the current tide of population migration, the tension between cultural soil "Stay" and actual requirements "immigrate" is a field of sociological concern. The traditional researches often analyze factors that encourage people to make the migration decisions and the impacts caused by movement of population from a qualitative or quantitative point of view.

Based on the existing research results, in this paper, we simulate the behaviors of population migration by constructing models and writing, and thus conduct research on movement of population, urban population size distribution and its generation mechanism. In the implementation process, on one hand, we tried to make the migratory behaviors conform to reality, on the other hand, we designed an evolutionary game model to observe the change of population rank size distribution and discover general principles from that.

The transformation of Chinese economic growth mode and social structure, regional disparity and regional tilt of investment, policy barriers of population migration dismantled by the government or undermined by the market reform, all of which have led to the increasing population migration and mobility.

The annual population migrations of developed countries are generally active, and compared with other developed countries in the world, the intensity of population migration in China has been at an apparently lower level over a long period of time. Nowadays, the main direction of China's migration is from rural areas to cities, and along with the deepening of reform and opening up and the promotion of urbanization, the population migration between cities, the population migration between towns and the population migration from towns to cities will be strengthened quickly, so the population migration will become more and more normalized.

Movement of population reflects the size of the difference between the residents' income, and focusing on movement of population can make us get to know the economic changes of the underlying residents. Combined with intelligent Agent technology, studying individual cognitions and psychological behaviors deeply is the main development trend of group behavior modeling. Modeling method based on agent (Agent-Based Modeling ABM), which is the representative result of this trend, have attracted extensive attentions of researchers. The design of population migration model must follow the principles as below: (1) the operation rules are required to reflect the current social migration situations; (2) friendly operation interfaces and results display interfaces are equipped, and computational results will be displayed intuitively; (3) a plurality of adjustable parameters which can be used to expressed different circumstances are equipped, so as to provide the basis of research discussion and results comparison.

2   Modeling Method Based on Agent

In recent years, not a few researchers try to study movement of population or migration evolution by introducing ABM method, including the agent-based spatial migration models on urban population which is used to study Zipf law formation mechanism, and prediction of urban population migration and distribution based on multi-agent and GIS [10]. However, the above researches either only draw attention to the analysis of the evolutionary process, characteristics and interactive elements of a single city, or only focus on the design of modeling simulation model, and furthermore, the research perspectives are based on analytic statistics and interpretations of historical events and lack of forecasts of future trends; besides, the research methods are mostly macro theoretical analysis or microscopic model design, which lack of synthetic study combined with actual management needs; the main research strategies are static analysis, lacking of dynamic continuous observation and simulation [11, 12].

Established in analyzing the evolution of population migration and based on the actual situations, in this paper, we put forward modeling methods of behaviors of immigrant group and designed the simulation model, revealed law of development and influencing factors of population migration behaviors entirely and systematically, explained the internal mechanism of migration and evolution, supported the predictions of migration trends and path, realized analysis of integrity and dynamics, and ultimately formed research findings which are of interpretability and reproducibility.

2.1 The Model is Closer to Reality

The past research models which use computer program to simulate population migration behaviors are relatively simple, in addition to drawing on the previous accumulation of research results, this paper added economic factors in the real world, and urban competitiveness, one of the important factors of affecting the population migration behaviors, was abstracted creatively to the model, divided the level of the subject and the city, so innovative results were obtained. These innovation points make the model closer to reality, and therefore, computational results are more convincing.

2.2 With Spatial Model

This paper established agent-based spatial model of urban population migration. Due to the improvement on the model and the strengthening of subjective characteristics, key characterizations in reality can be simulated, some of them can be undergone long-term evolution, and as a result, some problems in the development of real society can be understood in due course.

3 Conceptual Model Design

Combined with concept of population migration, this paper introduces related theories to analyze the influence of different levels of individuals taking different actions in different situations on the development trend of population migration and construct Agent conceptual model, among which migration settings are reflected in the design of game rules of population migration, and the overall model framework is discussed as below.

The most distinguishing feature of the current domestic migration/immigration in China is: (1) The majority group of domestic provincial population migration includes ordinary migrant workers or technical workers, and highly-educated talents, so the corresponding model is constructed, hereinafter it is called microscopic characterization model, which will be compared with traditional macro models.

The turtle in the model represents personnel of some area that can adopt left behind strategy or migration policy, are called simply as migrants and those leave behind, need to look for better living environments with the limited resources in order to achieve the goal of survival and reproduction. In interprovincial population migration model, the turtle refers to population individuals of each province, and they can move between different provinces, move means the individual move from one province to another, the population individual of each province is shown by different symbols, such as triangle, circle, star and so on; the tile represents provincial subject in the model, the size and color of each province are different, and the position is close to the actual position of China's provinces, which means a region of a specific color represents a province; so observers can observe and control the operation of the simulation world.

3.1 Rule Creation

According to the external representation of the current domestic population migration and the hypothesis made to simplify modeling, we translate it into a series of rules.

Rule 1. Different provinces included in the model have their own intrinsic characteristics, e.g. Guangdong which has a large population and a big economic gross, is the big province that the populations flow into; Henan which has a large population but a small economic gross, is the big province that the populations depart from. Besides, a number of neighboring cities may constitute urban agglomerations that are mutually complementary in economy, such as Pearl River Delta, Yangtze River Delta and so on.

Rule 2. Population migrations in different regions are of different preferences, and the current general trend is to flow into economically developed areas near the hometown, for instance, the populations of the Hunan area tend to flow into the Pearl River Delta over the Yangtze River Delta.

Rule 3. The economic growth rates of different regions are not the same. In general, cities with high growth rates are more likely to attract talents (for higher wages, benefits, and job opportunities); conversely, the economic growth rates of cities with high quality talents will increase; each city has a certain capacity, and any more outsiders will no longer be taken over when reaching the maximum capacity.

Rule 4. There are different types of individuals in each region, for instance, the initialization of Guangdong province is set as: 20% of individuals are farmers, 30% are workers, 20% are engineers, 20% are small business owners, 10% are owners of large enterprises, different types of individuals have different probabilities of possessing various grades of wealth and migrating to any other provinces (There are also preferences of migration sites), for example, the preference probability of Guangdong's big owners of enterprises and senior engineers migrating to Shanghai is 30%.

3.2 Individual Property Settings

Hypothesis 1. Considering the individuals under the age of 20 have little effect on the economy, the individuals are over 20 years old; Hypothesis 2. Each individual can only migrate once at most a lifetime. Hypothesis 3. The migrations take place at the end of each year; Hypothesis 4. The national birth rate, mortality rate, tax rate and any other policies that have effects on economy will remain unchanged.

Synthesizing the above hypothesis, the individuals in the model contain the following attributes. (1) The name of province of birth, and we will accordingly distribute colors for individuals (a province will be shown by the same color); (2) The name of province after migration, which is exactly the same with province of birth initially; (3) Migration preference list, which is the probability of migrating to a province, and all probabilities added up to 1 (Immigration preferences of populations engaged in the same occupation in the same province are the same, while immigration preferences of populations living in different provinces or engaged in different occupations are different); (4) The number of immigrations that is allowed, is 1 initially; (5) The occupation, engaged in different occupations, people have different capitals, earning capacities, immigration preferences, which are showed as different shapes in the diagram; (6) The age, is 20 years old when a person was born, and there is a possibility of death when reaching 50 years old; (7) A child and a father, the property can be divided when the individual dies; (8) Property; (9) Talent.

Each province in the model contains the following attributes. (1) The total population, which will be counted once a year; (2) The ratio of population occupation, which is used for those born to choose an occupation according to the ratio; (3) The ratio of the total economy, economic growth rate, total tax, tax rate, tax in public facilities and the ratio of tax revenues in welfare have an impact on the economy; (4) The birth rate, mortality rate, the proportion of property inheritance to their children (set to 1), (the same in all provinces); (5) The largest number of immigrants annually; (6) The location and color show in the diagram; (7) The information of longitude and latitude, which is used to calculate the distance; (8) The average wage.

In order to simplify the model, we made the following assumptions: (1) Because the individuals under the age of 20 have little effect on the economy, the individuals are over 20 years old; (2) Taking into account the situations of most people, each individual can only migrate once at best; (3) The migrations take place at the end of each year; (4) Over a long period of time, the national birth rate, mortality rate, tax rate and any other policies that have effects on economy will remain unchanged.

The modified push-pull model is formulated as a formula (1). And $t(i)$ indicates the unemployment rate in i, $b(j)$ indicates the average revenue of j, $M(i, j)$ represents the attraction of j to i.

$$M(i,j) = \{k1 * t(i) + k2 * b(j)\}/D(i,j) \tag{1}$$

The gravity model is formulated as a formula (2), among which $p(i)$ indicates the population of i, $D(i, j)$ indicates the distance from i to j, $M(i, j)$ represents the attractions of these two places.

$$M(i,j) = k * p(i) * p(j)/D(i,j) \tag{2}$$

"Cost-benefit" model is formulated as a formula (3), and $p(i)$ indicates the population of i, $D(i, j)$ indicates the distance from i to j, $M(i, j)$ represents the attractions of these two places.

$$M(i,j) = [1 - t(j)] * b(j) - b(i) \tag{3}$$

3.3   Model Flow

Step 1. Initialize each province and individual, the specific number of individuals of different occupations will be generated according to each province's attributes, various grades of random amount of money will be given according to occupations, the age of an individual is a random number, between 20 to 70, and the attributes of children and father are −1.

Step 2. Each individual earns money based on the economic growth rate of the province, pays tax according to the provincial tax rates (currently the tax rates are the same in all provinces), and the monthly total economy and total tax rate of the whole province can be calculated, the next month's economic growth rate can be calculated according the total economy of last month, and then the money used for welfare and public facilities will be calculated. The welfare is returned to the people (deemed to be the extra amount of money earned next month), and the money used for public facilities will make a positive effect on economic growth rate (to some extent).

Step 3. Repeat Step 2, perform calculation once a month and until the end of a year, then execute Step 4.

Step 4. The number of newborn can be obtained by the provincial birth rate, the newborn chooses an occupation according to the national occupation ratio, and then choose father in accordance with occupation normal distribution (For instance, if the son is an engineer, there is a great probability that his father is an engineer, and a small probability that his father engaged in other occupations). According to the number of deaths obtained by provincial mortality rate, we can add up the number of people over the age of 50 in a country and allocate these individuals a mortality rate (Every individual in the same province is the same), and the individuals will die on the basis of mortality rate, then their properties will be divided equally by their sons and daughters.

Step 5. Individual immigrant preferences to different provinces are generated according to the strategies, and three kinds of dynamic strategies are realized for choice: (1) The attraction model; (2) The push and pull model; (3) The cost-benefit model.

Step 6. Each individual that has never gone through immigration will choose the migration area with the corresponding probability according to the immigration preference. If it is the province, the movement is not needed, and if not, then migrate.

Step 7. Repeat executing Step 2 to Step 6.

By operating the population migration mode, we can have more specific and intuitive experience about various conditions affecting population migration in the evolution process.

Figure 1 shows the diagram of the domestic migration model. It can be divided into three sub parts. Left part shows the total wealth of different provinces; middle part shows the agents; right part shows the total population of different provinces. It can be seen that the numbers of population vary very slow with the time, also many provinces have little changes, only some special provinces have obvious rise or false, so we omit the data. We don't know if this is normal or some parts of the model need to be revised. Future research will discuss the problem in detail.

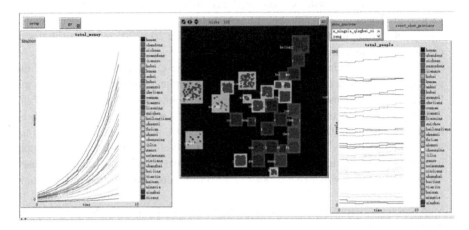

Fig. 1. The diagram of the domestic migration model.

In Table 1, total wealth of different provinces is exported from above model, which evolves under different initial value (referenced to real world value) with the growth of the year.

Table 1. GDP (Wealth) of different provinces evolve with the time (undergo 7 ticks).

Province	GDP (Billion yuan)							
Shandong	273119	281670.4	291816.4	300741.6	311223.1	321027	331346.5	341066.5
Henan	245014	253566.1	261362.4	269950.9	278605.4	287214.1	296232.2	304973.7
Sichuan	196823.2	202622.9	208481.5	215712	222118.5	228978.7	236359.5	244231.3
Guangdong	283786.2	291762.8	301369	310143	320651	330159.1	339566.2	349360.4
Jiangsu	208166.4	214490.6	221120.7	227936.3	234174.3	241351.3	248590.7	256826.4
Zhejiang	152190.5	157183.3	161991.9	166828.3	172068.6	178079.9	184038.1	190093.7
Jiangxi	108769.1	112195.6	115979.2	119464.9	122778.6	126686.3	131329.1	135722.2
Hubei	142392.1	146814.5	151626.4	155944	160867.2	166050.6	170942.7	175753.2
Hunan	190824.6	196664.8	201325.1	206956.7	213836.3	219908.3	227752.1	234664.8

It is observed that evolutionary results are affected by the initial values within a certain range, so it is very important to select the suitable development strategies for some relatively backward provinces to speed up development, such as subsidizing excellent talents and simplifying the settling procedures, etc.

4   Conclusions and Prospects

Faced to the actual demands of population migration management and based on the migratory population, the individual is represented as Agent in this paper, with the foundation of individual economic conditions, the interactive rule design model based on group behaviors and multi Agent technology is proposed, and we carry out modeling and simulation by combining actual data, through analyzing the closed loop process of "cognition - decision - behavior - feedback" evolved in the group behaviors of migration, reveal the law of development and influencing factors of population migration behavior entirety and systematically, explain the internal mechanism of migration and evolution and finally obtain maneuverable rules and models.

Through the operation and analysis of the simulation model, we discover that the factors including the higher average wage level, the shorter distance from home, the greater investment of city construction and so on can significantly boost the domestic migration behaviors. Through practical teaching use, some experiences of applying modeling methods in sociology are summarized.

(1)   Netlogo modeling provides a powerful tool and environment for teaching of social science courses. In the process of studying social phenomena, social model is constructed by combining the personal knowledge structure and practical rules. Application model can achieve immediate results, and while using models to predict and reason for solving problems, the thinking process is crystallized and the complex concepts are materialized.

(2)   With the help of the model's deductive function, researchers make use of logical and quantitative thinking to solve problems based on methods of analysis and reasoning. Nowadays, the domestic researches about social science disciplines tend to use conceptual representation, and seldom refer to microscopic characterization or deduction, intuitive and maneuverable research means are needed.

(3)   The model supports the construction of exploratory and visual environment, provides immediate feedback and validation, supports the deep exploration from phenomenon to essence, and by putting forward a series of related issues about population migration theory, building hypothesis independently, combining with the population migration model, observing the computing results to verify the hypothesis after having changed parameters of the model, and giving full play to the initiative, other members of the research team can modify and extend the model while combining their own personal thoughts.

In view of the problem that the current sociological researches lack of intuitive nature, the research activities are carried out by applying multi-agent modeling. Through the population migration model based on NetLogo simulation, the quantitative analysis of many factors affecting population migration is realized: with the elapse of time, we probe into various possible results of the particular individuals in multiple regions migrating between different regions under the condition of different parameters.

The future research can be developed from the following aspects: (1) Research on network group's cognitive behavior rules, study cognition process, cognitive steps, learning and memory, problem solving and other issues of network group in the evolution process of emergencies and network public opinions; (2) Research on behavior conversion rules of network group, study the mental state transition process, the conversion rules and its influence on group behavior of network group in the evolution process of emergencies and network public opinions; (3) Research on behavioral decision of network group, study the relationships of the action's purposes, standards, preference set, and Agent's emotional state of network group in the evolution process of emergencies and network public opinions; (4) Research on behavioral motive of network group, study "belief - desire - intention" of network group in the evolution process of emergencies and network public opinions, and then analyze the cognitive behaviors, reasoning behaviors, decision behaviors, learning behaviors of the group.

Acknowledgment. This research was financially supported by 2015 annual discipline construction project in Philosophy Social Sciences "12th Five-Year" planning of Guangdong Province (GD15XSH05), Natural Science Foundation of Guangdong Province, China (No. 2017A030313401, 2014A030313632, S2011040004387) and National Natural Science Foundation of China (No. 61375006, 11401223, 61402106).

References

1. Gilbert, S.W.: Model building and a definition of science. J. Res. Sci. Teach. **28**(1), 73–79 (1991)
2. Chen, K., Chen, B., Zhou, H.: The evaluation and reflection of teaching cases of chemical modeling based on Netlogo. Chin. Audiovis. Educ. **1**, 94–97 (2010)
3. Liu, X.: NetLogo platform based realization of dynamic model of public opinion. Inf. Documentation Serv. **1**, 55–60 (2012)
4. Geng, L., Li, Y.: Research on NetLogo computer simulation of cooperation behavior. Mod. Distance Educ. **1**, 66–69 (2011)
5. Ruyu, L., Zhu, W.: The simulation study of community innovation performance involved with multi agent. Prog. Technol. Strategy **30**(21), 06–10 (2013)
6. Riggs, W.W.: Agent-based modeling as constructionist pedagogy: an alternative teaching strategy for the social sciences. In: Reeves, T., Yamashita, S. (eds.) World Conference on E-Learning in Corporate, Government, Healthcare, and Higher Education 2006, vol. 1, pp. 1417–1423. Association for the Advancement of Computing in Education, Montreal (2006)
7. Macal, C.M., North, M.J.: Successful approaches for teaching agent-based simulation. J. Simul. **7**(1), 01–11 (2013)
8. Gammack, D.: Using NetLogo as a tool to encourage scientific thinking across disciplines. J. Teach. Learn. Technol. **4**(1), 22–39 (2015)
9. Frank, K.A., Lo, Y.J., Sun, M.: Social network analysis of the influences of educational reforms on teachers' practices and interactions. Zeitschrift für Erziehungswissenschaft. **17**(5), 117–134 (2014)
10. Jing, N.: Analysis of city population distribution prediction based on multi-agent and GIS. The doctoral dissertation of graduate school of Chinese Academy of Sciences (2007)
11. Wang, Z.: Some Agent-based model about spatial population transfer: researches on the dynamic mechanism of zipf's law. Master's thesis of Zhejiang Gongshang University (2007)
12. Peng, W., Yang, S., Zhang, J., Gao, Q.: Agent-based modeling and simulation of evolution of netizen crowd behavior in unexpected events public opinion. Mod. Librar. Inf. Technol. **7–8**, 65–71 (2015)

Application of Plant-Derived Anti Repellents in Prevention and Cure of *Parasaissetia Nigra* Nietner

Lihe Zhang, Bin Du, Baoli Qiu$^{(\boxtimes)}$, and Hui Wang

College of Agriculture, South China Agricultural University,
Guangzhou 510642, Guangdong, China
750446620@qq.com

Abstract. This study was aimed to find out whether the effects of natural plant-derived ant repellents on *Parasaissetia nigra* are good, long-persistent and safe to the natural enemies plant source ant repellent. Three botanical pesticides were used in repellent tests. It was found the effect of Arise ant repellent on the ants which had mutually beneficial relationship with *P. nigra* was the best, followed by Langsen keling and Capsaicint repellents. Meanwhile, the three plant-derived ant repellents were fatal to 1st and 2nd instar larva of *P. nigra*, and the highest integrated control effect was imposed by Arise ant repellent. It was found that the Arise ant repellent, when used to spray the plants around and 10 cm above the ground, not only repelled mutual-beneficial ants, but also caused the removed the nest around the plants. Mutual benefit of ant repellent significantly reduced the population of *P. nigra*, but significantly increased parasitoid *P. nigra* parasitic rate.

Keywords: Plant-derived ant repellents · *Parasaissetia nigra* Nietner
Mutual-beneficial ant · Parasitoid · Parasitic rate

1 Introduction

With the development of modern agriculture and people's growing emphasis on eco-logical environment, the use of chemical pesticides will continue to decline and gradually be replaced by plant-based pharmaceuticals. Different from chemical agents, botanical source to our production is life-beneficial, environmental-friendly and degradable, and does not undermine biological diversity or easily cause resistance (Zeng et al. 2009; Wang 2000). Plant repellents are plant-derived pesticides (Yang et al. 2006). So far, plant repellents have been studied extensively. Abroad, Hubert and Wiemer (1985) used infrared spectroscopy to determine the repellent activity of a ryegrass-isolated rye lactone on leafcutter ants. Capron and Wiemer (1996) studied the repellent effects of three cinnamic acid derivatives, isolated and purified by plants, on leafcutter ants. Plant pollen volatiles have a good repellent effect on ants (Willmer et al. (2009); Marinho et al. 2008). Moreover, β-eucalyptol in eucalyptus roots interferes with the recognition system and thus effectively repels leafcutter ants (Marinho et al. 2005; 2008). Sun et al. (2010) found most of the 22 tested plant extracts had a repellent effect on little black ants, with the cinnamon extract having the strongest repellent

© Springer Nature Singapore Pte Ltd. 2018
K. Li et al. (Eds.): ISICA 2017, CCIS 873, pp. 383–393, 2018.
https://doi.org/10.1007/978-981-13-1648-7_33

effect. Han et al. (2008) measured the repellent activities of 7 bridged terpenoids from *Monomorium pharaonis* and found they very actively repelled small yellow ants. They determined the repellent activities of *M. pharaonis* terpenoids, hydrogenated nopol, corresponding ethers and ester derivatives. The repellent rate of *M. pharaonis* was very high (Zheng et al. 2008; Zhao 2012).

Parasaissetia nigra Nietner, belonging to Hemiptera, Coccidae and *Parasaissetia*, is an important agricultural pest mainly distributed in South China. It can endanger garden plants and crops and seriously impact the rubber industry (Wang 2001). Ants secrete honey dew to maintain hygiene and to keep scale insects from the germs (Way 1963; Cud-joe et al. 1993; Gullan and Kosztarab, 1997). Ants can even repel scale-spread to other branches (Ho and Khoo 1997). Field investigation showed the insects had a mutual-beneficial relationship with the ants, called mutual-benefit ants, and the insects fed on the ant-secreted honey dew and helped to spread *P. nigra*. As a result, *P. nigra* increased in population and made the field increasingly more endangered. At present, the prevention and control of *P. nigra* are still dominated by chemical agents. Under the background of environmental protection and biodiversity conservation, an inevitable trend is to use plant-derived agents to repel ants and indirectly control *P. nigra*. To study the application of plant-derived ant repellents into the control of *P. nigra*, we conducted field tests with several plant-derived repellents in Danzhou, Hainan Province, aiming to provide reference for control of *P. nigra* and application of plant-derived agents in the field.

2 Materials and Methods

2.1 Test Location

The tests were conducted in July 2015 at Shatin, Treasure Island Village, Danzhou, Hainan. The investigation point was for private rubber plantation, and the variety of rubber was "hot research 7-33-97". The planting specification was 3×7 m^2, with 1-year grafted seedlings. During the test period, no pest control measures were taken in rubber plantation, and the rubber plantation fields were not managed, such as weeding, dead branches, twigs and weak branches.

2.2 Test Agents

The test botanical repellents were Langsen Kelin biological repellent (Beijing Harvest Sunshine Environmental Technology Co., Ltd.), Taiwan Tsui Yun natural plant repellent alternate name capsaicin (Shanghai Lee King Landscaping Co., Ltd.), and Germany Ariseant natural plant essential oils (Organic Products Co., Ltd.). The above test agents were all finished pharmaceuticals and directly sprayed without dilution.

2.3 Test Methods
2.3.1 Botanical Pharmacy Screening
In Shatian Team, Baojiaoxin, Danzhou, Hainan, a plantation field of rubber seedlings damaged by *P. nigra* was selected and randomly divided into four groups.

Three treatments were repeated for a total of 12 plots. Since the afternoon of August 11, 2015, the entire plants and 10 cm within the plants were sprayed every 7 days. The numbers of ants, *P. nigra* and parasitoids were recorded.

2.3.2 Effects of Ant Repellents on *P. nigra* and Parasitoids

At the peak of *P. nigra*, we selected 60 rubber tree seedlings and divided them into two groups (30 plants/group). Seedlings in one group were sprayed with Arise plant-derived ant repellent 10 cm above the ground and 10 cm around the basal part of the seedlings. Seedlings in the other group were given no treatment as a control. The numbers of ants, *P. nigra* and parasitoids in both groups were investigated on 7 d, 14 d, 21 d and 28 d. From the top of the rubber tree seedlings down 60 cm, *P. nigra* erased. During the investigation, the parasitic *P. nigra* population was statistically analyzed according to the parasitic death, individual color, appearance characteristics, and parasitic holes.

2.4 Data Processing

Data were processed using Excel, and variance analysis was performed on SPSS 20.0.

$$\text{Parasitization } (\%) = \frac{\text{Number of parasitizated } parasaissetia\ nigra\ Nietner}{Total\ number\ of\ parasaissetia\ nigra\ Nietner} \times 100$$

$$\text{Repellent } (\%) = \frac{Number\ of\ CK\ \text{population} - Number\ of\ \text{treatment population}}{Number\ of\ CK\ \text{population}} \times 100$$

$$\text{Dropping rate of insect } (\%) = \frac{Number\ of\ before\ treatment\ population - Number\ of\ after\ treatment\ population}{Number\ of\ before\ treated\ population} \times 100$$

$$\text{Larvae corrected mortality } (\%) = \frac{population\ decline\ rate\ of\ treatment - population\ decline\ rate\ of\ CK}{100\% - population\ decline\ rate\ of\ CK} \times 100$$

3 Results

3.1 Botanical Pharmacy Screening

3.1.1 Effect of Botanical Repellent on the Mutual-Beneficial Ant

We performed repelling test on the ant having a mutual-benefit relationship with *P. nigra* (Table 1). (1) Results after 3 replications of each botanical agent at 7 days showed that the average repellent results of Langsenkeling and Ariseant repellents were 50%, but the average repellent of capsaicin was only 20.68%. The repellent effect ranked as B>A>C. (2) The repelling result of Arise ant repellent was 85.98% after 14 d, but those of Langsen keling and capsaicin repellent began to decline in the order of A>B>C, indicating that Arise on 14 d was higher than the control on 7 d. The results of Langsen Keling were the opposite. (3) After 21 days, the results of the three medicaments still ranked as A>B>C, but the repellent results all decreased to some degrees. The repelling result of capsaicin was negative, indicating the repelling of capsaicin lasted for short time and disappeared after 21 d. (4) After 128 d, the repellents ranked in the order of A>B>C, which was the same order as those after 14 d and 21 d. The repelling results decreased sharply from 70.54% to 23.6% in the Langsen Keling repellent, with a decline of 66.54%. The repelling result of capsaicin was −26.38%, indicating no repelling effect on symbiotic ants. Field investigations showed hevea rubber tree seedlings during the period of 14 to 28 d, and the number of ants greatly increased. Although the results decreased, the repellent still exceeded 50% after 28 d. The results of 7, 14, 21 and 28 d show Ariseant repellent is good for both mutual-beneficial ants and durable effects.

Table 1. Comparison of repellent effects of three plant-derived ant repellents on mutual-beneficial ants in the filed

Treatment	Repeat	Initial population/head	7 days after drug use		14 days after drug use	
			Ant/head	Repellent %	Ant/head	Repellent %
A:Ariseant repellent	I	424	136	61.14	51	84.73
	II	489	175	43.18	65	80.48
	III	399	94	81.16	32	92.74
	Average	–	–	**61.83a**	–	**85.98a**
B:Langsenkeling	I	318	95	72.86	150	55.09
	II	348	67	78.25	147	55.86
	III	945	197	60.52	219	50.34
	Average	–	–	**70.54a**	–	**53.76ab**
C:capsaicin	I	429	280	20.00	393	−17.66
	II	529	402	−30.52	450	−35.14
	III	199	137	72.55	183	58.50
	Average	–	–	**20.68a**		**1.90b**
D:CK	I	208	350	–	334	–
	II	229	308	–	333	–
	III	310	499	–	441	–
Treatment	Repeat	Initial population/head	21 days after drug use		28 days after drug use	
			Ant/head	Repellent %	Ant/head	Repellent %
A: Ariseantrepellent	I	424	64	84.50	161	61.20
	II	489	93	78.37	213	53.19
	III	399	84	83.78	198	65.86
	Average	–	–	**82.22a**	–	**60.08a**

(*continued*)

Table 1. (*continued*)

Treatment	Repeat	Initial population/head	21 days after drug use		28 days after drug use	
			Ant/head	Repellent %	Ant/head	Repellent %
B:Langsenkeling	I	318	243	41.16	269	35.18
	II	348	165	61.63	256	43.74
	III	945	312	39.77	627	−8.10
	Average	–	–	**47.52ab**	–	**23.60ab**
C:capsaicin	I	429	506	−22.52	716	−72.53
	II	529	543	−26.28	664	−45.93
	III	199	278	46.33	352	39.31
	Average	–		**−0.82b**		**−26.38b**
D:CK	I	208	413	–	415	–
	II	229	430	–	455	–
	III	310	518	–	580	–

3.1.2 Effect of Botanical Repellents on *P. nigra* Nietner

Table 2 shows that Ariseant, Langsenkeling and capsaicin had a certain fatal effects on *P. nigra* 1st and 2nd nymphs. (1) After 7 d of repellent application, Ariseant and Langsenkeling were both very efficient on 1st and 2nd instar nymphs of *P. nigra*. The control efficacy was above 50% from both repellents, especially Langsenkeling (80.61%). The control effect of Capsaicin (47.23%) was worse than the first two repellents. The effectiveness of the three agents ranked as B>A>C. (2) Ariseant repellent and Langsenkeling increased the control efficiency by 97.01% and 85.09% respectively after 14 d, but that of capsaicin dropped significantly from 47.23% to 10.34%. According to the order of anti-efficiency A>B>C, Ariseant repellent was more effective on 14 d than 7 d and that of Langsenkeling was the opposite. (3) The order was A>B>C, but the control effect declined to some extent for all repellents. The repelling result of capsaicin is negative, indicating capsaicin has a short duration of control effect, and the effect on 1st and 2nd instar nymphs disappeared after 21 d. (4) After 28 d, the corrected dropping rate of insects ranked in the order of A>B>C. The order was not changed after 14 or 21 d. The corrected dropping rate of insects decreased with Langsenkeling repellent from 80% to 70%, and to −8.18 with capsaicin, without effect on the 1st and 2nd instar nymphs of *P. nigra*. The number of *P. nigra* began to increase after 21 and 28 d. This result is in line with the large increase in the ant number during the field investigation that the rubber tree seedlings were applied with capsaicin after 14 to 28 d. The control efficacy of Ariseant declined slightly after 28 d, but the number of revised insects reduced rate still exceeded 90. Ariseant, Langsenkeling, and capsaicin were significantly different at 7, 14, 21 and 28 d. Ariseant showed good control effect and durable effect on the 1st and 2nd instar nymphs of *P. nigra*.

Table 2. Influence of three plant-derived ant repellents on *P. nigra* in the field.

Treatment	Repeat	Initial population number/head	7 days after drug use			14 days after drug use		
			P. nigra/head	Dropping rate of insect %	Larvae corrected Mortality %	*P. nigra*/head	Dropping rate of insect %	Larvae corrected mortality %
A:Ariseant repellent	I	406	155	61.82	71.04	13	96.80	98.04
	II	401	147	63.34	65.54	18	95.51	96.16
	III	197	33	83.25	87.34	9	95.43	96.81
	Average	–	–	–	**74.64a**	–	–	**97.01a**
B:Langsen keling	I	138	62	55.07	65.92	59	57.25	73.87
	II	738	116	84.28	85.22	97	86.86	88.77
	III	852	105	87.68	90.68	90	89.44	92.63
	Average	–	–	–	**80.61b**	–	–	**85.09b**
C:capsaicin	I	341	234	31.38	47.94	585	−71.55	−4.84
	II	339	188	44.54	47.87	335	1.18	15.55
	III	183	131	28.42	45.89	209	−14.21	20.31
	Average	–	–	–	**47.23b**	–	–	**10.34b**
D:CK	I	22	29	−31.82	–	36	−63.64	–
	II	47	50	−6.38	–	55	−17.02	–
	III	127	168	−32.28	–	182	−43.31	–
Treatment	Repeat	Initial population number/rats	21 days after drug use			28 days after drug use		
			P. nigra/head	Dropping rate of insect %	Larvae corrected mortality %	*P. nigra*/head	Dropping rate of insect %	Larvae corrected Mortality %
A:Ariseant repellent	I	406	26	93.60	96.29	29	92.86	96.43
	II	401	25	93.77	95.12	41	89.78	94.66
	III	197	13	93.40	95.68	34	82.74	89.31
	Average	–	–	–	**95.70a**	–	–	**93.47a**
B:Langsen keling	I	138	84	39.13	64.76	129	6.52	53.26
	II	738	108	85.37	88.54	154	79.13	89.10
	III	852	128	84.98	90.17	208	75.59	84.88
	Average	–	–	–	**81.15b**	–	–	**75.75b**
C:capsaicin	I	341	663	−94.43	−12.56	681	−99.71	0.15
	II	339	495	−46.02	−14.38	674	−98.82	−3.83
	III	183	264	−44.26	5.56	357	−95.08	−20.86
	Average	–	–	–	**−7.13b**	–	–	**−8.18b**
D:CK	I	22	38	−72.73	–	44	−100.00	–
	II	47	60	−27.66	–	90	−91.49	–
	III	127	194	−52.76	–	205	−61.42	–

3.2   Effect of Plant-Derived Ant Repellents on Mutual-Beneficial Ants, *P. nigra* Nietner and Parasitoids

3.2.1   Plant-Derived Ant Repellents the Repelling Effect of Mutual-Beneficial Ants

After drug screening, the mutual-beneficial ants were repelling with Arise, and the ant populations on rubber tree seedlings were investigated 5 times (Fig. 1). Before

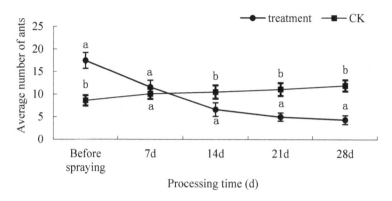

Fig. 1. Influence of Arise on mutual-benefiting ants.

repelling, the average ant population density in the test group surpassed that in the control group (17.5 ± 1.75 vs. 8.63 ± 1.16 heads/plant). After 7 days, the ant population on rubber tree seedlings dropped sharply in the test group, but did not change in the control group (Fig. 1). The ant population increased gradually after 14 days in the test group, but further decreased in the control group. After the control group began to decline slowly the number of anti-drive, the ant population still rose until the end of the experiment. This result fully suggests the Arise ant repellent greatly affects the population of mutual-beneficial ants. This agent has a good repelling effect and durable effect on rubber tree seedlings that have a mutual-benefit relationship with *P. nigra*.

The field experiments suggest the presence of mutual-benefit relationship ants with *P. nigra*, such as *Tapinoma melanocephalum* and *Anoplolepis gracilipes*. Most of formicaries were built around rubber trees, so during the experiments, chemicals should be sprayed within the range of 10 cm around the rubber trees. The number of formicaries within 10 cm around the trees was calculated. The survey showed no significant difference between groups before administration of Ariseant repellents ($P > 0.05$; Fig. 2). After application, the numbers of formicaries were significantly different between the test group and control group ($P < 0.05$). After 7 d of application, the formicary number of the test group dropped sharply, as most of the ants were affected by the Arise repellent; the formicary numbers after 14 or 28 d changed basically smoothly. The number of formicaries of the control group slowly increased until the end of the experiments.

3.2.2 Repelling Impact of Mutual-Beneficial Ants on *P. nigra* Nietner

After application of Arise repellent, the number of symbiotic ants declined, while the number of *P. nigra* varied. The population of *P. nigra* declined continuously after applying Arise repellent, whereas the number of *P. nigra* increased on the hevea repeats (Fig. 3). *P. nigra* population decreased after 7 d of treatment, but not significantly different from the control; ant populations at 14, 21 and 28 d after treatment were significantly different from the control group. This result suggests the damage level of *P. nigra* indirectly decreased after the use of Arise repellant.

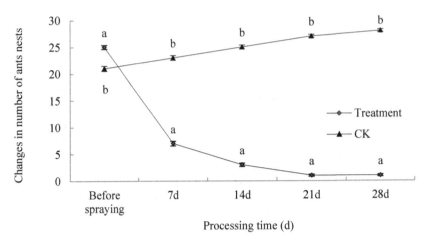

Fig. 2. Change in nest population on treatment and rubber seedlings after application of Arise.

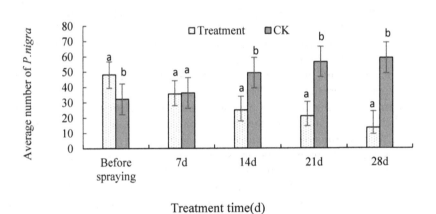

Fig. 3. Change in population of *P. nigra* Nietner on treatment and rubber seedlings.

3.2.3 Repelling Impact of Mutual-Beneficial Ants on Parasitoids

The proportions of parasitization on rubber tree seedlings were not significantly different between the test and control groups before treatment of Arise ant repellent (0.4 ± 0.1% vs. 0.2 ± 0.1%, P > 0.05; Fig. 4). No significant difference between groups was found at 7 or 14 d after treatment. But after 21 and 28 d, the proportions of parasitoids on the rubber tree seedlings were both significantly higher than the control group (P < 0.05), and parasitization reached 24 ± 0.9% on 28 d, while lower level of parasitization occurred on the rubber tree seedlings of the control group. Thus, the presence of ants modestly affected the parasitoids, and parasitoids after botanical repellant use significantly enhanced parasitization of *P. nigra*.

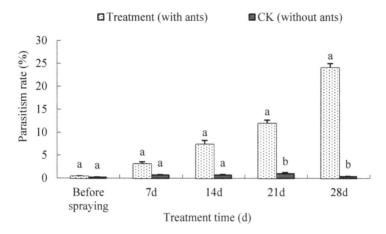

Fig. 4. Parasitism rate of Parasite wasps on treatment and check rubber seedlings.

4 Conclusions and Discussion

4.1 All Three Plant-Derived Ant Repellents Have Repelling Effects on Mutual-Beneficial Ants

The results rank in the order of Ariseant repellent>Langsenkeling>capsaicin. Ariseant repellent had good repelling effect and high durable effect, which were not significantly different from Langsenkeling (P > 0.05), but both were significantly different from capsaicin (P < 0.05). The experiment showed the three plant-derived ant repellents had certain fatal effects on the 1st and 2nd instar nymphs of *P. nigra* at the spraying sites without erasing *P. nigra*. The corrected dropping rates of insects of three ant repellents ranked in the order of Ariseant repellent>Langsenkeling>capsaicin, with significant differences (P < 0.05). The results suggest Ariseant repellent has the best combined effect and the most stable pharmacological effect, whether on mutual-beneficial ant or *P. nigra*.

4.2 Ariseant Repellent with the Mutual-Beneficial Ants Erased *P.* Nietner

It was proved the agent dose had a good repelling effect on mutual-beneficial ants. The population of *P. nigra* also decreased after the repelling of mutual-beneficial ants, and the parasitization of *P. nigra* by parasitoids also increased greatly. This conclusion is consistent with the ant isolation experiments by Zhang et al. (2009). It is fully demonstrated the mutual-beneficial ants help with the transmission of *P. nigra* and modestly protect against parasitism. Thus, plant-derived ant repellents are of great importance in the control of *P. nigra*.

4.3 All Three Repellents Are Botanical Agents and Botanical Pesticides

People use various vegetable oils, smoke and tar to disperse insects, forcing insects to leave (Yang et al. 2006). These applications can be traced back to 2000 years ago, as "Zhou Li" recorded "Shendan palm in addition to watermelon to attack Ying to Mangcao smoked." Chemical insecticides kill and repel insects, thus protecting bio-diversity and the human environment. Botanical repellents are derived from natural plants and are environment-friendly and degradable. Botanical repellents have become a hotspot and inevitable development trend (Wang et al. 2006; Su et al. 2009; Sun et al. 2010). Botanical repellents also show rapid and durable effects. In the future, *Parasaissetia nigra* Nietner can be controlled by botanicals that control pests and protect the environment and biodiversity.

References

Zeng, L., Yuan, Q., Hou, X., Yao, T.: Studies on the efficacy of plant source mixture in the control of western flower thrips. Proceler **35**(06), 164–166 (2009)

Wang, X.: Research and application of plant pesticide. World Agric. **02**, 30–32 (2000)

Hubert, T.D., Wiemer, D.F.: Ant-repellent terpenoids from Melampodium divaricatum. Phytochemistry **6**(24), 1197–1198 (1985)

Capron, M.A., Wiemer, D.F.: Piplaroxide, an ant-repellent piperidine epoxide from piper tuberculatum. J. Nat. Prod. **59**(8), 794–795 (1996)

Willmer, P.G., Nuttman, C.V., Raine, N.E., et al.: Floral volatiles controlling ant behavior. Funct. Ecol. **23**, 888–900 (2009)

Marinho, C.G.S., Della Lucia, T.M.C., Guedes, R.N.C., et al.: β-Eudesmol induced aggression in the leaf-cutting ant Attadensdens rubropilosa. Entomol. Exp. Appl. **117**, 89–93 (2005)

Marinho, C.G.S., Della Lucia, T.M.C., Ribeiro, M.M.R., et al.: Interference of β-eudesmol in nestmate recognition in Atta sexdens rubropilosa (Hymenoptera: Formicidae). Bull. Entomol. Res. **98**, 467–473 (2008)

Sun, M., Chen, Q., Song, G., et al.: Screening of plant-derived antrepellent. Chin. J. Vector Biol. Contr. **21**(6), 554–557 (2010)

Han, Z., Wang, Z., Jiang, Z., et al.: Synthesis of bridged ring terpenoids and their ant repell activity. J. Jiangxi Agric. Univ. **30**(4), 586–591 (2008)

Zheng, W., Jiang, Z., Han, Z., et al.: Study on antrepellent screening of terpenoids. Chin. J. Health Insecticidal Tools **14**(2), 84–86 (2008)

Zhao, L.: Hydrogenation of nopol and its derivatives synthesis and biological activity. Jiangxi Agricultural University, Nanchang (2012)

Wang, Z.Q.: Chinese zodiac, Insecta (volume 22), Homoptera, Scales Division, pp. 392–394. Science Press, Beijing (2001)

Way, M.J.: Mutualism between ants and honeydew-producing Homoptera. Ann. Rev. Entomol. **8**, 307–344 (1963)

Cudjoe, A.R., Neuenschwander, P., Copland, M.J.W.: Interference by ants in biological control of the cassava mealybug *Phenacoccus manihoti* (Hemiptera: Pseudococcidae) in Ghana. Bull. Entomol. Res. **83**, 15–22 (1993)

Gullan, P.J., Kosztarab, M.: Adaptation in scale insects. Ann. Rev. Entomol. **42**, 23–50 (1997)

Ho, C.T., Khoo, K.C.: Partners in biological control of cocoa pest: mutualism between *Dolichoderus thoracicus* (Hymenoptera: Formicidae) and *Cataenococcus hispidus* (Hemiptera: Pseudococcidae). Bull. Entomol. Res. **87**, 461–470 (1997)

Zhang, Z., Li, G., Zhou, M., Hongchang, A., et al.: Effects of yellow ants on the population dynamics of rubber *Acrocephalus kawakami*. Insect Knowl. **46**(6), 888–891 (2009)

Yang, C., Jiang, S., Xu, H.: Advances in botanical repellent research. Plant Prot. **06**, 4–9 (2006)

Wang, P., Zhou, H., Zhang, Z., et al.: Observation of two repell mosquitoes on plants with repellents. Chin. J. Trop. Med. **6**(10), 1791–1792 (2006)

Su, Y., Yang, C., Hua, H., et al.: Study on bioactivity of ethanol extracts from 13 species of plants to BPH. Chin. Agric. Sci. Bull. **25**(1), 198–202 (2009)

Research on Resource Trust Access Control Based on Cloud Computing Environment

Jun Nie[✉] and Dongbo Zhang

Software Department, Guangdong University of Science and Technology,
Dongguan, Guangdong, China
13739149@qq.com, 317710990@qq.com

Abstract. In this paper, a method of access control based on resource trust is proposed to solve the problem of trust access between cloud users and cloud resources. This method introduces trust into the attribute access control model in cloud environment, and proposes an optimization technology of attribute access control based on trust evaluation in cloud computing environment. By introducing the evaluation credibility, entity familiarity and evaluation similarity, the trust degree is calculated through this method, and the comprehensive trust degree of users or resources is achieved through direct trust, indirect trust and recommendation trust, so as to achieve the purpose of improving the security of attribute access control.

Keywords: Cloud computing environment · Resource trust · Access control

1 Introduction

How cloud users identify whether cloud service providers or cloud resources are real or maliciously false? How to select more secure services or identify false malicious sites from a large number of cloud service providers or cloud resources? Thus, cloud users' trust in cloud service providers is created [1]. On the other hand, how do cloud resources or cloud service providers prevent users from malicious attacks or conspiracy attacks, resulting in cloud resources and cloud service providers' trust in users [2]. In this paper, by introducing trust into the attribute access control model in cloud environment, we propose an optimization technology of attribute access control based on trust evaluation in cloud computing environment [3]. By introducing the evaluation credibility, entity familiarity and evaluation similarity, the trust degree is calculated through this method, and the comprehensive trust degree of users or resources is achieved through direct trust, indirect trust and recommendation trust, so as to achieve the purpose of improving the security of attribute access control.

© Springer Nature Singapore Pte Ltd. 2018
K. Li et al. (Eds.): ISICA 2017, CCIS 873, pp. 394–404, 2018.
https://doi.org/10.1007/978-981-13-1648-7_34

2 Related Definition

Definition 1 (cloud entity set): Suppose e_1, e_2, e_3 ... e_n is an entity in the cloud computing environment, and constitutes an entity domain $E = \{e_1, e_2, e_3 ... e_n\}$; $UE = \{u_1, u_2, u_3, ... u_n\}$ constitutes a collection of cloud n user entities, set for cloud service entities in the cloud computing environment. $SE = \{s_1, s_2, s_3 ... s_n\}$ is a collection of cloud service entities in the cloud computing environment and has no intersection with the cloud computing environment.

Definition 2 (trust): Trust is an evaluation of the credibility of an entity based on its own experience, and its evaluation results are represented by the degree of trust T.

Definition 3 (trust condition): Trust condition is a condition that the trust between entities must be satisfied. When a entity trusts a certain trust condition, the entity trusts another entity and vice versa.

Definition 4 (direct trust degree): The direct trust degree of the entity is related to the value of the trust evaluation. The higher the evaluation value, the higher the entity credibility, and the lower the vice. For two entities that have never been interacted, direct trust is zero.

Definition 5 (service satisfaction): Evaluation of the service satisfaction of an entity after multiple interactions with another entity.

Definition 6 (trust evaluation value): Evaluation of the entity after an entity interacts with another entity.

3 Process Design of Trust Computing Model

The main body in the cloud computer environment can determine whether the resource is a malicious cloud resource when accessing the cloud service resources according to the degree of trust [4]. If the principal is able to gain a higher degree of trust when the subject is accessed, the main body combines the policy center with the decision-making authority [5]. Make cloud services or cloud resources more secure and reliable for access trust in dynamic changes that are beneficial to the subject. Trust computing management module is mainly responsible for evaluating the degree of trust of cloud users and cloud resources [6]. According to the historical interaction records between cloud users and cloud owners, and combining information similarity and reliability, we can evaluate and decide the degree of trust [7]. Then the result of trust degree is submitted to the decision center, and the decision result is passed to the authentication and authorization module by the decision center [8]. After that, the corresponding access control policy is obtained after the authentication module is processed [9]. The access trust process for cloud users and cloud resources as shown in Fig. 1.

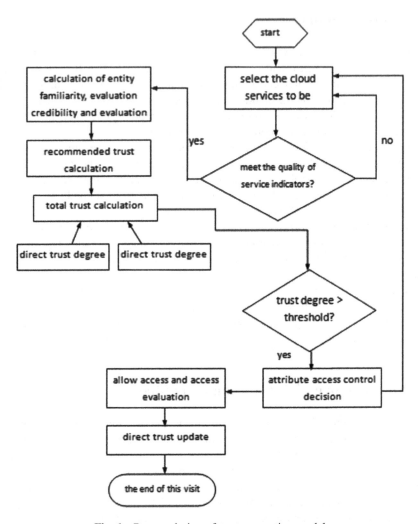

Fig. 1. Process design of trust computing model

In cloud computing environment, the overall trust degree of cloud user entity e_m to cloud service or cloud resource e_n is calculated by direct trust degree, indirect trust degree and recommendation trust degree [5]. The weight assignment between the two trust degrees directly affects the total trust degree and the accurate calculation [10].

(1) Calculation of Direct trust

According to the history of interactive records and based on whether the record is full of direct confidence, the formula is as follows:

$$T_{\text{Direct}}^{mn} = \sum_{k=1}^{N} \frac{1}{N} H(k) T_{mn}^{k} \tag{1.1}$$

In formula (1.1), the trust of a resource visitor to a service provider is calculated by using the historical interactive records of both parties. Where, T_{Direct}^{mn} is defined as the direct trust degree of entities to service providers after many interactions. N is the most effective number of historical records in the most recent interaction, and $H(k)$ is a time attenuation function, and $0 < H(k) < 1$.

(2) Calculation of the degree of indirect trust.

If there is no historical interaction window between the two entities, the calculation of the trust degree can be calculated indirectly through the transfer of trust.

$$T_{\text{indirect}}^{mn} = \sum_{k=1}^{N} (W(k) * T_{\text{Direct}}^{k} * \frac{1}{\sum_{i=1}^{n} W(k)} \tag{1.2}$$

In formula (1.2), T_{indirect}^{mn} represents indirect trust degree. T_{Direct}^{k} represents the trust degree of the entity of the evaluated entity in the Kth path, and $W(k)$ represents the weight in the recommended path of Kth.

(3) Calculation of Recommended trust

The overall evaluation credibility of user entities is represented by ET. It is based on the user satisfaction in each historical access window and the collection of cloud service entities that interact with users. $s(i)$ represents the degree of satisfaction of each evaluation, $\sum_{i=1}^{n} s(i)$ represents the credibility of cloud users' evaluation of a cloud service or cloud resource, and its comprehensive evaluation credibility is defined as follows:

$$ET = \sum_{t=1}^{n} \sum_{i=1}^{n} s(i)|P| \tag{1.3}$$

(4) Cloud users' overall trust calculation for cloud services

$$T(e_m, e_n) = \begin{cases} \frac{T_{mn}}{T_{mn} + |R|} T_{direct} + \frac{|R|}{T_{mn} + |R|} CRT(e_m, e_n) \\ T_{indirect} \quad T_{mn} = 0 \cap |R| = 0 \end{cases} \qquad (1.4)$$

In formula (1.4), $T_{m,n}$ is the number of historical interactions between cloud users and cloud services, and $|R|$ is the number of recommended entities. If there is no interaction and no recommendation, the overall degree of trust is the degree of indirect trust $T_{indirect}$. If the history interaction window is smaller than the recommended entity number, it shows that the more familiar users are in the cloud computing environment, the more the recommended, the heavier the weight of the trust is. Otherwise, the direct trust evidence is more sufficient, the weight of direct trust is increased and the weight allocation will be constantly adjusted with the renewal of trust. $CRT(e_m, e_n)$ represents the recommendation trust of the cloud user entity e_m for direct or indirect entities of cloud services or cloud resource en.

4 Correlation Algorithm for Trust Degree Calculation

The main function of trust computing management module is to calculate trust degree for cloud users to access cloud resources, and trust degree is combined with other attributes as a trust attribute, which is integrated by the access control decision center. The algorithm of trust computation is as follows:

TotalTrust(){//the trust of Cloud user entity m in cloud resources or cloud service entity n

Begin

If(Cloud user entity m has no historical direct interaction records for cloud resources or cloud service entity n, and n has no recommended entities)

Query(Indirect trust tree)，Weight of each path;

$W(j) = \prod_{m=0}^{j} T_{Dire}^{m}$;// Calculating Path weighting factor $W(j)$

The direct trust of the computing entity to the j path of the cloud resource is T_{Direct}^{j};

$T_{indirect}^{mn} = \sum_{j=1}^{k}(W(j)) * T_{Direct}^{j} * \frac{1}{\sum_{m=1}^{k}(W(j))}$;   //Calculation of the indirect trust of cloud user entity m to cloud resources or cloud resource entity n

return;

}

If(Cloud user entity m has an interactive history of cloud resources or cloud service entity n, and cloud user m does not recommend entities){

Query(A historical record of interaction)，Computer corresponding interactive history；

$H(j) = \begin{cases} 1 & j = N \\ H(j-1) + \frac{1}{N} & 1 \le j < N \end{cases}$//Computing time error function $H(j)$;

Calculation weight factor Wh;

$T_{mn}^{j} = \sum_{h=1}^{k} W_h (\frac{1}{|Q_{mn}^{iter,h} - Q_{h}^{claim}|})$;//Calculating the reliability of an interaction T_{mn}^{j}

$T_{Direct}^{mn} = \sum_{j=1}^{N} \frac{1}{N} H(j) T_{mn}^{j}$;//Calculating direct trust degree T_{Direct}^{mn}

return;

}

If(Cloud user entity m has no historical interaction record for cloud resources or cloud service entity n, and cloud user entity m has a collection of recommended entities){

$X(t) = \sum_{m=1}^{k} s(m) / |X|$;//Calculation of evaluation satisfaction $X(t)$;

calculate$|P|$;Calculate the number of entities that have historical interactive records of cloud resource or cloud service entity n at the cloud user entity m

$YT = \sum_{t=1}^{k} s(t)/|P|$;//Calculation of overall evaluation reliability $Y(t)$;

calculating$T_{direct,m}$ and $T_{direct,n}$;// The direct trust degree of cloud user entity m to cloud resources or cloud service entity n.

Calculate the number of entities that have historical interaction with cloud users m and cloud resources ;

$XY(m, n) = \frac{1 - \sqrt{\sum_{m,n=1}^{k}(|T_{direct,m} - T_{direct,n}|)^2}}{|P|}$;//Computational evaluation similarity $XY(m,n)$

Set the value of ∂ with a range of 0 to 0.65;

$EE(U_j = \partial \times YT + (1 - \partial) \times XY(m, n)$;//Calculation of recommended credibility EE(j)

Query(Indirect recommendation of the number of entities n);

$R_{indirect}$ (n);//Degree of indirect trust

EE(n);//Indirect recommendation of entity reliability

T_{mn} (j);//Historical interactive record of a cloud user m and an entity

Determine the δ parameter value;

$\delta \times T_{mn}$ (j);// Get the value of the entity familiarity

$CRT(m, n) =$

$$\frac{\sum_{i,j}^k (\partial \times T_{m,n} \times T_{direct}\ (i) + EE(i) + (1-\partial) \times T_{m,n}(j) \times R_{indirect}\ (j) \times EE(j)}{|P|}$$;//Calculation of comprehensive recommended trust

Return;

$$T(e_m, e_n) = \begin{cases} \frac{T_{mn}}{T_{mn} + |R|} T_{direct} + \frac{|R|}{T_{mn} + |R|} CRT(e_m, e_j) \\ T_{indirect}\ T_{mn} = 0 \cap |R| = 0 \end{cases}$$;// Computing overall

trust degree

return;

}

return;

End

}

5　Simulation Experiment and Analysis

Experimental hardware environment: virtual platform VMware, CPu with 8 core, primary frequency 3.0 GHz, memory configuration of 64 GB, storage capacity of 10 TB.

In the experiment, 2 entities are set, one is a cloud user entity, the other is a cloud resource or a cloud service entity [11]. They are divided into three categories: A, B and C. Cloud resource entities or cloud service entities are classified as follows:

Class A is a normal cloud resource entity and a cloud user entity that can always provide a real and effective service or service evaluation.
Class B is a semi normal cloud resource entity or a semi normal cloud user entity that can not fully provide a real service or service evaluation.
Class C is a malicious cloud resource entity, and the services provided are all unreal.

Cloud user entity and cloud service entity are relatively independent. An entity can be both a visitor and a service provider [6]. Its identity is relatively independent and does not affect each other [7].

Experimental software environment: the development platform is My Eclipse10, which is simulated with CloudSim, and the experimental parameters are shown in Table 1

Table 1. Experimental parameters

Parameter	Parameter values	Explain	Parameter	Parameter values	Explain
NT	2500	Normal cloud resource service entity	TN	40	The number of services provided by the entity
XN	8000	Entity scale	W	400	Entity interaction number
ST	700	Semi normal cloud resource service entity	δ	0.85	Weight of entity familiarity
HT	600	Malicious cloud resource service entity	∂	0.45	Weight of evaluation reliability
NM	2500	Normal cloud user entity	HM	690	Malicious cloud user entity
SM	600	Semi normal cloud user entity			

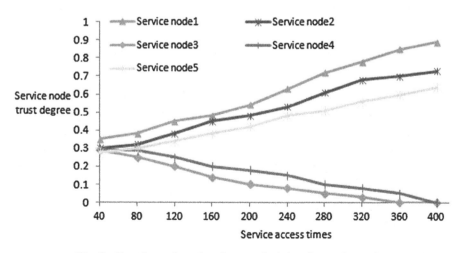

Fig. 2. Experimental results of trust calculation for service nodes

Experiment 1: the calculation experiment of malicious service node and trust degree

The experiment randomly selects 5 cloud service nodes to calculate their trust degree. You can see from Fig. 2, the service node 1, node 2 and node 5 are increasing with increasing number of access times. It is possible to determine that these two nodes are reliable service providers, while service node 3 and service node 4 gradually decline as the number of visits increases.

Experiment 2: cloud user evaluation reliability simulation experiment

As can be seen from Fig. 3, A class a cloud user entity is always reliable because of its evaluation, and its evaluation credibility is basically maintained at about 1. The B class cloud user entity is not entirely true service evaluation given by the cloud entities accessed, and its evaluation credibility is around 0.45. C class cloud user entities provide reliable or unreliable services because their evaluation credibility is below 30%, and it decreases to 1% of the untrustworthy state as the number of experiments increases.

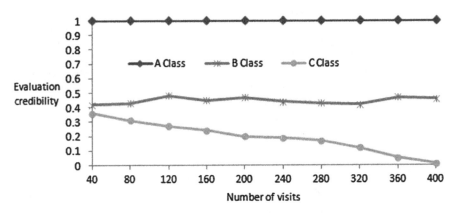

Fig. 3. Evaluation of experimental results for reliability calculation

Experiment 3: Validity test of trust computing management module

The experimental results show that when the ratio of C malicious service entities is below 35%, the success rate of our model and the CDTM model proposed in the document [12] is higher than that in the initial stage, reaching more than 85. But with the increasing proportion of malicious service entities in C, the success rate of model service is better in this paper, which indicates that the accuracy of this model is better in terms of total trust and recommendation trust (Fig. 4).

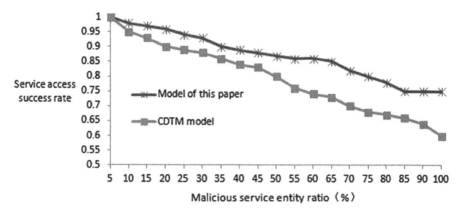

Fig. 4. Experimental results of access success rate under different scale malicious service entities

6 Conclusions

In this paper, the process design of the trust degree computing access control model of the entity in the cloud environment presents based on the study of access control technology. Then the algorithm of trust calculation is introduced in detail. Finally, the simulation experiment is used to verify the trust model. The result shows that the model can distinguish the malicious service entity better, and greatly improve the security of access control and the success rate of the service entity.

Acknowledgements. The work is supported in part by Department of Education of Guangdong Province under Grant 2015KQNCX193.

References

1. Wang, X.M., Fu, H., Zhang, L.C.: Research progress on attribute-based access control. Acta Electron. Sin. **38**(7), 1660–1667 (2010)
2. Feng, D.-G., Zhang, M.: Study on cloud computing security. J. Softw. **22**(1), 71–83 (2011)
3. Resch, J.K., Plank, J.S.: AONT-RS: blending security and performance in dispersed storage systems. In 9th USENIX FAST (2011)
4. Yingxum, F., Shengmei, L.: Survey of secure cloud storage system and key technologies. J. Comput. Res. Dev. **50**(1), 136–145 (2013)
5. Jinbo, X., Zhiqiang, Y.: A composite document model and its access control scheme in cloud computing. J. Xi'an Jiaotong Univ. **48**(2), 25–31 (2014)
6. Wang, Y.-D., Yang, J.-H.: Survey on access control technologies for cloud computing. J. Softw. **26**(5), 1129–1150 (2015)
7. Feng, C., Qin, Z.: Key techniques of access control for cloud computing. Acta Electron. Sin. **02**(43), 312–319 (2015)
8. Wang, X., Zhao, Y.: A Task-role-based access control model for cloud computing. Comput. Eng. **38**(24), 9–13 (2012)

9. Xu, J., Zhang, D., Shou, Y.: An improved parallel K-means algorithm based on MapReduce. Int. J. Embed. Syst. **9**(3), 275–280 (2017)
10. Liang, F., Yin, L.-H., Guo, Y.-C.: A survey of key technologies in attribute-based access control scheme. Chin. J. Comput. **07**(40), 1668–1698 (2017)
11. Hong, L., Jie, Y.: Exploration on access control technologies for the internet of things (Nat. Sci. Ed.). J. Southwest Univ. National. **06**(42), 665–670 (2016)
12. Yu, Y., Guo, Z.: Design and implementation of cloud computing access control system based on trust. Mod. Electron. Tech. **03**(40), 62–64 (2017)

A Statistical Study of Technological Innovation Factors in Beijing's Low-Carbon Economic Growth

Xiaofan Wang[(✉)] and Li Zhou[(✉)]

Beijing Wuzi University, Beijing, China
sunnyday1010@163.com, zhouli@bwu.edu.cn

Abstract. The current mode of economic development in China is still in the "economic transition period". Technological innovation has an irreplaceable position in all kinds of innovation, and the low-carbon economy has been a closely watched economic form in recent years. Therefore, in Beijing, for example, starting from the development condition and the mechanism of technological innovation, to measure its contribution to the Beijing low-carbon economy. This study has a certain practical significance, coincide not only with calls for the "innovation driven development" strategy in our country, but also with the United Nations stressed the "green economy" of the "environmental protection". In this paper, multiple regression models are used. Firstly, the comprehensive research directions are determined on the basis of related research at home and abroad, defining a low-carbon economy and technological innovation and, at the same time building index system, eliminating the cause of multicollinearity variables, thus establishing multivariate regression equation. Through empirical study, measuring the contribution of technology innovation in low carbon economy and the contribution of technological innovation in environmental protection, economic development, and residents living conditions, putting forward the correspondence strategy after the empirical analysis. This article runs the analysis under R Studio.

Keywords: Low-carbon economy · Technical innovation · Multiple regression

1 Introduction

1.1 Research Context

In recent years, more and more scholars have paid more attention to the definition of economic development in a low-carbon economy. China gradually into the key period of urbanization, but malignant problems also grow subsequently, our environment is worsening, shortage of non-renewable resources, increasingly, China's economic development but also greatly increase the emissions of carbon dioxide, at present, China's CO2 emissions are more than the United States, to become the world's first place with the international community to take measures to reduce carbon emissions in China calls for more and more high, so the development of low carbon economy for Beijing and China are imminent, the technology innovation has become the indispensable important part in developing low carbon economy.

© Springer Nature Singapore Pte Ltd. 2018
K. Li et al. (Eds.): ISICA 2017, CCIS 873, pp. 405–419, 2018.
https://doi.org/10.1007/978-981-13-1648-7_35

1.2   Research Significance

Technological innovation has an irreplaceable position in all kinds of innovation, and a low carbon economy is also a form has attracted much attention in recent years, compared with GDP, pay more attention to low carbon economy is the social development and environmental protection, coordinated development of various aspects, such as residents' life. In addition, compared with other cities, Beijing is the economic and cultural center of China and the ancient civilization of the world, which has irreplaceable status.

1.3   Research Ideas and Framework

According to the above questions, this article takes Beijing data from 2000–2015 as sample, through defining the concept of technological innovation and low carbon economy, analyzing the contribution of the technology innovation in Beijing environmental protection, economic development and residents' living environment, as well as the technological innovation is how to promote low-carbon economic growth, and put forward corresponding countermeasures (The technical framework is shown below).

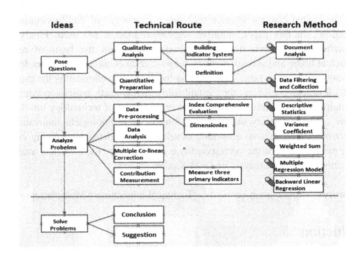

2   Research on Technological Innovation and Low-Carbon Economy

2.1   Theoretical Viewpoint of Low-Carbon Economy

On the concept of low carbon economy, "our energy future: creating a low carbon economy", the book points out that low carbon economy is through the use less natural resources and produce less pollution of the environment, to create more economic output and social wealth continuously, low-carbon economy at the same time also created more opportunities to develop, the use of advanced technology and communication.

Bloomberg [1] think that evaluation of low carbon economy should begin from the resource utilization and environmental protection, economic development and residents' life four aspects into account, the key to involve resources, ecological environment, economy, life, society, technology, etc. Wang [2], on the basis of resource environment constraint, there is a mutual promotion relationship between low-carbon economy and technological innovation.

This paper argues that a low carbon economy adhering to the concept of sustainable development, through the way of innovation, reducing emissions of greenhouse gases such as CO_2, achieve economic and social development, ecological environment protection, make full use of resources, residents living environment improve the win-win development of the form. The common point of the concept of low-carbon economy is to consume less natural resources and uphold the principle of reducing pollution in the process of creating value.

2.2 Theoretical Viewpoint of Technological Innovation

In the library of congress defined technical innovation to "technology innovation is a new idea to produce, research and development of the whole process, this process is thought to produce new products, the market application.

Yang [3] defined technical innovation to the enterprise or the relevant units in order to obtain more profits, by adopting new technology, new technology, new method such as new management organization, in the production and business operation system for the first time making new or improved products (or services) to research, design, development, production and marketing process." Luo [4], based on the six aspects of funding and human input, research and development and manufacturing capacity, economic benefits and technological achievements, she has evaluated the technological innovation ability of Anhui province. Zhu [5] emphasizes that technology innovation is not only to follow the principle of sustainable cycle of ecological, but more important is to people-oriented, so on the basis of system and technical indicators, she join the culture index into studying.

2.3 Selection and Determination of Indicators

Due to low carbon economy and the concept of technology innovation is not deterministic, does not have specific indicators like general concept can be evaluated directly, so we need according to certain principle and the method to select and determine the low carbon economy and the index evaluation system of technology innovation. Let's to select evaluation index of low carbon economy and technological innovation, on the basis of reference literature must be according to the principle of scientific, feasibility, selected low carbon economy by the three level indicators and nine secondary indicators. The selected technical innovation is evaluated by six indexes.

2.3.1 Indicators of Low-Carbon Economic

Visible by selecting index of negative index, namely, energy consumption per unit GDP, per unit of GDP water consumption, under normal circumstances, we will only take minus into plus indicators for comparison and calculation, the commonly used method is to take inverse method, so the two minus indicators take reciprocal method is adopted to improve the positive change (Table 1).

Table 1. Indicators of low-carbon economic evaluation

First grade indexes	Secondary index	Index direction
Environmental protection	Garbage disposal rate of living garbage (%)	Plus
	Sewage treating rate (%)	Plus
	Forest coverage rate (%)	Plus
Economic status	Energy consumption per unit of GDP (RMB/ton standard coal)	Minus
	Water consumption per unit of GDP (10,000 yuan/m3)	Minus
	GDP per capita (yuan/person)	Plus
Residents' life	Urban green coverage rate (%)	Plus
	Per capita park green area (square meter/person)	Plus
	Urban and rural disposable income (10,000 yuan)	Plus

2.3.2 Indicators of Technological Innovation

According to the theoretical research results of technological innovation, based on other scholars' research indexes, according to data availability. Build technical innovation indicators from the research and development (R&D) funds, research and development (R&D) test personnel, patent grant and the technology contract transaction number, new product sales, technology trading amount implementation rate of modeling to evaluate these six aspects.

3 A Statistical Description of the State of Beijing's Technological Innovation and Low-Carbon Economy

In the case of access to the raw data, first of all, making simple statistical description of low-carbon economy, mainly drawing a scatter diagram in order to make clear the development trend of the indicators, Due to the selection of nine indicators to evaluate the development status of a low-carbon economy with a low carbon economy, so the comprehensive value of low-carbon economy needs to be evaluated. Firstly, I'm going to make the data dimensionless, the weight is determined by variation coefficient method, then the weighted sum determines the final evaluation value, thus the low-carbon economy in Beijing can be seen at a glance.

After the basic descriptive of low carbon economy dependent and independent variables technology innovation, establishing a multiple regression model, detecting the serious multicollinearity of the model. Under normal circumstances, we can use correction model multicollinearity in the stepwise regression method, it chooses backward stepwise regression, eliminating the cause of multicollinearity variables, the multivariate regression equation is established.

The evaluation method of weight is divided into: subjective empowerment law and objective empowerment law. This paper selects the variation coefficient method in objective weighting method to evaluate the low carbon economic indicators.

The data in this article is from Beijing statistical yearbook 2016.

3.1 Assessment of the Development Trend of Low-Carbon Economy

First, we carried out simple descriptive statistics on the indicators of low-carbon economy, a simple trend analysis of nine indicators of the raw data of low-carbon economy. But in order to be able to see a change in the low-carbon economy over the years, we need to use comprehensive evaluation method to calculate the comprehensive value of low carbon economy based on the index system of construction. In this paper, the variation coefficient method is used to calculate the weight of the nine indicators. First, the mean value of each index is calculated, and the variation coefficient can be calculated to calculate the weight of each index. The weight of each index is obtained, and the data corresponding to the need to multiply can obtain the comprehensive evaluation value of low-carbon economy.

3.1.1 Descriptive Statistics of Low Carbon Economic Indicators

(1) Environmental protection status
The variable x11 in the graph represents the sewage treatment rate (%) and x12 represents the harmless disposal rate (%) of household garbage, X13 represents forest coverage (%) (Fig. 1).

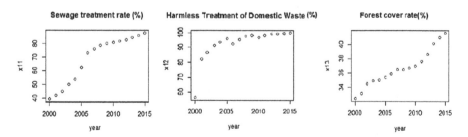

Fig. 1. Development trend of three secondary indicators in environmental protection

As can be seen from the figure, the relationship between the two variables is varying degrees. The figure said between 2000 and 2015, as the change of time, the Beijing municipal sewage treatment (%) of the (x11), hazard-free treatment rate of waste

(%) (x12), the forest coverage rate (%) (x13) ratio increased year by year, Beijing's environmental protection is also getting better and better, sewage disposal capacity is constantly improving, especially the sewage treatment, From the initial 39.4 percent governance rate to the later period up to 87.9 percent, the rate of harmless disposal of household garbage has risen from less than 60% in 2000 to 99.8% in 2015.

(2) situation of economic development.

The variable x21 in the figure shows energy consumption per unit of GDP (ten thousand yuan/ton standard coal), x22 indicates the water consumption per unit of GDP (ten thousand yuan/m^3), and x23 represents the per capita GDP (yuan/person) (Fig. 2).

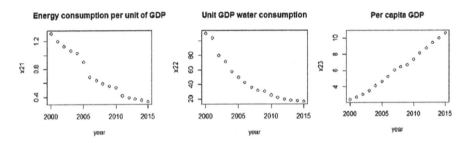

Fig. 2. Development trend chart of three secondary indicators in economic development

We can be seen from the graph, the two negative pointer unit GDP energy consumption and per unit of GDP water consumption in decline year by year, suggesting that the consumption of production per unit of GDP energy and water less and less, the unit GDP energy consumption from 1.311 in 2000 (ten thousand yuan/tons of standard coal) fell to 0.338 (ten thousand yuan/tons of standard coal), nearly fell by 4 times the amount of; The water consumption per unit of GDP fell from 110.98 (yuan/m^3) in 2000 to 16.6 (10,000 yuan/m^3), which was down nearly 6.68 times. At the same time, per capita GDP rose by 10.6497 yuan (yuan/person) from 2.412744 yuan (yuan/person) in 2000, basically a five-fold increase.

(3) Living Status of Inhabitants.

The variable x31 in the graph represents the green coverage of the city (%) and x32 represents per capita park green area (square meter/person) and x33 represents per capita disposable income (10,000 yuan) (Fig. 3).

As can be seen from the figure, the three positive indexes are on the rising trend, with the urban green coverage rate (%) rising to 48.4% from 36.5% in 2000. Per capita park green area (square meter/person) rose from 9.66 m^2 per person in 2000 to 16 m^2 per person. Per capita disposable income (10,000 yuan) rose from 0.4687 million yuan in 2000 to 2.0569 million yuan.

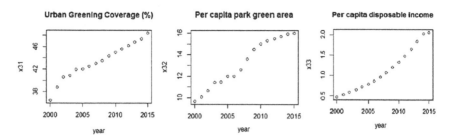

Fig. 3. Development trend of three secondary indexes in environmental protection

3.1.2 The Development of Low Carbon Economy in Beijing

The index of low carbon economy in Beijing from 2000 to 2015 was treated with variation coefficient method.

According to the variation coefficient method, the results of variation coefficient and weight are shown as follows:

Table 2. The variation coefficient and weight value of each index of low carbon economy

Objective layer	Criterion layer	Index layer	Variable coefficient	Weight
The situation of low-carbon economy development in Beijing	Environmental Protection	Garbage disposal rate of living garbage (%)	0.25075462	0.096778
		Sewage treating rate (%)	0.11742306	0.045319
		Forest coverage rate (%)	0.07166877	0.02766
	Economic Status	Energy consumption per unit of GDP (RMB/ton standard coal)	0.44563134	0.17199
		Water consumption per unit of GDP (10,000 yuan/m3)	0.55640643	0.214743
		GDP per capita (yuan/person)	0.43498249	0.16788
	Residents' life	Urban green coverage rate (%)	0.07471167	0.028835
		Per capita park green area (square meter/person)	0.16856596	0.065057
		Urban and rural disposable income (10,000 yuan)	0.47089202	0.181739

Standardize the original data according to the z-statistic formula, it's a process of dimensionless. Based on the known weights, the results are shown as follows (Table 2):

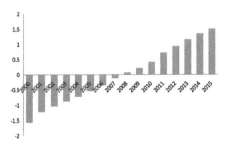

Fig. 4. The trend of development of the comprehensive evaluation of low-carbon economy in Beijing

It can be seen from the figure that the development trend of the low carbon economy in Beijing from 2000 to 2015. On the basis of 2008 and the measure, it can be seen in Beijing before 2008 low carbon economy evaluation index is negative, but in the process of increasing constantly, back in 2008, Beijing low-carbon economy on the basis of based on the positive (Fig. 4).

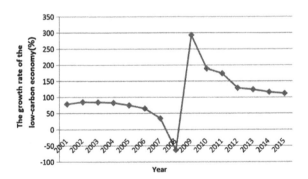

Fig. 5. Trends of low carbon economic growth rate in Beijing

The figure above shows Beijing's low carbon economy growth rate from 2001 to 2015. We can clearly see that, in the 2000–2007 growth rate basically maintained at between 0 and 1, growth rate plummeted in 2008, however, appeared the phenomenon of negative growth, this may is mainly affected by the financial crisis in 2008, the growth rate in 2009 after the financial crisis in 2008 recovered rapidly even reached nearly 300%, from 2009 to 2013, growth rate is over 100%, again in the future two years growth rate will gradually to maintain balance, there will be no too big trend of fluctuations (Fig. 5).

3.2 A Multivariate Regression Model of Low-Carbon Economy and Technological Innovation

In order to better explore the technology innovation influence the magnitude of the low carbon economy, this paper build a low-carbon economy and the multiple regression model of technology innovation, the use of low carbon economy value with the six indexes to establish a model of technology innovation, through the inspection we found that the model exists serious multicollinearity, the backward stepwise regression method is adopted to cause the multicollinearity of variables to weed out to overcome the multicollinearity in the model, using stepwise regression variables of technological innovation and low-carbon economy to establish regression model.

3.2.1 Explore the Relationship Between Low-Carbon Economy and Technological Innovation

The above table is the six measures chosen for technological innovation, which are analyzed after the economic data is usually taken from the logarithm. Take the logarithm we nature itself does not change the data and relationship, not only can eliminate heteroscedasticity, to some extent at the same time, the exponential, the meaning of economic variables with elastic, so the general argument the exponential form.

After reading the data, the relationship between independent variables and dependent variables can be explored by means of descriptive statistics. The output results are as follows (Fig. 6):

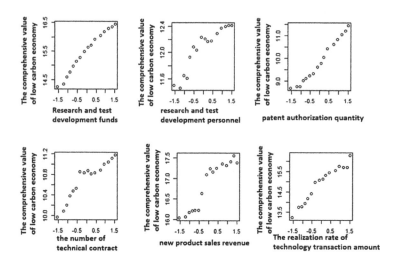

Fig. 6. Technical innovation index and low carbon economic scatter diagram

As shown in the above, by rendering technology innovation the scatterplot of dependent and independent variables and the low carbon economy, it can be seen that each independent variable and dependent variable in a relatively smooth straight near fluctuations, shows that there is relationship between two variables, and between each explained and interpreted variables of Pearson correlation coefficient is 0.987, 0.958, 0.917, 0.967, 0.951 and 0.934, and each independent variable and the low carbon economy of technological innovation has strong correlation between the dependent variable. Therefore, we can explore the relationship between technological innovation and low carbon economy by establishing multiple regression models.

3.2.2   A Multivariate Linear Regression Model Is Established

The comprehensive evaluation value of low carbon economy as the dependent variable and the technological innovation 6 indicators as the independent variable. The model of multiple linear regression model is established, and its basic expression is:

$$Y_i = \beta_1 + \beta_2 \cdot X_{2i} + \ldots + \beta_k \cdot X_{ki} + e_i, \, i = 1, 2, \ldots, n \qquad (1)$$

After reading the data, the following results can be obtained (the confidence level is 95%):

Table 3. The results of technical innovation data and low carbon economic regression

	Ln x1	Ln x2	Ln x3	Ln x4	Ln x5	Ln x6
P value	0.0017	0.0017	0.1174	0.653	0.9319	0.4305
Results	Passed	Passed	Unpassed	Unpassed	Unpassed	Unpassed

The explanation of the output results: after adjustment for the whole R is 0.9813, that equation fitting very well, F test P value is 2.929 e−08, far less than the confidence level, linear relationship of regression equations is notable. But we can see that the coefficient of significant t test failed, in addition to the variables X1, X2, the rest of the variables are not significant, and that the model, there are serious multiple linear index method to draw conditions index value was 8566603, according to conditions in general believe that conditions index CI value between the (0, 10), x is no multi-collinearity; The CI value is between (10, 100), and it is considered that X has strong multiple co-linear. When CI is greater than 100, it is considered that multiple collinear is very serious, and the larger the value, the more serious the multiple co-linear. We can see that the results of CI values mean that the model has serious multiple collinear. Therefore, the stepwise regression method is used to repair the multiple collinear of the model (Table 3).

3.2.3   Model Multiple Co-Linear Repair

Due to the severe multiple collinear in the model, the regression method is used to eliminate the variables.

The results show that the output results show that the first step is eliminated by the first step backward step by step. The second step removes the x4, because the AIC is at least −62.1. The third step is to get rid of the x3 variable, because at this time AIC is at least −64.07; The fourth step removes the x5 variable, because at this time AIC is at least −65.65; The fifth step is to not eliminate the variables at optimal time, i.e. the final

Table 4. Regression variables and corresponding AIC values are eliminated step by step

Steps	Step 1	Step 2	Step 3	Step 4
Eliminate variables	X6	X4	X3	X5
AIC value	−60.1	−62.1	−64.07	−65.65

model includes x1 research and test development (R&D) personnel, x2 R&D fund two independent variables. Therefore, the final model is obtained (Table 4):

$$Y = -15.2853 + 1.7287x1 - 0.9495 \, x2 \qquad (2)$$

Through texting: P value of the intercept 1.97e−06, variable X1 P value is 4.28e−08, variable X2 P value is 0.0144, three sig values are less than 0.05, shows that the equation of the regression coefficient by T test. In addition, F test statistic is 507.5, two degrees of freedom are 2 and 13 respectively, and the corresponding P value is 4.596e−13 is less than 0.05, which indicates that the whole model is significant and passes the F test.

The results of the model show that in the case of other variables, the research and test development (R&D) personnel increased by 1 unit, and the integrated value of low carbon economy increased by 1.7287 units. The total amount of R&D expenditure increased by 1 unit, and the total value of the low carbon economy decreased by 0.9495 units.

4 The Measurement of Contribution of the Technological Innovation in Beijing to the Components of Low-Carbon Economy

In this chapter, we will study the technical innovation in low-carbon economy to the contribution of measuring the level of evaluation system, the purpose is to explore how technology innovation index affect environmental protection, economic development and people life so as to influence the overall level of low carbon economy.

The variation coefficient method is used to assign weights to each level index, and the contribution of environmental protection, economic development and the comprehensive evaluation value of residents' life are respectively discussed.

4.1 Technological Innovation Estimates the Contribution of Environmental Protection

4.1.1 Environmental Protection and Development

You can see by the above, from 2000 to 2015, the environmental protection index of development, overall for the rising trend, the 2000–2005 growth rate faster, from 2006 to 2015, growth is slowing and there is no excessive growth, the slope stability (Fig. 7).

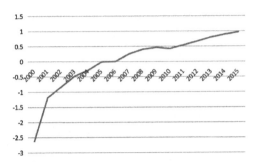

Fig. 7. Development trend of environmental protection indicators in Beijing's low-carbon economy in 2000–2015

4.1.2   The Contribution of Technological Innovation to Environmental Protection Is Calculated

Establish environmental protection and technology innovation of regression model, using the evaluation of technological innovation six indicators to with environmental protection index regression and eliminate multicollinearity, after stepwise regression to overcome the multicollinearity, we can get the regression equation of the environmental protection and low-carbon economy. The regression results: adjusted coefficient of determination $R^2 = 0.8706$, P-value is far less than 0.01 illustrates the sample regression effect is very good, strong linear relationship between the two. The specific equation is:

$$Y = -26.7311 - 0.1183 \, x \, 3 + 2.6089 \, x \, 4 \tag{3}$$

After testing: P value of the intercept 1.14 e−05, variable X3 P value is 0.018792, variable X2 P value is 0.000263, three sig values are less than 0.05, shows that the equation of the regression coefficient by T test. Moreover, the statistic of F test is 51.89, the two degrees of freedom are 2 and 13 respectively, and the corresponding P value is 6.348e−07 < 0.05, which indicates that the whole model is significant and passes the F test.

The result shows that, assuming the other variables are invariable, the patent authorization in the independent variable is increased by 1 unit, and the comprehensive value of environmental protection decreases by 0.1183 units. The number of technical contract transactions increased by 1, and the total environmental protection value increased by 2.6089 units;

4.2   The Measurement of the Contribution of Technological Innovation to Economic Development

4.2.1   Economic Development and Development

As can be seen from the figure above, the economic development between 2000 and 2015 is generally a gradual upward trend (Fig. 8).

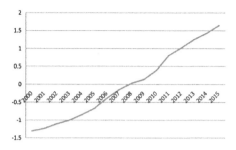

Fig. 8. Development trend of economic development indicators in Beijing's low carbon economy from 2000 to 2015

4.2.2 The Measurement of the Contribution of Technological Innovation to Economic Development

Establish the regression model of economic development and technology innovation, use of evaluation of technological innovation on the back and economic indexes of development index regression and eliminate multicollinearity, after stepwise regression to overcome the multicollinearity for economic development and the regression equation of the low carbon economy in the regression results: adjusted coefficient of determination $R^2 = 0.9946$, P-value is far less than 0.01, illustrates the sample regression effect is very good, strong linear relationship between the two. The specific equation is:

$$Y = -12.90586 + 0.91853 \times 3 + 0.35935 \times 4 \qquad (4)$$

After testing, the P value of the intercept is 1.27e−09, the P value of the variable X3 is 3.50e−11, the P value of the variable X2 is 0.00832, and the three sig values are less than 0.05, indicating that the regression coefficient of this equation is passed. In addition, the F test statistic is 1383, the two degrees of freedom are 2 and 13 respectively, and the corresponding P value is 7.18e−16 is less than 0.05, which indicates that the whole model is significant and passes the F test.

The result shows that, assuming the other variables are unchanged, the number of patent authorization increases by 1 unit, and the total value of economic growth increases by 0.91853. The number of technical contract transactions increased by 1 unit, and the economic growth composite value increased by 0.35935 units.

4.3 The Measurement of the Contribution of Technological Innovation to Residents' Life

4.3.1 Residents' Living conditions

As can be seen from the figure above, the economic development between 2000 and 2015 is generally a gradual upward trend. The 2014–2015 growth rate region remains unchanged (Fig. 9).

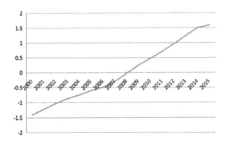

Fig. 9. The development trend of residents' living indicators in Beijing's low carbon economy from 2000 to 2015

4.3.2   The measurement of the contribution of technological innovation to residents' life

To establish the regression model of economic development and technology innovation, use of evaluation of technological innovation to residents and indexes of life index regression and eliminate multicollinearity, after stepwise regression to overcome the multicollinearity get the regression equation of the economic development and residents' life can be seen from the results of the adjusted coefficient of determination $R^2 = 0.9941$, P-value is far less than 0.01 illustrates the sample regression effect is very good, strong linear relationship between the two. The specific equation is:

$$Y = -12.14352 + 0.94252 \, x \, 3 + 0.26592 \, x \, 4 \tag{5}$$

After the test, the P value of the intercept is 3.99e−09, the P value of the variable X3 is 3.81e−11, the P value of the variable X2 is 0.0443, and all three sig values are less than 0.05, indicating that the regression coefficient of this equation is passed. Moreover, the F test statistic is 1271, the two degrees of freedom are 2 and 13 respectively, and the corresponding P value is 1.236e−15 < 0.05, which indicates that the whole model is significant and passes the F test.

The result shows that, assuming the other variables are invariable, the number of patent authorization increases by 1 unit, and the total value of residents' life is increased by 0.94252 units. The number of technical contract transactions increased by 1 unit, and the total living value of residents increased by 0.26592 units.

5   Conclusion

In this study, using stepwise regression to eliminate relevant indexes from technology innovation, getting the x3 (patent grant) and x4 (the number of technical contract) two indicators to represent the development status of technology innovation, according to the technological innovation and environmental protection, economic development and residents' life model, technology innovation play a vital role in environmental protection, economic development and residents' life, the most main is to promote economic development and residents' life, thereby giving impetus to the development of low-carbon economy.

From the results of the study, 2000–2015, the Beijing municipal level of low-carbon economy is to improve year by year, showing that the Beijing municipal government and relevant units or high levels of attention to the development of low-carbon economy, by adopting relevant measures to promote the development of low-carbon economy, finally also achieved fruitful results.

Acknowledgement. This study is supported by the science training program of Information School of the Beijing Wuzi University in 2017.

References

1. Peng, B.: The Construction and Empirical Study of Low Carbon Economic Evaluation Index System. Hunan University, Hunan (2011)
2. Wang, J.: Interaction Mechanism of Low-Carbon Economy and Technological Innovation. Shandong University of Finance and Economics (2012)
3. Yang, D.: Understanding and research on the concept of technological innovation. J. Harbin Inst. Technol. (Soc. Sci. Ed.) **2**, (2010)
4. Luo, Y.: Empirical Analysis of Technological Innovation Capability of Large and Medium-Sized Industrial Enterprises in Anhui Province. Anhui University, Hefei (2007)
5. Zhu, W.: Research on the Path Selection of Technology Innovation Under the Low-Carbon Economic Paradigm. Nanjing University of Aeronautics and Astronautics (2012)

Cloud Computing and Multiagent Systems – Cloud Engineering

Study on Critical Lines Identification in Complex Power Grids

Yi Wang[1], Zhiping Tan[1], and Yanli Zou[2(✉)]

[1] Guangdong University of Science and Technology, Dongguan 532000, China
[2] Guangxi Normal University, Guilin 541000, China

Abstract. Some key lines in power system have important influence on the stability of a power grid operation. If the key lines can be identified, we can take some protection against these key lines to reduce the probability of blackouts. A comprehensive evaluation index which considers local information and global information is proposed for the identification of critical lines in power grids in this paper. From the point of view of network structure and synchronization dynamics, we compare the ranking results obtained by the comprehensive index method and the edge betweenness centrality method based on global information in IEEE39 and IEEE118. The simulation results show that the proposed index is better than edge betweenness centrality index on the recognition effect of the importance of edges.

Keywords: Power grid · Critical line identification
Synchronization dynamics · Complex network

1 Introduction

Many blackouts occurred in power grids at home and abroad, which have caused huge economic losses [1–7]. One of the largest blackouts in eight states in the eastern United States and parts of Canada paralyzed traffic and industry in August 2003 [6, 7]. The accident caused great shock in all countries. In order to prevent the occurrence of blackouts and maintain the stable operation of power grids, the researchers are aware that it is necessary to study the internal causes of blackouts. In fact, one component failure causes other elements to fail through lines, which is the cascading failure, thus find key lines and carry on supervision and protection to these key lines can improve the reliability of the power system and reduce the probability of blackout. With the rapid development of complex network theory, more and more researchers used complex network theory to identify key lines of power system and achieved a series of results [8–18]. This article quantified the impact of small world characteristics on cascading failure propagation based on complex network theory. The study found that high node betweenness centrality and edge betweenness centrality to fault propagation play a role in fueling [14]. From the point of view of network topology structure, the authors put forward the route betweenness centrality index using the number of the shortest distance between any two points through the line, which is used to measure edge importance. The simulation results show that the betweenness centrality index can find the key lines in power grids [15–17]. Considering the distribution of power flow

© Springer Nature Singapore Pte Ltd. 2018
K. Li et al. (Eds.): ISICA 2017, CCIS 873, pp. 423–431, 2018.
https://doi.org/10.1007/978-981-13-1648-7_36

before and after disconnection of the line, a key line identification method was proposed. Using this identification method to identify the critical lines in a cascading failure process constructed by OPA blackout model. The experimental results show that the proposed identification method can identify the key lines in cascading failures, which can verify the effectiveness of the proposed method [9]. Key line identification methods based on cascading failure network diagram (CFG) were proposed respectively from node degree and different stages of cascading failures according to the propagation characteristics of cascading failures. Using the proposed key line identification methods to do simultaneous attack and timing attack experiment in IEEE118 node system. The simulation results show that the proposed critical line identification methods can not only reflect the vulnerability of power transmission lines, but also can reflect fault propagation relationship between lines, so the line identification methods have certain practical significance [18]. The above literatures all use single network feature index to identify key links in power grids.

The key line evaluation index is constructed from the network topology based on complex network theory in this paper. This index takes into account global information of edge betweenness centrality and local information of node degree, which overcomes the problem that a single index can not describe edge importance completely. In order to illustrate the effectiveness of the proposed index, the edge ranking result is compared with the results obtained by betweenness centrality method based on global information in IEEE39 and IEEE118 systems.

2 Network Model and Key Lines Identification Method

2.1 The Power Grid Model

The data of IEEE test system is the most recognized power grid data in the world at present, so power system researchers usually use IEEE standard test data for experimental simulation, such as IEEE14, IEEE30, IEEE39, IEEE57, IEEE118. Using graph theory to abstract topology structure of complex power system, it is convenient for us to do research on power grids. Generally, each component (generator, substation, user) is regarded as a node, and a power transmission line is regarded as an edge, so that a complex power grid is abstracted into a graph by node and edge. IEEE14 system has 5 generators, 9 loads, and 20 lines. There are 6 generators, 24 loads, and 41 lines in IEEE30 system. IEEE39 system consists of 39 nodes and 46 edges, including 10 generators and 29 loads. IEEE118 standard test system has 118 nodes and 179 edges, including 54 generators and 64 loads. The topology diagram of complex power networks is shown in Fig. 1. In this paper, we do the experiment simulation using unweighed and undirected networks.

2.2 The Power Grid Dynamic Model

The power grid dynamic model is the two order Kuramoto-like oscillator model [19–21], which is widely used in synchronous performance analysis and stability research. The model is described as follows:

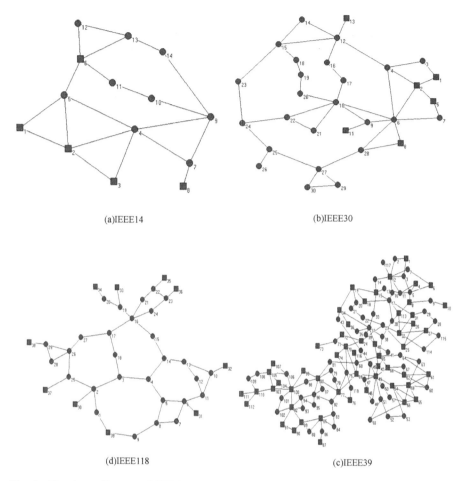

(a)IEEE14   (b)IEEE30

(d)IEEE118   (c)IEEE39

Fig. 1. Topology diagram of IEEE system: square nodes represent generator nodes, ellipse nodes represent load nodes

$$\ddot{\theta}_i = P_i - \alpha\, \dot{\theta}_i + K \sum_{j=1}^{N} a_{ij} \sin(\theta_j - \theta_i) \tag{1}$$

In order to facilitate numerical simulation, the above two order differential equation is rewritten into two first order differential equations:

$$\begin{cases} \dot{\theta}_i = \omega_i \\ \dot{\omega}_i = P_i - \alpha\omega_i + K \sum_{j=1}^{N} a_{ij} \sin(\theta_j - \theta_i) & i = 1, 2, \ldots, N \end{cases} \tag{2}$$

Where θ_i is phase offset of node i, P_i is power of node i, α is loss parameter, K is coupling strength between nodes, $A = (a_{ij})$ is adjacency matrix of network connection which describes topological structure of network. If there is a link between node i and node j in unweighed and undirected networks, so $a_{ij} = a_{ji} = 1$, otherwise $a_{ij} = a_{ji} = 0$.

It can be seen from formula (2) that node phase and node frequency jointly determine motion state of node i. When a power network works normally, all nodes run at the frequency of 50 Hz or 60 Hz and the power grid is in synchronization state.

We use phase order parameter to measure synchronization ability of network [22, 23]. Order parameter is defined as follows:

$$r(t)e^{i\varphi(t)} = \frac{1}{N}\sum_{j=1}^{N} e^{i\theta_j(t)} \tag{3}$$

Where $\theta_j(t)$ represents phase of oscillator j, $\varphi(t)$ represents a mean value of all oscillator phases in a network, N represents number of oscillators. It can be seen from formula (3) that order parameter $r(t)$ describes synchronization situation of all oscillators at a given moment. Steady order parameter r_∞ represents a mean value of order parameter of one system in a stationary state. The concrete formula of steady order parameter r_∞ is as follows:

$$r_\infty = \lim_{t_1 \to \infty} \frac{1}{t_2} \int_{t_1}^{t_1+t_2} r(t)dt \tag{4}$$

2.3 The Key Line Identification Method

Considering node degree and edge betweenness centrality, comprehensive evaluation index about importance of edges is put forward as follows:

$$NL_{ei-j} = \frac{(k_i + k_j)}{N * \langle k \rangle} * \frac{B_{ei-j}}{\max(B)} \tag{5}$$

Where k_i is degree of node i, k_j is degree of node j, N is node number, $\langle k \rangle$ is average degree of network, so $N * \langle k \rangle$ is sum of all node degree. $\frac{(k_i+k_j)}{N*\langle k \rangle}$ reflects proportion that local degree information which is composed of two nodes i and j connecting one edge occupies whole network node degree. Edge betweenness centrality B_{ei-j} is number of shortest paths between any two nodes passes edge e_{i-j}, which describes edge importance in the global information transmission. B is all edge betweenness centrality set. From the above analysis, it can be seen that comprehensive evaluation index considers not only local characteristics of nodes, but also global characteristics of edges.

According to the definition of each index, we calculate the index value of each edge in IEEE39 and IEEE118 standard test system, then we arrange edge in accordance with the index value from large to small order. Due to the limitation of length in this paper,

Tables 1 and 2 lists the edge numbering of top 30 of each index in IEEE39 system. The higher an edge sorting position is, the more important it is.

Table 1. Edge betweenness centrality sort table

Order number	Edge	Order number	Edge	Order number	Edge
1	e_{14-15}	11	e_{2-25}	21	e_{26-29}
2	e_{15-16}	12	e_{25-26}	22	e_{21-22}
3	e_{16-17}	13	e_{26-27}	23	e_{23-24}
4	e_{2-3}	14	e_{16-21}	24	e_{5-8}
5	e_{4-5}	15	e_{16-24}	25	e_{10-13}
6	e_{4-14}	16	e_{3-18}	26	e_{1-39}
7	e_{16-19}	17	e_{5-6}	27	e_{8-9}
8	e_{3-4}	18	e_{17-18}	28	e_{6-11}
9	e_{13-14}	19	e_{19-20}	29	e_{12-13}
10	e_{17-27}	20	e_{1-2}	30	e_{2-30}

Table 2. Comprehensive index sort table

Order number	Edge	Order number	Edge	Order number	Edge
1	e_{16-17}	11	e_{25-26}	21	e_{5-8}
2	e_{15-16}	12	e_{16-21}	22	e_{10-13}
3	e_{16-19}	13	e_{16-24}	23	e_{19-20}
4	e_{2-3}	14	e_{5-6}	24	e_{21-22}
5	e_{4-5}	15	e_{26-27}	25	e_{23-24}
6	e_{14-15}	16	e_{17-27}	26	e_{6-11}
7	e_{4-14}	17	e_{26-29}	27	e_{8-9}
8	e_{3-4}	18	e_{3-18}	28	e_{26-28}
9	e_{13-14}	19	e_{17-18}	29	e_{1-39}
10	e_{17-27}	20	e_{1-2}	30	e_{12-13}

3   Comparison of Line Identification Methods Based on Network Structure

We compare the sorting methods about edge importance from the perspective of network structure. Network efficiency $C(e_{i-j})$ is defined as follows [24]:

$$C(ei - j) = N^R_{ei-j} \Big/ N \tag{6}$$

Where N^R_{ei-j} says node number in maximum connected subnet of network after the removal of the edge e_{i-j}, N says primitive network node number. N^R_{ei-j}/N represents the vulnerability of the network after the failure of the edge. The smaller the value of $C(e_{i-j})$ after removing edge e_{i-j} is, the more important of edge e_{i-j} to a network connectivity is from formula (6). According to the index value, an edge is removed in order of large to small, and the relationship between the number of removing edges and network efficiency is studied in each power grid mode. The simulation results are shown in Fig. 2.

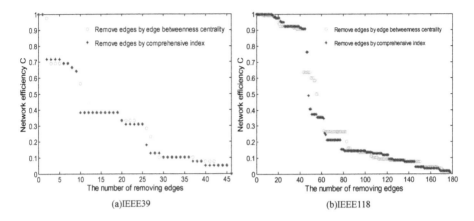

Fig. 2. Network efficiency under different ways to remove edges

In view of Fig. 2, it is found that removing edges number is inversely proportional to network efficiency value. The more edges are removed, the smaller network efficiency value is in each way, which indicates that connected network is divided into several scattered small connected region or isolated nodes and thus it reduces node number on the largest connected branch. If one edge is deleted, the number of nodes on the largest connected branch is reduced faster, then it can be explained that the deletion of the edge is more destructive to the network. In view of Fig. 2(a), when deleting the first 2 edges, network efficiency value is greatly reduced using comprehensive index method. At this time, network efficiency $C_{comprehensive\ method} < C_{betweenness\ centrality\ method}$ which shows that the destruction of a network is greater with the removal of edges by comprehensive method. According to the sorting result of Tables 1 and 2, it is found that the first 2 edges contain e_{15-16} and the other edge is different, which shows that e_{16-17} is more important than e_{14-15}. In addition, when removing the third edge of e_{16-17} by betweenness centrality method, we found $C_{comprehensive\ method} > C_{betweenness\ centrality\ method}$ which shows that the removal of edge e_{16-17} is more destructive to connectivity of a network. The above analysis illustrates that comprehensive index method is more reasonable in edge importance ranking. When removing the tenth edge by comprehensive index method, network efficiency $C_{comprehensive\ method}$ drops faster. According to edge importance ranking result using comprehensive index method, we found that this edge correspond to e_{2-25} while taking e_{2-25} in eleventh place in betweenness centrality method. The higher the sorting location is, the more important the link is. From this point of view, comprehensive index can find key edges of a network more effectively than betweenness centrality method. Combining Tables 1 and 2, we observe network efficiency of removing the twenty-first edge and the twenty-sixth edge in different ways, and we also find that the comprehensive method is more effective in identifying important edges. Figure 2(b) can also obtain the above conclusions.

4 Comparison of Line Identification Methods Based on Dynamics

When an important line fails and is removed from a power grid in the operation of power system, synchronous capability of a power grid decreases, which affects the stable operation of a power grid. We usually use critical synchronous coupling strength k_c to measure synchronization capability of the power grid.In general, critical synchronization coupling strength k_c of a power grid will increase with the decrease of synchronization capability. We use two order Kuramoto-like model as the power grid dynamics model to study synchronous dynamic behavior of a power grid. This paper studies synchronization ability of power grids when deleting a link in different ways, and the simulation results are shown in Fig. 3.

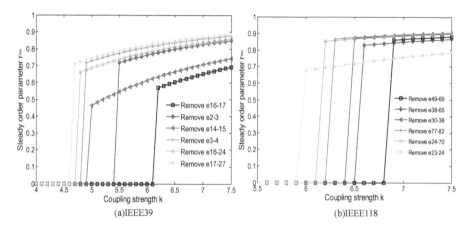

Fig. 3. Power grid synchronization capability when deleting a link

We use the edge importance index $IM_{ei\text{-}j}$ to indicate importance of edge.The results of critical synchronization coupling strength k_c of a power grid when links are removed in Fig. 3(a) is as follows: $k_{c,\text{remove } e16\text{-}17} > k_{c,\text{remove } e2\text{-}3} > k_{c,\text{remove } e14\text{-}15} > k_{c,\text{remove } e3\text{-}4} > k_{c,\text{remove } e16\text{-}24} > k_{c,\text{remove } e17\text{-}27}$, and these links importance is sorted as follows: $IM_{e16\text{-}17} > IM_{e2\text{-}3} > IM_{e14\text{-}15} > IM_{e3\text{-}4} > IM_{e16\text{-}24} > IM_{e17\text{-}27}$. The ordering result of these edges importance by betweenness centrality method is as follows: $IM_{e14\text{-}15} > IM_{e16\text{-}17} > IM_{e2\text{-}3} > IM_{e3\text{-}4} > IM_{e17\text{-}27} > IM_{e16\text{-}24}$. These edges importance ranked by comprehensive method is as follows: $IM_{e16\text{-}17} > IM_{e2\text{-}3} > IM_{e14\text{-}15} > IM_{e3\text{-}4} > IM_{e16\text{-}24} > IM_{e17\text{-}27}$.

In addition, the ranking result of edge importance obtained by experimental simulation in Fig. 3(b) is as follows: $IM_{e49\text{-}69} > IM_{e38\text{-}65} > IM_{e30\text{-}38} > IM_{e77\text{-}82} > IM_{e24\text{-}70} > IM_{e23\text{-}24}$. The order of importance of these edges determined by betweenness centrality is as follows: $IM_{e38\text{-}65} > IM_{e30\text{-}38} > IM_{e49\text{-}69} > IM_{e24\text{-}70} > IM_{e23\text{-}24} > IM_{e77\text{-}82}$. The order situation of importance of these links obtained by comprehensive method is as follows: $IM_{e49\text{-}69} > IM_{e38\text{-}65} > IM_{e30\text{-}38} > IM_{e77\text{-}82} > IM_{e24\text{-}70} > IM_{e23\text{-}24}$. By comparing simulation results and edge importance ranking results in different ways, it is found that the ranking results of edge importance obtained by comprehensive method and the experimental simulation are consistent. It shows that comprehensive method is more effective than betweenness centrality method in identifying the importance of links. Figure 3(b) can also get the same conclusion.

5   Conclusion

A comprehensive evaluation index is proposed in this paper, which considers local information of node degree and global information of edge betweenness centrality. Comparing the identification effect of comprehensive index method and betweenness centrality method on important edges from network structure and synchronization dynamics in IEEE39 and IEEE118 systems. The simulation results show that comprehensive method is more effective than betweenness centrality method to find key links in power grids.

Both edge betweenness centrality method and comprehensive index method all need to calculate betweenness, so the amount of calculation is large. From this point of view, these two methods are usually used for key line identification in small networks, which are not very suitable for large networks.

Acknowledgements. The work of this paper was supported by National Natural Science Foundation of China: Research on Synchronization, Stability and Cascaded Faults of Power Network Based on Complex Network Theory (Grant No. 11562003), and Multi-source Information Mining and Safety Key Laboratory of Guangxi (Grant No. 13-A-02-03).

References

1. Ding, M., Han, P.: Research on power system cascading failure with complex system theory. J. Hefei Univ. Technol. (Nat. Sci.) **28**(9), 1047–1052 (2005)
2. Bai, W., Wang, B., Zhou, T.: Brief review of blackouts on electric power grids in viewpoint of complex networks. Complex Syst. Complex. Sci. **2**(3), 29–37 (2005)
3. Pourbeik, P., Kundur, P.S., Taylor, C.W.: The anatomy of a power grid blackout-root causes and dynamics of recent major blackouts. IEEE Power Energy Mag. **4**(5), 22–29 (2006)
4. Lu, Z.: Survey of the research on the complexity of power grids and reliability analysis of blackouts. Autom. Electr. Power Syst. **29**(12), 93–97 (2005)
5. Ding, L., Cao, Y., Liu, M.: Dynamic modeling and analysis on cascading failure of complex power grids. J. Zhejiang Univ. (Eng. Sci.) **42**(4), 641–646 (2008)
6. Yang, T., Wang, Y., Wang, S.: Research review of safety risk of complex power system construction. J. Shanghai Univ. Electr. Power **28**(5), 457–462 (2012)
7. Yin, Y., Guo, J., Zhao, J., Bu, G.: Preliminary analysis of large scale blackout in interconnected north America power grid on August 14 and lessons to be drawn. Power Syst. Technol. **27**(10), 8–11 (2003)

8. Cao, Y., Chen, Y., Cao, L., Tan, Y.: Prospects of studies on application of complex system theory in power systems. Proceed. CSEE **32**(19), 1–9 (2012)
9. Zeng, K., Wen, J., Cheng, S., Lu, E., Wang, N.: Critical line identification of complex power system in cascading failure. Proceed. CSEE **34**(7), 1103–1112 (2014)
10. Liu, G., Liu, Y., Gu, X.: Identification of critical lines in restoration scheme for transmission network. Autom. Electr. Power Syst. **35**(1), 23–28 (2011)
11. Ding, L., Liu, M., Cao, Y., Han, Z.: Power system key lines identification based on hidden failure model and risk theory. Autom. Electr. Power Syst. **31**(6), 1–5 (2007)
12. Cai, Z., Wang, X., Ren, X.: A review of complex network theory and its application in power systems. Power Syst. Technol. **36**(11), 114–121 (2012)
13. Zhang, G., Li, Z., Zhang, B., Halang, W.A.: Understanding the cascading failures in Indian power grids with complex networks theory. Phys. A: Stat. Mech. Appl. **392**(15), 3273–3280 (2013)
14. Ding, M., Han, P.: Small-world topological model based vulnerability assessment algorithm for large-scale power grid. Autom. Electr. Power Syst. **30**(8), 7–10 (2006)
15. Dwivedi, A., Yu, X., Sokolowski, P.: Identifying vulnerable lines in a power network using complex network theory. In: Proceedings of IEEE International Symposium on Industrial Electronics, 5–8 July 2009, Seoul, Korea, pp. 18–23 (2009)
16. Chen, G., Dong, Z.Y., Hill, D.J., Zhang, G.H.: An improved model for structural vulnerability of analysis of power networks. Phys. A Stat. Mech. Appl. **388**(19), 4259–4266 (2009)
17. Chen, X., Sun, K., Cao, Y.: Structural vulnerability analysis of large power grid based on complex network theory. Trans. China Electrotech. Soc. **22**(10), 138–144 (2007)
18. Wei, X., Gao, S., Li, D., Huang, T., Pi, R., Wang, T.: Cascading fault graph for the analysis of transmission network vulnerability under different attacks. Proceed. CSEE **29**, 1–11 (2009)
19. Motter, A.E., Myers, S.A., Anghel, M., Nishikawa, T.: Spontaneous synchrony in power-grid networks. Nat. Phys. **9**(3), 191–197 (2013)
20. Filatrella, G., Nielsen, A.H., Pedersen, N.F.: Analysis of a power grid using a Kuramoto-like model. Phys. Condens. Matter **61**(4), 485–491 (2008)
21. Maistrenko, Y., Popovych, O., Burylko, O., Tass, P.A.: Mechanism of desynchronization in the finite-dimensional Kuramoto model. Phys. Rev. Lett. **93**(8), 084102 (2004)
22. Rohden, M., Sorge, A., Witthaut, D., Timme, M.: Impact of network topology on synchrony of oscillatory power grids. Chaos **24**(1), 279–312 (2014)
23. Rohden, M., Sorge, A., Timme, M., Wittaut, D.: Self-organized synchronization in decentralized power grids. Phys. Rev. Lett. **109**(2), 1–7 (2012)
24. Liu, J., Ren, Z., Guo, Q., Wagn, B.-H.: Node importance ranking of complex networks. Acta Phys. Sinica **62**(17), 178901,1–10 (2013)

A Review of Multi-sensor Data Fusion for Traffic

Xue Zhao and Dongbo Zhang[✉]

Guangdong University of Science and Technology, Dongguan, Guangdong, China
77188649@qq.com

Abstract. Data fusion is the process of integrating multi-sources data to obtain more consistent, accurate, and beneficial information than that provided by any single data source. This paper sum the state of the data fusion field and describes the most relevant studies. We first explain data fusion and multiple sensors Then, data fusion in traffic are reviewed.

Keywords: Data fusion · Multiple sensor · Traffic state estimation

1 Introduction

1.1 Data Fusion

The general concept of multi-sensor data fusion is analogous to the manner in which humans and animals use a combination of multiple senses, experience, and the ability to reason to improve their chances of survival [1] In particular, the brain fuses information from our surrounding environment and attempts to derive knowledge, draw conclusions or inferences from the fused information [2]. For example, consider how many sensors are used by a human being when eating. Assessing the quality of an edible substance may not be possible using only the sense of vision; the combination of sight, touch, smell, and taste is far more effective [3].

While there is not one commonly referenced definition of data fusion, there is a general consensus of what fusing data means. [1] suggests that multi-sensor data fusion is "the theory, techniques and tools which are used for combining sensor data, or data derived from sensory data, into a common representational format in performing sensor fusion our aim is to improve the quality of the information, so that it is, in some sense, better than would be possible if the data sources were used individually." [3] propose "data fusion techniques combine data from multiple sensors and related information to achieve more specific inferences than could be achieved by using a single, independent sensor." [2] provides the simplest definition, stating that "fusion involves the combination of data and information from more than once source." As can be seen from these three definitions, there is a common understanding that data fusion encompasses a wide variety of activities that involve using multiple data sources. Unfortunately, the universality of data fusion has engendered a profusion of overlapping research and development in many applications. A jumble of confusing terminology (Fig. 1) and ad hoc methods in a variety of scientific, engineering, management, and educational disciplines obscures the fact that the same ground has been covered repeatedly [3].

© Springer Nature Singapore Pte Ltd. 2018
K. Li et al. (Eds.): ISICA 2017, CCIS 873, pp. 432–444, 2018.
https://doi.org/10.1007/978-981-13-1648-7_37

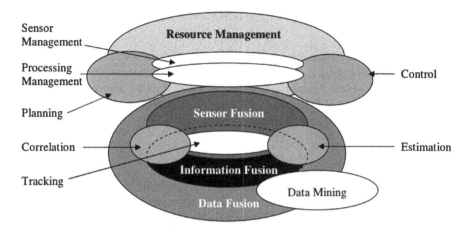

Fig. 1. (Con)fusion of terminology – adapted from [3]

There are many reasons why we need data and information fusion systems as noted by [1–5] Reliability/Robustness/Redundancy: A system that depends on a single source of input is not robust in the sense that if the single source fails to function properly, the whole system operation will fail. However, the system fusing several sources of data has a higher fault-tolerance since multiple sensors providing redundant information serve to increase reliability in the case of sensor error or failure.

- Accuracy/Certainty: Combining readings from several different kinds of sensors can give a system more accurate information. Combining several readings from the same sensor makes a system less sensitive to noise and temporary glitches. Therefore, multiple independent sources of data can not only help improve accuracy, but can also add certainty by removing ambiguity in the data.
- Completeness/Coverage/Complementarity: More data sources will provide extended coverage of information on an observed object or state. Extended coverage is particularly relevant in spatial and temporal environments for the sake of completeness. Sometimes information from multiple sensors is complementary and allows features in the environment to be perceived that are impossible to perceive using just the information from each individual sensor operating separately (see Sect. 2.3.3).
- Cost effectiveness: To build a single sensor that can perform multiple functions is often more expensive than to integrate several simple and cheap sensors with specific functions.
- Representation: Another problem that sensor fusion attempts to address is information overload. The amount of time required to make a decision increases rapidly as the amount of information available increases. Sensor fusion is necessary to combine information and clearly present the best interpretation of the sensor data to allow for a well informed and timely decision.
- Timeliness: More timely information may be provided by multiple sensors due to either the actual speed of operation of each sensor, or the processing parallelism that

may be possible, as compared to the speed at which it could be provided by a single sensor.

Of course, all of these benefits hinge on the assumption that there is no single perfect source of information. This assumption is well made since all sensors have a few things in common: every sensor device has a limited accuracy, limited coverage, is subject to the effect of some type of noise, and will under some conditions function incorrectly. Hence, there is no single perfect source of information.

1.2   Multiple Sensors

[6] first classified a multi-sensor data fusion system according to its sensor configuration. The typology proposed by this author gained popularity and is now widely used in the data fusion research community. The three basic types of configurations are: complementary, competitive, and cooperative. While these divisions are defined by the functionality of the sensor network, they are not necessarily mutually exclusive.

(1)   Complementary

A sensor configuration is called complementary if the sensors do not directly depend on each other, but can be combined in order to give a more complete image of the phenomenon under study. Complementary sensors help resolve the problem of incompleteness. As a simple example, Fig. 2 shows a temperature monitoring system that consists of several thermometers each covering a different region. This configuration is complementary because each thermometer provides the same type of data but for a different geographic region. In general, fusing complementary data is intuitive and easy.

Fig. 2. A complementary sensor network may consist of several thermometers, each covering a different geographical region (note there is no overlap in coverage)

(2)   Competitive

A sensor configuration is competitive if each sensor delivers an independent measurement of the same property. Since they provide what should be identical data, the sensors are in competition as to which reading should be believed by the system in the case of discrepancies. Competing sensors can be identical or they can use different methods of measuring the same attribute. The aim of competitive fusion is reduce the effect of uncertain and erroneous measurements, provide greater reliability, and/or add fault tolerance to a system. Figure 3 shows three thermometers partially surveying the same region (shaded darker). Note that this type of configuration would still be able to function for the joined region if one of the thermometers were to cease functioning.

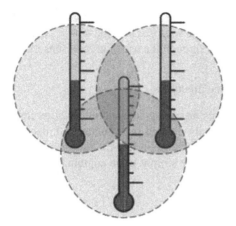

Fig. 3. Competitive thermometers would all return information regarding the same region (note the overlap in coverage)

(3) Cooperative

A cooperative sensor configuration uses the information provided by two or more independent sensors to derive information that would not be available from the single sensors alone. Figure 4 shows four thermometers that measure the temperature at different points along a line. Not only can they be used as complementary sensors to provide temperature

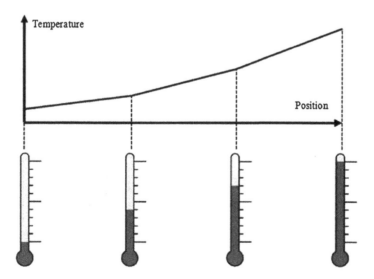

Fig. 4. Thermometers separated by equal distance along a line provide information about temperature. They can also be used cooperatively to find the rate of change of temperature

information over a combined area (as in Sect. 2.3.1), but they can also be use coopera-tively to determine the rate of change of temperature along the line: rate of change can be estimated as the difference between two readings divided by the distance between two thermometers. Note that the temperature change along the line could never be determined by using only one sensor. Thus, the aim of cooperative sensor networks is to derive new information through the use of several sensors.

1.3   Fusion System Architectures

There are several ways of classifying data and information fusion system architectures but they are most commonly divided into: centralized, decentralized, and distributed architectures. Occasionally, there is reference to other types such as hierarchical or hybrid architectures, which are simply some combination of the three aforementioned architectures.

(1) Centralized

In centralized fusion architectures, the fusion unit is located at a central processor that collects all of the raw data from the various sensors as shown in Fig. 5. All processing and decisions are made at this node and instructions or task assignments are given out to the respective sensors.

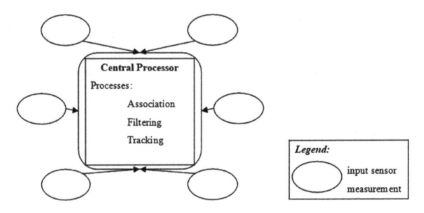

Fig. 5. Centralized architecture with a central processor – adapted from [3]

(2) Decentralized

Decentralized fusion architectures consist of a network of nodes, where each node has its own processor. There is no central fusion or central communication center. Fusion occurs at each node on the basis of local information and information from neighboring nodes. Additionally, nodes have no knowledge of the global network architecture of which they are a part. Decentralized fusion architectures could be further categorized as fully connected (as shown in Fig. 6) or partially connected (not shown).

(3) Distributed

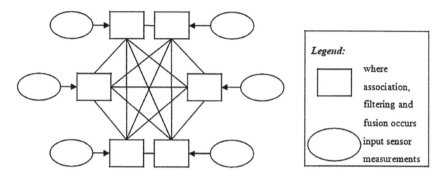

Fig. 6. Decentralized fusion architecture – adapted from [3]

Distributed fusion architectures are an extension of the centralized fusion architecture, where each sensor' s measurements are processed independently before sending the estimate (often referred to as a "track") to a central processor for fusion with other distributed sources of input. A distributed fusion architecture is shown in Fig. 7.

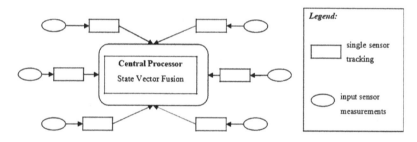

Fig. 7. Distributed fusion architecture – adapted from [3]

The next chapter provides an introduction to data fusion applications in transportation engineering and a comprehensive literature review on data fusion studies conducted in traffic flow and travel state estimation.

2   Data Fusion in Transportation Engineering

Transportation management centers are continuously motivated to obtain reliable information for traffic monitoring and control operations. Basic traffic data is typically obtained from sensors embedded in the road pavement, namely loop detectors. These fixed sensors are very useful, but they fail in measuring spatial characteristics of traffic. That is, the loop detector represents only the specific location of the traffic. A proliferation of new measurement devices (cameras, cell phones, Bluetooth, GPS, etc.) mean that other sources of data are becoming increasingly available to complement the information provided by conventional loop detectors. Although these technologies vary, they promote a common trend: probe vehicle data collection. In this sense, cars on the road act as a moving sensor, continuously providing information about traffic conditions. Therefore, a wide spectrum of data and heterogeneous sources of information are now becoming available for traffic monitoring. As a result, many applications in transportation engineering involve a data fusion problem [7].

Though various reviews of data fusion have been conducted, [8] was the first to specifically examine data fusion technology with an eye to its application in Intelligent Transportation Systems (ITS). Table 1 provides a synopsis of the data fusion techniques and key ITS projects reviewed by [8]. Note that the year given in column two of Table 1 represents the date the article was published and not necessarily the date the ITS project was completed. Therefore, one must keep in mind that some of the data fusion techniques listed in Table 1 may not have actually been implemented in the final version of the ITS project cited.

Table 1. Summary of fusion techniques applied to ITS – adapted from literature [11]

Project (Author)	Year	Technique(s)	Purpose
PROMETHEUS (Behringer et al.) (Martinez et al.)	1992 1990	Kalman filter	Constructs 4-D position estimates from autonomous driving
		Expert system	Decomposes a driving task into independent subtasks
		Neural network	Allocates one neural net for each driving subtask
Brainmaker (Change)	1992	Neural network	Pattern-matches current traffic situations with historical situations
IGHLC (Niehaus, Stengel)	1991	Kalman filter	Determines vehicle position
		Bayesian	Deals with traffic uncertainty
TravTek (Summer)	1991	Expert system	Models concept of Worse-Case Decision Making
		Fuzzy logic	Permits traffic conditions to be described with qualitative measures rather than simple "yes-no" responses
DRIVE (Martinez et al.)	1990	Expert system	Decomposes a driving task into independent subtasks
		Neural network	Allocates one neural net for each driving subtask
PRODYN (Kessaci et al.)	1989	Kalman filter	Estimates traffic-turning movements
		Bayesian	Estimates traffic-state variables, e.g., queues and saturation
(Harris)	1988	Fuzzy logic	Effectively controls AGV's lateral motions in real time

Similar to the previous study, [9] investigated Data Fusion for ITS but with a particular focus on delivering Advanced Traveler Information Services (ATIS). The purpose of ATIS is to provide practical and timely information to aid travelers in an integrated, multi-modal environment. Their study includes a literature review of ATIS data fusion practices, the development of an appropriate ATIS data fusion model, and general guidelines on the development of an ATIS data fusion system. Table 2 shows some of the data fusion techniques assessed by [10], the applicability of which is bound by the specific task for which the data fusion model is being developed. In other words, each method cannot be applied to all data fusion problems, but rather solve a particular problem through data fusion. For example, Dempster-Shafer theory cannot be used for all data fusion problems.

Table 2. The relative merits of level 1 data fusion techniques–adapt from literature[12]

Relative scalable performance	Relative scalable performance		Maintenance	
Parametric based				
Classical inference	Excellent	Excellent	Excellent	Excellent	Excellent
Bayesian inference	Good	Poor	Good	Poor	Poor
Dempster-shafter	Good	Poor	Good	Poor	Poor
GEP (Generalized evidence processing)	Poor	Poor	Poor	Poor	Poor
Non parametric based				
Parametric templates	Poor	Good	Good	Poor	Poor
Neural nets	Good	Good	Poor	Poor	Poor
Clustering	Good	Excellent	Good	Good	Good
Voting	Good	Excellent	Excellent	Good	Excellent
Figure of merit	Good	Good	Good	Good	Good
Correlation measures	Excellent	Excellent	Good	Good	Excellent
Pattern recognition	Good	Poor	Poor	Poor	Poor
Cognitive based				
Logical templates	Poor	Good	Poor	Poor	Good
Knowledge-based	Poor	Poor	Poor	Good	Poor
Fuzzy set techniques	Good	Good	Good	Good	Good

3   Data Fusion for Traffic Speed and Travel Time Estimation

The purpose of this section is to provide a comprehensive overview of data fusion research for traffic speed and travel time estimation. For the sake of readability, the research projects are presented in chronological order under one of the following subheadings: statistical approaches, Kalman filter applications, neural network models, evidence theory (Dempster–Shafer theory), and other contributions.

(1)   Statistical Approaches

Tarko & Rouphail (1993) [13] proposed one of the earliest data fusion applications to travel state estimation for the *ADVANCE* (Advanced Driver and Vehicle Advisory Navigation Concept) project. The *ADVANCE* project in the Chicago metropolitan area was one of several ATIS operational tests that were underway in the USA and abroad in the early nineties. The basic data fusion concept involved the following steps:

1. Evaluate the expected link travel time from detector data (*EDTT*) using a regression model developed off-line (these detectors did not measure speed),
2. Compute the mean probe travel time (*EDTT*) from probe data received during the last interval,
3. Fuse EDTT with EPTT in order to obtain the on-line link travel time estimate (*EOTT*),
4. Obtain the final link travel time (*EFTT*) by fusing *EOTT* with a historical travel time estimate (*ESTT*).

In step 1, the regression model chosen for detector data conversion is used to establish a relationship between detector occupancy and expected travel time since loop detectors measuring speed were not available. In the following steps, the estimates are fused based on the squared estimation error similar to that of the simple convex combination described in Sect. 4.1, only the weights of the estimates are additionally influenced by the sample size of the estimator.

El Faouzi (2004b) [14] explores the use of aggregative strategies in which all competing estimators are aggregated to form a single estimate. This author assumes the intuitive strategy of using a weighted average of individual estimators that minimizes the overall estimation error. In other words, the optimal weights are proportional to the quality of the set of estimators. Although the terminology is not explicitly used by the author, this research is essentially an investigation of the simple convex combination and the Bar-Shalom/Campo combination (Sects. 4.1 and 4.2).

In either case, the research shows combining two estimators consistently outperformed the individual estimators and provided substantial improvement under the root of mean squared error (RMSE) criterion. El Faouzi (2004b) [14] notes the main advantage of aggregative strategies is that errors from different estimators may cancel out one another, but it is not easy to determine the expected improvement of fusion for a given configuration. Moreover, it is not straightforward how to determine which combination of estimators can improve the overall estimation reliability.

(2) Kalman Filter Applications

Guo et al. (2009) [15] sought to develop an accurate speed estimation methodology for single loop detectors under congested conditions. Based on this requirement their research team developed a method using the Kalman filter technique (Sect. 4.3), based on an empirical investigation into the relationship among single loop measurements. In their study, the unknown speed is treated as a hidden state, and the common instrument of assuming a random walk model for the state transition is used In order to evaluate the proposed methodology, the estimated speeds (determined only using the flow rate and occupancy data from the stations) were compared to the dual loop measured speeds that served as the ground truth. The proposed algorithm was tested using data from two urban regions in Northern Virginia and Northern California. The empirical evaluation showed that their proposed algorithm can produce acceptable speed estimates under congested traffic conditions. The comparison of the proposed approach with the g-factor approach (adopted for comparison in their study) shows that the proposed method consistently outperformed the traditional g-factor approach. Overall, their study shows the Kalman filter is capable of estimating speed accurately in an online fashion.

Peng et al. [16] present another Kalman filter based traffic measurement fusion method to solve the problem of monitoring road sections without GPS traffic data by using neighbouring roads that are monitored by GPS devices. The Kalman filter algorithm is verified with the real-world GPS traffic data from the city of Hangzhou in China, and the fusion results are compared to the results obtained by using historical speed data to replace the current estimated speed. In the first experiment, the relative mean error (RME) of the fused results is 19.61%, while the relative mean error of using historical speed data to replace the current speed is 44.44%. In the second experiment, the RME of the fused results is 19.84%, while the RME of using historical speed data to replace the current speed is 64.75%. Thus, for road sections without GPS sampling signals, the fusion results of the speed information of associated road sections are better than the results obtained by using historical speed data to estimate the current speed. The authors note that while the Kalman filter algorithm is simple and easy to implement, and the computation cost is very small, there is a need for optimization or identification methods to determine the parameters to calibrate the model.

Byon et al. [17] Any single sensor that uses multiple data sources to evaluate present traffic conditions is active applying single-constraint-at-a-time (SCAAT) . This method addresses the need for a versatile technique that can fuse data from not only loop detectors and probe vehicles, but also other available data sources, which may not necessarily have the same frequency or accuracy. Moreover, some information may have a different unit, such as that based either fully or partially on human judgment, for instance the information provided by websites of traffic departments and by radio stations (e.g. "moving well", "moving slowly", "extremely slow" or "not moving"). For the initial application and evaluation of the proposed SCAAT Kalman filter, different types of data are collected from 4 different sources: a floating car survey using a GPS unit, 40 loop detectors across Highway 401 in Toronto, radio broadcasting from AM 640 Chopper Traffic channel and the on-line freeway management system of the Ontario Ministry of Transportation (MTO). The results indicate the SCAAT estimated travel times are reasonably close to what a road user actually experiences. Furthermore, a micro simulation package is applied for having access to the true traffic conditions of a simulated context that has been calibrated for a particular road section in Toronto. Then, the performance of the developed SCAAT filters are compared with the true traffic conditions under different sampling strategies with varying number of probes and varying sampling frequencies of sensors. The conclusion from the microsimulation study is that the data should be combined only if there are data gaps from the most accurate sensor. Fusing the lower quality loop detectors in the presence of other better sensors would only increase the error measures. Overall, the major advantage of adopting the SCAAT data fusion method for traffic monitoring is that any change in the sampling rate or addition/removal of any new/old sensor can be handled with no additional major modifications to the filtering framework. Thus, the flexible nature of SCAAT filtering can enable robust and easy-to-implement traffic monitoring systems.

(3) Neural Network Models

Nelson & Palacharla [18] introduced the idea of using neural networks (see Sect. 4.7) for tacking the travel state estimation problem. Their paper describes the ability of a Counter propagation neural network model to classify input traffic flow patterns and output present travel state estimates. Neural networks are enable learning how to classify and connect input/output patterns, making them fit for tacking problems like estimating present travel times from traffic flow patterns received from several different sources. The outputs of the counterpropagation neural network developed in this study are statistical averages of the travel time outputs that are best representative of each traffic flow pattern class. The authors experimented using the model for a group of ten consecutive links where some links have loop detectors in all lanes, other links have detectors in a few lanes, and some links have no detectors at all. Altogether, the ten links have 27 input sources (loop detectors). The network was trained using 4 input traffic flow patterns (at four different times of the day), and following completion of training, the counterpropagation neural network functions as a look up table. Thus during normal operation of the neural network, traffic flow patterns are collected during each interval from all sources, and the Kohonen weight vector that is closest to the input traffic flow pattern is selected (i.e. the closest traffic pattern stored in the neural network), and the corresponding Grossberg weight vector is given as the present travel time output (i.e. the associated travel time of that traffic pattern stored in the neural network).

Cheu et al. [19] proposed a more advanced neural network based arterial speed estimation model using data from mobile probe vehicles and inductive loop detectors. The input layer has three neurons receiving: (1) average speed estimated by the loop detector modules; (2) average speed estimated by the probe vehicle module; and (3) probe vehicle sample size. The latter gives an indication on the accuracy of the average speed calculated from probe vehicle data. The authors decided to base their model development and testing on simulated data from a microscopic traffic simulation model capable of simulating probe vehicles and loop detectors. To evaluate the performance of the data fusion model, the error between the fused speed estimates and all vehicle link speeds is computed from a subset of data reserved for this purpose. The root mean squared error (RMSE) of speed estimates is reduced to 1.08 km/h. This is an improvement from using either probe vehicle estimates or detector estimates alone. Park & Lee [20] developed a neural network with one input layer, one hidden layer, and one output node. The inputs to the neural network in this study were four outputs obtained from a dual-loop detector: average speed, average occupancy, flow, maximum occupancy. The evaluation of their model is based on probe data collected on Whasan-ro in Jeonju, South Korea, on the 23rd and 24th of April, 2003, and dual-loop detector data collected in the same period. Whasan-ro is an arterial road consisting of 6 links. Probe vehicle data are collected by recording all of the passing license plates last four digits at the 7 intersections on Whasan-ro. To simulate the effect of probe vehicles as learning data, two vehicles are randomly sampled in a given link during each time interval and taken as probe vehicles. Furthermore, two versions of the neural network were developed: one uses the average of all observed vehicles speeds as a target value for training, and the second uses only the probe vehicles (the average of two sample cars speeds) as target values. The performance of the two neural networks is almost the same, except the neural network trained on probe vehicle data can show some biases generated by the characteristics of probe vehicle type (an intuitive result).

4 Conclusions

This paper reviews some of the methods and techniques applied in traffic speed and travel time estimation draw heavily from other disciplines and fields of study; this is a reminder of Figure 2-1, which shows how a number of disciplines, areas of study, and techniques contribute to the field of data fusion in general.

The paper reviewed all share the common objective of fusing multiple traffic sensors, with a particular focus on data from conventional loop detectors and a small sample of travel time measurements from probe vehicles obtained during the same time period. Although traffic state estimation have been addressed from different perspectives, the preferred fusion approach is also undetermined because of some major loophole in bibliographical information. First of all, none of the experts made an attempt to compare their proposed technique with competing techniques. Instead, each researcher proposed a fusion method and then compared their method with mono-sensor approaches. Also, these single researches cannot be correctly compared due to the lack of a common measure of effectiveness. Many researches did not even use quantitative measures to evaluate their proposed approach, and when they did, the same quantitative measure was not used among experts.

References

1. Alpaydin, E.: Combined 5 × 2 cv F test for comparing supervised classification learning algorithms. Neural Comput. **11**(8), 1885–1892 (1999)
2. Beale, M.H., Hagan, M.T., Demuth, H.B.: Neural network toolbox 7 User's guide. The MathWorks Inc., Natick (2010)
3. Berkow, M., Monsere, C.M., Koonce, P., Bertini, R.L., Wolfe, M.: Prototype for data fusion using stationary and mobile data. Transp. Res. Rec. **2099**, 102–112 (2009)
4. Bouchon-Meunier, B. (ed.): Aggregation and fusion of imperfect information. Physica-Verlag, New York (1998)
5. Brooks, R.R., Iyengar, S.S.: Multi-sensor fusion : Fundamentals and applications with software. Prentice Hall PTR, Upper Saddle River (1998)
6. Byon, Y., Shalaby, A., Abdulhai, B., Elshafiey, S.: Traffic data fusion using SCAAT kalman filters. In: TRB 89th Annual Meeting Compendium of Papers DVD, Washington, DC. (2010)
7. Chang, K.C., Saha, R.K., Bar-Shalom, Y.: On optimal track-to-track fusion. IEEE Trans. Aerosp. Electron. Syst. **33**(4), 1271–1276 (1997)
8. Cheu, R.L., Lee, D., Xie, C.: An arterial speed estimation model fusing data from stationary and mobile sensors. In: 2001 IEEE Intelligent Transportation Systems Conference Proceedings, Oakland, CA, pp. 573–578 (2001)
9. Choi, K., Chung, Y.: A data fusion algorithm for estimating link travel time. J. Intell. Transp. Syst. Technol. Plann. Oper. **7**(3–4), 235–260 (2002)
10. Chong, C., Mori, S.: Convex combination and covariance intersection algorithms in distributed fusion. In: FUSION 2001 Proceedings, Montréal, Quebec, Canada (2001)

11. Dailey, D.J.: ITS data fusion No. Research Project T9903, Task 9: ATIS/ATMS Regional IVHS Demonstration. Washington State Transportation Commission, Washington (1996)
12. Keever, D., Shimizu, M., Seplow, J.: Data fusion for delivering advanced traveler information services. Fhwa Report (2003)
13. Tarko, A., Rouphail, N.: Travel time data fusion in advance. In: Pacific Rim TransTech Conference@sVolume I: Advanced Technologies. ASCE (1993)
14. El Faouzi, N.: Data-driven aggregative schemes for multisource estimation fusion: a road travel time application. In: Proceedings of SPIE, vol. 5434, pp. 351–359 (2004)
15. Guo, J., Xia, J., Smith, B.L.: Kalman filter approach to speed estimation using single loop detector measurements under congested conditions. J. Transp. Eng. **135**(12), 927–934 (2009)
16. Peng, D., Zuo, X., Wu, J., et al.: A kalman filter based information fusion method for traffic speed estimation. In: International Conference on Power Electronics and Intelligent Transportation System, pp. 374–379. IEEE (2009)
17. Byon, Y.J., Shalaby, A., Abdulhai, B., et al.: Traffic data fusion using SCAAT Kalman filters. In: Transportation Research Board 89th Annual Meeting (2010)
18. Nelson, P., Palacharla, P.: A neural network model for data fusion in ADVANCE. In: Proceedings Pacific Rim TransTech Conference, Seattle, Washington, vol. 1 (1993)
19. Cheu, R.L., Lee, D.H., Xie, C.: An arterial speed estimation model fusing data from stationary and mobile sensors. In: Proceedings of the Intelligent Transportation Systems, vol. 2002, pp. 573–578. IEEE (2001)
20. Park, T., Lee, S.: A Bayesian approach for estimating link travel time on urban arterial road network. In: Laganá, A., Gavrilova, M.L., Kumar, V., Mun, Y., Tan, C.J.K., Gervasi, O. (eds.) ICCSA 2004. LNCS, vol. 3043, pp. 1017–1025. Springer, Heidelberg (2004). https://doi.org/10.1007/978-3-540-24707-4_114

Research on Evaluation Method of Service Quality

Yan Zhao, Xiaxia Niu[✉], and Li Zhou

School of Information, Beijing Wuzi University, Beijing, China
605671232@qq.com, 15136212624@163.com, zhoulibit@126.com

Abstract. With the rapid development of service economy and information technology, the information age has brought great changes in consumer attitudes and service needs. In addition to price and speed, quality has become the most important factor. At present, the experts and scholars for the courier service quality evaluation methods have not yet formed a unified standard, which seriously restricts the express delivery service quality certification development. Based on this, this paper summarizes the existing methods of service quality evaluation and evaluation methods in domestic and foreign literature as much as possible, summarizes and compares the disputes and unresolved questions of scholars, and aims to do the research on service quality certification of service industry collecting work. The purpose is to express the quality of service evaluation method as the basis for other industries to provide guidance on the quality of service evaluation.

Keywords: Service quality · Evaluation method · Express delivery service

1 Evaluation Methods and Classification of Service Quality

The service quality evaluation model and evaluation method have always been the core of service quality research at home and abroad, and also the forefront topic in the field of global enterprise management. Enterprises need to provide satisfactory service for customers, dynamically all-the-way tracking service quality, in a timely manner to the entire service chain evaluation at various stages of decomposition, so as to find the crux of the problem as soon as possible, and in the first place to develop a response plan, these is very important to the development of the enterprise [1]. A lot of scholars on service quality evaluation model and the extension of concept, research on express service quality evaluation of service quality from the traditional field of FedEx and respectively, the view from the objective of research, also is the enterprise perspective and customer perspective both in determining the logistics service quality dimensions, the representative the significance of the model [2]. This article summarizes the existing evaluation models and evaluation methods of service quality in domestic and foreign literature, summarizes and compares scholars' disputes and unresolved issues, and aims to do a good job of literature research for service quality certification in service industry. Many scholars at home and abroad put forward innovative theoretical models and evaluation methods in this field. Foreign scholars have Christian Gronroos, PZB, Lee, del Stella Di Vick, Soh [3] uses the triangular fuzzy AHP to analyze the third-party logistics

© Springer Nature Singapore Pte Ltd. 2018
K. Li et al. (Eds.): ISICA 2017, CCIS 873, pp. 445–455, 2018.
https://doi.org/10.1007/978-981-13-1648-7_38

enterprises and gives the results of sensitivity analysis, Medjoudj [4] applies AHP to customer satisfaction surveys in the power industry etc., domestic scholars such as Professor Lang Zhizheng, Professor Cui Lixin. Common service quality evaluation models and evaluation methods include: Geluluosi customer perceived service quality model, Li Ya Del and Stella Di Vic's relationship quality model, PZB customer perceived service quality model, the gap analysis model and its included A series of evaluation methods, Q matrix - multidimensional evaluation model for measuring service quality, comparative evaluation model based on psychology judgment standard, value curve evaluation model, system dynamics method, multidimensional fuzzy language information, service blueprint method, shadow customer survey Method, vague language information method, interview method and so on. In view of space limitations, this article describes several representative models and evaluation methods.

1.1 Gap Analysis Model

(1) Service Quality Model
In the mid-1980s to the early 1990s, the PZB academic community proposed a service quality gap model, also known as "5GAP model". The 5GAP model is a conceptual framework for research services, which can effectively evaluate the performance and capabilities of organizational services, and has been applied in many enterprises and industries. The gap between 1 and 4 is the difference between the expectation of the customer, the wrong service quality standard, the failure to provide service standards, the delivered service and the propaganda service. The gap 5 is the gap between customer expectations and perception, which may be caused by one or more of the above gaps.

The five mainstream evaluation methods are SERVQUAL measurement method, SERYPERF measurement method, weighted performance evaluation method (Adequacy-Importance measurement method), attribution model and non-difference (Non-Difference) evaluation method.

(2) SERVQUAL Measurement Method
SERVQUAL first appeared in the article "SERVQUAL: a multivariate customer perceived service quality measurement method" published by PZB academic team in 1988. The PZB academic team presents ten elements that determine the quality of service, after merged into five elements, and developed a questionnaire that includes 22 questions, and later scholars call the "SERVQUAL" method (see Table 1). The measurement method is based on the basis of customer service quality perception, the difference between the measurement and calculation of customer expectations of service and the actual perception, and the calculation results as the judgment basis of service quality. Application of SERVQUAL measurement method, can better understand the customer's expectations and perception process, help enterprises to find the dimensions which influence the customer perceived service quality, find the key problems affecting the quality of service, so as to make the quality improvement decision better. As the PZB academic team draws conclusions from the studies of four industries including Walk, Credit Card Company, Equipment Maintenance and Long Distance, its general applicability may not be established. They think, SERVQUAL measurement is a dynamic

quality of service evaluation method, in the specific application should be based on different cultural backgrounds, industry characteristics of service quality dimensions and questions to make appropriate adjustments, so as to ensure the SERVQUAL

Table 1. SERVQUAL scales.

Ten elements	Five elements	To form a project
Tangible	Tangible	1. The service facilities provided are modern
		2. To provide services to attract public facilities
Accessibility		3. Staff clothing and appearance more neat
		4. Services that can be promised through an enterprise service facility
Reliability	**Reliability**	5. Enterprises can be timely to complete the commitment to the customer thing
		6. Businesses can help customers with difficulties
Ability		7. For customers, the business is reliable
		8. When the customer to provide the commitment to the business when the business is very punctual
		9. Enterprises can ensure that the service records are correct
Responsiveness	Responsiveness	10. The staff will tell the customer the exact time to provide the service
		11. Enterprises will provide customers with timely service
Communicate		12. Enterprise staff has always been willing to help customers
		13. employees will not be too busy to meet the needs of customers
Credibility	Guaranteed	14. Can trust business staff
		15. Customers are relieved about the transactions with the business
Courtesy		16. Business staff are very polite to customers
		17. Companies encourage employees to provide better service to customers as much as possible
Understanding	Empathy	18. Enterprises for different types of customers to provide targeted services
		19. Corporate staff will give customers individual care
Safety		20. Enterprise staff take the initiative to understand the needs of customers
		21. Enterprises in the decision-making will give priority to the interests of customers
		22. The customer believes that the time of service provided by the enterprise is in line with it

Note (1) Source: Parasuraman, A., Zeithaml, V.A., Berry, L.: SERVQUAL, a multiple-item scale for measuring consumer perceptions of service quality. J. Retail. **64** (1), 12–40 (1988).

measurement science Sexuality and effectiveness. Quality of Service (SQ) = Service Awareness (P) - Service Expectation (E). When P > E, the customer's perception of the service is more than the expectation of the service, the customer can feel the high quality service; when P = E, When P = E, the customer's perception of the service is equal to the expectation of the service, and the customer feels that the service quality is acceptable. When P < E, the customer's perception of service below expectations, customer service feel low quality.

In the five dimensions of Table 1, tangibles include the actual facilities, equipment, and the appearance of the service personnel. Reliability means the ability to reliably and accurately fulfill the performance pledge. Responsiveness refers to the enterprise to help customers and quickly improve the level of service desire; Assurance refers to the staff has the knowledge, etiquette and express the ability to confident and credible; Empathy is concerned and provide customers with the ability and desire to personalize services.

Although the PZB team believes that SERVQUAL can be widely used in all service industry quality of service evaluation. However, empirical studies have shown that, when SERVQUAL transplanted to different industries, the evaluation of the questionnaire and evaluation indicators may need to be adjusted. Finn and Lamb [5] for the quality of service, the development of existing measurement tools is mainly based on the definition of perceived service quality and the gap model, that is, the customer according to a series of quality elements, through the service expectations and service experience comparison services Quality perception found in the study of retail stores, SERVQUAL scale five dimensions of the measurement architecture in the retail industry does not apply [6]. Therefore, the proposed for different enterprises and industry service quality measurement, SERVQUAL scale should be based on its characteristics to a certain degree of modification.

1. Modified SERVQUAL model

Because the quality of service is a kind of customer perceived quality, with strong subjectivity, so it is very difficult to measure effectively, SERVQUAL method of birth to better solve this problem. However, any new theory and method in the early stages of application, there will be some flaws. Many scholars have tested various aspects since the SERVQUAL model was born and put forward many suggestions and criticisms for improvement. As the PZB from the 20th century, 90 years, the SERVQUAL evaluation method has been modified several times, in 1991 proposed amendments SERVQUAL evaluation method. In the revised SERVQUAL evaluation method, the number of questionnaires was greatly increased, and the negative question in the original questionnaire was changed to the positive question, and the tone of the question was changed, so that the revised SERVQUAL evaluation method the reliability and validity are better than the original SERVQUAL. It is worth mentioning that this approach proposes that there is a cross between the dimensions of service quality. Therefore, in-depth analysis of the relationship between the dimensions of service quality is a very important area of future service quality evaluation methods.

2. SERVPERF measurement method

Cronin and Taylor (1992) to overcome the shortcomings of the SERVQUAL measurement model empirical study, proposed SERVPERF measurement method. SERVPERF performance measurement service quality measurement method refers to the use of a service performance variables to measure customer perceived quality of service. The measurement method has a great dependence on SERVQUAL, there is no substantial difference between the two in the questionnaire design, and does not involve weighting problem in the measurement process. So Cronin and Taylor think SERVPERF is a relatively simple and practical evaluation method.

3. Adequacy-Importance measurement method (weighted performance evaluation method)

Mazis (1975) and others argue that different customers have different evaluation criteria for quality of service, and different weights can be set for different factors by means of weighting, indicating that different customers have different preferences. In the 1970s, the study of perceived service quality had not yet begun, and the limitations of the level of service management research at that time were doomed to be imperfect. This method places too much emphasis on perceived service performance and basically neglects the service expectation.

4. Attribution model

Bkner (1990) formally introduced attribution to service management research. The so-called attribution refers to consumers experience a spontaneous inquiry when the service results and expectations inconsistent and adjust the relationship between performance perception and expectations of a reasonable state. Bkner has conducted a large number of empirical studies on attribution models and found different meanings of attribution models from other measurement methods. One is to add more logical and reasoning variables within the framework of the comparison of differences. The second is to emphasize the service provider's marketing mix the formation of consumer's last reasoning has an important role.

5. Non-difference evaluation method

Brown et al. (1933) argue that customer perceived quality of service can be judged by measuring the difference between the actual feelings and service expectations of customers. From a specific operational point of view, the use of non-differential evaluation method requires the use of ERVQUAL scale, and only 22 data, more simple to operate.

1.2   Comparative Evaluation Model Based on Psychological Judgment Criteria

The purpose of the comparative evaluation model based on the criteria of psychological judgment is twofold: one is to establish three sets of optional evaluation criteria for measuring the quality of service on the basis of the psychological judgment criteria to solve the unresolved methodological problems; The second purpose is to combine the extended concept of expectation into an alternative evaluation criteria for measuring the quality of service, resulting in another evaluation model of customer perceived service quality - a comparative evaluation model based on the criteria of psychological judgment. This model has a certain reference for measuring the quality of service and understanding customer perception.

(1) Three Sets of Questionnaires

Three sets of selectable questionnaires were selected, one with a combination of differential scores, and the other two were a direct measure of the quality of service. Each of the three sets of questionnaires also incorporates the expanded concept of expectation to be simultaneously defined by us. These three sets of questionnaires are given separately for the required, lowest and perceived service scores, which are designed side by side under the same evaluation criteria. It needs to calculate the perceived perceptual value separately, and perceive the appropriate difference values to quantify the measure of merit (MSS) and the appropriate service measure (MSA), respectively. Thus, although the project team is not repeated in the SERVQUAL evaluation model, the procedure for measuring the quality of service is similar to the SERVQUAL evaluation model.

In sum, evaluating the perceived values, combined with the lowest and required service expectations, helps to accurately diagnose service deficiencies and take appropriate improvements through innovation. In order to achieve these goals, the three-column questionnaire provides more detailed and accurate data than the other two questionnaires.

(2) Conclusions and Effects

The overall model of the results tells us that there are many trade-offs in choosing the most appropriate method of measuring the quality of service. These three questionnaires have relative strengths and weaknesses, which emphasize the trade-offs between each other and the impact on practice and theory.

1. Practical impact

Three questionnaires and two questionnaires provide a measure of quality service and minimum service, and three columns require three separate values, so it takes more time for the respondent. However, the simpler completion of the three-column questionnaire alleviates the time it takes to provide additional information while increasing the time spent. The two questionnaires have obvious complexity, and the respondents believe that such questionnaires have greater difficulty and have lower confidence in responding to this questionnaire, so that although it only requires two numerical sets of values, it will take more time to accomplish it.

As for the value of the information obtained through the measurement of the quality of service, the three-column questionnaire is better than the two-column questionnaire, and the two-column questionnaire is superior to the questionnaire. However, managers who like to measure the perceived relative to expectation may want to measure only MSS through a questionnaire.

What's important here is that you should decide which method to measure based on your business goals. If the goal is to predict the quality of service, then only the perceived evaluation method is the best. However, if identifying critical service defects is the primary goal, then the three-column questionnaire appears to be the most useful; and this form of questionnaires also provides independent perceived values for the forecast.

In short, the evaluation industry should consider adopting a service quality measurement system that can independently measure the minimum service, required services and perceived services. Substituting the existing measurement system may not be an easy task. Especially those with inherent thinking of the same courier business, but the conclusions of the study can also affect such a courier business. Only the express delivery system of the perceived measurement system should consider at least one required service measurement method to improve the current system to be able to more accurately identify service defects. If it is not possible to add a new measure, consideration should be given to making the perceived measure only a direct measure of the difference between perceived and required service expectations. Moreover, managers should be soberly aware of the importance of dynamic observation, if the manager to track the value of the entire time, will be more valuable data, that is, with the passage of time, the performance of each feature is improved, Deterioration or to maintain the same level, then you can take accurate and effective action.

2. Theoretical influence

The conclusion of the study, while expanding the knowledge about the quality of service and its measurement methods, also suggests additional issues that require further study.

Although the three questionnaires have a superior use value, there may be practical difficulties in the full implementation, especially in the telephone survey, or when more specific terms relating to the actual company are attached to the 22 basic terms, increased practical difficulties. Therefore, the need for an effective method of additional research, this method can explore the implementation of the reliability, soundness, but also get a full use of value. Research on the reasons for the differences between the predicted differences and the actual results can give us a better understanding of which factors are truly important in the measure of quality of service. Such research is particularly urgent. It is necessary to combine the standards of practice with the traditional evaluation model, which is dominated by psychology standards.

For example, the information needed to create a chart of tolerance can be obtained by the following "incomplete" method.

First, get the appropriate service values and perceived values from half of the samples, and get the desired service values and perceived values from the other half;

Second, the appropriate service values, the desired service values and the perceived values are obtained from the three comparable samples, respectively;

Thirdly, the whole sample is divided into five sub-samples that can be compared, and all three values of only one of the five aspects of SERVQUAL are obtained from each sample. Studies are also needed to evaluate the reliability and effectiveness of these methods, to evaluate the statistical significance of the results, and to evaluate the management of the entire questionnaire relative to the overall sample.

1.3   System Dynamics Method

(1) Basic Concepts of System Dynamics Modeling

The emergence of system dynamics began in 1956, the founder is professor Jay w. Forrester of the Massachusetts institute of technology. In the late 1950s, the system dynamics gradually developed into a new discipline, which is an important branch of system science. From the birth, the system dynamics has an independent theoretical system and scientific methods, and the method of system dynamics research is a combination of systematic analysis and comprehensive reasoning, qualitative analysis and quantitative analysis. Zhang Qiongfang [7] in the master's thesis "system dynamics based evaluation of the quality of express logistics service research", detailed description of how the system dynamics to evaluate the quality of courier logistics services.

(2) Basic Steps in System Dynamics Modeling

System dynamics is a simulation laboratory of complex systems in society. Different strategies can be used to verify the different results obtained by system dynamics model simulation. The basic steps of system dynamics modeling and simulation are:

1. Defining modeling purposes

Before you build a model, make sure you understand the system goals, which are modeling purposes. The ultimate goal of modeling is to analyze and solve problems in real life through the built model, such as predicting future trends or analyzing the causes of what has happened. The first thing to be clear in modeling is the socioeconomic phenomenon to be studied. The ultimate goal of the study is to solve the controllable problems in these phenomena and what effect to achieve.

2. Defining system boundaries

System dynamics is adapted to the study of complex systems. By defining the system under study and its system environment, the boundaries of the system can be delineated between the two systems. The system environment is a collection of things that are outside the system that are associated with the system. As far as possible to narrow the boundaries of the system, if some variables for the study of the problem does not matter that it should not be considered.

3. To build a causal diagram

All the factors in the system interact with each other and affect the overall system behavior. Therefore, if we want to understand the system structure and behavior characteristics, we should start with constructing the causal relationship between the factors. According to the purpose of modeling, we will build a system dynamics causal diagram on the basis of fully understanding the research questions and all relevant information such as relevant experience, knowledge and existing research results.

4. To establish a system flow chart

According to the system dynamics causality diagram constructed above, the nature and type of the basic factors are differentiated, the flow variable, the flow rate variable, the auxiliary variable and other variables are determined, the system dynamic flow diagram is constructed and the relevant variables are written System dynamics Equation.

5. Computer simulation

According to the system flow diagram model, the dynamic model of the system is simulated by means of computer, using simulation software and editing the variable Equation.

6. Simulation results analysis

According to the results of the simulation software output, the behavior characteristics of the research system can be analyzed, and the policy effect can be evaluated through the policy adjustment to provide scientific basis for decision-making.

(3) The Principle of System Dynamics Model Construction

1. Combined with express delivery service quality evaluation index system

The index system summarizes the indicators applied in the express service quality evaluation system model from the evaluation dimension. In constructing the system dynamics model, the factors that can cause the change of these indexes can be added to establish the causal relationship diagram, so as to analyze the influence of express logistics service quality Factors and the relationship between the factors. Through the construction and simulation of the system dynamics model, express delivery enterprises can find the control points and improvement measures of their logistics service quality.

2. Combined with customer perceived service quality model
The customer perceived quality of service model clearly describes the logical relationship between the factors of service quality evaluation. It is the result of the expectation of perceived - perceived gap to measure the level of service from the perspective of psychological cognition. It is the result of banner research in the field of service quality research.

1.4   Existing Customer Perceived Service Quality Assessment Method Has not yet Solved the Problem

Many scholars at home and abroad have conducted in-depth research on the evaluation methods of service quality, resulting in more evaluation methods, especially SERVQUAL and SERVPERF. However, there are different opinions about the comparison of the two methods. The empirical research shows that the evaluation methods of the two customer perceived service quality evaluation methods are consistent and have accurate dimension division, high reliability and validity, and variation Explanatory ability, but SERVPERF than the SERVQUAL method easier to answer and have a higher reliability, validity and explanatory power of variation [8]. Although SERVQUAL has been criticized by many scholars, but as a pioneering study of customer perceived service quality evaluation methods, other scholars on this issue are basically based on the basis of SERVQUAL, the vast majority of scholars in the empirical study, The use of the questionnaire, the question in the questionnaire and even the quality of service elements to determine, basically based on SERVQUL made. But it can't be denied that, whether it is SERVAUAL or SERVPERF, or other evaluation methods, there are still many unresolved issues.

(1) The Quality of Service "Qualitative" and Its Relationship with Customer Satisfaction

In the process of creating SERVAUAL PZB on the quality of service "qualitative problem" does not seem to have much to grasp. At the same time, they try to distinguish between quality of service and customer satisfaction research, the method used is to use the difference analysis method to measure customer perceived service quality.

Oliver's relationship between attitudes and customer satisfaction has been defined, and he argues that attitudes are a relatively stable view and inclination of a product or service, and satisfaction is that when consumers consume the product or service, Comparison of the emotional response, so the attitude of the measurement is based on a relatively broad basis, less affected by the random. Wojciechowska thinks that the quality of logistics service is reflected in the degree of customer satisfaction. It is determined by the function of goods being delivered on time, and on time and by additional services [9].

Until today, this debate is still continuing, there is still no authoritative conclusion on the relationship between customer perceived service quality and customer satisfaction. Especially for customer perceived service quality, customer satisfaction and customer repurchase will relationship between the different scholars of the conclusions are still different, it is difficult to reach an agreement.

(2) Reliability and Validity of the Problem

Cronin and Taylor conducted a comparison of SERVQAL and SERVPERF for three different industries. The conclusion of the study is that SERVPERF is superior to SERVQUAL in terms of reliability or validity, since the former have passed the test and the latter has failed to pass the test.

Su Yunhua through the SERVAUAL, revised SERVQUAL, SERVPERF, modified SERVPERF and modified Non-Difference comparative study, the conclusions and Cronin and Taylor are basically the same, in terms of reliability, the best performance correction SERVPERF, followed by the amendment Non-Difference, and finally is to amend SERVQUAL, the same, in terms of validity, the sort of the same situation and reliability.

However, PZB questioned Cronin and Taylor's research methods. For example, for the problem of coefficient, in the dimension of the problem analysis, its role is far enough, must be combined with other statistical methods to study in order to draw more objective and scientific conclusions.

(3) On the Customer Perceived Quality Evaluation Method "Cross-Cultural" Applicability

Almost all of the customer perception of quality evaluation methods, scholars almost all ignore the cultural background, whether it is SERVAUAL or SERVPERF, are produced in the European and American cultural background. Cultural factors in the customer perception of service quality formation and evaluation process in the end play a big role, whether it will affect the reliability and validity of these evaluation methods. Malhotra has studied the quality of service determinants of SERVQUAL in a detailed study of developed and underdeveloped countries, concluding that there are differences in the way customers perceive quality of service in different countries; How to evaluate the service contact problem between American and Japanese customers, the result is a model and a specific measure of the service quality of customers in two countries. In measuring the quality of service, the problem of cultural difference should be included in the research field. However, there is no authoritative conclusion on a series of issues such as how cultural differences affect the perceived quality of service, the degree of influence, how to measure and detect these effects.

(4) The "One-Sidedness" of the Evaluation Method

We refer to the "one-sidedness" in two aspects: First, from the perspective of service quality, the vast majority of the existing quality of service evaluation methods are limited to the evaluation of functional quality, and technical quality point of view of the service

Second, the existing evaluation is basically the quality of service contact evaluation, and the quality of service providers and customer relationship evaluation is no corresponding way.

Some scholars have been aware of this problem. For example, Johnny and Terry have made it clear that SERVQUAL measures only the quality of service process, but does not involve the quality of service results, which is one-sided, and incomplete. Almost all of the existing evaluation methods focus on the quality of the service process, such as SERVQUAL, SERVPERF and Non-Difference evaluation methods, and from the technical quality point of view is almost blank, which undoubtedly make the existing evaluation methods More or less have a certain degree of one-sidedness [8].

2 Conclusion

Services now occupy an important place, so the quality of service research has become increasingly important, if there is a standard in all sectors of service quality evaluation methods, to enhance the quality of service will be of great help. The next will focus on express delivery industry service quality evaluation methods, and strive to express the industry can have a uniform standard of service quality evaluation methods.

Acknowledgement. The study is supported by Research and application of logistics service certification scheme that is the sub project of national key research and development plan (2016YFF0204105-1), and Beijing the Great Wall scholars program (No. CIT&TCD20170317).

References

1. Cui, L.: Quality of Service Evaluation Model. Economic Daily Press, May 2003
2. Thai, V.V.: Logistics service quality: conceptual model and empirical evidence. Int. J. Logist. Res. Appl. **16**(2), 114–131 (2013)
3. Soh, S.: A decision model for evaluating third-party logistics providers using fuzzy analytic hierarchy process. Afr. J. Bus. Manag. **4**(3), 339–349 (2010)
4. Medjoudj, R., Laifa, A., Aissani, D.: Decision making on power customer satisfaction and enterprise profitability analysis using the analytic hierarchy process. Int. J. Prod. Res. **50**(17), 4793–4805 (2012)
5. Finn, D.W., Lamb Jr., L.C.W.: An evaluation of the SERVQUAL scales in a retailing setting. Adv. Consum. Res. **18**, 483–490 (1991)
6. Zheng, B.: B2C Network Store Logistics Service Quality and Its Relationship with Customer Loyalty. Dalian University of Technology, Dalian (2008)
7. Fang, Y.: An empirical comparison of evaluation methods of customer perceived service quality - revisiting SERVPERF and SERVQUAL. J. Ningbo Univ. Technol. **24**(4), 53–57 (2012)
8. Du, J.: Enterprise and Customer Interaction Mechanism Research. Kunming University of Science and Technology, Yunnan (2004)
9. Werbinska-Wojciechowska, S.: On logistics service quality evaluation-case study. Logist. Transp. **2**(13), 45–46 (2011)

Research on Traffic Data Fusion
Based on Multi Detector

Suping Liu[(⊠)]

Department of Computer Science,
Guangdong University of Science and Technology, Dongguan, China
457789090@qq.com

Abstract. In order to alleviate urban traffic congestion, it is necessary to obtain roadway network traffic flow parameters to estimate the traffic conditions. Single-detector data may not be sufficient to obtain a comprehensive,effective, accurate and high quality traffic flow data. The neural network and regression analysis data fusion methods are employed to expand data sources as well as improve data quality. The multi-source detector data can provide fundamental support for traffic management. An empirical analysis is conducted using Beijing urban expressway traffic flow parameters acquisition technology. The results show that the proposed data fusion method is feasible and can provide reliable data sources.

Keywords: Data fusion · Neural network method · Regression analysis
Multi-source data

1 Introduction

With the increase in car ownership, the urban traffic congestion situation also increasingly serious [1]. By obtaining traffic status, it can reflect the intersection of cities. Traffic flow speed is the best way to reflect traffic flow characteristics, and a traffic flow parameter that is easier to obtain.

Obtaining traffic flow speed can well reflect the traffic status, thus providing reliable parameter data for traffic managers. After the data fusion, the multi-source traffic information can be more authentic and reliable traffic information [2], which provides strong support for traffic operation and management. In order to better serve the traffic management, control and guidance, it is necessary for the multi detector to detect the traffic flow speed.

In order to better serve the induction in traffic management, control and multi detector on traffic flow detection has become inevitable. Using information fusion method source detector collected traffic flow data with the data fusion are more abundant and high quality traffic information, improve traffic mobility, safety and orderly. By using the information fusion technology, the traffic flow velocity data collected by multi-source detectors can be obtained by means of data fusion, and more abundant and high-quality traffic information can be obtained, so as to improve the mobility, safety and orderliness of traffic.

© Springer Nature Singapore Pte Ltd. 2018
K. Li et al. (Eds.): ISICA 2017, CCIS 873, pp. 456–467, 2018.
https://doi.org/10.1007/978-981-13-1648-7_39

The processing object of data fusion is the speed of data from different sources, and the core of data fusion is to coordinate data optimization and comprehensive treatment. It is necessary to carry on traffic flow speed acquisition integration, the advantages of data fusion, the characteristics of the intelligent transportation system is the key technology and traffic flow and joint decision. Data fusion technology can get more accurate traffic information through comprehensive processing of traffic data from different sources.

As the center of the city within the scope of the road network in the main framework, Beijing city fast road network system without traffic lights, although the length of the total length of road network in the city accounted for just 8% days, but the traffic volume in the future to bear at least accounted for the city's total traffic volume 50% [3]. In the city of Beijing Expressway Development in city traffic plays an important function, the quality of its operation, operation will directly affect the overall road network in Beijing City, the supply capacity of expressway is a freeway traffic flow characteristics reflected, whether in traffic planning or control the daily traffic management, how to correctly analyze determine the traffic flow characteristics of the expressway, will determine the running state of city traffic network.

2 The Model of Data Fusion

2.1 B-P Model

The advantages of B-P algorithm: The research theory is mature, has the rigorous derivation process, the fault-tolerant ability is strong, the versatility is good, so far is the main algorithm of the forward network learning. The disadvantages of B-P algorithm: (1) the learning efficiency is low, the training time is long, the convergence speed is slow, and the increase of the sample dimension will make the network performance worse. (2) Greedy algorithm is especially easy to form local minimum, so that it can't get the global optimum. (3) The selection of hidden nodes in network is lack of theoretical support. (4) In the course of training, learning the new sample will forget the trend of the previous samples [4, 5].

In the application of B-P model to solve practical problems, better data normalization method can be used to avoid convergence or slow convergence, and to a large extent improve the performance of the network.

2.2 The Method of Speed Fusion Based on Linear Regression

Regression analysis is a statistical analysis method that determines the degree of influence or interdependence between the changes of one or more variables to another specific variable. In this paper, the relationship between the velocity data and the microwave velocity data (independent variable) and travel time velocity data (dependent variable) from the same cross section is studied, and the relation is expressed by regression method.

2.3   The Model of Two Dimensional Regression and Regression Equation

Set 2 independent variables are x_1, x_2, and the dependent variable is y. The equation describing the relationship between the dependent variable, the independent variable and the error term ε is called bivariate regression model. Its general form as follow:

$$y = \beta_0 + \beta_1 x_1 + \beta_2 x_2 + \varepsilon \tag{1}$$

Where, β_0, β_1, β_2 are parameters of data fusion Model, ε is the error term.

Formula (1) indicates that y is the sum of the linear functions of x_1, x_2 and error term。 The error term ε reflects the influence of random factors other than independent variables on y, which is not explained by the linear relationship between independent variables and dependent variables. The parameter β_0, β_1, β_2 in the regression equation is unknown. β_0, β_1, β_2 are estimated by sample statistic $\hat{\beta}_0$, $\hat{\beta}_1$, $\hat{\beta}_2$, and finally the bivariate regression equation is obtained. Its general form as follow:

$$\hat{y} = \hat{\beta}_0 + \hat{\beta}_1 x_1 + \hat{\beta}_2 x_2 \tag{2}$$

Where, $\hat{y}$ is the estimate of the dependent variable y; $\hat{\beta}_0$, $\hat{\beta}_1$, $\hat{\beta}_2$ f is the estimated value of parameter β_0, β_1, β_2. $\hat{\beta}_1$ means the average mean amount of change of dependent variable y that when x_2 remains unchanged, x_1 changes 1 units. $\hat{\beta}_2$ means the average mean amount of change of dependent variable y that when x_1 remains unchanged, x_2 changes 1 units.

2.4   Least Square Estimation of Parameters

The $\hat{\beta}_0$, $\hat{\beta}_1$, $\hat{\beta}_2$ in the regression equation is obtained by the least square method, The purpose is to minimize the sum of squares of residuals [7].

$$E = \sum_{i=1}^{n} (y_i - \hat{y}_i)^2 = \sum_{i=1}^{n} (y - \hat{\beta}_0 - \hat{\beta}_1 x_1 - \hat{\beta}_2 x_2)^2$$
$$= \sum_{i=1}^{n} (y_i - \hat{\beta}_0 - \hat{\beta}_1 x_{1i} - \hat{\beta}_2 x_{2i})^2$$

The standard equations for solving $\hat{\beta}_0, \hat{\beta}_1, \hat{\beta}_2$ can be obtained.

$$\begin{cases} \frac{\partial E}{\partial \beta_0} \Big|_{\beta_0 = \hat{\beta}_0 = 0} \\ \frac{\partial E}{\partial \beta_1} \Big|_{\beta_i = \hat{\beta}_i = 0} \end{cases} \quad where, \ i = 1, 2 \tag{3}$$

3   Data Acquisition and Preprocessing of Beijing Network

This paper mainly studies and analyzes the data from 1 directions of 1 license plate detectors in express way, namely Guanghui Nanli district to SanKuaiban village. Beijing expressway traffic flow detector mainly refers to the microwave detector and the license plate recognition detector. According to statistics, there are 829 microwave detectors and 231 license plate detectors in the expressway.

3.1   Selection of Fusion and Forecast Periods

In this paper, the traffic flow velocity data of the whole day are fused and forecasted through 3 periods of morning peak, flat peak and evening peak.

To study the fusion of the 3 detector speeds, data for the 3 detectors must be present for one day's data. In accordance with this principle, the data obtained from the Beijing Traffic Management Bureau screening, to find eligible only on June 20, 2011, 21, and 22 days of data more appropriate. Because there is a large number of data missing in the license plate recognition detector, the data of early peak, flat peak and evening peak can be used for fusion and prediction. The data for these 3 periods are predicted by the morning peak model, using data from June 22, 2011 5:00–10:00, and model validation using 10:00–11:00 data. The model of predictions of flat peak using June 20, 2011 12:00–17:00 data, model validation using 17:00–18:00 data. Model predictions for evening peaks were trained using data from June 22, 2011 15:00–20:00, and model validation using 20:00–21:00 data.

3.2   Acquisition of Raw Data

In this paper, the data of license plate recognition detector which locate from Guanghui Nanli district to Sanbankuai village are fused.

The data used include floating vehicle detector data, microwave detector data, and license plate detector data, all from the Beijing Traffic Authority Database. In the choice of data, based on to the integrity of the data, the license plate recognition detector data is large enough basis for data screening. In data time selection, data fusion of morning peak and flat peak periods is performed separately. Research on the road is a highway between Guanghui Nanli district to Sanbankuai village. In this section, the expressway is crossed with the East Third Ring Expressway.

The road is Hebei highway road Tonghui between X cell and y a, on the road, Hebei road and East sanhuan Tonghui Expressway Interchange. Study on the road between Guanghui Nanli district to Sanbankuai village section of the road. The main reason for choosing this section is that, on the one hand, the coverage length of the fixed detector can be guaranteed. On the other hand, the number of floating vehicles on the road can also be satisfied. After comparing the original data of the expressway, the data of the license plate recognition detector is relatively sufficient to ensure that the following fusion can be carried out smoothly. The attributes of the data are shown in Table 1.

Table 1. Date and time of investigation and peak type

Date	2017-06-02	2017-06-22	2017-06-22
	(Monday)	(Wednesday)	(Wednesday)
Time	12:00–18:00	5:00–11:00	15:21:00
Type	Flat peak	Morning peak	Evening peak

Method for obtaining data and collating data

(1) Requirements are floating vehicles larger amount of data, license detector data missing less sections, in accordance with this requirement to find the road section.

(2) In order to make the speed of the road closer to the real value and reflect the real-time traffic condition, the data in the basic time interval of different types of detectors are processed differently. The weighted average of the speed of the different moving vehicles in the same direction is weighted equal to the reciprocal of the flow, which is equivalent to the license detector section. The speed of different lanes detected by the coil detector is weighted by the reciprocal of the flow, and the average speed of the detector is obtained by weighting the average speed of all lanes detected by the detector. Then, the average speed of different

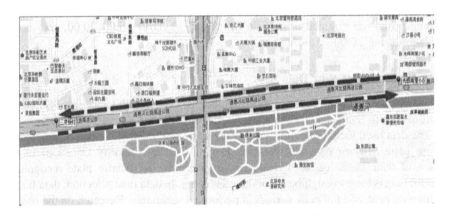

Fig. 1. Expressway section

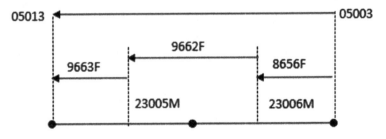

Fig. 2. Detector distribution of section

detectors in the same direction is weighted by the reciprocal of the corresponding flow, and weighted average to the license detector section. The average value is calculated after the bad value is removed from the travel time speed within the basic time interval of 2 min. Figure 1 is a screenshot of a quick section on Baidu maps. Figure 2 shows the distribution of detectors in the study section.

3.3 Data Preprocessing

The data fusion and analysis carried out in this paper are based on 2 min as the basic data time interval.

3.3.1 Microwave Data Preprocessing

The raw data obtained by microwave detector include flow rate, lane number, time occupancy, speed, etc. The road section studied in this paper is a section with license plate recognition as the basic unit. It includes a microwave detector with lanes numbered 23005 and 23006. The preprocessing of microwave data needs to be done in 2 steps.

The first step: obtaining weighted average speed data for each lane speed of a sub detector with an interval of 2 min.

Because of the reciprocal $1/k$ of the vehicle traffic density, Q can be used as the weight to weigh each lane speed to obtain the weighted velocity data of the microwave detector number 23005 (or 23006). The process is as follows:

$$\bar{v}_{kl} = \frac{\sum\limits_{i=1}^{n_l} \frac{v_{kli}}{q_{kli}}}{\sum\limits_{i=1}^{n_l} \frac{1}{q_{kli}}} \quad i = 1, 2, \ldots, n_l(i = 11, \ldots, n_l);\ l = 23005,\ 23006 \tag{4}$$

where, The $\bar{v}_{kl}$ represents the weighted average of the speed of each lane of the Lth microwave detector within the kth interval. v_{kli} indicates the speed of the ith Lane detected by the Lth microwave detector within the kth time interval. q_{kli} indicates the ith Lane flow detected by the Lth microwave detector within the kth time interval.

The second step: Obtain weighted average speed of equivalent license plate recognition section at 2 min intervals. The average speed of the 2 detectors is weighted by the reciprocal of the flow at the microwave detector. The processing method is as follows:

$$\bar{v}_k = \frac{\sum\limits_{l=23005}^{23006} \frac{v_{kl}}{\sum\limits_{i}^{n_l} q_{kli}}}{\sum\limits_{l=23005}^{23006} \frac{1}{\sum\limits_{i}^{n_l} q_{kli}}} \quad i = 1, 2, \ldots, n_l(i = 11, \ldots, n_l);\ l = 23005,\ 23006 \tag{5}$$

$$\bar{v}_k = \frac{\sum_{i=1}^{3} \frac{v_{ki}}{q_{ki}}}{\sum_{i=1}^{3} \frac{1}{q_{ki}}} \quad i = 1, 2, \ldots, n_l (i = 11, \ldots, n_l); \; l = 23005, 23006 \tag{6}$$

where, $\bar{v}_k$ represents the weighted average of vehicle speed detected by 2 microwave detectors within Kth time interval. $\bar{v}_{kl}$ represents the weighted average of the speed of each lane of the Lth microwave detector within the Kth time interval. q_{kli} indicates the ith Lane flow detected by the Lth microwave detector within the kth time interval.

3.3.2 Data Preprocessing of Floating Vehicles

Average traffic, average speed, and average occupancy rate are the original data of the floating vehicles obtained in this paper. The data feature is data that has been standardized to be 2 min intervals. The road section "Guanghuinanli district-SanbanKuai village" studied in this paper is a section with license plate recognition as the basic unit. It is divided into 3 floating vehicle sections according to the direction of traffic flow, in which the number of 3 floating vehicles in the direction of "Guanghuinanli district-SanbanKuai village" is 86569662 and 9663 respectively. Because of the $1/k = 0$ of the traffic density, $1/Q$ can be used as the weight to weigh the speed of the 3 floating vehicles in each direction, and the weighted average speed data of the floating vehicle of the section "Guanghuinanli district-SanbanKuai village" is obtained on the average. The preprocessing procedure is as follows:

$$\bar{v}_k = \frac{\sum_{i=1}^{3} \frac{v_{ki}}{q_{ki}}}{\sum_{i=1}^{3} \frac{1}{q_{ki}}} \quad i = 1, 2, 3 \tag{7}$$

Where, $\bar{v}_k$ represents the weighted average value of the 3 floating vehicles in the same direction in the Kth interval. v_{ki} represents the detection speed of the ith floating vehicle section in the kth time interval. q_{ki} represents the detection flow of the ith floating vehicle section within the kth time interval.

3.3.3 The Preprocessing of the License Plate Identification Detector Data

The original data obtained by the license plate recognition detector mainly include vehicle entry time and exit time, and the average speed of the road can be obtained with the length of the section. Its data characteristics are data that has not been standardized processed for 2 min intervals.

The section "Guanghuinanli district-SanbanKuai village" studied in this paper is the section with the license plate recognition as the basic unit. The data preprocessing method eliminates the large deviation value of the data in 1 unit time intervals (2 min) with the confidence of $a = 0.05$, and then obtains the travel time data of the 2 min intervals on the remaining data. Since the data obtained within 2 min are generally less than 20, are small samples that can be considered as t distributions subject to the degree of freedom of n − 1 (N is the sample size of travel time data for the first k intervals), i.e. V-T. The confidence interval of travel time speed can be obtained under the condition that the significance level is a = 0.05.

The average speed in a Kth interval:

$$\bar{v}_k = \frac{\sum_{i=1}^{n_k} v_{ki}}{n_k} \qquad (8)$$

Confidence interval of velocity in the first k interval.

$$\left[\bar{v}_k \pm t_{\alpha/2}(n_k - 1)\frac{s_k}{\sqrt{n_k}}\right] \qquad (9)$$

where, $\bar{v}_k$ represents the average speed in the Kth time interval; v_{kl} indicates the average speed of the ith cars detected by the license plate identification method within the kth time interval. n_k represents the sample size of the license plate identification method detected within the kth time interval. s_k indicates the speed standard deviation detected by the license plate recognition method within the Kth time interval. In the confidence interval, external velocity data should be eliminated. Velocity data within the confidence interval should be retained, and velocity data outside the confidence interval should be eliminated.

4 Example Analysis

4.1 Fusion of B-P Neural Network Method

The model training is divided into 2 parts according to the different research periods: rush hour and peak hour. The rush hour is divided into early rush hour and evening rush hour. A total of 3 model training and 3 validity judgments were tested. Considering the average speed of the road section, it is possible to stagger the morning peak and the evening peak of the traffic to a certain extent. The solution of the model is forward 1 h and pushes backward 2 h on the basis of flow early peak and evening peak. That is, the data time for model training is 5 h, and the time of data verification is 1 h.

Taking into account the non coincidence of the peak period of speed and flow, the data come from June 22, 2011 5:00–10:00 were used for trained; and the data come from 10:00–11:00 were used for validation in morning peak stage. The data come from June 20, 2011 12:00–17:00 were used for trained; and the data come from 17:00–18:00 were used for validation in flat peak stage. The data come from June 22, 2011 15:00–20:00 were used for trained; and the data come from 20:00–21:00 were used for validation in evening peak stage.

The morning peak period fusion and prediction results are shown in Figs. 3 and 4 respectively.

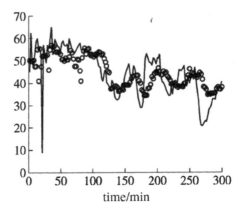

Fig. 3. Training results of morning peak

At the same time, the data of early morning peak, flat peak and evening peak are used for training and prediction,and the predicted fusion value is compared with the accurate value, as shown in Table 2. Table 2 is a comparison of fusion values with floating vehicle data and microwave detection data. Table 2 shows the following conclusions:

(1) The fusion value LSE is smaller than that of the floating vehicle LSE and the microwave detector LSE, so the model is valid.

(2) The difference between the floating vehicle LSE and the fusion value LSE is less than the difference between the microwave detector LSE and the fusion value LSE, which shows that the floating vehicle detector is more accurate than the microwave detector in speed detection. The reason may be that for the whole section, the microwave detector only detects the average speed of the 2 points, and is not representative of the average speed of the whole road; The floating vehicle detected by the floating car detector is basically distributed evenly on the road section. The average speed of multiple points along the direction of the road is detected, and the average speed of the road is more apparent.

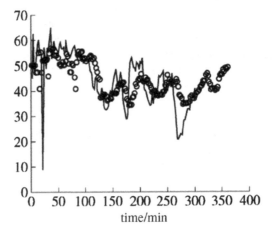

Fig. 4. Predicating results of morning peak

(3) Compared with LSE in 3 periods of morning peak, flat peak and evening peak, it is concluded that the accuracy of the detection data is higher in the flat peak period than in the morning peak period.

Table 2. Comparison of fusion values and two type of detector of expressway

Type of peak	LSE of floating car	Microwave detector LSE	LSE of Fusion values
Morning peak	0.219	0.656	0.147
Flat peak	0.133	0.470	0.040
Evening peak	0.176	0.560	0.062

4.2 Regression Analysis for Data Fusion

The regression equation was established with the speed of license plate recognition detector as dependent variable, the speed of floating vehicle and the speed of microwave detector as independent variables.

(1) The data come from morning peak June 22, 2011 5:00–10:00 were regressed to obtain the regression equations.

$$y = 15.82389 + 0.552847x_1 - 0.00597x_2$$

Because of the significance $F = 1.39 \times 10^{-16} < \alpha = 0.05$, the linear relationship is obvious.

(2) The data come from flat peak June 20, 2011 12:00–17:00 were regressed to obtain the regression equations.

$$y = 60.13448 - 0.13127x_1 - 0.01601x_2$$

Because of the significance $F = 0.022544 < \alpha = 0.05$, the linear relationship is obvious.

(3) The data come from evening peak June 222, 2011 15:00–20:00 were regressed to obtain the regression equations.

$$y = 44.12753 - 0.088128x_1 - 0.009443x_2$$

Because of the significance $F = 0.017131 < \alpha = 0.05$, the linear relationship is obvious.

The data of morning peak, flat peak and evening peak are trained and predicted, and the predicted fusion values are compared with the accurate ones. The results are shown in Table 3.

Table 3. Comparison of two detectors and fusion values of LSE in expressway

Type of peak	LSE of floating car	Microwave detector LSE	LSE of Fusion values
Morning peak	0.219	0.656	0.141
Flat peak	0.133	0.472	0.044
Evening peak	0.176	0.560	0.062

(1) The fusion value LSE is smaller than the floating vehicle LSE and the microwave detector LSE, so the model is valid.

(2) The difference between the floating vehicle LSE and the fusion value LSE is less than the difference between the microwave detector LSE and the fusion value LSE, which shows that the floating vehicle detector is more accurate than the microwave detector in speed detection.

(3) Compared with the 3 peaks of morning peak, flat peak and evening peak LSE, it is concluded that the accuracy of the detection data is higher in the flat peak period than in the rush hour period.

(4) From the analysis of the fusion effect, the fusion effect of the flat peak period is better than that of the rush hour period.

From the above analysis, we can get the same result by the neural network method and the linear regression method. The 2 models have been tested by the effectiveness, and achieved good results.

5 Conclusions

In this paper, the neural network fusion model and the two element regression model are deeply studied. Based on the 2 models, the 3 stages of the early peak, flat peak and evening peak of the traffic flow in the fast section of Beijing are tested. To judge the validity of the model, the fusion value of LSE was less than the floating car LSE and microwave detector LSE, the value of flat peak period is more close to the real value of speed of the fusion seriously. At the same time, the difference between the floating vehicle LSE and the fusion value LSE is less than the difference between the microwave detector LSE and the fusion value LSE, indicating the floating vehicle detector is more accurate than the microwave detector in speed detection. Two models are designed to predict the road speed value, and the results are better. The LSE between the fusion value and the accurate value is smaller than the LSE value between the unit detector and the accurate value, whether in early peaks or flat peaks.

References

1. Bu, Y., Howe, B., Balazinska, M., Ernst, M.D.: HaLoop: efficient iterative data processing on large clusters. Proc. VLDB Endow. **3**(1–2), 285–296 (2010)
2. Foschini, L., Hershberger, J., Suri, S.: On the complexity of time-dependent shortest paths. In: ACM-SIAM Symposium on Discrete Algorithms, vol. 68, pp. 327–341. Society for Industrial and Applied Mathematics (2011)

3. Frigioni, D., Marchetti-Spaccamela, A., Nanni, U.: Fully dynamic algorithms for maintaining shortest paths trees. J. Algorithms **34**(2), 251–281 (2000)
4. Gao, M.X., He, G.-G.: An arc labeling algorithm for shortest path problem considering turn penalties and prohibitions at intersections. J. Lanzhou Jiaotong Univ. (2011)
5. Choi, K., Chung, Y.S.: Travel time estimation algorithm using CPS probe and loop detector data fusion. In: 80th Transportation Research Board Annual Meeting [CD], Washington D.C. (2001)
6. Klein, L.A., Yi, P., Teng, H.L.: Decision support system for advanced traffic management through data fusion. Transp. Res. Rec. 173–178 (2002)
7. Hu, X., Huang, X.: Solving traveling salesman problem with characteristic of clustering by parallel genetic algorithm. Comput. Eng. Appl. **40**(35), 66–69 (2004)
8. Zhi-Heng, H., Xiaoping, R., Cheng-Wei, D., Xian-Feng, S., Jing, W.: A heuristic bidirectional hierarchical path planning algorithm based on hierarchical partitioning. Open Cybern. Syst. J. **9**(1), 306–312 (2015)
9. Shi, H., Spencer, T.H.: Time–work tradeoffs of the single-source shortest paths problem ✰. J. Algorithms **30**(1), 19–32 (1999)
10. Ivan, J.N.: Neural network representations for arterial Street incident detection data fusion. Transp. Res. Part C: Emerg. Technol. **5**(2–3), 245–254 (1997)

A Recommend Method of Hotspots Knowledge Based on Big Data from Evolving Network

Yi Zhao[1], Zhao Li[2(✉)], and Jun Wu[2]

[1] Computer School of Wuhan University, Wuhan 430072, China
ivwepriu@sina.com
[2] China Three Gorges University, Yichang 443002, China
zhaoli@ctgu.edu.cn

Abstract. Because of the rapid growth of all kinds of professional knowledge on the internet. There are three characteristics about the big data in the Internet: the amount of information is huge, the type of information is changing quickly, the requirements of people are diverse. How to help users discover hotspots knowledge to meet their personalized are becomes a challenging and hot issue. Thus, in this paper, we propose the evolving network combine with word2vec. To address this issue. first, we construct the evolving network and analyze the data relationships based on Wikipedia knowledge tree and software engineering open source community (e.g., Stack Overflow) in real time; second, the evolving network can merge similarity and synonym of hotspots terms by Word2vec. Finally, we can calculate the weight of each hotspots, the nodes with the greatest weight in each domain are recommend knowledge. We evaluated stability, recall rate, precision rate, and F-Measure through an experiment and the results showed that our method is more accurate than existing approaches.

Keywords: Evolving network · Word2vec · Hotspots knowledge

1 Introduction

IN recent years, knowledge-oriented meaning is emerging as a new promising research issue, which has a significant impact on software engineering and social science. At the same time, users' requirements on the Internet are usually drive the direction of knowledge development. In order to provide hotspots knowledge with high quality and facilitate the realization of on-demand knowledge recommend. It is required to organize and recommend hotspot knowledge effectively in the user-centric mode.

Thus, we propose then new method that can automatic draw evolving network by user questions on the Stack Overflow. It is well known that the real word exists in many physical complex networks, The knowledge evolving network model represents a transition between power-law and exponential scaling, while the Barabási–Albert [1, 2] scale-free model is only one of its special (limiting) cases. We used that this knowledge evolving network model can maintain the robustness of scale-free networks and can improve the network reliance against intentional attacks, which is the inherent fragility of most scale-free networks. In the environment of big data, user's diverse and individualized requirements make sense in the two aspects. On one hand, there exists great

© Springer Nature Singapore Pte Ltd. 2018
K. Li et al. (Eds.): ISICA 2017, CCIS 873, pp. 468–479, 2018.
https://doi.org/10.1007/978-981-13-1648-7_40

difference of the hot topics (Quality of knowledge) values among knowledge, such as number of click by user, hot degree, attention degree, and so on. On the other hand, a user may get different knowledge in different situations, and user attention will charge over time, so we must use weighted network express User change demand.

To address the first problem, many knowledge clustering methods have been proposed: 1. Hotspots knowledge mining method based on probability [3], there are main methods is LDA-hotspots meaning; 2. Hotspots knowledge mining method based on complex network [1, 4, 5], it construct the knowledge network to represent association relation. But the first method disadvantage is that the article is a natural language structure with semantics. If hotspots analysis is not combined with context, a lot of synonyms related hotspots knowledge is ignored.

These approaches described above do not take to knowledge semantic and the inheritance relationship of knowledge, they just simply calculates statistical tendency by means of the keyword search; and the knowledge structural relations between lexical, such as lexical hyponymy, are not involved in the work; Moreover, user search history might be exaggerated deliberately, which cannot represent the real user requirements. These methods mainly organize knowledge from the aspects of knowledge product process constraint relationship, knowledge hierarchy and so on. In order to solve the above shortcoming, we propose three improve method:

1. We need to construct weight network express hotspots knowledge phenomenon, weight value is users' requests hot degree for knowledge.
2. Oriented Stack Overflow website hotspots knowledge data, we should first determine which type of network distribution of hotspots knowledge. There are mainly three type of the network degree distribution, it includes random network, small-world networks, scale-free network. If most of nodes are paralyzed from random networks, it will inevitably lead to the division of the network. If most of the nodes are removed in the network, they will not be able to communicate with each other and end up with being isolated islands of information. However, simulation results of scale-free networks showed a different situation. Even removing more than 80% of randomly selected Internet routers in the node, still the rest of the routers can form a complete cluster and certificate path between any two nodes.
3. After constructing scale-free network, we find the hottest knowledge through weighted path, and use Word2vec to search synonyms of hot knowledge based on context.

2   Relate Work

In the era of big data, it possesses the 4V characteristics of knowledge: volume, velocity, variety, veracity [4]. Moreover, Evolving Science System network [5] has been propose to comply with Mapping the Network Architecture of Knowledge production. A local-world evolving network model is proposed and testified by Li and Chen [1]. The model proves that the real network is scale-free model; the interpersonal relationship and knowledge exchange are in accordance with the typical characteristics of scale-free network.

BA network is extend that the A comprehensive weighted evolving network model [6]. Because extract streaming data and construction knowledge system is significant importance for big data research. But compared with other data types, Internet big data has following characteristics: there are a huge amount of information, knowledge change quickly, hotspots isn't easy to preserve. Thus, In view of the above problems, there are the new methods has proposed, it includes A weighted local-world evolving network model with aging nodes [7], self-organized criticality and partial synchronization in an evolving network [8], consensus problem in multi-agent systems with physical position neighborhood evolving network [9]. In 2007, the Turing Award winner Jim Gray, described intensive data discovery, which he called fourth paradigm [10–14], and put forward the unification of theory and method of the fourth paradigms. He emphasized it is important that building big data oriented domain knowledge and stored data in the computer. In 2012, Li pointed out that big data of scientific research should be separated from third paradigm (computer simulation), evolving network–simulation study [15–19]. And the reason is that the required research methods are different from traditional mathematical based method. This is not only a transformation of science research method, but also the change of thinking more about big data. Then, We present a structured approach and the Word2vec approach to constructing a description of the fourth paradigm of big data. This paper presents a new method to process stream data.

Moreover, we provide the new knowledge nodes merge approach; it can accelerate the time to find answer. Hotspots trend recommendation can enable developers to obtain the frontier knowledge.

3   Overall Architecture of Our Framework Modeling Evolving Knowledge Network

As shown in Fig. 1, we propose a framework of knowledge organization and recommendation in the orientation of hotspots knowledge. The framework includes the following several modules: evolving network modeling, Merging of similarity knowledge clustering, on-demand user answer, multi-granularity knowledge recommendation and evolution of organized knowledge. In this section, we first elaborate the BA knowledge network construction; we then briefly introduce characteristics of BA knowledge graph, which is used to the scale free characteristics construction evolving network.

(1) Knowledge Domain Modeling According to the users' common needs in specific answer on the Stack Overflow, Basic knowledge tree structure are obtained through modeling by domain experts(e.g.: Java domain, and Android domain). The definitions of the Wikipedia models and all the symbols will be introduced in Sect. 3.1. These models are the basis of knowledge organization. Clustering According to the specific type of knowledge, knowledge which realizes similar functions but have different hotspots degree values is clustered to form specific knowledge clusters. How to cluster knowledge according to the knowledge type is introduced in Sect. 3.2, which is not detailed in this paper.

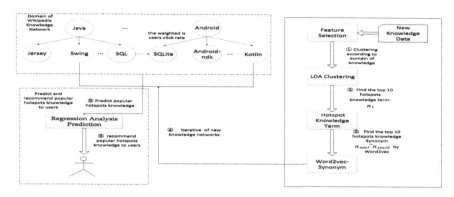

Fig. 1. Hotspot knowledge prediction and recommendation model

(2) On-demand knowledge top 10 terms are organized according to users' query or click number. Knowledge is organized in the orientation of specific domain problem. In Fig. 1, n_t denotes the hotspots knowledge.

(3) $n_{syn1} \sim n_{syn10}$ denote the top 10 hotspots knowledge synonym by Word2vec method.

(4) Evolution of Organized knowledge Based on the knowledge being organized, the new knowledge node grows on the original knowledge network. These operations mainly include addition, configuration, and deletion.

(5) Multi-granularity hotspots knowledge Recommendation According to users' different expression modes, and predict popular hotspots knowledge to users.

3.1 BA Knowledge Network Construction

Stack Overflow was founded by Jeff Atwood and Joel Spolsky, it has 16M unique visitors per month, more than 72M Page Viewer. Similarly, according to statistics, the volume of Wikipedia has grown over time. We analyze the Stack Overflow community data, and obtain original question and answer text data, which is an edit tool developed with Python, and this tool is suitable for analyzing complex network modeling. Besides, because of evolving network must Wikipedia knowledge tree and complex network analysis algorithm, it can carry out complex network data analysis compared with WebFetch.

In a community network, each node is a conceptual entity, which is connected with some other nodes. The number of connections is called the degree of node, and the web in a randomly selected node, its degree is the probability distribution, which is called the degree distribution of nodes. Community in the network of knowledge evolution is not in a Poisson distribution (Poisson), because experiments have shown that network node connection possibility deviates from the specified value probability, and it is not exponential decline. Such as the Stack Overflow community, this is based on the software engineering domain. Most question web pages have no more than 10 answers, or even unanswered; only very few issues have more than 3000 answers. While for some core issues, they have more than 10000 answers or even more answers had not been adopted. We can see that such community networks may not be consistent with the Poisson distribution, but in a power-law distribution [12]. That is to say, node degree is random in the random network, node degree probability is:

$$P(d = k) \propto \frac{1}{k^\gamma} \tag{1}$$

$d = k$ is probability of proportional. The greater value is k, the lower probability is $d = k$, but in a scale-free network decreasing speed is polynomial. Barabás and Albert proposed two characteristics of scale free network in 1999 [13]:

1. Evolving Network Growth: a smaller Knowledge Tree G_0 started, (the network has nodes N_0, edges E_0), it is self-growth, according to the time step, each time you add a new node has at least one side of the line, and this line is connected to the already exists the other node.
2. Preferential attachment: assuming the original network has nodes $(n_1, n_2 \ldots n_i)$. Time t_i add a new node n_{i+1}, the original node connected with it written $s_j (1 \leq j \leq i)$, the original nodes in the network degree written d_j, the probability of a new node connected with original nodes.

$$p_j = \frac{d_j}{\sum_{k=1}^{i} d_k} \tag{2}$$

As the time after t times, the new network has $n_0 + t$ nodes, there are $E_0 + mt$ edges.

However BA model has limitations, and Dorogovtsevetal, proposed the DMS model (Dorogovtsev-Mends-Samukhin). The probability of each new node and the original node i connected probability depend on k_i and β_i, and obey the following rules:

$$\prod = \frac{k_i + \beta_i}{\sum_j (\beta_j + k_j)} \tag{3}$$

k_i is the degree of node i, and β_i is the attractive factor of node i, $\sum_j (\beta_j + k_j)$ is the sum of the rest of the nodes degree and the attraction factor.

3.2   Characteristics of BA Knowledge Network

There are many evolving networks in real world, which are both weighted networks with different weight and directed networks. Therefore, we propose weight and direction evolving network based on the topology of Wikipedia knowledge, and propose attractive factor concepts to measure the index of in-strength and out-strength. Find the maximum node degree by the number of user answer and question.

Model includes the following steps:

(1) Starting nodes is s_0, e_0 is the edge from s_0 of directed networks.
(2) At each time interval t, randomly select node set S as an initial network and looping execute the following two steps:

    A. Randomly selected node set S_i from the initial network as an evolving network Ω, each time t, add a new node, it is connected with the original network Ω, select the node in Ω based on priority strategy of in-degree or out-degree:

$$\prod i_{(in)} = \frac{S}{s_0 + t} \frac{k_{i(in)}}{\sum\limits_{j \in \Omega} k_j} \tag{4}$$

OR

$$\prod i_{(out)} = \frac{S}{s_0 + t} \frac{k_{i(out)}}{\sum\limits_{j \in \Omega} k_j} \tag{5}$$

    B. Because new node s has attractive factor, so the connection weight between s and S_i conforms to the following rules:

$$\prod i_{(in)} = \frac{S}{s_0 + t} \frac{k_{i(in)} + \Gamma_i}{\sum\limits_{j \in \Omega} (k_j + \Gamma_j)} \tag{6}$$

OR

$$\prod i_{(out)} = \frac{S}{s_0 + t} \frac{k_{i(out)} + \Gamma_i}{\sum\limits_{j \in \Omega} (k_j + \Gamma_j)} \tag{7}$$

$\prod i_{(in)}$ is the connection weight based on in-degree to select a node in the local network.

$\prod i_{(out)}$ is the connection weight based on out-degree size to select a node in the local network. $k_{i(in)}$ is in-degree of the node i, $k_{i(out)}$ is i out-degree of the node. Γ is attractive factor of the node i $\sum\limits_{j \in \Omega} (k_j + \Gamma_j)$ is the sum of degree and attractive factor of other node in the network. The calculation method of Attracting factor Γ expressed as $\Gamma = \frac{n}{\Delta t}$, Δt is unit time, n is linking number in unit time. Unlike BA model, the probability of new node with the original network connected is determined by node degrees and attractive factor.

3.3 Evolving BA Knowledge Graph

In Formulas (6) and (7), due to the effects of Γ, the node connections are considered in three different situations.

(1) When k_i and Γ_i are in high value, the node has strong attraction to new node. New node tends to connect the node. This node will gradually evolve into distributed node. This situation is shown in Fig. 1, node A point degree is high, and attractive factor is high.

(2) When k_i is higher, Γ_i is lower and, this is equivalent to an original node decline, although degree is high but the attractive factor is low, so there is not much advantage to attract other nodes.

(3) When k_i is lower, Γ_i is higher and, this situation is very favorable to the new node, even if its degree is very low, but the value is high, It will get a large number of connections in a short time.

As shown in Fig. 2, node A is accorded with the first case, node A has high value, attractive factor is also high. B accords with the second case, although node B and node A are the initial nodes in the original network, but node B own attract factor is smaller than node A and node C, so node B degree is decreased continuously, because node A attractive factor is higher than node B (Java knowledge is not the new technology at present, but Android knowledge is popular). Node C enters the network later, although degree is less, but attractive factor is very high. Then node C will be able to obtain new nodes, thus it degree value to more than original nodes (A and B).

4 Mining Synonyms of Hotspots Knowledge Words Using Word2vec Algorithm and Predictive

The main objective of the Word2vec system is to perform the three main tasks of Aspect Based Hotspots knowledge Analysis at the same time. That is, to mining synonyms of hotspots knowledge into the same domain aspects using Word2vec and LDA method. In addition, our system separates unless sentiment-terms from opinion-words without requiring additional resources or supervision. The system at its core consists of an LDA-based topic model, extended with biased topic modeling hyper-parameters based on continuous word embedding's and combined with an unsupervised pre-trained classification model for sentiment-term and knowledge-word separation.

4.1 Mining Synonyms Knowledge Model

Figure 3 describes the process to obtain the mining synonyms model. First, the news hotspots knowledge is clustered by LDA method. Then, the occurrences of these words are bootstrapped from the unlabeled domain synonyms according to wrod2vec methods, [−2, +2], context window. Next, context words are replaced by their corresponding Brown cluster to build each training instance. Finally, domain aspect hotspots terms sorting, it finds the top 100 knowledge terms.

The core of the system is a LDA-based topic model, extended to include hotspots-term and synonyms-word (Shorthand: Syn) defined domain aspect. While synonyms-word merge to generate new evolving network nodes is guided method as described in Sect. 2, topic (i.e., domain aspect) and LDA-Word2vec modeling are done by biasing

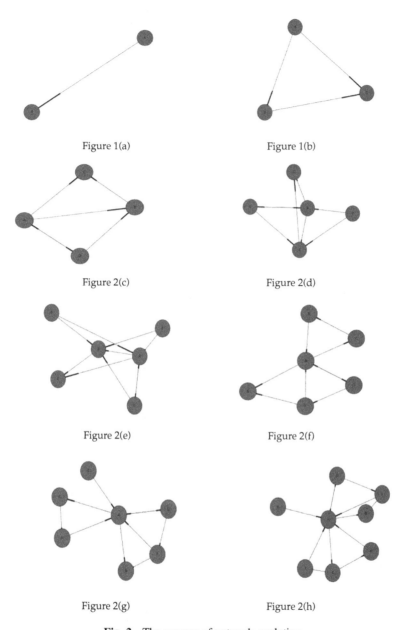

Figure 1(a)

Figure 1(b)

Figure 2(c)

Figure 2(d)

Figure 2(e)

Figure 2(f)

Figure 2(g)

Figure 2(h)

Fig. 2. The process of network evolution

certain hyper parameters according to the given domain aspects configuration. Synonyms are added to the subject domain to improve the theme.

LDA-Word2vec algorithm clustering the problem of the new nodes (user question) with following steps:

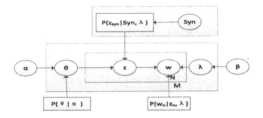

Fig. 3. LDA-Word2vec synonyms mining algorithm

The δ hyper-parameters control the polarity distribution for each document, and they are calculated for each document as shown in (8).

$$p(w|\alpha, \beta, \theta, \lambda, S_d)$$

$$= \prod_{i=1}^{V_t} p(w|\alpha, \beta, \theta, \lambda, Syn)$$

$$= \prod_{i=1}^{V_t} \sum_{r=1}^{R} \sum_{k=1}^{K} p\left(w_{ti}, z_{syni} = k, r_t = Syn|\alpha, \beta, \theta, \lambda, Syn_h\right)$$

$$= p(\theta|\alpha) \prod_{i=1}^{V_t} \sum_{r=1}^{R} \sum_{k=1}^{K} p\left(w_{ti}|z_{syni} = k, \theta\right) \cdot p(z_{ti} = k|r_t = Syn, \lambda) \cdot p(r_t = Syn|Syn_h))$$

$$= p(\theta|\alpha) \prod_{i=1}^{V_t} \frac{1}{S_d} \sum_{r=1}^{R} \sum_{k=1}^{K} \lambda_{wk}\theta_{kt}$$

In these formulas, Syn is the Synonyms for hot knowledge vocabulary, Syn_d has been assigned to hotspots-term h; w_{ti} is the number of words in the document d assigned to topic t; λ are the pre-trained hotspots-term model weights for word-type u; and z_{syni} is the feature vector for Syn_h, n, composed by the Brown clusters of the context words. Analogously, w_{ti} has been assigned to topic t and polarity $q \in \{P, N\}$.

4.2   Recommend Analysis

The BA scale-free network evolution model solves the new question about the unstructured data and diversified data analysis. In this paper, we use regression analysis method, which is based on the relationship between independent variable and dependent variable, to analyze various phenomena. The regression equation is established between the variables, and the regression equation as the predictive model, according to the number change of independent variable in prediction period to predict the dependent variable relationship.

Therefore, we choose Unary Linear Regression method. Let X as the independent variable, Y as dependent variable, there is a Unary Linear relationship between X and Y, and a Unary Linear Regression method is show in Formula (9):

$$y_i = a + bx_i + \delta \ i = 1, 2, \ldots, n \tag{8}$$

In formula 9, δ is sum of various random effects factors. Thus,

$$\delta = (0, \phi); y = N(a + bx, \phi^2) \tag{9}$$

$$y_i' = a + bx_i \tag{10}$$

According to Generalized Least Squares [14]:

$$K = \sum_{i=1} (y_i + y_i')^2 = \sum_{i=1} (y_i - a - bx_i)^2 \tag{11}$$

According to the principle of extreme value, to make K has a minimum value, a' and b' have partial derivative function:

$$a' = \frac{\sum y_i}{n} - b' \frac{\sum x_i}{n} \tag{12}$$

$$b' = \frac{n \sum x_i y_i - \sum x_i \sum y_i}{n \sum x_i^2 - (\sum x_i)^2} \tag{13}$$

Formula (9) is used to calculate value of y', and y' is the predicted value, and then we use Formula (9), (10), (11), and (12) to verify the prediction results. The verification results can be found in Sect. 4.

5   Experimental Evaluation

5.1   Regression Analysis Prediction vs. Degree Distribution Prediction

Stack Overflow website is a typical knowledge exchange community. We need to solve how to provide the better answer to user's question, and inform the user the current hotspots knowledge and predict the future hotspots knowledge. If we want to know that each user's demand, we have to understand that which category the user's problem is belongs to.

According to parameters of Formula (9), (10), (11), and (12) are a', b', R, S_y.the above two prediction models, we selected degree of randomly nodes, for example use Regression Analysis Prediction model to calculate "php" (to show Table 3) node in the 2006. $a' = 37.11$, $b' = 26.13$; $R^2 = 0.95$; $S_y = 9.93$.

However, Degree Distribution Prediction [15] is based on the physical model, did not consider the human factor, thus, the big data of real world relies on three dimensional worlds: man, machine and things.

Because S_y is standard error estimate. Thus, inaccuracy is smaller, results is better.

As shown in Figs. 4 and 5, we found that the amount of sample data has increased with time. The standard error estimation of Degree Distribution becomes large. But in this work, knowledge integration Regression Analysis model S_y is not a big change, because knowledge integration algorithm can choose the best dependent variable.

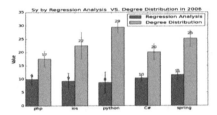

Fig. 4. S_y value of regression and degree distribution

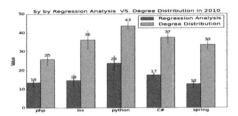

Fig. 5. S_y value of regression and degree distribution

Acknowledgements. This work was supported in part by the National Key Research and Development Programs of China (2016YFC0802503, 2016YFB0800403), by the PI Project of Hubei Provincial Collaborative Innovation Center for New Energy Microgrid (CTGU) (8006116), by the Open Foundation of Hubei Key Laboratory of Intelligent Vision Based Monitoring for Hydroelectric Engineering (2015KLA03), and by the Open Foundation of Key Laboratory of Geological Hazards on Three Gorges Reservoir Area (Ministry of Education) (2015KDZ05).

References

1. Li, X., Chen, G.: A local-world evolving network model. Phys. A Stat. Mech. Appl. **328**(1), 274–286 (2003)
2. He, K., Li, B., Ma, Y., Huang, Y.: The key technology of software engineering of the era of big data. China Comput. Commun. Fed. Beijing **3**(10), 8–18 (2014)
3. Albert, R., Barabási, A.-L.: Statistical mechanics of complex networks. Rev. Mod. Phys. **74**, 47–97 (2002)
4. Bahr, P.R.: Double jeopardy: Testing the effects of multiple basic skill deficiencies on successful remediation. Res. High. Educ. **48**, 695–725 (2007)
5. Zelman, A.G.: Mediated communication and the evolving science system: mapping the network architecture of knowledge production. Relig. Educ. Off. J. Relig. Educ. Assoc. **3**, 86–91 (2002)
6. Li, C., Chen, G.: A comprehensive weighted evolving network model. Phys. A Stat. Mech. Appl. **343**(1), 288–294 (2004)
7. Zhang, Y., Lo, D., Xia, X., et al.: Multi-factor duplicate question detection in stack overflow. J. Comput. Sci. Technol. **30**(5), 981–997 (2015). https://doi.org/10.1007/s11390-015-1576-4
8. Barabási, A.-L., Albert, R.: Emergence of scaling in random networks. Science **286**(5439), 509–512 (1999)

9. He, K., He, Y.: Ontology Meta Modeling Theory Method and Application. China Science Publishing, Beijing (2008)
10. Newman, M.E.J.: Power laws, Pareto distributions and Zipf's law. Contemp. Phys. **46**(5), 323–351 (2005)
11. Makowiec, D.: Evolving network–simulation study. Eur. Phys. J. B Condens. Matter Complex Syst. **48**(4), 547–555 (2005)
12. Christopher, D.M., Prabhakar, R., Hinrich, S.: An Introduction to Information Retrieval, vol. 181. Cambridge University Press (2009)
13. Yi, Z., Evolution knowledge tree for services computing domain in wikipedia. J. Wuhan Univ. (Nat. Sci. Edition). (04), 331–338 (2015)
14. Wang, C.-Y.: The fusion of support vector machine and multi-layer fuzzy neural network. Mach. Learn., June 2012
15. Yan, X.: Linear Regression Analysis: Theory and Computing, pp. 1–2. World Scientific (2009). ISBN: 9789812834119
16. Meade, N., Islam, T.: Prediction intervals for growth curve forecasts. J. Forecast. **14**(5), 413–430 (1995)
17. Hansen, C.B.: Generalized least squares inference in panel and multilevel models with serial correlation and fixed effects. J. Econ. **140**(2), 670–694 (2007)
18. Dean, J., Ghemawat, S.: MapReduce: Simplified Data Processing on Large Clusters. Accessed 2004
19. Park, S.Y., Bera, A.K.: Maximum entropy autoregressive conditional heteroskedasticity model. J. Econ. 219–230 (2009). Accessed 02 June 2011

Everywhere Connectivity – IoT Solutions

Improved Location Algorithm Based on DV-Hop for Indoor Internet of Things

Qian Cai[✉]

Jinagxi University of Science and Technology, Ganzhou, China
caiqian1980@163.com

Abstract. Information perception is the basic function of the Internet of things and the basic means which an of Things (IoT) information system is capable of comprehensive perception. In this paper, the DV-Hop localization algorithm is improved by modifying the average jump distance and weighting factor to improve the information localization and perception accuracy of the Internet of Things. The simulation results show that the algorithm meets the requirements of the information sense of the Internet.

Keywords: Internet of things · DV-Hop · Information perception
Weighting factor · Average jump distance

1 Introduction

Wireless Sensor Networks (WSNs) are the key technology underlying IoT networks [1, 2]. WSNs can obtain the information they need at any time, at any stage and under any environmental conditions. This is an important support technology for IoT information sensing systems, and it also lays the foundation for the development of the IoT. Because WSNs have advantages, including self-organization, rapid deployment, high fault tolerance and strong concealment, they are suitable for a variety of applications, such as battlefield target location [3], physiological data collection [4], intelligent transportation systems [5] and ocean exploration [6].

With the development and innovation of WSN technology for the IoT, research on WSNs has gradually extended from local WSN perception to wide-ranging material interconnection. In 2006, some universities and well-known enterprises in the United States proposed an effective solution to WSNs. At the same time, the US government vigorously promoted the family of intelligent energy systems based on WSNs. By 2020, 60% of US households are expected to install intelligent energy systems. At the same time, other countries are also stepping up the deployment of WSN-related development strategies and gradually promoting the construction of WSN network infrastructure. At present, China's WSN technology and industrial supply chain is still in the early stages of development. Related institutions working on WSN theory and applications of in-depth research and exploration are described in the "National Medium and Long-term Science and Technology Development Plan (2006–2020)," which regards "sensor network and intelligent information processing" as a priority area, and establishes the

© Springer Nature Singapore Pte Ltd. 2018
K. Li et al. (Eds.): ISICA 2017, CCIS 873, pp. 483–491, 2018.
https://doi.org/10.1007/978-981-13-1648-7_41

priority "to give policy support," which will play a large role in promoting WSN technology architecture and the formation of industrial models [7–10].

Information perception provides a source of information for IoT applications and is the basis for these applications. The most basic form of information perception is data collection, that is, when nodes transmit perceived data over the network to the sink section. However, the original sensing information obtained through wireless sensor networks might contain significant uncertainty and a high degree of redundancy. The uncertainty of information is mainly manifested in the lack of uniformity, inaccuracy, discontinuity, and incompleteness [11–15]. Therefore, precise node positioning algorithms in WSNs have a very important role in IoT systems. In recent years, many experts and scholars in wireless sensor node positioning algorithm research have also put forward a number of positioning algorithms. Positioning algorithms mainly include range-based localization algorithms and range-free localization algorithms [16–18]. Due to the characteristics of the application environment, range-free localization algorithms are more suitable for IoT perception systems. This paper employs an improved DV-Hop localization algorithm to achieve accurate perception.

2 Related Work

At present, many experts and scholars have proposed a variety of improved algorithms for non-ranging positioning. Chakchai et al. proposed Hybrid Fuzzy centroid with MDV-Hop BAT localization algorithms. In this algorithm, the hybrid scheme utilizing the key characteristics of Centroid and DV-Hop uses an extra weight with signal normalization derived from a fuzzy logic function in the Centroid. The result is that the distance between the unknown node and the beacon node is more accurate and the positioning accuracy of the unknown node is improved [19]. Guozhi Song et al. proposed two novel DV-Hop localization algorithms. A hyperbolic location method and the selection of appropriate beacon nodes are used to improve the location accuracy compared with traditional DV-Hop localization. Simulations proved that this localization algorithm improves localization accuracy [20]. Zhang proposed an improved normalized collinearity DV-Hop algorithm. This localization algorithm employs a hop count threshold and collinearity to improve the localization accuracy, and the proposed algorithm can achieve better positioning accuracy in both homogeneous and anisotropic networks [21]. In this paper, the DV-Hop localization algorithm is improved in two respects to improve the positioning accuracy [22–24]. First, the beacon node is selected using the beacon node's power control to transmit the beacon signal with different levels of transmitting power, and the optimal beacon node is modified with its neighbor. The number of hops between nodes is determined, with the number of hops allowed to be a decimal number, and the average jump distance is corrected. The second improvement is the use of the number of hops and the number of different representatives of the beacon node on the unknown node to determine the weighting factor.

One of the important features of IoT is the exchange of information between objects. Each object is represented as an object. Therefore, as it is responsible for sensing and documenting information, the underlying IoT network must be able to reflect the characteristics of each object. This function is realized by the node positioning technology in the wireless sensor network. The WSN architecture is shown in Fig. 1. The Internet of things is realized based on WSN technology [25]. IoT information exchange is very different from traditional human-computer interaction in several respects. (1) The "user" of IoT information interaction is ubiquitous. "Object interconnection" supports information exchange among things everywhere, which will provide interactive information exchange, expanding the network to all people, machines, and objects. (2) IoT information exchange is an active mode of interaction. In the application of things, the information exchange is not a passive response to the interaction, but the network node on-demand actively accesses information and self-consciously addresses the process of sensing information. (3) The IoT information exchange process is highly complex. Many factors give rise to this complexity, including the participation of a large number of heterogeneous network nodes, the distributed existence of mass information in the network, the dynamics and instability of the wireless network, and the limitations of the node resources. Due to the above characteristics, the accuracy of WSN node positioning technology seriously affects IoT performance. Therefore, this paper optimizes the hop count between nodes to improve the localization accuracy of the DV-Hop localization algorithm.

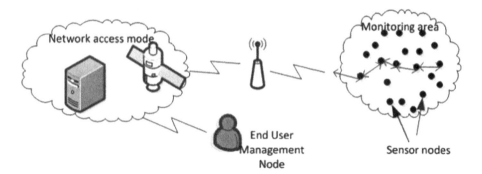

Fig. 1. WSN architecture

3   Improving the DV-Hop Algorithm

In this paper, to improve the localization accuracy and information sensing accuracy of WSNs in IoT, the traditional DV-Hop localization algorithm is improved by optimizing the hop count between nodes and by promoting the hop count accuracy. Therefore, this paper improves the DV-Hop localization algorithm in two respects. First, a beacon signal of different power is transmitted by the power control of the beacon node, and the hop count between the beacon node and its neighbor node is corrected. Second, the average jump distance is corrected.

3.1 Traditional DV-Hop Localization Algorithm

DV-Hop localization algorithm is the most widely used localization method in APS algorithm. Its localization process does not depend on the range finding method. It uses multi hop beacon node information to participate in node localization, and the location coverage is larger. Distance vector routing mechanism of DV-Hop algorithm is very similar to the traditional network. In this localization mechanism, firstly, the unknown nodes and beacon nodes calculate the minimum number of hops; secondly, estimating the average hop distance; thirdly, calculating distance of between unknown nodes and beacon node using the minimum number of hops multiplied by the average hop distance; lastly, calculating the unknown node coordinates using three trilateration or maximum likelihood estimation method.

DV-Hop localization algorithm can be divided into the following 3 stages:

(1) Calculating the minimum number of hops for unknown nodes and each beacon node.
    The beacon node broadcasts its own location information to the neighbor node, including the hop field, initialized to 0. The receiving node records the minimum number of hops to each beacon node, ignoring the number of hops from the same beacon node of the grouping. The hop value is then incremented by 1 and forwarded to the neighbor node. Through this method all nodes in the network can record the minimum number of hops to each beacon node.
(2) Calculating the actual jump distance between the unknown node and the beacon node
    Each beacon node uses the following formula to estimate the actual distance per hop based on the position information and the number of hops of the other beacon nodes recorded in the first stage:
(3) Obtaining the location of unknown node

The unknown node uses the truncated distance of each beacon node recorded in the second stage to calculate the self coordinate by the trilateral measurement method or the maximum likelihood estimation method.

As shown in Fig. 2, the distance and the number of hops between the beacon nodes L_1 and L_2 and L_3 can be calculated by the first and second stages. The beacon node L_2 calculates the corrected value (i.e., the average distance per hop) is $(40 + 75)/(2 + 5) = 16.42$. Assuming that the unknown node A obtains the correction value from L2, it is associated with three beacon nodes The distance between them is L_1: 3×16.42, L_2: 2×16.42, L_3: 3×16.42, and finally the position of node A can be determined by trilateral measurement method.

The traditional DV-Hop algorithm uses the average distance per hop to estimate the actual distance, the node hardware requirements are low, to achieve a simple. The disadvantage is the use of jump distance to replace the straight line distance, there is a certain error.

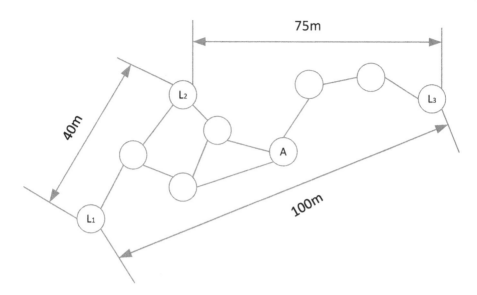

Fig. 2. The traditional DV-Hop localization algorithm

3.2 Traditional DV-Hop Localization Algorithm

In the first stage of the traditional DV-Hop algorithm, the distance between nodes is too large, leading to an increase in the localization error. In this paper, the hop count between nodes is weighted, reducing the localization error problem. The calculation process is as follows:

(1) The beacon nodes propagate data tuples $\{ID, x_i, y_i, Hops_i, RSSI_i\}$ to the surrounding nodes in a flooded way. *ID* represents the label of the node; x_i and y_i are the coordinate values of the nodes; $Hops_i$ is the hop count, with an initial value set to 0; $RSSI_i$ is the signal strength of the receiving node, and the initial value is set to 0.

(2) After the node adjacent to the beacon node receives this data tuple, the value of $Hops_i$ is set to 1 and the *RSSI* value is saved in the $RSSI_i$. This data tuple continues to be transmitted to the indirect neighbor node.

(3) For the indirect neighbor nodes, assume that the beacon nodes have k-level transmission power, that is, $R = \{R_n \mid n = 1,2,3,L, k \boxplus R_1 < R_2 < R_3 < L < R_k\}$. The received direct neighbor node RSSI value is saved in R_1 and divided by R_i as the weighted hop count of the receiving node, that is:

$$L = \frac{R_1 \times \delta_1}{R_i \times \delta_i} \qquad (1)$$

Where δ_i is the power regulation factor, $\delta_i = (k/i)^\beta (i = 1, 2, 3, \cdots, n)$, n is the beacon node signal transmission time, R_i is the maximum power, and β is the loss factor.

At present, the *Hops$_i$* value in the data tuple is updated to:

$$Hops_i = Hops_i + L \tag{2}$$

(4) The *ID* packet received by the node is compared with the hop count value *Hops$_i$* of the same *ID* in the known data table, and the data tuple with the smallest number of hops is reserved.

3.3 Correcting Average Jump Distance

In the traditional DV-Hop localization algorithm, the unknown node will observe the most recent and optimal principles when selecting the average jump distance, but this will inevitably lead to an increase in the error in the specific application environment where the nodes are randomly distributed. Suppose that d_i (average distance per hop) is estimated by the beacon node i and that the minimum hop count between beacon node i and unknown node j is h_{ij}. The weighting factor is expressed by α and expressed by the following formula:

$$\alpha_{ij} = \frac{1}{h_{ij}} \tag{3}$$

After the average jump distance is modified, the average jump distance is expressed as follows:

$$d_j = \frac{\sum_{i=1}^{n} \alpha_{ij} \cdot d_i}{\sum_{i=1}^{n} \alpha_{ij}} \tag{4}$$

3.4 Proposed Algorithm

In this paper, the accuracy of the distance estimation in the DV-Hop localization algorithm is improved by modifying the average jump and hop to improve the positioning accuracy. This paper develops the algorithm as follows:

(1) The beacon nodes broadcast at different power levels and the neighbor nodes record different power numbers, taking into account the limited energy of the nodes, and the neighbor nodes do not forward the broadcasts. After the time interval T, the beacon node transmits power increases in turn, and when the beacon node transmits power increases to R_i, the received beacon node information is broadcast in flooding mode. In this way, the node information and the minimum hop count for each beacon node are recorded in all other nodes in the network.

(2) Formula (2) is used to calculate the average distance per hop.

(3) Formula (3) is used to obtain the weight factor, and corrected according to (4) based on the average distance per hop.
(4) The maximum likelihood estimation method is used to calculate the coordinates of the node to be positioned.

4 Simulation

The experiment is performed with MATLAB software. First, the best k value is experimentally selected. Second, the best k value for the beacon node ratio and the communication radius of the two different circumstances is determined based on a simulation. Two localization algorithms from the literature [20, 21] and the localization algorithm developed in this paper are applied to the simulation environment set up in this paper, and their positioning accuracy is analyzed and compared.

The simulation environment and parameters are set as follows:

(1) This paper uses the MATLAB platform for simulation experiments based on a simulated networking system with an indoor length of 100 m and a width of 100 m.
(2) Nodes are randomly placed in the simulated indoor model, and all nodes can communicate with other nodes within the communication range. The beacon nodes have multiple levels of transmission power, corresponding to a limited communication radius. Unknown nodes have the same radius of communication, and all nodes have symmetrical communication capabilities.
(3) Unknown nodes are randomly distributed in the indoor walls, and these nodes are placed again for each experiment, so that each experimental network will have a different topology. The reported experimental results represent the average of 100 individual experimental results.
(4) The error between the improved algorithm and the original algorithm is analyzed based on the experimental results. The formula is as follows:

$$e_i = \sqrt{(x_i - x)^2 + (y_i - y)^2} \qquad (5)$$

Where, (x_i, y_i) is the estimated position of the unknown node and (x, y) is the actual position of the unknown node. If the number of unknown nodes is n, then the average error is $\sum e_i / n$.

The Influence of the Number of Anchor Nodes on the Algorithm, in the next experiment, k = 8, the communication radius R = 15, the total number of sensor nodes is set to 100, and the proportion of beacon nodes is 12%, 14%, 16%, 18%, 20%, 22%, 24%, 26%, 28% or 30%. The experimental results shown in Fig. 3 show that the algorithm in this paper greatly improves the positioning accuracy compared with localization algorithms previously reported in the literature [20, 21]. The localization accuracy of the algorithm developed here is higher when there are fewer beacon nodes.

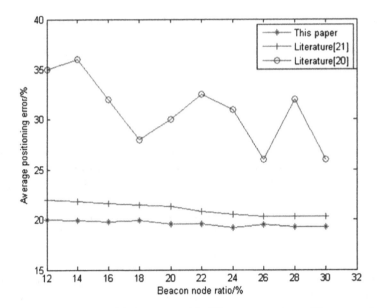

Fig. 3. The average positioning error with different proportions of beacon nodes

5 Conclusion

Aiming to fulfill the exact requirements of IoT sensing systems, this paper proposed an improved DV-Hop localization algorithm. The power of the beacon node was controlled using many power levels to transmit the beacon signal. The optimal transmission level was selected based on the beacon signal area, thereby improving the accuracy of the hops. This paper introduces a weighting factor to distinguish the distance between the beacon nodes from the distance to the positioning node. The experimental results show that the best value does not need to be selected for specific applications, but rather the environment can be set at the optimal value based on the proportion of sub-beacon node changes and changes in the size of the communications range. The experimental results show that the algorithm designed in this paper meets the requirements of system accuracy and that the positioning accuracy is greatly improved compared with the traditional algorithm.

References

1. Liu, Q., Huang, X.H., Leng, S.P.: Deployment strategy of wireless sensor networks for internet of things. China Commun. **8**(8), 111–120 (2011)
2. Zhou, G.D., Yi, T.H.: Recent developments on wireless sensor networks technology for bridge health monitoring. Math. Probl. Eng. **2013**(3), 1–33 (2013)
3. Viani, F., Rocca, P., Oliveri, G., et al.: Localization, tracking and imaging of targets in wireless sensor networks: an invited review. Radio Sci. **46**(5), 1161–1166 (2011)

4. Emeka, E.E., Abraham, O.F.: A survey of system architecture requirements for health care-based wireless sensor networks. Sensors 11(5), 4875–4898 (2011)
5. Fernando, L., Antonio-Javier, G., Felipe, G., et al.: A comprehensive approach to WSN-based ITS applications: a survey. Sensors 11(11), 10220–10265 (2011)
6. Cristina, A., Pedro, S., Andrés, I., et al.: Wireless sensor networks for oceanographic monitoring: a systematic review. Sensors 10(7), 6948–6968 (2010)
7. Ni, M., Liu, Y., Zhu, Y.: China's national research project on wireless sensor networks. IEEE Wirel. Commun. 14(6), 78–83 (2007)
8. Qian, Z., Wang, Y.: IoT technology and application. Acta Electron. Sin. 40(5), 1023–1029 (2012)
9. Sun, Q., Liu, J., Li, S., et al.: Internet of things: summarize on concepts, architecture and key technology problem. J. Beijing Univ. Posts Telecommun. 33(3), 1–9 (2010)
10. Akyildiz, I.F., Tommaso, M., Kaushik, R.: Wireless multimedia sensor networks: a survey. IEEE Wirel. Commun. 14(6), 32–39 (2007)
11. Simplício, M.A., Barreto, P., Margi, C.B., et al.: A survey on key management mechanisms for distributed wireless sensor networks. Comput. Netw. 54(15), 2591–2612 (2010)
12. Gracanin, D., Eltoweissy, M., Wadaa, A., et al.: A service-centric model for wireless sensor networks. IEEE J. Sel. Areas Commun. 23(6), 1159–1165 (2005)
13. Davis, T.W., Xu, L., Miguel, N., et al.: An experimental study of WSN power efficiency: MICAz networks with Xmesh. Int. J. Distrib. Sens. Netw. 2012(14), 1–12 (2012)
14. Atzori, L., Iera, A., Morabito, G.: The internet of things: a survey. Comput. Netw. 54(15), 2787–2805 (2010)
15. Pantazis, N.A., Nikolidakis, S.A., Vergados, D.D.: Energy efficient routing protocols in wireless sensor networks. IEEE Commun. Surv. Tutor. 15(2), 384–387 (2012)
16. Han, G., Xu, H., Dong, T.Q., et al.: Localization algorithms of wireless sensor networks. Telecommun. Syst. 52(4), 2419–2436 (2013)
17. Liu, H., Darabi, H., Banerjee, P., Liu, J.: Survey of wireless indoor positioning techniques and systems. IEEE Trans. Syst. Man Cybern. Part C Appl. Rev. 37(6), 1067–1080 (2007)
18. Robles, J.J.: Indoor localization based on wireless sensor networks. Int. J. Electron. Commun. 68(7), 578–580 (2014)
19. Chakchai, S., Katekaew, W.: Hybrid fuzzy centroid with MDV-Hop BAT localization algorithms in wireless sensor networks. Int. J. Distrib. Sens. Netw. 2015(1), 1–18 (2015)
20. Song, G., Tam, D.: Two novel Dv-Hop localization algorithms for randomly deployed wireless sensor networks. Int. J. Distrib. Sens. Netw. 2015(1), 1–9 (2015)
21. Zhang, Y., Xiang, S., Fu, W., et al.: Improved normalized collinearity DV-Hop algorithm for node localization in wireless sensor network. Int. J. Distrib. Sens. Netw. 2014(11), 1–14 (2015)
22. Adewumi, A.O., Arasomwan, M.A.: On the performance of particle swarm optimisation with (out) some control parameters for global optimisation. Int. J. Bio-Inspir. Comput. 8(1), 14–32 (2016)
23. Wang, H., Wang, W., Sun, H., Rahnamayan, S.: Firefly algorithm with random attraction. Int. J. Bio-Inspir. Comput. 8(1), 33–41 (2016)
24. Jia, Z., Duan, H., Shi, Y.: Hybrid brain storm optimisation and simulated annealing algorithm for continuous optimisation problems. Int. J. Bio-Inspir. Comput. 8(2), 109–121 (2016)
25. Xu, Z., Unveren, A., Acan, A.: Probability collectives hybridised with differential evolution for global optimisation. Int. J. Bio-Inspir. Comput. 8(3), 133–153 (2016)

A Design of the Shared Farmland System Based on the Internet of Things Technology and IMS

Na Chang[✉] and Junhua Ku

College of Information Engineering, Hainan Institute of Science and Technology,
Haikou, Hainan, China
474263408@qq.com, 370634001@qq.com

Abstract. In foreign countries, farmer owners have long been able to make what work needs to be done on their farmland public in the internet, so that clients can make appointments to experience farmland lives, which brings to the farm-owners millions of benefits. Over the past years, the domestic also set off a "shared farm" boom, and Hainan Province recently held "to develop 'shared farm' as the starting point for the construction of pastoral complex and beautiful village" training promotion, making the "shared farm" become a project that government attaches much importance.

In the past researches, due to the immaturity of real-time communication equipment, as well as the camera and other equipment is expensive, researchers have encountered much resistance. In this system, in order to solve those problems, Raspberry Pi and PGIO interface information collecting technology are applied to reduce cost, making this research feasible.

In the research and practice of Raspberry Pi and sensor hardware connection and access to the Internet, we have found that using python 3 language brings not only high speed but also the scalability, making the connection with the sensors smoothly successful.

Keywords: GPS · Sensors keyword · Web camera · Raspberry Pi · Shared

1 Introduction

The sharing of farms is a hot issue in both domestic and foreign researches. It has great research value in land, agricultural machinery and logistics sharing. The project has put forward the proposal of solving these problems and practiced the key technology.

This paper mainly studies the Internet of things and the self-developed IMS system. In the aspect of sensor data collection technology, this project chooses the python 3.5 as the main development voice, which has rich modules. Considering the cost of project, we have adopted the Raspberry Pi as the sensor server, and studied the PGIO interface programming technology of Raspberry Pi. In terms of improving users' experience, the system has adopted the HTML5 technology to program the App and an android App has been developed.

This paper first introduces the basic technical modules used in the system, and then explains the main technical designs, and gives the screenshots of the research results.

© Springer Nature Singapore Pte Ltd. 2018
K. Li et al. (Eds.): ISICA 2017, CCIS 873, pp. 492–498, 2018.
https://doi.org/10.1007/978-981-13-1648-7_42

2   The Introduction of Basic Technologies

2.1   The GPS Technology

GPS is the English abbreviation of Global Positioning System (quoted from Baidu encyclopedia), used for converting geographical location to coordinate data. In this project, the farmers' land is distributed sporadically around in the city, and with the combination of the technique of GPS and Baidu map, it is convenient for users to find their goals and drive to there to manege the land.

2.2   Web Camera Technology

The webcam can transmit what the camera films to the users through the network. The users thus get to know the crop growth situation and the land condition, and decide what the next agricultural operations should be carried out.

2.3   Sensors

Modern agriculture management [4] needs to analyze the soil, weather, moisture so on and scientific adjustment to achieve the best crop growth environment, increase production and reduce energy waste. The sensors can digitalize the situation of the environment, and pass it through the interface to the software to deal with data, and the sensors commonly used in agricultural production are sensors such as temperature, humidity, rain, direction, soil nutrient.

2.4   Raspberry Pi

Raspberry Pi [3] (abbreviated as RPi, (or RasPi/RPI), a microcomputer as large as a credit card, its system is based on Linux, using python as the main programming language (quoted from Baidu encyclopedia). Raspberry Pi demands a strong interface ability, and can easily realize the network connection, camera connection, various sensors via PGIO interface. Furthermore, it is cheap and suitable for large-scale deployment so that the whole cost can be reduced.

3   Design of the Project

This system is designed to guide agricultural production, provide information about land and labor with the basis of modern information management technology and convenient information communication conditions, at the aim of farmland sharing.

Html5 has the features of browser adaptive, local offline storage and other new features, which is very suitable for the project, so the system uses the html5 as the front-end development language of the project. Because python 3.5 enjoys many mature hardware operation modules, which can save time and increase efficiency, the system uses it as the back-end language of the project. Considering the performance of the system, the deployment and the use of the project, mongodb [2] has been chosen. In addition, the front-end also uses the bidirectional binding data technology–vue.js framework. The back-end uses the excellent tornado framework, which when used together with nginx, can be optimized to the response speed of 8000 times/second by reasonable optimization.

3.1 The Technology Design of GPS

When farm-owners use the App to register the land information, they can use GPS [9] to mark the positions of their farms in the system by Baidu map [5], which can show the position of the farmland and help to guide the route to the farmland.

3.2 The Technology Design of Web-Camera

With USB camera and Raspberry Pi, web camera function can thus be achieved.

Install motion and connect the USB camera with Raspberry Pi with motion. When finished, the pictures captured by the camera can be viewed by browsers.

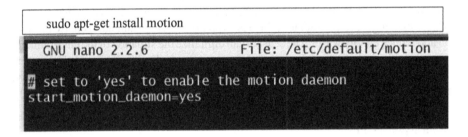

```
sudo apt-get install motion
```

```
GNU nano 2.2.6                    File: /etc/default/motion

# set to 'yes' to enable the motion daemon
start_motion_daemon=yes
```

3.3 The Technology Design of Sensors

The sensors [6] used in the system are temperature sensors, soil moisture sensors, wind humidity sensors, etc., and the PGIO [8] interface of the Raspberry Pi will store the data in the database.

The numbering ways of Raspberry Pi:

Functional Physics Pins
From left to right, top to bottom: odd number at left, even number at right:: 1–40

BCM
This kind of numbering focuses on the CPU register, according to the GPIO register number of BCM2835.

wpi
This kind of numbering focuses on logic, numbering from zero, which is convenient for programming, as shown in Fig. 1:

HCM encoding	wpi encoding	function			Physical interface				function		wpi encoding	HCM encoding
BCM	wPi	Name	Mode	V	Physical	V	Mode	Name	wPi	BCM
					B Plus						
		3.3v			1 || 2			5v		
2	8	SDA.1	ALTO	1	3 || 4			5v		
3	9	SCL.1	ALTO	1	5 || 6			0v		
4	7	GPIO. 7	IN	1	7 || 8	0	ALTO	TxD	15	14
		0v			9 || 10	1	ALTO	RxD	16	15
17	0	GPIO. 0	IN	0	11 || 12	0	IN	GPIO. 1	1	18
27	2	GPIO. 2	IN	0	13 || 14			0v		
22	3	GPIO. 3	IN	0	15 || 16	0	IN	GPIO. 4	4	23
		3.3v			17 || 18	1	OUT	GPIO. 5	5	24
10	12	MOSI	ALTO	0	19 || 20			0v		
9	13	MISO	ALTO	1	21 || 22	1	OUT	GPIO. 6	6	25
11	14	SCLK	ALTO	1	23 || 24	1	ALTO	CE0	10	8
		0v			25 || 26	1	ALTO	CE1	11	7
0	30	SDA.0	ALTO	1	27 || 28	1	ALTO	SCL.0	31	1
5	21	GPIO.21	IN	1	29 || 30			0v		
6	22	GPIO.22	IN	1	31 || 32	0	IN	GPIO.26	26	12
13	23	GPIO.23	IN	0	33 || 34			0v		
19	24	GPIO.24	IN	0	35 || 36	0	IN	GPIO.27	27	16
26	25	GPIO.25	IN	0	37 || 38	0	IN	GPIO.28	28	20
		0v			39 || 40	0	IN	GPIO.29	29	21
BCM	wPi	Name	Mode	V	Physical	V	Mode	Name	wPi	BCM

Fig. 1. GPIO interface connection

3.4 Roles for Users

There are three roles designed in the system: administrators, farmers and claimants.

The administrators are in charge of the operation of the system. Their main job is to manage the cameras and the relationships of land-binding, to keep the camera sensors, temperature sensors, and humidity sensors under good running, to adjust the labor price and to back up file.

Farmers can apply land, manage land and provide labor services (Fig. 2).

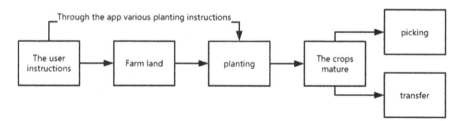

Fig. 2. The flow chart

Claimants are the main participants of the system. They can purchase land, plant crops and publish online labor or picking activities through the system.

3.5 The Design of Data Model and Use Case

(See Figs. 3 and 4).

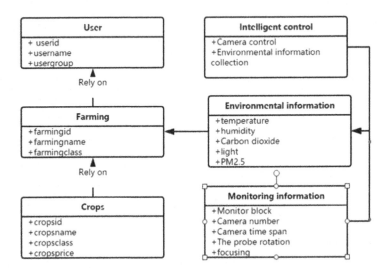

Fig. 3. The data model

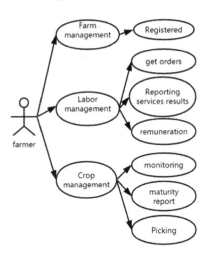

Fig. 4. Farmer' use cases

3.6   The Design of Users' Interfaces

(See Fig. 5).

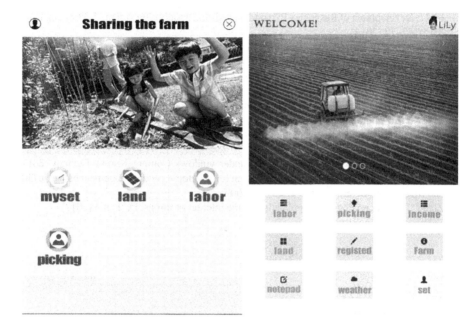

Fig. 5. Users' interfaces

4 Conclusion

In summary, the design of a reasonable IMS system and measurement sensor networking technology [10], the collection and analysis of farm data, are the keys to achieving farm sharing [1].

The client and off-stage management used in the system are programmed in html5 language, which has taken users' experience into full consideration. The use of the low-cost Raspberry Pi can reduce the cost of the whole project. The adaptation of python 3.5 to the sensors and the obvious advantages of mongodb [7] will bring a new kind of developmental pattern to farmland management and farmland sharing.

Acknowledgment. This research was supported by the natural science foundation project of Hainan province (614248), and the key science research project of Education Department of Hainan province (Hnky2017ZD-20).

References

1. Bai, C.: A study of the enterprise managerial pattern of farmlands for leisure. Mod. Agric. Sci. Technol. **1**, 265–267 (2017)
2. Chen, W.: Data storage design of the internet of things open platform based on MongoDB. J. Xi'an Univ. Posts Telecommun. **21**(2), 78–82 (2016)

3. Li, L.: Development of Raspberry PI based on real-time monitoring system. J. Minjiang Univ. **35**(5), 67–72 (2014)

4. Lu, Y.: An analysis of the application of the internet of things in modern agriculture planting and management. Agric. Netw. Inf. **3** (2017)

5. Ajwaliya, R.J.: Web-GIS based application for utility management system. J. Geomat. **11**(1), 86–97 (2017)

6. Zhanpeng, S.: The system of inner-environment monitoring by multi-sensors. Transducer Microsyst. Technol. **36**(1), 87–90 (2017)

7. Tong, Y.: Design and implementation of information integration system based on MongoDB. Inf. Technol. **2**, 125–129 (2015)

8. Ys, Z.: Simulating I2C bus with GPIO pins under windows. Comput. Knowl. Technol. (2017)

9. Zheng, S.: The information integration system for developing and management based on GIS and MIS technologies. Off. Inf. **1**, 28–29 (2009)

10. Zhu, L.: The technology and application of the internet of things. PC Fan **3** (2017)

Personalized Recommendation Algorithm
Based on Commodity Label

Yuehua Dong and Xuelei Liang[✉]

Jiangxi University of Science and Technology, Ganzhou 341000, China
347505278@99.com

Abstract. Aiming at the fact that the traditional collaborative filtering recommendation algorithm is insufficient in the number of users' implicit feedback, and the user interest preference model is too rough, a collaborative filtering recommendation algorithm with the importance of tags is proposed. Type and frequency of use of the label reflect user preferences and preferences, in order to establish a new user preferences model for better mining and use implicit user feedback data will affect the degree of the label on the user to quantify, to establish a new method for similarity computation. The experimental results show that the proposed algorithm has obvious advantages, improves the recommendation accuracy and alleviates the cold start problem.

Keywords: Collaborative filtering · Recommendation algorithm
Implicit feedback · Similarity measure

1 Introduction

Recommendation algorithm mining user's interests and hobbies based on the customer's feedback data, the user's feedback data refers to the customer in the process of purchasing goods produced a series of data, such as: the user of its purchased goods subjective evaluation, rating or the goods Browse, plus purchase, share operations and so on. Feedback data can be divided into explicit data and implicit data. Users' subjective ratings or evaluation of goods are explicit feedback data, which can directly reflect the user's hobbies, but this feedback data is less, the data sparse more serious. Most of the implicit data from the user's data on the operation of the product, these data can reflect the current needs of users, it's more and easy to get. The two feedback data have their own advantages and disadvantages when mining the user's interest needs. Therefore, this paper focuses on the combination of two kinds of feedback data, complement each other and make their own efforts to establish a new recommendation model and calculation method of similarity to improve the quality of recommendation.

With the rapid development of recommendation systems in recent years [1], different opinions have been put forward in the past when digging user implicit feedback data. Literature [2] first excavates user's display and implicit data and applies it to recommendation system. The literature [3] separately calculated the user's two kinds of feedback data with different ratios and established the matrix decomposition model to improve the recommendation quality. Literature [4] proposed the migration learning

© Springer Nature Singapore Pte Ltd. 2018
K. Li et al. (Eds.): ISICA 2017, CCIS 873, pp. 499–512, 2018.
https://doi.org/10.1007/978-981-13-1648-7_43

algorithm to improve the recommendation quality. GAIL [5] put forward the fusion of two kinds of feedback data applied to the personality. Luk et al. [6] applied the implicit feedback data trust mechanism to the recommendation system; Lu et al. [7] compared all kinds of implicit feedback recommendation methods and gave various evaluation indexes; Jawaheer et al. [8] summarize the advantages and disadvantages of implicit and explicit feedback applications in recommendation systems. However, there are few researchers on heterogeneous feedback data, and the models that describe the preferences of users are inaccurate and do not make full use of user's feedback information, it decreased the calculation precision of similarity. As a result, the reliability of the nearest-neighbor set of similar users obtained is poor, and the recommendation result is not accurate enough.

In view of the user's interest preference model is too rough, the method of calculating the similarity is not accurate and other issues. This paper gives the following innovations to improve the accuracy of the recommendation system:

1. Establish the relationship between user – commodity - label. Through the observation and analysis of the relationship between the establishment of user – product - label, the user needs to buy the goods, so explicit and implicit feedback data can be digging when the user find the goods, including implicit data tag related Information, the process is the user, product, the link between the tags, followed by the establishment of the relationship between the three.
2. Quantification of labels for the first time. On the basis of the relationship between user-product-label, different kinds of labels are given different amounts through a series of calculations. For example, the object has the weight and the charge has the same amount of electricity, and the user's interest preference is more accurately obtained through the calculation of the amount and the degree of preference.
3. Establish a new user interest preference model. The label is the implicit feedback data of the user. The type and frequency of the label used by the user reflect the preference of the user to a certain extent and establish a new similarity calculation method according to the model.
4. Establish a new similarity calculation method. The importance of tags between users is proportional to the frequency difference between the tags they use.

2   Traditional Collaborative Filtering Recommendation Algorithm

User-based Collaborative Filtering is the most widely used nominate algorithm in the recommendation system. It recommended to target users based on the principle "Birds of a feather flock together", for example: when li Ming buys headphones, recommendation system will find li Ming's similar users, and then put the similar user has bought and Ming are likely to be interested in working through the calculations of this headset recommended to him.

2.1 The Calculation Process of Traditional Similarity Degree

Based on the above principle, UBCF algorithm follow the three steps below:

1. Data Initialization: The initial input user's rating matrix A(n × m) refers to n users and m projects, and x_{nm} is the user's rating of the product m.
2. Find the nearest neighbor: Calculate the similarity $sim(u_k, u), (1 \leq k \leq K)$ between the marked user u and other users u_k, and get the user's neighbor set.
3. The forecast score produces the recommended results: the user nearest neighbor set $N(u) = \{u_1, u_2, \cdots, u_K\}$, and the forecast score of the project is $P_{u,i}$, and the prediction formula is:

$$P_{u,i} = \frac{1}{\|N(u)\|} \sum_{v \in N(u)} R_{v,i} \tag{1}$$

$P_{u,i}$ is the prediction score, $\|N(u)\|$ is refers to the number of close neighbors and recommendation is made on the basis of $P_{u,i}$.

2.2 Traditional Similarity Calculation Method

At present, the methods for calculating the similarity between users of UBCF mainly include the following:

(1) Cosine similarity

$$sim(t, o) = \cos\left(\overrightarrow{R_t}, \overrightarrow{R_o}\right) = \frac{\sum_{i \in I_{to}} R_{ti} R_{oi}}{\sqrt{\sum_{i \in I_{to}} R_{ti}^2} \sqrt{\sum_{i \in I_{vto}} R_{oi}^2}} \tag{2}$$

(2) Correct cosine similarity

$$sim(t, o) = \frac{\sum_{i \in I_{to}} (R_{ti} - \bar{R}_t)(R_{oi} - \bar{R}_o)}{\sqrt{\sum_{i \in R_i} (R_{ti} - \bar{R}_t)^2} \sqrt{\sum_{j \in R_j} (R_{oj} - \bar{R}_j)^2}} \tag{3}$$

(3) Pearson Coefficient correlation

$$sim(t, o) = \frac{\sum_{i \in I_{to}} (R_{ti} - \bar{R}_t)(R_{oi} - \bar{R}_o)}{\sqrt{\sum_{i \in I_{to}} (R_{ti} - \bar{R}_t)^2} \sqrt{\sum_{j \in I_{to}} (R_{oj} - \bar{R}_J)^2}} \tag{4}$$

Above three kinds of method to calculate the similarity of each have advantages and disadvantages, but the recommended performance is not precise enough in a word. In order to improve the recommendation accuracy, this paper given a similarity calculation method of label important degree, choose the highest similarity calculation results of a number of users, or above a certain threshold number of users as the target user's neighbor [9, 11] to recommend.

3   Personalized Recommendation Algorithm

3.1   Establish the Relationship Between User - Merchandise - Labels

There are much large e-commerce platforms at home and abroad, when the user browsing the goods in the platform, the choice of their favorite project random browse, add to cart, purchase, or share, etc. These operations directly reflect the user's interests.

The operation of the commodity as a credible element that reflects the user's interest preference, It can express user's interest preferences directly [12], and also makes it easier for users to share resources. When the user to manipulate, the goods between the user, the user follow operation and goods between and among forms a network relation, the relation expression between the user and the user through the common implicit feedback data together, make possible collaborative filtering recommendation algorithm, as shown in Fig. 1

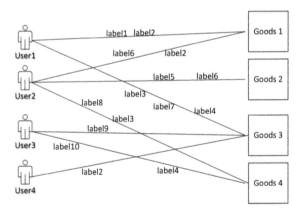

Fig. 1.   Relationship diagram example

There are labels 1 and label 1 between the user 1 and the merchandise 2 in the picture. When the user 1 is browsing, sharing or purchasing the goods 1, the label 1 and the label 2 are used to mark the goods, that is, the user 1 uses the label 1 and the label 2 to mark the commodity 1, and the other elements in the picture are the same.

3.2   Establish a New Similarity Calculation Model

In order to dig the user's implicit feedback data in the electric business platform, The system monitors the user's clicked event to obtain the click object and matches the corresponding word according to the attribute of the clicked object. Just as many people like to go shopping, they will come up with their own favorite products, and then they will select the products they are interested in (for example, share, browse, add to cart, purchase, etc.), When the user is operating on the commodity, the use of the commodity itself with similar attributes of the label (for example: watches have mechanical watches, electronic watches, quartz watches, etc. Skirts have long skirts, skirts, dresses, etc.), system to choose matching words to help users create a product tag library. This process is considered the user is using the label.

Definition 1. Product label, Commodity tag. In order to help a user organize, manage and categorize commodity resources, convenient for users to share resources, and an appropriate word or phrase is selected to mark the commodity when the user operates the commodity. For example, when browsing, purchasing or sharing a short skirt product, the system selects the word "skirt" to mark the product, and the word or phrase to mark the product is called the product label.

Definition 2. Preference goods. The article considers that the goods that are viewed, Shared and purchased by the user are preferred commodities, which are the items that appear in Table 1.

Table 1. User commodity matrix

User	Item			
	I_1	I_2	...	I_m
u_1	x_{11}	x_{12}	...	x_{1m}
u_2	x_{21}	x_{22}	...	x_{2m}
...	...	...	...	...
u_n	x_{n1}	x_{n2}	...	x_{nm}

Definition 3. User-commodity score Matrix R (user-item Matrix). N users and m projects form an n by m matrix.

The row represents the user, the column represents the item, that is the product, when the user I have purchased the product j and the score of the product j is x, the $r_{ij} = x$, otherwise $r_{ij} = 0$ (Table 2).

Table 2. User-commodity label frequency matrix

User	Label-F			
	t_1	t_2	...	t_g
u_1	a_{11}	a_{12}	...	a_{1g}
u_2	a_{21}	a_{22}	...	a_{2g}
...	...	...	...	...
u_n	a_{n1}	a_{n2}	...	a_{ng}

Definition 4. User-commodity label Frequency matrix A (user-commodity Tag Frequency Matric). n users and g commodity labels, constitute an n by g matrix as shown in Table 3:

Table 3. Commodity-label frequency matrix

Commodity	Label-F			
	t_1	t_2	...	t_g
I_1	q_{11}	q_{12}	...	q_{1g}
I_2	q_{21}	q_{22}	...	q_{2g}
...	...	...	...	...
I_m	q_{m1}	q_{m2}	...	q_{mg}

Wherein, the row represents the user, the column represents the frequency of the commodity tag used by the user, and u_i uses x commodity tags t_g, then a_{ij} = x, that is, the number of the user's commodity tags; otherwise, a_{ij} = 0.

Definition 5. Commodity-Tag Frequency Matrix Q (Commodity Tag -item Frequency Matric). If there are g labels $T = (t_1, t_2, \cdots t_g)$ and m items $I = (i_1, i_2, \cdots i_m)$, this forms a matrix of g × m, which behaves as a product label.

When there are y items labeled t_g item t_g, q_{gj} = y, that is, the number of items labeled the item, otherwise q_{gj} = 0.

Because people aspire and love each other, the same type of tag has a different degree of importance for different users. In order to better calculate the similarity, tags are given an average importance independent of different users. Each user has their own preferred product, a product can be marked by multiple labels, a label can be marked a variety of goods, for example: a dress can be "shirt", "round neck", "summer" and other labels, So User u_n has i favorite product, The ith favorite item has g tags, calculate the importance of the user I the first product according to the importance of tags to, and then establish the importance of all the preferences of the user a Dimension vector. In the paper, the similarity of two users is considered as the inverse of the difference between the users' preference of all goods and the square of the difference between them using the common label. Finally, the prediction score and the recommendation of Top n are given. The model is as follows:

$$sim(T, I) = k\frac{M(u_t, I_m) \cdot M(u_i, I_m)}{d_{ti}^2} \quad (5)$$

Where sim(T, I) is the similarity of the target user u_t and other users u_i, k is the adjustment factor used to adjust the step size between two users, $M(u_t, I_m)$ represents the degree of importance of the favorite item of the user u_t, $M(u_i, I_m)$ represents the degree of importance of the favorite item of the user u_i.d_{ti} is the Euclidean distance of the label j used by the user u_i and the user u_t.

3.3   Calculation of Similarity of Label Importance

For a certain label, it may mark the same preference product of many users, but this label may have different importance among different users. For example, both A and B users use the same "round neck" label to mark the same clothes. User A may pay more attention to the layout of the clothes, and user B may pay more attention to the collar of the clothes. Therefore, since the label "round neck" is of different importance to user A and user B, it is necessary to stipulate here that the same label is independent of The importance of different users, that is, the average degree of importance $\overline{m(t)}$. In calculating the average importance level, the labels appearing in more users 'favorite items are less important than those appearing in fewer users' preferred items, for example, the labels t_i and t_j both appear in the preferred product of users A and B, however, the label t_i appears in 1000 users 'preference items, and the label t_j appears in 10 users' preference items. In measuring the similarity of user A and user B, the label t_j is more important than t_i, label A may appear in the 1000 or 10000 user's preference product may be because label A is universal, for example: jacket label, whether it is summer, winter or round neck, V-neck can be called a jacket. The labels that appear in less preferred items of the user are personalized and targeted, such as jackets, self-cultivation, limited edition, etc. Therefore, in the article, it is considered important that the labels appearing in the less-preferred items of the user have a greater importance, The degree of importance is calculated as (6) below.

$$\overline{m(t_g)} = \log(\frac{n}{t_g}) \tag{6}$$

$\overline{m(t_g)}$ is the average importance of label g, n is the sum of users in the system, and t_g is the total number of labels g in the system. The importance of the label g to the user u_m is calculated as shown in (7).

$$m(T_g, u_n) = w(t_g, u_n) \times \overline{m(t_g)} \tag{7}$$

$m(T_g, u_n)$ is the importance of the tag t_g in the preferred item of user u_n, $w(t_g, u_n)$ is the important parameter of the preferred item. The preferred product importance parameter is calculated by multiplying the word frequency by the inverse document frequency [15, 16]. The term frequency can be used to indicate how often a tag appears in a preferred product, and the greater the frequency with which a particular tag is used for different users, the more preferred the user is to the tagged product; reverse file frequency indicates the type of tag distinguishing ability, the category can reflect to some extent the user's preference for goods, such as Korean version, Slim, sweater and other labels, if the system has g label, Then the importance of the formula parameters such as:

$$w(t_g, u_m) = \frac{s_n^{t_g}}{s_n^T} \times \log(\frac{n}{s_{t_g}}) \tag{8}$$

$s_n^{t_g}$ is the number of labels u of the user u_n in the preferred item, s_n^T is the total number of labels of the user u_n in the preferred item, n is the total number of users, and s_{t_g} is the total number of preferred items containing the label g. The degree of importance of the tag g on the preference item of the user u_n is calculated as follows:

$$m(T_g, u_n) = \frac{s_n^{t_g}}{s_n^T} \times \log(\frac{n}{s_{t_g}}) \times \overline{m(t_g)} \tag{9}$$

We can calculate the importance of the user u_n preferred product, If user u_n has I preferred the product, then the ith preferred product contains g labels, and each label can calculate the degree of importance of the user's preferred product. Then, the total importance degree of all preferred products of user u_n is calculated as follows:

$$M(u_n, I_m) = \sum_{i=1}^{m} (\sum_{j=1}^{g} m(T_g, u_n)_j)_i \tag{10}$$

$M(I_m, u_n)$ is the overall degree of importance of all the preferred products of user u_n, user u_n has a total of m preferred products, and the ith preferred product has a total of g tags.

When calculating the similarity between users, the degree of importance of the two users' preference commodities is inversely proportional to the square of the frequency difference between them. The larger the difference is, the less similar the two users are. In this paper, the Euclidean distance is used to calculate the difference between two user tags.

$$d_{ti} = \sqrt{\sum_{j=1}^{g} (q_{tj} - q_{ij})^2} \tag{11}$$

d_{ti} is the Euclidean distance of the target user u_t and other users u_i using the label j, g is the number of the label types used by the user u_t and the user u_i, j is a label in g, q_{tj} is the frequency of the target user u_t using the j label, q_{ij} is the frequency at which the target user u_i uses the j tag.

The similarity calculation formula of the marked user u_t and other users u_i, and substituting Eqs. (10) and (11) into Eq. (5) is as below:

$$\text{sim}(T, I) = k \frac{\sum_{i=1}^{m}(\sum_{j=1}^{g} m(T_g, u_t)_j)_i \cdot \sum_{i=1}^{m}(\sum_{j=1}^{g} m(T_g, u_i)_j)_i}{\sqrt{\sum_{j=1}^{g}(q_{tj} - q_{ij})^2}^2} \tag{12}$$

In this paper, the similarity calculation method first calculates the average importance degree of all the users' tags in the platform and calculates the importance degree of the users' preference products according to the average importance degree of the tags so as to obtain the importance of the users' total preference products. When calculating the similarity between two users, In calculating the similarity between two users, the paper considers that the product of the total product importance of two users is inversely proportional to the square of the frequency difference between two users, and the difference is calculated using the Euclidean distance. Finally, the target user similar to other users Degree calculation formula was got.

3.4 Collaborative Filtering Recommended Algorithm Description for Label Importance

Combined with the traditional similarity calculation method, the importance of label is calculated by using the inverse text frequency. Finally, the similarity of label importance degree is calculated, and the description of the algorithm and the analysis of time complexity are given.

The calculation steps are as follows

Input: user item - product matrix $R_{(n,m)}$, user-label frequency matrix $A_{(n,g)}$, product - label frequency matrix $Q_{(m,g)}$;

Output: Top-N items recommended for the target user, namely, neighbor sets.

Step1: Filter out the target user U_t and other users U_i to use the common tag set $I_{(t,i)}$, the total number of tags in the system t_g and the total number of users in the system n, $U_t \neq U_i$;

Step2: Calculate the average importance of tags independent of different users $\overline{m(t_g)}$;

Step3: Calculate the importance parameter $w(t_g, u_m)$ using the inverse text frequency;

Step4: Calculate the importance of the tag g to the preference U of the user U m $m(T_g, u_n)$;

Step5: Calculate the total importance $M(u_n, I_m)$ of the tag to the user U according to $m(T_g, u_n)$ in Step4;

Step6: Use Euclidean distance to calculate the difference between g kinds of label frequency d_{ti};

Step7: Calculate the similarity sim(T, I) between the target user U_t and other users U_i;

Step8: Filter out the U_t nearest neighbors of the target user and predict the score of the U_t user who has not been rated by the target user U_t in the user-product matrix R;

Step9: The prediction score Top-N recommended to the target user U_t;

Time complexity calculation: Assuming that the total number of users in the system is N, the number of labels that the user labels the most goods by using the label is M and the maximum number of similar users is K and $K < N$, the complexity of the label importance degree algorithm is as follows: $T(n) = NM^2 + NM + M + M^2K = O(NM^2)$.

4   Experimental Results and Analysis

4.1   Data Sets and Evaluation Criteria

The experiment uses MovieLens dataset because it is widely used in the experiments and tests of various recommender systems, so it is more persuasive. More importantly, the dataset contains social tags, and the social tags are user- Favorite movies are tagged with words randomly. These are phrases such as martial arts, action, and so on. In this article, the user's product tag is tagged with a phrase such as a sock, a skirt, or an electronic product. Therefore, using a social tag Instead of the user's product label, the movie in the dataset replaces the product. The total number of users, movies, and labels in this dataset is 2113, 10197, and 13222, respectively.

In order to achieve a better measure of accuracy, this article analyzes the mean absolute error (MAE), precision [21, 24, 25, 27] and recall rate [28].

The MAE measures the accuracy by predicting the absolute distance between the score and the actual score. The larger the absolute value of the weighted average, the lower the accuracy, as shown in Eq. (13) [17, 18, 20]:

$$MAE = \frac{\sum_{i=1}^{n} |P_i - R_i|}{n} \tag{13}$$

P_i is the recommended prediction score, R_i is the actual score of the user, n is the size of the set, and the larger the MAE value, the smaller the accuracy.

The accuracy is the ratio of correct items in the recommended list, the higher the accuracy is, the better the recommended performance of the algorithm is, as shown in Eq. (14):

$$prs = \frac{M_{simil}}{M_p} \tag{14}$$

M_{simil} is the number of user preferences in the referral list, M_p is the number of items on the referral list.

Recall rate is the proportion of projects in the test set that are correctly recommended, the higher the recall rate, the better the recommended performance. As shown in (15):

$$Rc = \frac{M_{simil}}{M_{test}} \tag{15}$$

M_{simil} is the number of user preferences in the recommendation list, M_{test} is the total number of items in the test set.

4.2   Experimental Results and Analysis

In order to test the validity of the collaborative filtering recommendation algorithm, Refer to a variety of experimental principles [10, 13, 14, 19], comparison methods [22] and models [23], the traditional user-based collaborative filtering recommendation algorithm and the recommendation of the collaborative filtering recommendation algorithm based on label group are compared with the performance of the proposed algorithm [26].

Randomly selected from the test set of 10 users as the target user, the average as the test results.

In the first test, we choose the scale of 10 to 60 and the step size of 10 from the neighbor set to test the influence of the neighbor size on the recommended performance, and compare with the common similarity calculation method, including cosine similarity, modified cosine similarity, Pearson similarity and synonym tag grouping recommendation algorithm, as shown in Fig. 2:

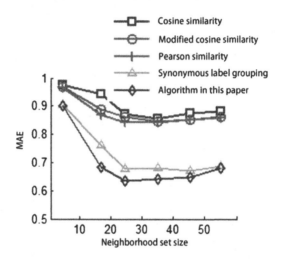

Fig. 2. Comparison of mean absolute error

The smaller the average absolute error is, the greater the accuracy of the recommended prediction. As can be seen from the figure above, the absolute value of the algorithm in the bottom of the curve, the performance of different algorithms is not obvious when the number of neighbors is small, With the increase of scale, the curve of recommended algorithm for collaborative filtering of tag importance declines rapidly, and the performance of the algorithm is obviously enhanced. When the number of neighbor users reaches 30, the performance of the algorithm reaches the maximum. As the number of neighbor users increases, the performance of the algorithm decreases, which indicates that too many neighbor users will increase the noise score, thus affecting the accuracy of the algorithm recommendation.

In the second test, we choose the cosine similarity, the modified cosine similarity, the Pearson similarity, the recommended algorithm of label grouping and the similarity calculation method in this paper to compare the performances using the best test result of the first test with the neighbor set of 30, As shown in Fig. 3:

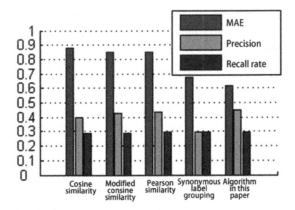

Fig. 3. Comparison of performance-testing

It can be seen from Fig. 3 that the accuracy, recall rate and average absolute error of collaborative filtering recommendation algorithm of label importance are better than other similarity metric algorithms, which shows that the algorithm has better recommendation performance.

5   Conclusion

In order to solve the problem of insufficient user, implicit feedback data mining and low recommendation precision in the existing recommendation system, a new user preference model and similarity calculation method are proposed in this paper. The model takes into account that user's implicit feedback data can directly react user's interest preferences, establish the relationship between user goods label three, users' frequency of labeling reflects the preference of users, and then improves the user preference model. The similarity calculation uses the importance of labels and improves the traditional similarity calculation method. It can be seen from the experimental results that the proposed algorithm has great advantages and the recommendation accuracy has been greatly improved.

The next step is to order the user's product label, combined with the user's own quantitative properties and other properties of the similarity reconstruction method.

References

1. Amatriain, X.: Mining large streams of user data for personalized reconnendations. ACM SIGKDD Explor. Newsl. **14**(2), 37–48 (2013)
2. Bell, R.M., Koren, Y.: Scalable collaborative filtering with jointly derived neighborhood interpolation weights. In: 2007 Seventh IEEE International Conference on Data Mining, ICDM 2007, pp. 43–52. IEEE (2007)
3. Liu, N.N., Xiang, E.W., Zhao, M., et al.: Unifying explicit and implicit feedback for collaborative filtering. In: Proceedings of the 19th ACM International Conference on Information and Knowledge Management, pp. 1445–1448. ACM (2010)

4. Pan, W., Xiang, E.W., Yang, Q.: Transfer learning in collaborative filtering with uncertain ratings, vol. 12, pp. 662–668. AAAI (2012)
5. Li, G., Chen, Q.: Exploiting explicit and implicit feedback for personalized ranking. Math. Prob. Eng. **2016** (2016)
6. Lu, K., Xie, L., Li, M.: Research on implied-trust aware collaborative filtering recommendation algorithm. J. Chin. Comput. Syst. **37**(2), 241–245 (2016)
7. Lu, Y., Cao, J.: Research status and future trends of recommender systems for implicit feedback. Comput. Sci. **43**(4), 7–15 (2016)
8. Jawaheer, G., Szomszor, M., Kostkova, P.: Comparison of implicit and explicit feedback from an online music recommendation service. In: Proceedings of the 1st International Workshop on and Scalable Location-Aware Recommender System (2013). IEEE Trans. Knowl. Data Eng. **26**(6), 1384–1399 (2013)
9. Herlocker, J.L., Konstan, J.A., Borchers, A., et al.: An algorithmic framework for performing collaborative filtering. In: Proceedings of the 22nd Annual International ACM SIGIR Conference on Research and Development in Information on Retrieval (1999)
10. Adewumi, A.O., Arasomwan, M.A.: On the performance of particle swarm optimisation with (out) some control parameters for global optimisation. Int. J. Bio-Inspired Comput. **8**(1), 14–32 (2016)
11. Resnich, P., Iacovou, N., Suchak, M., et al.: GroupLens: an open architecture for collaborative filtering of netnews. In: Proceedings of ACM Conference on Computer Supported Cooperative Work, pp. 175–186 (1994)
12. Wang, G.: Collaborative filtering recommendation algorithm based on user's gravitation. Appl. Reaction Res. Comput. **33**(11), 3329–3333 (2016)
13. Wang, H., Wang, W., Sun, H., Rahnamayan, S.: Firefly algorithm with random attraction. Int. J. Bio-Inspired Comput. **8**(1), 33–41 (2016)
14. Jia, Z., Duan, H., Shi, Y.: Hybrid brain storm optimisation and simulated annealing algorithm for continuous optimisation problems. Int. J. Bio-Inspired Comput. **8**(2), 109–121 (2016)
15. Ren, Q.: The Research of Algorithm about Social Network Recommendation Service based on Hadoop. Jilin University, Changchun (2013)
16. Yang, S.: A method of patent retrieval based on automatic query expansion. Zhe Jiang University (2013)
17. Yanan, H.A.N., Han, C.A.O., Liangliang, L.I.U.: Collaborative filtering recommendation algorithm based on score matrix filling and user interest. Comput. Eng. **42**(1), 36–40 (2016)
18. Ailin, D.E.N.G., Yangyong, Z.H.U., His, B.A.: Collaborative filtering recommendation algorithm based on item rating prediction. J. Softw. **14**(9), 1621–1628 (2013)
19. Xu, Z., Unveren, A., Acan, A.: Probability collectives hybridised with differential evolution for global optimisation. Int. J. Bio-Inspired Comput. **8**(3), 133–153 (2016)
20. Xijun, Y.E., Yue, G.O.N.G.: Study on diversity of collaborative filtering recommendation algorithm based on item category. Comput. Eng. **41**(10), 42–46 (2015)
21. Nan, Z., Qiudan, L.: A recommender system based on tag and time information for social tagging systems. Expert Syst. Appl. **38**, 4575–4578 (2011)
22. Cai, X., Gao, X.Z., Xue, Y.: Improved bat algorithm with optimal forage strategy and random disturbance strategy. Int. J. Bio-Inspired Comput. **8**(4), 205–214 (2016)
23. Gálvez, A., Iglesias, A.: New memetic self-adaptive firefly algorithm for continuous optimisation. Int. J. Bio-Inspired Comput. **8**(5), 300–317 (2016)
24. Sarwar, B., Karypis, G., Konstan, J., et al.: Analysis of recommendation algorithms for e-commerce. In: Proceedings of the 2nd ACM Conference on Electronic Commerce (2000)
25. Su, X.Y., Khoshgoftaar, T.M.: A survey of collaborative filtering techniques. In: Advances in Artificial Intelligence, pp. 1–19 (2009)

26. Srivastava, P.R.: Test case optimisation a nature inspired approach using bacteriologic algorithm. Int. J. Bio-Inspired Comput. **8**(2), 122–131 (2016)
27. Yang, X., Yu, J., Tu, E., Yi, B., et al.: Collaborative filtering model fusing singularity and diffusion process. J. Softw. **8**, 1868–1884 (2013)
28. Hailing, X.U., Xiao, W.U., Xiaodong, L.I., et al.: Comparison study if internet recommendation system. J. Softw. **20**(2), 350–362 (2009)

Mobile Node Localization Based on Angle Self-adjustment with Mine Wireless Sensor Networks

Wangsheng Fang, Hui Wang$^{(\boxtimes)}$, and Zhongdong Hu

School of Information Engineering,
Jiangxi University of Science and Technology, Ganzhou, China
minishift@163.com

Abstract. In a mining disaster, flooding or a collapse can isolate portions of underground tunnels, severing wired communication links and preventing radio communication. In that condition, wireless sensor networks are an important technology in mine disaster first aid, and mobile node localization is a key technology in WSNs. Compared to other proposed non-self-adjustment localization algorithms, the dynamic self-adjustment localization scheme only uses the time interval adaptive adjustment, which makes it simple to implement and robust to variations in the underground propagation environment. In this paper, a localization algorithm is proposed which improves localization accuracy of mobile node and simplifies mobile node localization algorithm in WSNs. The localization algorithm is based on angle self-adjustment method and the anchor node selection mechanism using the Gauss elimination method. The proposed algorithm has the advantages of requiring little prior information and lower power consumption. The simulation results show that the proposed algorithm is better than SAL in terms of the localization accuracy.

Keywords: Wireless sensor networks · Self-adjustment localization
Mobile node localization · Least square method · Gauss elimination method

1 Introduction

In the safety production process of mine, it has been a crucial task to monitor working environment, miner and vehicle in mines. Many factors exist, including the amount of gas and dust, need to be monitored in underground laneways (which are usually long and narrow with lengths of tens of kilometers and widths of several meters) [1–6]. Therefore, wireless sensor networks need to have full coverages as well as flexible layouts. Removing, installing or rearranging monitoring points are very common. Wireless sensor network localization technology is used to address these problems.

Wireless sensor networks (WSNs) localization algorithm primarily refers to two kinds of localization algorithms. The first one is static localization algorithm which has been maturely developed. It composes of range-based [7] algorithms and range-free [8] algorithms. Range-based algorithms employ measurements, such as RSSI [9], while range-free algorithms employ minimum hops between the unknown node and beacon node and the hop distance to localize the position of unknown node, such as DV-Hop

© Springer Nature Singapore Pte Ltd. 2018
K. Li et al. (Eds.): ISICA 2017, CCIS 873, pp. 513–520, 2018.
https://doi.org/10.1007/978-981-13-1648-7_44

[10]. The second algorithm is dynamic localization [11] algorithm. The dynamic localization mode can be categorized as active localization and passive localization. Steps of active localization are described as follows: the mobile nodes transmit signals to beacon nodes, and then the beacon nodes receive the signal and estimate distances. In this paper, we focus on the active localization method.

Several active algorithms have been proposed to estimating the position of mobile node. One is trajectory prediction algorithm based on Gaussian mixture model was proposed [12]. The algorithm uses the data centralized trajectory model that is more complex and various, which is difficult to describe in a single Gauss process, and the prediction time is long. Hao Chu and Cheng-dong Wu proposed a Kalman framework based mobile node localization in rough environment using wireless sensor network [13]. The algorithm only uses linear regression prediction method to predict the trajectory, and there is no clustering analysis of different trajectory patterns. The prediction error is in hence larger. Therefore, it is not suitable for the location prediction of mobile nodes in underground mine. In this paper, by analyzed the reasons of localization error caused by offset angle estimation and the degraded state when using the least square, a localization method for mobile node is proposed deriving from moving predictor and anchor node selection mechanism. We compared our proposed scheme with some existing solutions via simulations and experiments.

2 Localization Algorithm for Mobile Node in Mine

In this section, self-adjustment localization (SAL) [14] algorithm and the proposed localization algorithm are introduced, respectively. SAL is the basis of the proposed localization algorithm.

2.1 SAL

SAL introduces the time interval self-adaptive in accordance with variation of localization error. The aim is to reduce the time-interval of real-localization when localization error surpasses the fixed localization error. SAL also decreases energy consumption.

The motion state of the mobile nodes in a small time period is slowly changing in underground laneway, which can be similar to that in the before and after. The time interval of the two times localization of the mobile node is moving in a uniform linear motion. Supposing at time $t(i)$, $V_{t(i)}$ is the motion speed of the mobile node, the direction angle is $\theta_{t(i)}$, location is $X_i = (x_{t(i)}, y_{t(i)}, z_{t(i)})$, the interval to next executive location $t(i+1)$ is $\Delta T_{t(i+1)}$, the location $(x_{t(i+1)}, y_{t(i+1)}, z_{t(i+1)})$ of mobile node at $t(i+1)$ can be forecasted from under motion model:

$$\begin{cases} x_{t(i+1)} = x_{t(i)} + V_{t(i)} \bullet \Delta T_{t(i)} \bullet \cos \theta_{t(i)} \\ y_{t(i+1)} = y_{t(i)} + V_{t(i)} \bullet \Delta T_{t(i)} \bullet \sin \theta_{t(i)} \end{cases} \tag{1}$$

Suppose at time $t(i)$, the last time is $t(i-1)$, location is $(x_{t(i-1)}, y_{t(i-1)})$, and estimation of motion speed can be gotten. The following is the equation of $V_{t(i)}$:

$$V_{t(i)} = \sqrt{\frac{\left(x_{t(i)} - x_{t(i-1)}\right)^2 + \left(x_{t(i)}, x_{t(i-1)}\right)}{\Delta T_{t(i)}}} \tag{2}$$

$$\Delta T_{t(i)} = \begin{cases} \arctan\frac{y_{t(i)} - y_{t(i-1)}}{x_{t(i)} - x_{t(i-1)}} & x_{t(i)} - x_{t(i-1)} \geq 0 \quad (a) \\ \pi + \arctan\frac{y_{t(i)} - y_{t(i-1)}}{x_{t(i)} - x_{t(i-1)}} & x_{t(i)} - x_{t(i-1)} \leq 0 \quad (b) \end{cases} \tag{3}$$

2.2   Proposed Mobile Node Localization Algorithm in Mine Tunnel

This localization model is mainly divided into two phases, that is, prediction phase and positioning phase. In this section, movement predictor method and localization method based anchor node selection mechanism will be introduced in detail.

(1) Prediction Phase
According to Node position information of the first two moments, the position coordinates of the current moment can be predicted by using formula (1), (2), (3a) and (3b).

(2) Localization Phase
The mine tunnel can be regarded as a three-dimensional long and narrow tunnel. Apart from a small part of the tunnel, most areas in the tunnel can be considered as long and narrow tunnels. These long and narrow tunnels may exist in curved parts. But most bending radius is larger so that they can be considered as a long and narrow tunnel. The algorithm considers the heterogeneity and mobility of the nodes in the long straight tunnel, and proposes a RSSI location algorithm which is suitable for the mine underground positioning accuracy.

(1) Propagation loss model and RSSI ranging model

In accordance with the theory of Fresnel zone, this propagation area consists of two kinds: multi mode propagation and waveguide propagation. The expression for calculating critical distance is obtained [15], as shown below:

$$d_{NF} = Maximum\left(\frac{h^2}{\lambda}, \frac{w^2}{\lambda}\right) \tag{4}$$

The fundamentals ranging based on RSSI (Received Signal Strength Indicator) is: according to the signal intensity of beacon node and the signal intensity which had been received by the receiving node, path loss can be calculated in transmitting process, and the distance is obtained from transmitting loss by applying a signal model of theory or experiences.

The relationship between path loss and distance d, between the transmitter and receiver follows a power-law relation and can be described by:

$$PL(d) = PL0 + n10\log_{10}\frac{d}{d_0} + X_\sigma \tag{5}$$

where d is the real distance, $PL0$ is the received signal strength at reference distance $d_0(d_0 = 1m)$, n is the path loss exponent, and X_σ is a zero-mean Gaussian noise with standard deviation σ. According to the overall loss, in the underground, the relationship between signal intensity attenuation and distance can be obtained by using RSSI ranging model, shown in the formula (6):

$$PL(d) = 20\log_{10}f + 14.04\log_{10}d + 32.89 \tag{6}$$

According to change of signal, the relationship between signal intensity attenuation and distance can be obtained. In underground roadway, we use RSSI distance channel model of underground roadway, propagation loss as shown in the formula (6), RSSI distance estimation expression for the underground roadway:

$$d = 10^{\frac{(PL(d_0) - PL(d))}{10n}} \tag{7}$$

(2) RSSI localization model

And we will use RSSI to measure the distance d_1, d_2, d_3, d_4 from be measured node (x, y, z) to four selected beacon nodes $((x_1, y_1, z_1), (x_2, y_2, z_2), (x_3, y_3, z_3), (x_4, y_4, z_4))$, and the equation will be established as follows:

$$\begin{cases} (x - x_1)^2 + (y - y_1)^2 + (z - z_1)^2 = d_1^2 & \text{(a)} \\ (x - x_2)^2 + (y - y_2)^2 + (z - z_2)^2 = d_2^2 & \text{(b)} \\ (x - x_3)^2 + (y - y_3)^2 + (z - z_3)^2 = d_3^2 & \text{(c)} \\ (x - x_4)^2 + (y - y_4)^2 + (z - z_4)^2 = d_4^2 & \text{(d)} \end{cases} \tag{8}$$

Simultaneous Eqs. (8a), (8b), (8c), (8d), we can obtain the Eq. (10):

$$AX = b \tag{9}$$

Using the least square method to estimate, it has the form of formula (14a):

$$X = (A^T A)^{-1} A^T B \tag{10}$$

(3) Selection model of anchor node

The plane $S_{2,3,4}$ is formed by anchor node Anchor2, Anchor3 and Anchor4, $H(x_H, y_H, z_H)$ is the pedal from Anchor1 to $S_{2,3,4}$. The distance from Anchor1 to $S_{2,3,4}$ is α.

When the distance between the selected four pairwise anchor nodes is very small, prismatic shape formed by four anchor nodes is also very small. As a consequence, the position error of anchor nodes greatly increases location error. Limiting the size of α is necessary, and thus we can obtain:

$$\sqrt{(x_H - x_1)^2 + (y_H - y_1)^2 + (z_H - z_1)^2} \geq \alpha \tag{11}$$

In accordance with characteristics of roadway, location of unknown node is mainly to estimate its radial position, and then we can obtain:

$$y_1 - \min(y_n) \leq \alpha \leq$$
$$\sqrt{(x_1 - \max(x_n))^2 + (y_1 - \max(y_n))^2 + (z_1 - \max(z_n))^2}(n = 2, 3, 4) \tag{12}$$

From Eq. (14a), we can see that the principle selecting the anchor node based on the RSSI location under the mine is that (1) the selected anchor nodes are in the different plane, and (2) the distance from one of the four anchor nodes to plane that is composed of other three anchor nodes is greater than or equal to α.

(3) The Self-adapting Adjustment for Mobile Nodes

(1) Time adaptive model for movement nodes

Now error between estimated value and measured location is calculated:

$$E = \sqrt{\left(x_{t(i+1)} - x'_{t(i+1)}\right)^2 + \left(y_{t(i+1)} - y'_{t(i+1)}\right)^2 + \left(z_{t(i+1)} - z'_{t(i+1)}\right)^2} \tag{13}$$

$$\Delta T_{t(i)} = \begin{cases} \omega_1 \bullet \Delta T_{t(i-1)} + \omega_2 \bullet \Delta T_{\min} & E \geq E_{th} \quad \text{(a)} \\ \omega_1 \bullet \Delta T_{t(i-1)} + \omega_2 \bullet \Delta T_{\max} & E \leq E_{th} \quad \text{(b)} \end{cases} \tag{14}$$

where, ω_1 and ω_2 are defined weighed value which can be adjusted to. It shows that ω_1 and ω_2 should be defined: $\omega_1 = \omega_2 = 0.5$.

(2) Angle adaptive model for moving of nodes

The positioning angle offset adjustment can reduce the positioning error, when $y_{t(i+1)} - y'_{t(i+1)} \geq 0$, then, the mobile node is moving in the direction of the forward, when $y_{t(i+1)} - y'_{t(i+1)} \leq 0$, The mobile node is moving in the direction of the back, that is:

$$\theta_{t(i)} = \begin{cases} \pi & y_{t(i+1)} - y'_{t(i+1)} \geq 0 \quad \text{(a)} \\ -\pi & y_{t(i+1)} - y'_{t(i+1)} \leq 0 \quad \text{(b)} \end{cases} \tag{15}$$

Where, $\theta_{t(i)}$ is the angle between the mobile node and the radial coordinate $y_{t(i+1)}$ or $y'_{t(i+1)}$.

(4) Localization Algorithm Process
In this section, we describe the proposed mobile localization method; the steps of the proposed mobile localization method are described in detail, as follows:

(1) When mobile nodes need to identify their own position, to around beacon node sends positioning request packet; When a beacon node receives a request from a mobile node, it returns a message packet to the mobile node: nodes ID, coordinates and its location information; after the mobile node receives a packet of beacon nodes information, recorded with the average value of the same one beacon RSSI value; When the mobile node receives N beacon nodes information that their signal intensity scale out the threshold, it will be in accordance with the RSSI value from large to small sort and its numbers is 1,2,…,N.

(2) Using N recorded the beacon nodes RSSI value and the path attenuation model, from formula (7) we can obtain distance between the beacon nodes and the mobile node.

(3) If numbers 1–4 corresponding to the anchor nodes do not meet the Eqs. (10) and (11), one can return this process to step 3, due to the sort of 1st and 2nd RSSI value is larger, and retain arranging the 1, 2 RSSI value of anchor nodes, in order to reduce the localization error caused by distance error from RSSI value of low recognition. Numbers 1–4 corresponding to the anchor node does not satisfy formula (10) and (11). Its reason is that any permutation and combination of three anchor nodes are collinear, and distances between anchor nodes are too close.

(4) The least square method is used to calculate the coordinates of the mobile nodes.

(5) Calculation of error E. Comparison between E and E_{th}, if $E \geq E_{th}$, from formula (14a) obtaining the next time interval $\Delta T_{t(i)}$, and when $y_{t(i+1)} \geq y'_{t(i+1)}$, from formula (14b) obtaining the orientation angle of the next moment, and when $y_{t(i+1)} < y'_{t(i+1)}$, from formula (15a) obtaining the orientation angle of the next moment, otherwise from formula (15b) obtaining the next time interval $\Delta T_{t(i)}$, and when $y_{t(i+1)} \geq y'_{t(i+1)}$, from formula (14a) obtaining the orientation angle of the next moment, and when, from formula (14b) obtaining the orientation angle of the next moment.

(6) Localization end, the next position, repeating first to fourth steps.

3   Simulation

Using MATLAB to simulate the tunnel environment, this algorithm is simulated and compared with the SALMN algorithm. The place about $300\,m \times 8\,m \times 6\,m$ is selected, in this space 30 anchor nodes are deployed, the maximum communication range of the anchor nodes is 30 m. Only one mobile node is selected to simulate localization for the purpose of observing test results. The result of multiple nodes is the superposition of a single simulation node, so we do not have to test repeat. Suppose if initial location of mobile node is (0, 4, 0), then moves for the Fig. 1, the simulation lasts 100 s.

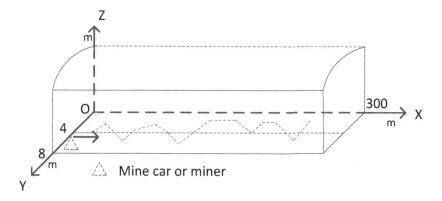

Fig. 1. Random motion trail of movement node

Figure 2 illustrates the localization error in each time index for this paper algorithm and self-adapting dynamic localization algorithm. In comparison with self-adapting dynamic localization algorithm, the proposed method owns the lowest localization error in most time indexes. The probability of locating error this paper falls within the range of 1 m is 90%, and the adaptive algorithm is only 62%. In general, in the application of underground personnel positioning, the localization accuracy of the algorithm is better than that of the self-adapting dynamic localization algorithm.

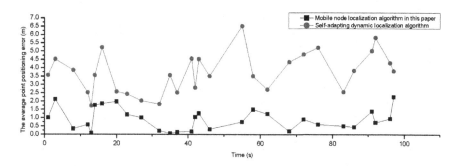

Fig. 2. Comparison of point position error of two algorithms

4 Conclusion

In this paper, an accurate mobile localization method has been proposed for wireless sensor network in mine environment using the angle self-adjustment prediction method and the anchor node selection mechanism. This paper firstly adopts the theory of Fresnel zone to determine wireless signal propagation area. Within identified region, empirical model is used to calculate the signal propagation distance, and anchor node selection mechanism selects anchor node is used to select anchor node, and self-adjustment prediction method is used in position predictor. Finally the least square method is used to estimate the position of mobile node.

Acknowledgment. This research was supported by the National Natural Science Foundation of China and funded by the Ministry of Education (61562038).

References

1. Bellazreg, R., Boudriga, N.: DynTunKey: a dynamic distributed group key tunneling management protocol for heterogeneous wireless sensor networks. EURASIP J. Wirel. Commun. Netw. **2014**(1), 1–19 (2014)
2. Bhattacharjee, S., Roy, P., Ghosh, S., et al.: Wireless sensor network-based fire detection, alarming, monitoring and prevention system for Bord-and-Pillar coal mines. J. Syst. Softw. **85**(3), 571–581 (2012)
3. Cheng, B., Zhou, P., Zhu, D., et al.: The complex alarming event detecting and disposal processing approach for coal mine safety using wireless sensor network. Int. J. Distrib. Sens. Netw. **2012**(6), 1319–1322 (2012)
4. Li, M., Liu, Y.: Underground coal mine monitoring with wireless sensor networks. ACM Trans. Sens. Netw. **5**(2), 777–781 (2010)
5. Li, M., Liu, Y.: Underground structure monitoring with wireless sensor networks. In: Proceedings of the 6th International Conference on Information Processing in Sensor Networks, 2007, pp. 69–78. Springer, Heidelberg (2007)
6. Zhang, S., Cao, J., Li, C., et al.: Accurate and energy-efficient range-free localization for mobile sensor networks. IEEE Trans. Mob. Comput. **9**(6), 897–910 (2010)
7. Aarthi, S., Sankaran, L.: A scalable geographic routing protocol for mobile ad hoc network with grid positioning system. Int. J. Comput. Appl. **88**(12), 12–15 (2014)
8. Qiao, S., Jin, K., Han, N., et al.: Trajectory prediction algorithm based on Gaussian mixture model. J. Softw. **26**(05), 1048–1063 (2015)
9. Chu, H., Wu, C.D.: A Kalman framework based mobile node localization in rough environment using wireless sensor network. Int. J. Distrib. Sens. Netw. **2015**(5), 1–9 (2015)
10. Emslie, A., Lagace, R., Strong, P.: Theory of the propagation of UHF radio waves in coal mine tunnels. Antennas Propag. IEEE Trans. **23**(2), 192–205 (1975)
11. Sun, Z., Akyildiz, I.F.: Channel modeling and analysis for wireless networks in underground mines and road tunnels. IEEE Trans. Commun. **58**(6), 1758–1768 (2010)
12. Zhang, Y.P., Zhang, W.M., et al.: A hybrid mode for propagation loss prediction in tunnels. Acta Electronica Sinica **29**(9), 1283–1286 (2001)
13. Forooshani, A.E., Bashir, S., Michelson, D.G., et al.: A survey of wireless communications and propagation modeling in underground mines. IEEE Commun. Surv. Tutor. **15**(4), 1524–1545 (2013)
14. Hrovat, A., Kandus, G., Javornik, T.: A survey of radio propagation modeling for tunnels. IEEE Commun. Surv. Tutor. **16**(2), 658–669 (2013)
15. Tan, Z., Zhang, H.: A modified mobile location algorithm based on RSSI. J. Beijing Univ. Posts Telecommun. **36**(3), 88–91 (2013)

Research on the Development Strategy of O2O e-Commerce in Traditional Retail Enterprises

Sisi Li[1(✉)] and Suping Liu[2]

[1] GuangZhou Huashang Vocational College, Zengcheng, Guangzhou City 511300,
Guangdong Province, China
2676501836@qq.com
[2] Department of Computer Science, Guangdong University of Science and Technology,
Dongguan City 523000, Guangdong Province, China
457789090@qq.com

Abstract. In recent years, China's traditional retail enterprises are looking for transition, to develop the O2O e-commerce, but because of enterprise resources, capital and experience are inadequate in the transformation, caused the risk of transformation is high, and the enterprise innovation ability is insufficient, mode is single and lack of experience. Therefore, this paper put forward that traditional enterprises in the process of transformation need to improve their management service level, choose differentiation strategy integration channel and improve the consumer experience to realize the transformation of enterprises.

Keywords: Traditional retail enterprises · O2O · New retail

With the rapid development of e-commerce many traditional retail enterprises face the difficulties in operation, the management costs of enterprises are raising, and consumer's habits change lead to the contraction in business, etc., traditional retail enterprises began to transition, focus on the development of O2O business. Jack Ma put forward the concept of "new retail" in 2016, which also provided ideas for the transformation of traditional retail enterprises.

1 O2O E-Commerce

1.1 The Background of E-Commerce

O2O [1]. Based on Electronic Technology and based on the theories of business. That is, under the open network environment of the Internet, the confirmation of the transaction information is completed. Based on various forms of display, the transaction mode of the consumer choice, transaction and other behaviors is finally completed under the line. The bottom information exchange can keep the advantages of e-commerce in time, accuracy and convenience, and the transaction link maintains the traditional trade in the reality of transactions, cash settlement and other characteristics, it is an important business model that is inevitable in the development of the concept of Internet of things. It is another wave of e-commerce after B2C and C2C.

© Springer Nature Singapore Pte Ltd. 2018
K. Li et al. (Eds.): ISICA 2017, CCIS 873, pp. 521–530, 2018.
https://doi.org/10.1007/978-981-13-1648-7_45

Across the globe, car rental, travel and information service companies are the first companies to use O2O. CTRP, founded in 1999, is the earliest O2O model company in china. Consumers can browse the latest travel information online for selection, booking, etc., and then pay offline and enjoy the service. Recently, "shake cart", "didi taxi" and other taxi information services also belong to the O2O model. The O2O model has begun to be favored by Bravo and Hem [2], and the traditional retail business. These enterprises are aware that online and offline e-commerce in favor of information, logistics, inventory and resource sharing, It is a good way out for online enterprises to get out of pure online trading and the transformation and development of traditional retail enterprises, It will become a promising and competitive e-commerce development model and represent the new direction of the future e-commerce development. This is why Sunning, Gome and other traditional retail enterprises to "online and offline coverage of the whole network" as the main direction of future development, and more and more enterprises began to try this new e-commerce model.

1.2   O2O Transaction Process

"O2O" model is a business model which combines online virtual economy and offline entity store operation. The heart of this model is to bring online consumers to the real store, consumers can choose goods and services, transactions, payment, settlement online, and then go to enjoy services offline (Zhang [3]). According to Ma [4] Online to offline refers to the definition, through wired or wireless Internet provide sales infor-mation, gather effective purchasing groups, and online payment, with various forms of credentials offline, That is, real world goods or service providers do their consumption. Obviously, in Online offline, Internet has become the front line of consumer transactions,

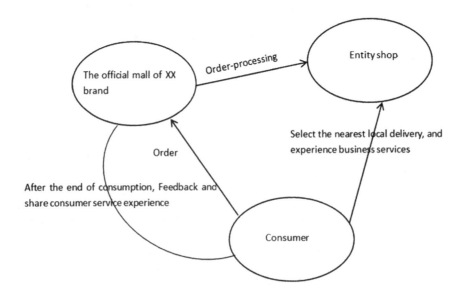

Fig. 1. The diagram of the O2O schema

so that offline services can showmanship online, and consumers can chose service online. The diagram of the O2O schema is shown in Fig. 1.

2 The Feasibility of O2O Transformation in Traditional Retail Industry

2.1 The Continuous Increase of Traditional Retail Costs Made the Traditional Physical Retail Industry Had to Transition

The cost of traditional retail industry keeps rising in China, especially the rising of labor costs more than the enterprise's growth, [5] China's the labor costs of retail enterprise rose by 4.2% in 2016, more than the increase in retail sales. In addition, along with the increase of labor costs, there is also the company's rent, management expenses, financial expenses and sales expenses. The rising costs have further squeezed the profit margins of traditional retail businesses, and even many enterprises have had to close stores in the face of business difficulties. For example, in 2014–2016, traditional retail enterprises closed 1,075 stores, and 833 stores were closed in 2015 alone. In 2016, many enterprises tried to transition, the number of shops closed narrowed, and the first half of 2016 closed 17 stores.

2.2 O2O the Development of Online Retail Industry Is Fast and Stable, and the Enterprises Both Online and Offline Are Tending to Develop O2O

In recent years, the online retail industry has developed rapidly. In 2016, online retail sales volume was RMB5.16 trillion in China, increased by 26.2% over 2015, but the growth rate has narrowed. In the online retail markets, the online retail sales of physical goods are RMB4.19 trillion, accounting for 2.6% of consumer goods. Overall, China's retail industry is still dominated by physical retail. With the development of the online retail market becoming more stable, e-commerce retail industry are looking for transformation, hoping to improve the service level and the experience feeling of consumers through the construction of O2O retail model [6]. At the same time, many physical retail enterprises are also seeking to cooperate with e-commerce enterprises to carry out O2O e-commerce business. Jack Ma pointed out that e-commerce will be disappeared in the next 10 years and 20 years, whether physical stores or e-commerce enterprises need to develop the business online and offline with logistics, that is, new retail.

This paper mainly studies the application of O2O electricity supplier Yonghui supermarket and rainbow mall.

A. O2O fusion of Yonghui supermarket

December 8, 2015, Yonghui supermarket and Jingdong to achieve successful business cooperation, with the help of the Jingdong mobile APP to carry out online business. First, by virtue of its own advantages in purchasing and warehousing, and combining with the Jingdong distribution network, resource sharing is realized. Second, Yonghui supermarket and Jingdong to share customer information in accordance with the level of orders. Through cooperation to achieve complementary advantages of resources, in the supply chain management business docking specifically see Fig. 2.

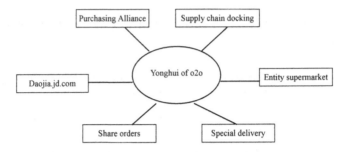

Fig. 2. O2O fusion status of Yonghui supermarket

B.   O2O fusion of rainbow supermarket

Rainbow Department store since 2010, the development of e-commerce on the line of the flow of resources, channel function, management system, organization structure and other aspects of the basic O2O fusion depth integration. At present, the stereoscopic e-commerce model of rainbow scarf + WeChat + micro products has been formed. Rainbow scarf self-platform (PC terminal, mobile terminal and mobile terminal) rainbow micro products at the same time using the WeChat platform, the establishment of public service, the rainbow, combined with the line of the rapid development of physical stores, has formed the basic scene of O2O fusion. Specifically see Fig. 3.

Fig. 3.   O2O fusion status of rainbow supermarket

2.3   Traditional Enterprises Have a Great Advantage in Carrying Out O2O Model

Online retail sales account for a very low percentage of retail sales across the whole society, but its growth began to slow, this is mainly due to online retail itself has some problems, such as product quality is not high, logistics cost is high, the logistics system is not perfect, consumer experience feeling is not good [7]. However, these problems of online retail are the advantages of physical retail enterprises. Therefore, traditional enterprises have a great advantage in carrying out O2O model, and traditional physical stores can enhance the experience feeling of consumers. Based on a physical store to carry out the O2O mode, can also solve the problem of logistics the last kilometer, consumers can easily come to pick up the goods or experience, also can take the stores as the center, to carry out the small range of distribution, this can save the inventory cost and delivery cost of enterprise to a certain extent, etc. In addition, traditional retail enterprises can obtain identification resources through the original supply chain, give full play to their advantages, and then it can easily carry out the O2O business with appropriate training.

3   Problems Existing in the Traditional Retail Industry in Carrying Out O2O

In recent years, many traditional retail enterprises have been transformed, but there are still some problems in the process of transformation.

3.1   The Enterprises Are Lack of Relevant Resources and Experience, the Risk of Transformation Is High

Since 2014, many traditional retailers have tried to transform to the O2O model in China; however, the success of the transformation is not much. It mainly depends on the domestic retail enterprises don't have resources and experience to carry out e-commerce business, in carrying out O2O mode, some enterprises simply used the online advertising, sent coupons, and failed to achieve O2O mode. In order to obtain the sales volume, some enterprises carried out group buying online, and neglected the most important customer experience in O2O model. Some enterprises could not handle the channel conflicts of traditional channels and e-commerce channels. Some enterprises caused many problems due to the service level of logistics can't keep up with O2O development. For example, [8] in 2015, Wanda group had to shut down 10 department stores because of its difficulties in carrying out O2O business, and it compressed 25 stores that were not operated well. Carrying out the O2O mode in traditional enterprises, often requires constant investment, not only need to do online marketing sales well, but also need to develop a complete set of offline products, logistics, service level and so on, many enterprises don't have specific related resources, service level and experience to carry out the O2O business, blindly imitate others to carry out O2O and trigger a series of transformation risk.

3.2 In the Process of Transformation, There Is Often Lack of Innovation and Model Is Single

Traditional retail enterprises carrying out O2O mode in our country, many of them are simple in electric business platform or conduct WeChat microblogging marketing, not fully understand the properties of these platforms and tools, simply copy others to do, lack of depth innovation, caused a series of problems, such as the companies' O2O pattern is similar, lack of competitiveness, etc. In the end, the traditional retail enterprises have no clear understanding of the business model and lack of innovation ability when they carry out O2O model.

3.3 There Are Still Problems in Users' Experience Feeling After the Transformation

Traditional retail enterprises compared with online retail, the most advantageous is the customer experience. However, many traditional retail enterprises transformed to O2O model, in order to save cost and solve the channel conflict to implement the online and offline differentiation strategy and ignore the user experience. The most typical is some stores provide online sales of goods, the customer come to store to experience, play at a discount, or poor quality of the product, or poor service levels, the results caused consumer dissatisfaction, and then affected online reviews, even influenced the traditional physical shop sales.

4 The Options with Transition Modes of Traditional Enterprise

Dongqiang Guo and Youcheng Chen (2015) point out that the development of O2O model is a business process that requires collaboration and interaction, this process is quite complex. O2O operating companies act as a bridge between customers and businesses within the region. We can use the method of analysis of customer consumption data and browsing, personal information and purchase intention are analyzed, which can be used to the online platform to achieve the purchasing behavior, and complete the online payment. Furthermore, Enterprises and local offline merchants to participate in order inquiry and preparation activities, at the same time, O2O enterprises can provide the corresponding marketing plan, thereby promoting its maximum commercial value, get bigger business opportunities. In addition, customers can purchase goods and service experience in the local offline business, at the same time, it can feedback the purchase experience and opinions and suggestions to the O2O enterprises through comments. The enterprise will be based on feedback in-depth analysis and tracking, and thus objectively evaluate the credibility of local businesses offline. At the same time, consumers can also through the online evaluation and reputation and other relate information, to achieve their own purchase experience accumulation, in order to provide a reference for the next shopping activities. According to the difference of service provided by this model and the degree of participation, the O2O model is divided into three modes: transaction service, online drainage and resource integration.

4.1 Trading Services

Enterprises can not only create independent e-commerce companies, but also can cooperate with the third party platform, so as to transform the enterprise to the O2O of transaction services. At present, in the field of Global trade, a number of commodities appear O2O trading platform. This kind of platform mainly realizes the profit through the transaction commission way to the merchant. In our country, the typical transaction service class O2O has Juhuasuan, the Meituan and so on. According to statistics, the number of users who use this service platform in China is less than 40%, which indicates that there is still a lot of room for development.

4.2 Offline Drainage

The model can effectively solve the problem of information asymmetry in the early development of the enterprise, the enterprise will not interfere with the advantage of traditional business, the enterprise can make full use of the characteristics of Internet spread widely, its products will be produced or services to more consumers, to achieve the next line of drainage. It is worth mentioning that, through this model, enterprises can promote customers to conduct shopping quickly, and can enjoy good service online. Normally as long as the products produced by enterprises can realize online sales, they can be sold through this mode, the mode transformation is relatively simple, and the applicability is stronger.

4.3 Resource Integration

When the development of O2O model matures, it will produce resource integration model, which is also an important symbol of the overall transformation of enterprises. The transformation mode usually occurs in the traditional retail enterprises, usually, the smaller the scale of the enterprise, the faster the transformation will be, and the corresponding losses will be smaller. It is worth noting that the key to the development of the model is to realize real-time and accurate docking between data offline and online, some traditional companies have set up electricity supplier company, not only the integration of the resources of a line of the unit, but also to its store for effective integration. The traditional business transformation O2O occurred in the food service industry, covering the general merchandise, service and other life service industry and other industries, the type represents the O2O mode in the future direction of development, which can provide the perfect solution for consumers, and can effectively integrate various resources and realize seamless docking, and thus O2O model into every aspect of the production of the sales process.

5   Countermeasures and Suggestions for Carrying Out O2O in Traditional Retail Industry

5.1   Improve and Strengthen Electronic and Information Construction of the O2O Retail Enterprise

The level of electronic and information construction is also an important aspect of O2O retail stores, because it directly affects the O2O model to create value for consumers. For example, you can use big data as information tool to achieve online and offline consumer preference information sharing. When consumers choose online shopping channels, Online channels can analyze their potential needs based on their previous consumption traces, and employees in offline stores can recommend what they might need; When consumers choose online shopping, according to the line of accumulated information for many years of operation, online to recommend products for consumers, to avoid consumers in the vast amount of information in the choice of goods. After the completion of the entire shopping process, consumers feel the intimate care of enterprises, so to improve the evaluation of retail enterprises, and retail enterprises have achieved improved performance [9].

5.2   To Cooperate with E-Commerce Enterprises, Use Their Respective Advantages, and Form Strong Joint Development Advantage

For some large retail enterprises, they can cooperate with e-commerce enterprises, use their respective advantages, and strengthen their own advantages to build a closed-loop O2O platform. Such as Walmart has many stores, large items, has incomparable advantages in the supply chain; it chooses to work with Jingdong, built the Jingdong-Walmart supermarket online. WalMart's advantages in goods and logistics make up for the shortage of the goods and the high cost of warehousing in Jingdong. At the same time, Jingdong has a large number of e-commerce users and wealth of e-commerce experience, which is what Walmart lacks. The cooperation between the two sides can maximize the development of customer value, reduce the logistics cost of warehousing distribution, and also can improve the customer's experience. Rationally integrate online and offline channels adopt differentiated strategies and avoid channel conflicts.

5.3   Rationally Integrate Online and Offline Channels Adopt Differentiated Strategies and Avoid Channel Conflicts

The integration of online and offline channels is the only way for traditional enterprises to carry out O2O, how to deal with the conflicts of channels is very important. This can adopt the differentiated strategy, including product differentiation, service differentiation, price differentiation and experience differentiation. But when implement the differentiation strategy, should pay attention to hold good degrees, cannot sacrifice user experience for cost savings, especially the product quality is not suitable for big differentiation, can carry out the differentiation in the function, components and services of the products. Such as a certain brand of baby diapers, product quality of online channels significantly worse in the physical shops, lead to part of the online customer experience feeling bad, and feel this brand baby diapers quality is bad, and change to buy other brands, this kind of differentiation strategy will only do more harm than good.

5.4 Strengthen the Management Level and Service Level of the Enterprises, Improve the User Experience Feeling

Traditional enterprises want to carry out O2O model; they should constantly improve their management level and service level, to improve the customer's experience. In particular, enterprises in conducting O2O mode need to perfect the original enterprise management system, strengthen the construction of informatization, train the personnel, put the O2O development into the goals of enterprise strategic development, improve the management and service level of enterprises, to enhance the user experience.

5.5 Pay Attention to Take the Customer as the Center of the Retail Concept

In essence, the traditional retail enterprise belongs to the services, so its essence is to manage customers. To grasp the initiative in the retail competition of fierce locusts, O2O retail enterprises only provide better customer experience and loyalty. Traditional retail mode focuses on commodities and stores as the center, most of the store, especially the department stores rely on rental site management, from the customer from [10], because it is not put the customer first approach, the traditional retail enterprises gradually lost the favor of consumers, so that it makes the rapid development of e-commerce growth. At present, many O2O enterprises have done a lot of efforts in online and offline integration, but they have not got rid of the original situation, the reason is that these O2O retailers do not have a customer centered concept that really goes through the practice and completely ignores how to manage customers. Therefore, the urgent task for O2O retail enterprises is to return to the customer centered concept. Specifically, the user stickiness can be improved through the following three approaches.

A. improve the quality of online goods
B. pay attention to the quality of service offline
C. to create a diversified and secure payment system.

6   Conclusion

All in all, traditional retail industry must be prepared for transformation in China, to greet the arrival of the new retail period, realize their development goals, improve their level of management and service, in the development of traditional O2O mode should also be integrated the traditional channel and e-commerce channel, reasonably use differentiation strategy, enhance the user experience, to ensure that the transformation from traditional retail enterprise to O2O successfully.

Acknowledgment. This work was partially supported by Key platform construction leap plan and major project and achievement cultivation plan project, characteristic innovation project of Department of Education of Guangdong Province of China (Grant Nos. 2014GXJK174). We would like to thank the anonymous reviewers for their valuable comments that greatly helped us to improve the contents of this paper.

References

1. Hu, L.: Analysis of electronic commerce O2O. Merch. Qual. (8), 31 (2011)
2. Bravo, R., Hem, L.E.: Form online to offline though brand extensions and alliances. Int. J. E-Bus. Res. **8**(1), 17–34 (2012)
3. Zhang, W.: Analysis of O2O business model. Chin. Electron. Commer. (3), 2–3 (2012)
4. Ma, Y.: Analysis of O2O e-commerce model in China. Chin. Electron. Commer. (4), 17–18 (2012)
5. Gao, Y.: Reconstruction mechanism and path of traditional retail O2O model in China. Commer. Econ. Res. (16) (2016)
6. Ministry of Commerce: China retail industry development report, July 2016–2017
7. Guo, X., Zhang, J.: The main models and countermeasures analysis on online and offline integration development for retail industry in China. J. Beijing Bus. Univ. (5) (2014)
8. Yang, H.: Feasibility research on O2O model of traditional retail enterprises. Commerc. Econ. Res. (1) (2015)
9. Zhao, Y.: Research on the development and Countermeasures of O2O e-commerce in traditional retail enterprises under the background of transformation. Commerc. Econ. Res. (03), 90–92 (2017)
10. Wang, L.Q.: Research on O2O development strategy of traditional retail enterprises in China. Fujian Normal University (2016)
11. Wang, L.Q.: O2O transformation of traditional retail enterprises in mobile Internet era. J. Fujian Commerc. Coll. (06) (2015)
12. Sang, X.Q., Fang, H.: SWOT analysis and countermeasures of O2O model of traditional retail enterprises in China. E-Bus. J. (12) (2015)
13. Dai, X.F., Zheng, S.R.: Mode selection and development strategy of O2O in retail industry. E-Bus. J. (31) (2015)
14. Wei, J.F.: Analysis and exploration of the application of O2O mode in retail industry in China. E-Bus. J. (22) (2015)

Implementation of Academic News Recommendation System Based on User Profile and Message Semantics

Weiling Li, Yong Tang, Guohua Chen$^{(\boxtimes)}$, Danyang Xiao, and Chengzhe Yuan

School of Computer Science, South China Normal University, Guangzhou, China
{weiling,ytang,chengh,danyangxiao,yuanchengzhe}@m.scnu.edu.cn

Abstract. The development of academic social network has changed the way researchers make access to academic news. In this paper, we propose a news recommendation system to provide user personalized recommendation based on their profile. Our system presents the generation of user profile: (1) extracts the user tag through TextRank; (2) incorporates the timeliness of the news and generates the user profile according to the users' behavior; (3) combines the semantic space vector of news, calculate the similarity of the above to make news recommendations. Finally, the experiment is carried out on the academic social network platform - scholat.com. The experimental results show that the news recommendation system can update the user profile in real time, and this recommendation achieves the due goal.

Keywords: Academic social network · News recommendation
Semantic computing · User profile

1 Introduction

In recent years, the rapid development of academic social network attracts the numerous scholars. Academic social platform is providing an important platform for scholars to exchange ideas, promote scholars' influence and dissemination. Different from traditional social network, the main users of academic social network is scholars, teachers and students. Through the academic social platform, scholars can publish real-time news, show their personal academic information, create teams and courses. On one hand, it can get closer the cooperation between scholars, on the other hand, the cooperation between scholars can be closer [1]. Nowadays, with rapid growth of news posted in academic social network, terabytes of data is generated daily. The overloaded information will bring a new selection challenge to users. In traditional news media, people will pay more attention to the mainstream news, while other valuable news will be always ignored. Improving non-mainstream news attention by exploring the long tail of news will bring unexpected benefits. News recommendation system is an effective way to solve the problem about overloaded information and long tail effect.

News recommendation system aims to find the users' need based on their interests, and provide customized services to users. Therefore, the design and implementation of

© Springer Nature Singapore Pte Ltd. 2018
K. Li et al. (Eds.): ISICA 2017, CCIS 873, pp. 531–540, 2018.
https://doi.org/10.1007/978-981-13-1648-7_46

news recommendation system has become an important research topic in the academic social platform.

Currently, there are some well-known online academic social platforms at domestic and abroad industries, such as Scholat, Baidu academic, Google Scholar, Mendeley, Research Gate etc. Among them, Scholat, as an academic social platform for researchers, covering the functions with academic information management, paper retrieval, academic information, system message, instant messaging, academic calendar and course management. The recommendation function of this platform will help the scholars with the same research interest to understand their latest developments and share academic achievements in real time. Based on Scholat, this paper is introducing the design and develop of news recommendation system for recommending more interested news to our users.

In this paper, Sect. 2 will introduce the related work and technology of current news recommendation. Then in Sect. 3, we will show the system architecture design, and discusses the news recommendation algorithm based on the Scholat. Last, Sect. 4 is the experiment result of Scholat and conclusion.

2   Related Work

Currently, news recommendation system [2, 3] already has a wide range of research at domestic and abroad inducstries. The first news recommendation was proposed by a foreign website named Reddit, which is a social news website. Users can score the published links according to their preferences, and the high score links will be placed on the front page by Reddit. Besides, many other foreign websites such as Yahoo! News, Google News, New York Times also use recommendation algorithms to select the most popular and latest news information. Same as well-known Chinese news websites, such as Tencent News, NetEase News Sohu News. They selectively recommend the news to users according to their user's subscriptions or browsing habits. Among them, NetEase news infers the users preferred tags based on the user's reading and collection of text, and then calculates the similarity according to the freshness and users' tags.

In 2006, Facebook launched News Feed [4]. News Feed used EdgeRank with repeated optimization and iteration to recommend dynamic trends for users. The first edition of EdgeRank, developed by the instruction of Serkan Piantino, scored out every new item with high scores from quantifying intimacy, edge weights and freshness. After 2011, Facebook improved News Feed by machine learning to deal with the rising number of users and advertisers.

At present, the common recommendation methods are collaborative filtering algorithm [5], including content-based collaborative filtering algorithm, model-based recommendation algorithm, and hybrid recommendation algorithm.

Collaborative filtering algorithm, also known as the algorithm based on neighborhood, exploring user's preference by their behavioral analysis with widely used personalized recommendation technology, and then recommending users with similar taste items according to their different preferences [6, 7]. There are two main types of collaborative filtering algorithms, one is item-based and other is use-based. With the user's historical behavior data and user-based collaborative filtering algorithm, we found the users with similar preferences then we will recommend the suitable things to them. According to the user's previous preferences and their behavior, the item-based collaborative filtering algorithm will calculate the similarity between commodities, and then recommend it to users.

Model-based recommendation algorithm [8] takes the existing user's preferences as training samples, then training a model which can predict users' preferences to recommend their goods. This method incorporates users' real-time or recent preference with trained models so as to enhance the accuracy of recommendations.

So far, there are so many recommendation algorithms have been proposed and applied in the industry. After a lot of practice proved, each method has its advantages and limitations. Hybrid recommendation algorithm [9] combines two or more different recommendation algorithms weighting is to generate the final sorting results of the recommendations. The most common method is the combination of content-based and collaborative filtering based, which is not only overcoming the shortcomings of both but also owing the advantages of both. In the famous international recommendation competitions held by Netflix Company, the winning team won by their high accuracy which benefit from the combination of varieties of weighted mixture models and algorithms.

In addition, Zhou et al. [10] propose a method to perform news recommendation based on the combination of semantic analysis and content, which combined the inverse document frequency of synonym sets with semantic similarity and calculated for similarity with WordNet synonym sets.

In this paper, we will talk about the news recommendation according to the collaborative filtering of contents. Firstly, it will adopt the TextRank method to extract keywords from users' information, then combining it with users' behaviors (the users' favorites, reviews, reprints, appreciations and browsing times) to generate users' profile and making use of the profile and news set to perform semantic calculations. At last, it will rank the news according to its semantic similarity to gain the list of recommendations.

3   System Architecture and Algorithm Design

Based on Scholat, we have implemented a news recommendation system. This system falls into 2 parts: user profile analysis module and real-time recommendation module. The architecture diagram of the system is as shown in Fig. 1.

In Fig. 1, we can find that Scholat server record users' browsing history and extract user tags by means of the TextRank as well as store them in database. Based on user tags, the recommendation system will search the similar news through similarity calculation, and recommend TOP-N the most similar news to users.

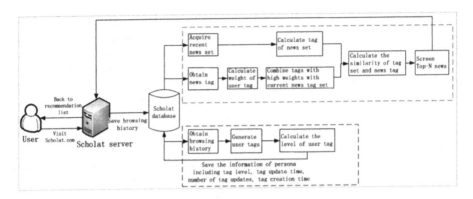

Fig. 1. System framework

3.1 Extracting User Tags Based on TextRank

In this paper, we use the TextRank to generate tags for users, a graph based sorting algorithm based on Google's PageRank [11, 12], which is mainly used to extract document keywords and document abstracts. The basic idea is that the text is divided into a plurality of words for part-of-speech tagging process, then the stop words and symbols will be filtered out and leave the nouns, adjectives, verbs and adverbs. These words will obtain a collection of words as candidate words which will be taken as nodes to build a graphical model, finally it will rank the important words in texts by using the voting mechanism.

The formula of TextRank is as follows.

$$ws(v_i) = (1 - d) + d \cdot \sum_{v_j \in In(v_i)} \frac{w_{ji}}{\sum_{v_k \in Out(v_j)} w_{jk}} ws(v_j) \tag{1}$$

In Eq. 1, $ws(v_i)$ is the score weight of the word v_i, and d is the damping coefficient, which ranges from 0 to 1, and the value is generally 0.85. w_{ji} is the weight between the nodes v_i and v_j in the graph, and $In(v_i)$ is the set of nodes pointing to the node v_i, and $Out(v_j)$ is the set of nodes that the node v_j points to. Before calculating the weight of each word by using the TextRank, each word should be given a default initial value, and the formula 1 is used to iteratively propagate the weight of each node until the convergence condition is reached. Finally, the top K most important words with the highest weights are selected as user tags.

Figure 2 is a flow chart for extracting user tags based on the TextRank.

In this system, each word's weight in each news is calculated by using the TextRank, and the first two words will be added to the user tags. However, the levels of user tags generated by behaviors, such as bookmarking, commenting on, reprinting, clicking the "like" button and general browsing news, carried out by scholars and users are different. Every behavior will correspond to a tag level coefficient a. The correspondence relation of users' behaviors over the news are as shown in Table 1.

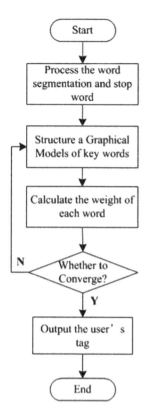

Fig. 2. User tag flowchart based on TextRank

Table 1. Correspondence table of user behavior levels.

User behaviors	Level	Descriptions
Bookmarking	$1.5 \cdot a$	Bookmarking shows users' preference over the article, of which the tag is more important for the user
Commenting on	Positive comment. $1.5 \cdot a$ Negative comment. $-0.3 \cdot a$	Users' comments are divided into two parts, positive or negative, which are judged by the system according to keywords matching. Plus points for positive comments and minus points for negative comments
Reprinting	$0.8 \cdot a$	Reprinting shows users' great appreciation on the articles that they want to share it to other people
Clicking the "Like" button	$0.5 \cdot a$	It shows users' great appreciation on the articles
Browsing	$0.15 \cdot a$	Whether or not reading shows users' favorableness to the articles

The extracted tags are compared with the user's existing tags. If the tag already exists, the level will be updated according to the level coefficient, or a new tag will be added. At the same time, the initial value of the level will be set according to the level coefficient.

In news recommendation, in addition to the level coefficient, we also need to use the creation time, update time, and the number of updates of the tags are saved. Algorithm 1 is an algorithm updating the levels of tags, where *newsTitles* is the headlines of browsing news, *userTags* is the user tag, *a* is the level coefficient of user tags, *updateTime* is the update time of a user's tag, and *count* is the number of update of a user tag.

Algorithm 1. Update Tag Level

```
Input: newsTitles, userTags,a.
Output: userTags, updateTime, count.
FOR ALL newsT in newsTitles Do
    TextRank(newsT) → userTag
    IF(userTag ∈ userTags) THEN
        Update the a, updateTime, count in the database
    ELSE
        1. Insert userTag into the database
        2. Initialize the a, updateTime and count
    END IF
END FOR
```

3.2   Calculate the Weight of a User Tag

For news recommendation, different users may have different needs, like:

1. The system keep on recording the user's browsing history, which will gradually increase as time goes by and the system will extract more and more user tags at the same time. How to select key tags from all extracted tags is a very significant issue.
2. Users' interests might get changed as time goes by, and the system may recommend the news which the user used to be interested in it.
3. Users may be more interested in other news relating to this current news.

This system will introduce weight coefficient for each tag, and the coefficients are related to user behaviors (i.e. tag levels generated by adding in favorite, commenting and other operations, etc.), tag creation time, tag update time, numbers of tag updates, scholar research direction and the tags of current browsing articles.

The storage structure of tags in the database is as shown in Table 2.

Table 2.   Storage structure of tags.

Field	Name
accId	Account number ID
tagName	Names of tags
level	Levels of tags
updateCount	Numbers of updates
updateTime	Latest update time
createTime	Creation time

In order to meet the needs of users, we adopt two methods to obtain user tags. One method is to recommend similar news relate to the current news, while the other is to recommend the news related to users' tags.

Due to there are some news browsing tags which are not existing among the user's tags, the weights of tags extracted from current browsing tags are relatively minor and hard to compare the tags in database when user tag weight calculation is adopted. The tags calculated by means of weight calculation which is not including the tags extracted from current browsing news. Also we want the news recommendation list can include the relative similar news, therefore, here will directly add the tags which is extracted from the current news to the calculation of the similar news.

In addition, we will use the method of weight calculation when we recommend news related to user tags, and the weight calculation formula for recommending news related to user tags are as follows.

$$W = T_l(1 + (C_t - L_t)x + \frac{y}{M_t - N_u} + S \cdot z)$$

(2)

In the formula 2, T_l is the tag level, C_t is current time, L_t is last update time, M_t is the maximum update times of all tags, N_u is numbers of updates, S is the similarity between tag and scholar research direction. x, y and z are all optional parameters, which are used to adjust the ratios of the latest update time, the number of updates and the direction of the scholars' research.

After calculating all the weights, the weights are ranked from the largest to the smallest for searching related news. With the two methods above, the current news tags are combined with the tags obtained by tag weight calculation, so as to meet more requirements of news recommendation.

3.3   Calculate the Similarity of User Tags and News

Vector Space Model is the most commonly used similarity calculation model. Each word is taken as one dimension of a vector, and the word frequency represents the value of this dimension. The similarity between two articles is the cosine of the vector composed by the keywords of the two articles. The closer the cosine is to 1, the closer of intersection angle of two vectors is to 0, and more similar the two articles are.

The similarity formula is as follows.

$$Sim(X, Y) = \frac{\sum_{i=1}^{n} (X_i \cdot Y_i)}{\sqrt{\sum_{i}^{n} (X_i)^2} \cdot \sqrt{\sum_{i}^{n} (Y_i)^2}}$$

(3)

where, X and Y are the spatial vectors of keywords of two articles, $Sim(X, Y)$ represents the similarity of vector X and vector, X_i is the No. i element of vector X, and Y_i, likewise, is the No. i element of vector Y.

The similarity between user tags and news articles is calculated by formula 3, and higher similarity means the user may be more interested in the article.

3.4 Generation of Recommendation

In consideration of the news is full of timeliness, users will tend to prefer the latest news. Therefore, this system will select the news happened in certain period of time as a candidate news set according to the requirements of users.

As the increasing of amount of news, the tags of the users will also increase. When we use user's tags to calculate the similarity of articles, the more tags are, the more calculating vector dimensions will be. On the contrary the less contribution to similarity calculation the user tags, the less weight will make. Therefore, when we calculate similar between tags, we only can use the top K tags with largest weights in the calculation. Based on the formula 3, we calculate the users' tag and the news' set's similarities separately and find there is a corresponding similar value in each news and tag.

Calculating the similarity between the news and user tags, then ranking the similarity from the highest to the lowest with their recommendation coefficients, we will find the recommended Top-N articles to public in the website.

4 Results of Experiment

To verity the arithmetic validity of this paper, user's scanning information data in Scholat has been referred, including the scholar's research situation, team dynamics, conference solicit articles, recruitment information and the instant academic news. During the statistic analysis of the experiment result, Precision and Recall are mainly adopted to evaluate the effect of recommended algorithm. The Precision means the proportion that the number of news recommended and appeal to user divides the total number of news items; the Recall means the proportion that the number of news recommended and appeal to user divides the total news that reader has scanned. The Precision and Recall rate are between 0 and 1. The closer the value is, the better the recommendation algorithm is. So the formula for Precision and Recall is as follows:

As shown in Fig. 3, the *precision* of the recommended news is changed with the number of recommended news. The more the number of recommendations, the lower the user's attention, so the *precision* will decrease.

As shown in Fig. 4, as the number of news items recommended to users increases, the number of recommended news items read by users promotes, and the *recall* of the recommended system also increase.

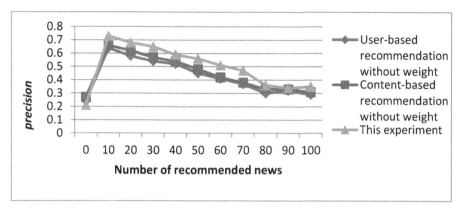

Fig. 3. Recommendation precision of the three algorithms

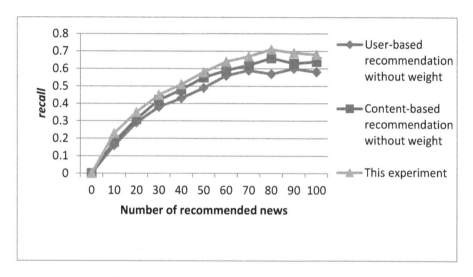

Fig. 4. Recommendation recall of the three algorithms

As can be seen from Figs. 3 and 4, the *precision* and *recall* based on the recommendation algorithm of this experiment is higher than that of traditional user based recommendation algorithms. There are two main reasons behind this: one is for scholar's news recommendation, they are more interested in the recommendation based on their personal browse record than others. The other is to give users more information about what they are interested in, after the weight is attenuated on the importance of the label. Based on this, the recommendation system proposed in this paper has higher *precision* and better recommendation ability compared with other recommendation systems.

5   Conclusion

This paper proposed a news recommendation algorithm for academic social network. It will record users' browsing news, then extract users' tags based on TextRank, for semantic computing by users' profile and news set. Finally gain the recommendation list from news sort according to its semantic similarity. Though the experiment on scholat.com, we proved this method can help user to get their interested news and improved the utilization of news.

Acknowledgement. This work was partially supported by Guangzhou Science and Technology Project (No. 201604046017).

References

1. Xia, Q.J., Li, W.C., Xue, J.J., et al.: New social networking platform for the academic fields: research networking systems. J. Intell. **33**(9), 167–172 (2014)
2. Hopfgartner, F., et al.: Benchmarking news recommendations: the Clef NewsREEL use case. ACM SIGIR Forum **49**(2), 129–136 (2016)
3. Lin, C., Xie, R., Guan, X., Li, L., Li, T.: Personalized news recommendation via implicit social experts. Inf. Sci. **254**(1), 1–18 (2014)
4. Hoffman, L., Glynn, C., Huge, M.: News Feed: A Profile of News Use on Social Networking Sites. Social Science Electronic Publishing (2011)
5. Bobadilla, J., Ortega, F., Hernando, A., et al.: Recommender systems survey. Knowl. Based Syst. **46**(1), 109–132 (2013)
6. Hofmann, T.: Latent semantic models for collaborative filtering. ACM Trans. Inf. Syst. **22**(1), 89–115 (2017)
7. Koren, Y., Bell, R.: Advances in collaborative filtering. In: Ricci, F., Rokach, L., Shapira, B., Kantor, P. (eds.) Recommender Systems Handbook, pp. 145–186. Springer, Boston (2011). https://doi.org/10.1007/978-0-387-85820-3_5
8. Zhang, H.: A random forest approach to model-based recommendation. J. Inf. Comput. Sci. **11**(15), 5341–5348 (2014)
9. Dong, X., Yu, L., Wu, Z.H., Sun, Y.X., Yuan, L.F., Zhang, F.X.: A hybrid collaborative filtering model with deep structure for recommender systems. In: AAAI, pp. 1309–1315 (2017)
10. Zhou, Y., Dai, W.H.: News recommendation technology combining semantic analysis with TF-IDF method. Comput. Sci. **40**(s2), 267–269 (2013)
11. Haveliwala, T.H.: Efficient computation of PageRank. Stanford Db Group Technical report (2016)
12. Balcerzak, B., Jaworski, W., Wierzbicki, A.: Application of TextRank algorithm for credibility assessment. In: IEEE/WIC/ACM International Joint Conferences on Web Intelligence, vol. 1, pp. 451–454. IEEE (2014)

Everywhere Connectivity – Wireless Sensor Networks

An Optimal Sink Placement for High Coverage and Low Deployment Cost in Mobile Wireless Sensor Networks

Qingzhong Liang[1,2] and Yuanyuan Fan[1,2(✉)]

[1] Hubei Key Laboratory of Intelligent Geo-Information Processing,
Wuhan, China
[2] School of Computer Science in China University of Geosciences,
Wuhan, China
{qzliang,yyfan}@cug.edu.cn

Abstract. Reliable communication quality and reasonable cost control are two of important goals of Sink node location problem in modern communication. How to achieve an optimization solution by balancing both the two goals is a difficult tradeoff, especially in mobile wireless sensor networks. Therefore, with the maximum signal coverage as the communication quality goal, a constrained multi-objective optimization model is proposed by optimizing cost and communication quality at the same time. In order to deal with the constraints, we improve the classic multi-objective algorithm NSGA_II and adopt the strategy of binary tournament based on CV violation value to select the parent, so that the superior individuals have a greater probability to participate in the genetic operation. Finally, the problem model and algorithm proposed in this paper are applied to a set of Sink node location problems with different cost and network parameters. Compared with the NSGA_II algorithm, experimental results show that the new algorithm can improve the signal coverage better without much increase in cost. At the same time, the variances of the optimization results obtained in new algorithm are smaller, which means its optimization is more stable.

Keywords: Wireless sensor network · Multi-objective optimization
Sink node placement

1 Introduction

Mobile Wireless Sensor Networks (MWSNs) are characterized by the mobile sensor nodes that continuously observe physical phenomenon [1]. MWSNs have more rapid topology changes due to their moving sensor nodes [2]. The placement of sink nodes is to select a set of requirements in a large number of candidate sink node locations, taking into account the factors such as placement cost, communication capacity and network coverage of sink nodes [3]. Therefore, the placement of sink nodes plays an important role in mobile wireless networks [4]. In order to ensure efficient network service and reliable communication quality, it is necessary to place the sink node reasonably considering the composition of the communication service and the density

© Springer Nature Singapore Pte Ltd. 2018
K. Li et al. (Eds.): ISICA 2017, CCIS 873, pp. 543–551, 2018.
https://doi.org/10.1007/978-981-13-1648-7_47

of the communication. In addition, the rational placement of the sink nodes will be beneficial to the mobility and scalability of the MWSNs.

Due to more than one sink node in a MWSN, each node needs to choose which sink node to communicate with, that has a great impact on the performance of the MWSNs. To extend the network life time of WSNs, early work aiming at minimize transmission delay and energy consumption is used to placing sinks in large-scale WSNs [5]. For protracting the lifetime of WSNs, a particle swarm optimization (PSO) based algorithm, PSO-MSPA, is proposed for placement of multiple-sink in WSNs. In this algorithm, an efficient scheme of particle encoding and novel fitness function is developed for the energy efficiency optimization considering Euclidian distance and hop count from the gateways to the sinks [6]. Since finding the optimal number of the sinks and their locations is an NP Hard problem, a genetic algorithm based optimum sink placement that find out the near optimal results for coverage with the minimum possible sinks has been adopted as a Sink Placement strategy [7].

In this paper, the network coverage and placement cost are considered in placing 4G sink nodes. Firstly, we analyze the placement problem of 4G sink nodes which can be modeled as a multi-objective optimization problem and the factors that should be paid attention to. This paper proposes a problem model that includes two optimization objectives of cost and coverage, and designs a CV_NSGA_II algorithm with the parent selection strategy by means of the binary tournament based on the CV violation value. Finally, the model and the algorithm are applied to the Sink node placement problem, and the eclipse software emulates the signal coverage. The results of multiple experiments show that the new algorithm can improve the signal coverage at the similar cost and achieve more stable optimization results.

2 System Model and Problem Statement

In this paper, we consider a scenario for sink nodes placement in a mobile wireless sensor network, as illustrated in Fig. 1. In this network, a variety of different sensor nodes collect data with their moving carrier such as buses and pedestrians. Then, they send the data to the server by communicating with a fixed sink node. To locate sink nodes reasonably, two factors including network coverage and placement cost are considered.

2.1 System Model

In this paper, sink node placement problem can be regarded as a Set Covering Location Problem (SCLP) [8] and Maximal Covering Location Problem (MCLP) [9]. SCLP are the optimization problems to select a minimal set of service stations, with which every demand point in the target area will be covered by at least one service station and the total cost of building the stations is minimized. Maximum Covering Location Problem (MCLP) or P-cover Problem is the optimization problem how to respond to the maximum demand by P service stations assumed that the number and the service radius of stations are all known.

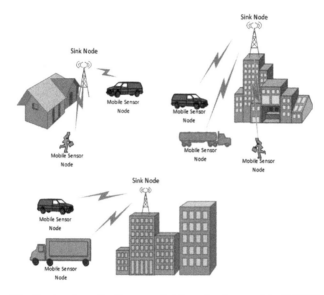

Fig. 1. A schematic diagram of sink node placement in MWSN

It is assumed that BS (Backbone Sinks) represents a set of N sink nodes and MS (Mobile Sensors) represents a set of M mobile sensor nodes. For mobile node i in MS and sink node j in BS, the description of the placement scenario can be expressed as formula (1) and (2).

$$x_j = \begin{cases} 1, & \text{Placing the sink node at j} \\ 0, & \text{else} \end{cases} \tag{1}$$

$$y_{ij} = \begin{cases} 1, & \text{Moble node i is connected to sink node j} \\ 0, & \text{else} \end{cases} \tag{2}$$

The coverage cost value function for the j-th sink node is expressed as formula (3).

$$C'_j = k_1 \sum_{i \in MS_j} P_i + k_2 \sum_{i \in MS_j} P_{ji} + C_j \tag{3}$$

Where, k_1 and k_2 is the cost weighting coefficients of the uplink and downlink power respectively. $\sum_{i \in MS_j} P_i$ is the total power of the mobile nodes covered by the j-th sink node. $\sum_{i \in MS_j} P_{ji}$ is the total transmission power of this sink node, where MS_j denotes the set of mobile nodes covered by the j-th sink node. C_j is the construction cost of this sink node.

The sink node placement problem based on the set-covering model can be described as that a set of sink nodes is selected from the BS so that it can cover all the mobile nodes. Moreover, the total amount of coverage costs calculated based on the number of mobile nodes covered by each sink node and the power consumed is minimized. It is expressed as formula (4).

$$Minimize\left(\sum_{j\in BS} C'_j x_j\right) = Minimize\left(k_1 \sum_{i\in MS_j} P_i + k_2 \sum_{j\in BS}\sum_{i\in MS_j} P_{ji}x_j + \sum_{j\in BS} C_j x_j\right) \quad (4)$$

In addition, in order to satisfy the maximum coverage model, the sink node placement problem can be described as follows. From the candidate sink node set BS, under the restriction of the fund budget (or sink node number), a subset including N0 nodes is selected for the purpose of maximizing the number of covered mobile nodes. It is as the formula (5).

$$Maxmizie\left(\sum_{i\in MS}\sum_{j\in BS} y_{ij}\right) \quad (5)$$

2.2   Problem Statement

In previous studies, the set cover model as shown in formula (4) required that all demand points be covered. However, in the sink node's placement problem, it is not necessary to calculate all mobile nodes within cost. Only when mobile node i is connected with sink node j, the power cost of mobile node i will be included in the total cost. Based on this judgment, the cost optimization objective function of this paper is improved as following formula.

$$Minimize\left(\sum_{j\in BS} C'_j x_j\right) = Minimize\left(k_1 \sum_{i\in MS_j} P_i y_{ij} + k_2 \sum_{j\in BS}\sum_{i\in MS_j} P_{ji}x_j + \sum_{j\in BS} C_j x_j\right) \quad (6)$$

Another optimization objective of the sink node placement problem is to maximize the number of mobile nodes covered. However, in actual networks, the coverage ratio can more directly reflect the signal coverage than the number of connected mobile nodes. Therefore, we improve this optimization goal to maximize the signal coverage, as shown in formula (5.4), where |MS| represents the total number of mobile nodes.

$$Maxmizie\left(\sum_{i\in MS}\sum_{j\in BS} y_{ij}/|MS|\right) \quad (7)$$

It can be seen from the above discussion that the Sink node placement problem includes two optimization objectives: minimizing the cost of placing sink nodes and maximizing the signal coverage, and the two are mutually contradictory. Therefore, the problem is a constrained multi-objective optimization problem.

2.3   Constraints

In order to describe constraints of sink node placement, we define the following variable symbols.

$P_{j,0}$ denotes the maximum dedicated channel transmit power of sink node j.

$P_{j,m}$ is the transmission power of sink node j to mobile node m.

P_m is the transmit power of mobile node m.

$g_{m,b}$ denotes the link gain between the mobile node m and the sink node b.

$P_{j,\eta}$ denotes the receiver noise power at the communication node j, where j can be a mobile node or a sink node.

$\bar{P}_b$ is the transmission power of the common channel allocated to sink node b.

$SIR^{\uparrow}_{(0,m)}$ represents the target SIR for the uplink of mobile node m.

$SIR^{\downarrow}_{(0,m)}$ represents the target SIR for the downlink of mobile node m.

$\bar{\alpha} = 1 - \alpha$ where α is the orthogonalization factor of the downlink.

C_j is the construction cost of a sink node j.

N' is the total number of sink nodes to be set up.

F_0 is the total capital budget.

A mobile node i can only connect to one sink node, so its constraint is as formula (8).

$$\sum_{j \in BS} y_{ij} \tag{8}$$

It is only possible for the mobile node i to connect with the sink node if it is placed at j, then this constraint is described as formula (9).

$$y_{ij} \le x_j \tag{9}$$

The sum of the transmit powers of all the mobile nodes connected by a sink node is smaller than the maximum transmits power of this sink node, described as formula (10).

$$0 < \sum_{i \in MS} P_{ib} y_{ib} < P_{b,0} \tag{10}$$

Constraints on node number and capital budget are described as formula (11) and (12).

$$\sum_{j \in BS} x_j = N' \tag{11}$$

$$\sum_{j \in BS} C_j x_j \le F_0 \tag{12}$$

In addition, for all mobile nodes m and candidate sink nodes b, they must satisfy the uplink and downlink transmit power constraints.

3   CV_NSGA_II Algorithm for Sink Node Placement

NSGA_II algorithm based on the Pareto optimal concept is one of the most famous, most frequently used and best performing multi-objective evolutionary algorithms. However, the algorithm does not consider the constraints in the selection process so as to probably produce infeasible descendants. Consequently, it will have a significant impact on the results of the latter part of the experiment. Therefore, a new selection strategy of tournaments is proposed, using the violation value of the constraint CV as an important selection condition to generate the parents who will participate in genetic operations.

The parent selection strategy by means of the binary tournament based on the violation value CV is as follows.

Two individuals are randomly selected from the t-th generation of population P_t and the better individual will be chosen as one of the parents.

(1) If both of them are infeasible, the individual with the smaller violation value of the constraint will be chosen as the parent.
(2) If both are feasible solutions, one of the individuals is randomly chosen as the parent.
(3) If one individual is feasible and the other is infeasible, the feasible one will be chosen as the parent.

The same strategy can be used to choose another parent. Then the offspring individuals are generated by the two parents via crossover and mutation. Assuming a population size of 2N, 2N offspring individuals can be generated according to this method to form the next generation population P_{t+1}. The algorithm controls the parental quality based on the parent selection strategy mentioned above so that the better individual has a higher probability to participate in the genetic operation. The algorithm CV_NSGA_II is shown in Algorithm 1, which is an improved algorithm NSGA_II with the parent selection strategy based on the violation value.

Algorithm 1. The framework of CV_NSGA_II.
1. BEGIN
2.   Initialize and evaluate P_0, t=0.
3.   while(not shutting down)   {
4.   for(i=1; i<=N; i++){
5.   Produce two patrons $v_i(t)$ and $u_i(t)$ with CV_ParentSelection.
6.   Generate new individuals $v_i(t+1)$ and $u_i(t+1)$ with crossover and mutation.
7. $P_t = P_t \, U\{v_i(t+1), u_i(t+1)\}$;
8.   }
9.   Using NSGA_II selection strategy to generate the next generation population P_{t+1}.
10. t=t+1. }
11. END

4 Performance Evaluation

In order to compare the new algorithm with the traditional NSGA_II algorithm, a number of experiments are designed and implemented in this paper. The Sink node placement problem in this paper is improved based on SCLP and MCLP, which includes two optimization objectives: the minimum sink node cost and the maximum mobile node coverage. The Sink nodes and the mobile nodes are evenly and randomly distributed in the plane of [0,100] * [0,100]. All the parameters in the experiments refer to the simulation parameters in Table 1. Some of the randomly generated parameters follow a uniformly random distribution. Two of the important parameters are the number of sink nodes N = |BS| and the number of mobile nodes M = |MS|.

Table 1. Parameter settings in the simulation experiments.

Parameters	Range	Number of variables
k_1	0.01	1
k_2	0.1	1
$\alpha = 0.4$	$\bar{\alpha} = 1 - \alpha = 0.6$	1
$g_{m,b}$	[0,15]	M * N
$P_{j,0}$	[30,70]	N
$P_{i,j}$	$[0, P_{j,0}]$	M * N
P_i	[1,15]	M
$P_{b,n}$	[−102,−100]	N
$P_{m,n}$	[−106,−104]	M
$\overline{P_j}$	[5,15]	N
C_j	[10,400]	N
$SIR_{0,m}^{\uparrow}$	[−6,−2]	M
$SIR_{0,m}^{\downarrow}$	[−8,−4]	M

Based on this problem model, two groups of different simulation experiments are conducted in this paper. The NSGA-II algorithm and the new CV_NSGA_II algorithm are respectively used for optimization on four different values of (N, M): (3,5), (5,18), (8,36), (10,60). On each value the two algorithms each run ten times to compare the average cost and coverage. The two algorithms have the same parameters: the population size N is 100, the crossover probability is 1.0, the mutation probability is 1/N, and the number of iterations is 10000.

We compare the average cost and coverage achieved by the two algorithms on the different values of (N, M) in Figs. 2 and 3. It can be seen that in the first three smaller dimensions, the variance of the optimization results by CV_NSGA_II algorithm is much smaller than that of NSGA_II algorithm, indicating that the optimization of the former are more stable.

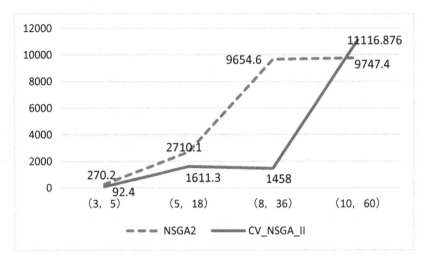

Fig. 2. Comparison chart for cost variance.

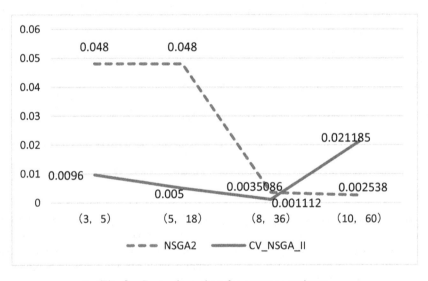

Fig. 3. Comparison chart for coverage variance

5 Conclusion

In this paper, aiming at the actual optimization of the Sink node placement problem, a multi-objective optimization model is proposed to optimize cost and coverage at the same time. The signal coverage instead of the number of connected nodes improves the optimization goal of signal coverage. And a multi-objective algorithm CV_NSGA_II algorithm is proposed, which select a parent through a binary tournament strategy based on the constraint violation value CV, so that the superior individuals have a higher probability of participating in the genetic operation. Experimental results show that the proposed algorithm can improve the signal coverage in the Sink node placement problem when the cost is not much increased. At the same time, the variance of the optimization results obtained by the new algorithm is smaller, which means the optimization of it is more stable.

Acknowledgment. This research was supported by the NSF of China (Grant No. 61673354, 61672474, 61402425, 61272470, 61305087,61440060,61501412), the Provincial Natural Science Foundation of Hubei (Grant No. 2015CFA065). This paper has been subjected to Hubei Key Laboratory of Intelligent Geo-Information Processing, China University of Geosciences, Wuhan 430074, China. It was also supported by Open Research Project of The Hubei Key Laboratory of Intelligent Geo-Information Processing (KLIGIP201603 and KLIGIP201607).

References

1. Munir, S.A., Ren, B., Jiao, W., Wang, B., Xie, D., Ma, J.: Mobile wireless sensor network: architecture and enabling technologies for ubiquitous computing. In: International Conference on Advanced Information Networking and Applications Workshops, pp. 113–120 (2007)
2. Hu, C.Y., Tian, D.J., Liu, C., Yan, X.: Sensors placement in water distribution systems based on co-evolutionary optimization algorithm. In: International Conference on Industrial Networks and Intelligent Systems, pp. 7–11 (2015)
3. Lee, J., Jang, J.Y., Lee, E., Kim, S.H.: Efficient sink location service for prolonging the network lifetime in wireless sensor networks. In: IEEE Consumer Communications and Networking Conference, pp. 290–291 (2016)
4. Kaur, D., Kaur, M.M.: An approach of mobile wireless sensor network for target coverage and network connectivity with minimum movement (2017)
5. Das, D., Rehena, Z., Roy, S., Mukherjee, N.: Multiple-sink placement strategies in wireless sensor networks. In: Fifth International Conference on Communication Systems and Networks, pp. 1–7 (2013)
6. Srinivasa Rao P, C., Banka, H., Jana, Prasanta K.: PSO-based multiple-sink placement algorithm for protracting the lifetime of wireless sensor networks. In: Satapathy, S.C., Raju, K.S., Mandal, J.K., Bhateja, V. (eds.) Proceedings of the Second International Conference on Computer and Communication Technologies. AISC, vol. 379, pp. 605–616. Springer, New Delhi (2016). https://doi.org/10.1007/978-81-322-2517-1_58
7. Ghosh, S., Snigdh, I., Singh, A.: GA optimal sink placement for maximizing coverage in wireless sensor networks. In: International Conference on Wireless Communications, Signal Processing and Networking (2016)
8. Šarac, D., Kopić, M., Mostarac, K., Kujačić, M., Jovanović, B.: Application of set covering location problem for organizing the public postal network. Promet - Traffic Transp. **28**(4), 403–413 (2016)
9. Stiffler, S.A.: Application of the maximal covering location problem to habitat reserve site selection. Int. Reg. Sci. Rev. **39**(1), 18–32 (2016)

Wireless Sensor Network Time Synchronization Algorithm Overview

Chunqiang Liu[1], Haijie Pang[1], Ning Cao[1(✉)], Xinze Li[2], and Dongchen Xu[1]

[1] Qingdao BinHai University, Qingdao 266555, China
`ning.cao2008@hotmail.com`
[2] Yunnan Open University, Kunming 650000, China

Abstract. Time synchronization is the key supporting technology for wireless sensor networks in many applications such as Internet of things, environmental monitoring and target tracking. On the basis of summarizing and analyzing the time synchronization algorithm of two-way message exchange and the distributed time synchronization algorithm, combining with the research status at home and abroad, the future research direction of time synchronization algorithm is given.

Keywords: Wireless sensor networks · Time synchronization
Bidirectional message exchange · Distributed consistent clock synchronization

1 Introduction

Since 1990s, wireless sensor networks have been developed rapidly, and are widely used in forest, architecture, military battlefield and other perception monitoring area [1]. The time synchronization technology is an important underlying wireless sensor network technology. Its purpose is to provide a common timestamp to local time nodes in the network, so as to realize the localization [2], node scheduling mechanism and node mode switching [3] network functions.

Prior to the emergence of wireless sensor networks, time synchronization has been extensively studied. Typical time synchronization is atomic clocks, such as GPS (Global Positioning System). Due to the energy consumption and price constraints of sensor nodes, it is obviously infeasible to install a GPS receiver for each sensor node. In addition, the reception of GPS satellite signals is affected by line of sight, and it can only be used outdoors. Because the hardware clock of sensor nodes is often designed simply, and there is a lot of clock drift in the application, the sensor network often needs complex time synchronization algorithm. Moreover, compared with wired networks, multi hop phenomenon of wireless sensor networks makes inter node time synchronization problem more complex, because it is not suitable for a simple C/S time synchronization method in wireless sensor networks.

With the research of time synchronization in wireless sensor networks, scholars have put forward many different algorithms, there are mainly centralized synchronization and distributed consistent time synchronization, in which the centralized synchronization algorithm adopts the method of selecting reference node to realize the time

© Springer Nature Singapore Pte Ltd. 2018
K. Li et al. (Eds.): ISICA 2017, CCIS 873, pp. 552–561, 2018.
https://doi.org/10.1007/978-981-13-1648-7_48

synchronization of all nodes in the network. The distributed consistent time synchronization does not need to select the reference node, and simultaneously realizes the clock synchronization to the entire network node. Bidirectional message exchange is the main mode of centralized synchronization algorithm. Based on the existing research results, this paper mainly discusses bidirectional message exchange time synchronization algorithm and the distributed consistent time synchronization algorithm.

2   A Bidirectional Message Exchange Algorithm Based on Sender-Receiver

At present, many scholars study the algorithm is a bidirectional message exchange synchronization algorithm based on sender-receiver, this is a classic communication mechanism proposed by Ref. [4], communication time for information between two adjacent nodes, and is widely used in the Ref. [5–7]. The two-way information exchange mechanism is shown in Fig. 1.

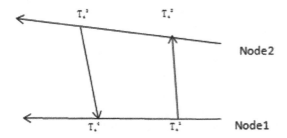

Fig. 1. Two-way information exchange mechanism

In the two-way information exchange process, in the first round of I in the process of information exchange, the reference node A sends a T_i^a include the time sent to the common node B, and the common node B records the access time in the time T_i^b, in the time T_i^c sends a confirmation message to the reference node A, this information contains in the T_i^b and T_i^c time stamp. Finally, the reference node A receives the confirmation information from the common node B at T_i^d, and the end of I round of information exchange is completed. This process repeats N times, and N indicates the number of synchronization required. In the process of clock synchronization information exchange, various communication delays affect the transfer of clock messages between nodes, which makes the clock synchronization process more complex. The communication delay can be divided into fixed and variable parts, for the delay of the variable part, many scholars use the probability density function to put forward the stochastic delay model in various wireless sensor networks, including the Gaussian distribution, exponential distribution, gamma distribution, to estimate the accuracy of clock synchronization.

Based on the bidirectional information exchange mechanism, the Ref. [8, 9] proposes a maximum likelihood estimation algorithm for the estimation of clock offset in Gaussian stochastic time delay model, and the obtained parametric mean square error is close to that of Cramer-Rao lower bound. In the literature about node localization, [10, 11] takes the fixed delay part as a parameter to estimate. The Ref. [11] combined minimum variance unbiased estimation and maximum likelihood estimation algorithm, by simplifying the existing bidirectional information exchange mechanism, estimating the clock offset and clock frequency drift, the algorithm improves the precision of clock parameter estimation.

Reference [12] converts the joint maximum likelihood estimation problem into a linear programming problem (LP) under the exponential stochastic delay model. Although the solution of the linear programming problem is to guarantee the global optimal solution, it is inefficient to solve the linear programming problem directly. With the geometric analysis of the new feasible domain, the equivalent maximum likelihood estimator of low complexity is proposed. Under the same delay model, Ref. [13] proposed a linear clock model that can work in both symmetric and asymmetric links, and does not require the exponential distribution of the variable delay part obeying the same parameter. The paper optimizes the mean square error for the joint estimation of clock offset and clock frequency drift, which is suitable for the case that the number of observation sets is limited or the stochastic delay is large.

The Jackknife deviation correction proposed by Ref. [14] corrects the maximum likelihood estimation under exponential distribution, and compared with the bootstrap deviation correction, the algorithm proposed in Ref. [15] has a lower deviation, but the variance is relatively large, and the larger mean square error is produced. In Ref. [15, 16], the accuracy of time synchronization is improved by the method of bootstrap deviation correction. Ref. [17] proposes a time synchronization algorithm of WSN, which combines bidirectional synchronization and single synchronization, and on the basis of clustering, two-way time message synchronization is adopted between cluster head node and base station node, and the time frequency drift and clock offset are corrected by parameter estimation and minimum variance unbiased estimation. In the cluster, one-way time message synchronization is used between the cluster head node and the member node, and the linear regression is used to estimate the time frequency drift and the clock offset. This method improves the accuracy of time synchronization, reduces the number of synchronizations, and reduces the energy consumption of nodes, has long synchronization period and the low communication cost. Similarly, the Ref. [18, 19] adopts a close sender-receiver message exchange synchronization algorithm and a loosely dispersed message time synchronization algorithm in the cluster, which supports the scalability of the network, reduces the error of time synchronization, enhances the efficiency of energy utilization and increases the network throughput. However, the Ref. [18] only supports the GCF algorithm, which is not suitable for other TDMA scheduling algorithms.

3   Two-Way Broadcast Message Exchange Time Synchronization Algorithm

Two-way broadcast message exchange (PBS) mechanism is made by Noh in the Ref. [20], this mechanism combines the sender-receiver and receiver-receiver two-way message exchange, in the proposed synchronization method, a subset of sensor nodes eavesdrop on the time synchronization information exchange of a pair of sensor nodes.

Due to power constraints, the sensor's communication range is strictly limited to the radius depending on the transmission power (radio geometry) in the circle. Each node within the inspection zone can receive messages from reference node P and Node A (a paired synchronization is performed between reference node P and Node A using two-way time information exchange). All nodes within the inspection area can then receive a series of timestamp information synchronization messages that contain pair synchronizations. Nodes in the region uses this information, and can also be synchronized to parent node P with an RBS-like approach, with no additional time message transmission. Node P and Node A can provide synchronous beacons for all nodes located near them, and are therefore called super nodes.

In order to select the appropriate reference node and eavesdropping node, Ref. [21] based on the PBS message propagation mechanism, a distributed protocol is proposed on the basis of the Multi-hop sensor network, which is inspired by the greedy algorithm. This protocol reduces the number of time message exchanges, saves energy more than TPSN and RBS, and supports network expansion and connectivity.

Albu and Labit in Ref. [22] mixed PBS and IEEE 1588 standard proposed a layered structure protocol, the simulation of the clock synchronization accuracy reached One-tenth microsecond level, and get effective energy saving, but this protocol does not support network scalability, anti-jamming is not high. In order to improve the efficiency of energy utilization, Ref. [23] proposes a protocol based on PBS, which is called improved pair broadcast synchronization (IPBS), which guarantees synchronization accuracy and is very efficient when exchanging with less messages. However, the algorithm can only work effectively within the range of two hops, and the limitation of network size is large.

For the clock synchronization of inactive nodes, Ref. [24] based on PBS time message exchange mechanism, using the maximum likelihood estimation algorithm under exponential stochastic delay and Gaussian mixed Kalman particle filter (GMKPF) algorithm to estimate the clock offset of inactive nodes. Because of the improved performance of GMKPF and the applicability of arbitrary and generalized non-Gaussian or arbitrary stochastic delay models, it is better than maximum likelihood estimation of clock offset, and can replace maximum likelihood estimation algorithm. The Ref. [25] uses joint maximum likelihood estimation (JMLE) and generalized maximum likelihood estimation (GMLLE) algorithm to estimate clock offset and drift, and the comparison shows that JMLE is usually better than GMLLE algorithm at the cost of larger computational complexity. JMLE also has lower computational complexity in achieving the same precision as LP linear programming. But at the same time, the fixed delay in the communication link is considered as a constant, which does not conform to the practical application.

The Ref. [26] based on PBS proposes a new scheme called back reference timestamp (SPIRT). SPIRT using the wireless sensor network with Double-layer network structure, the nodes in the network are divided into disjoint clusters, each cluster has a cluster head node, its aggregated data comes from its members. Each cluster header exchanges a synchronized clock with a reference node message in the network. Because cluster members can overhear the cluster header transmission, SPIRT adjusts the clock of the cluster member with this synchronous communication advantage. SPIRT needs to attach a reference timestamp to the cluster header information, and the cluster members can estimate their clock adjustments. This cuts energy consumption and increases SPIRT synchronization efficiencies over other competitive schemes. Similar to the literature 25, this algorithm does not apply to multi-hop networks. Using the cluster structure of similar literatures 24, Ref. [27] proposes a time synchronization (FBTS) method based on feedback to adjust the resynchronization process in real time. According to the difference between synchronous precision requirement and synchronization error, the group resynchronization cycle is adjusted in real time and evaluated as much as possible, this method can effectively reduce the energy consumption caused by time synchronization.

4   Distributed Time Synchronization Algorithm

Two-way time synchronization algorithm based on send-receive and two-way broadcast message exchange algorithm cannot solve the problem of network topological dynamic change caused by joining, invalidation and moving of nodes in wireless sensor network, and it cannot reduce the cumulative synchronization error in network. In order to solve this problem, the distributed Uniform Time Synchronization protocol emerges. In a distributed consistent time synchronization algorithm, it is generally not necessary to synchronize nodes, but simultaneously synchronize all sensor nodes. Instead of estimating the clock offset between the two nodes and the relative clock frequency drift, the clock offset and drift of each independent node is estimated.

Solis first proposed a distributed algorithm for precise time synchronization in large-scale multi hop wireless networks in Ref. [28]. The core is to use a large number of global constraints in a multi hop network, and must satisfy the same time regulation, that is, the prototype of the distributed consensus time algorithm.

The Ref. [29–31] proposes a method based on belief propagation. Under the assumption of Gaussian distribution stochastic delay, a belief-based algorithm is proposed to estimate the clock offset. The Ref. [29–31] estimates the clock offset and drift. In this algorithm, each node exchanges Information with a neighbor node, estimates its clock offset and drift in a fully distributed and asynchronous way. The algorithm does not require any centralized information processing or coordination, and has the scalability of network scale.

Only the clock offset estimation is considered in the exponential delay, the Ref. [32] pioneered a new algorithm of network based on factor graph representation and maximum product information updating, and a completely distributed algorithm is proposed on the basis of exponential delay.

The Ref. [33] proposes a distributed two order consistent time synchronization, which is a completely localized distributed asynchronous implementation method, the main idea is to realize global synchronous offset compensation and offset compensation.

The algorithm does not modify the local clock directly, so the virtual clock is a compensation parameter based on the local clock. Each node implements a virtual consensus clock by periodically updating the compensation parameters.

Because most of the consensus algorithms are based on "node", that is, each node adopts traversal consensus algorithm to achieve the synchronization state. This increases overall information complexity, congestion and latency in the network, and high-energy consumption. The Ref. [34] proposes a general topology optimization strategy, which accelerates the clock synchronization protocol based on the consensus of the concept of delay equalization topology. In an energy-constrained environment, the reference proposes to select a subset of a priori sensor node and a finite number of adjacent sensor nodes to minimize the complexity of the message and reduce the energy consumption. In addition, the selection of "subset" sensor nodes must ensure the connectivity of the consistent propagation to achieve synchronization of the network. Adjacent sensor nodes must be allocated to minimize latency, thus gaining a faster convergence of consensus. The whole problem is transformed into a delay equilibrium topology (CDSDBT) problem based on a connected dominating set. In order to make the problem easy to deal with, the reference proposes a stochastic weighted genetic algorithm (RWGA) to deal with the trade-off between the objective functions and select the Pareto optimal solution based strategic topology. With this strategy, the convergence speed is obviously improved, the global synchronization error is reduced, and the message quantity of energy consumption is optimized.

Because the accuracy of timestamps is a significant constraint on the achievable synchronization scenarios, most existing synchronization schemes do not consider propagation latency. The Ref. [35] proposes a method to estimate and compensate network propagation latency. Based on a custom cumulative propagation delay encoding, use constructive transmission disturbances to send similar packets to build a spanning tree network that does not need to be maintained. It can be used to improve the delay compensation based on the flood-general synchronization scheme; even the microsecond-level clock synchronization is obtained in the network with significant propagation latency.

5   Conclusion

Through the above analysis, it can be found that the existing time synchronization algorithm has its own defects, most of which need to consider the energy loss of the node, to ensure the low energy consumption while improving the time synchronization accuracy. Due to the external environment and the impact of the nodes' own hardware to generate communication delay, time synchronization calculation and precision become more complex. In addition, most of the existing time synchronization protocols need to select the reference node, in order to avoid energy of reference nodes loss too fast due to relatively more communication, the neighbor node is not synchronized, it is necessary to study the appropriate selection of reference nodes, so that time synchronization algorithm becomes more complex. The recently proposed distributed consistent time synchronization algorithm avoids the choice of reference nodes and reduces the complexity of the algorithm. In the future study, the following three research directions are mainly considered:

(1) For the bidirectional message exchange time synchronization algorithm, the PBS algorithm combined with the broadcast mechanism obviously has lower energy consumption than the sender-receiver message exchange time synchronization algorithm, it reduces the exchange times of time synchronization messages, and in future studies, it mainly considers from the angle of statistical signal processing, the time synchronization parameter clock offset and the clock frequency drift are estimated, the fixed delay and the variable delay of the nodes are taken into consideration, and the parameter estimation accuracy and the error range can be reduced while reducing the energy consumption of nodes.

(2) Whether it is a two-way message exchange time synchronization algorithm or a PBS algorithm, the topology-spanning tree or cluster initiated by the root node or the reference node is used first. In the synchronization phase, the node can only be synchronized with the previous level node, can not be directly synchronized with the root node, it will inevitably lead to the impact of single hop accumulation, the whole network topology is poor, the whole network synchronization convergence speed is slow. With the increasing scale of wireless sensor network and the shrinking of node size, the application of these algorithms is restricted more and more. The distributed synchronization mechanism has the characteristics of expansibility, good survivability, and so on, through the information fusion of neighbor nodes, the time of node is synchronized to a virtual time, and the problem of the whole network resynchronization is not caused by the root node failure, and the convergence speed is fast.

(3) The main advantage of distributed consistent time synchronization algorithm is to cancel the selection of reference nodes, so many node selection algorithms for solving the problem of energy consumption of reference nodes no longer need to be considered, thus there will not be an uneven distribution of energy consumption. However, the distributed consistent time synchronization algorithm synchronizes the nodes in the whole wireless sensor network, and the periodic and continuous propagation of the synchronous signal can lead to the increase of all kinds of delay, and the accuracy of the clock synchronization is decreasing continuously. Therefore, in the future, we can consider using multi-hop method to synchronize the clock in the distributed consistent time synchronization algorithm and optimizing the network topology, reducing the network congestion and message delay caused by traversal, and on the basis of reducing the energy consumption and improving the clock synchronization precision, the communication delay can be further used as the influence factor to study the parameter estimation of the clock synchronization. In addition, the characteristics of global traversal of distributed consistent algorithm need to consider the convergence speed of the whole network, so how to improve the convergence speed of the network is also a key direction to be researched in the future.

References

1. Corke, P., Wark, T., Jurdak, R., et al.: Environmental wireless sensor networks. Proc. IEEE **98**(11), 1903–1917 (2010)
2. Peng, Y., Wang, D.: Wireless sensor network positioning technology overview. Electron. Measur. Instrum. **25**(5), 389–399 (2011)
3. He, C.: Wireless sensor network time synchronization technology research. South China university of technology (2010)
4. Ganeriwal, S., Kumar, R., Srivastava, M.B.: Timing-sync protocol for sensor networks. In: Proceedings of the 1st International Conference on Embedded Networked Sensor Systems (SenSys 2003), pp. 138–149 (2003)
5. Rucksana, S., Babu, C., Saranyabharathi, S.: Efficient timing-sync protocol in wireless sensor network. In: Proceedings of International Conference on Innovations in Information, Embedded and Communication Systems (ICIIECS), Coimbatore, pp. 1–5 (2015)
6. Seth, G., Harisha, A.: Energy efficient timing-sync protocol for sensor network. In: Proceedings of the International Conference on Computing and Network Communications (CoCoNet), Trivandrum, pp. 912–916 (2015)
7. Wang, H., Zeng, H., Wang, P.: Clock skew estimation of listening nodes with clock correction upon every synchronization in wireless sensor networks. IEEE Sig. Process. Lett. **22**(12), 2440–2444 (2015)
8. Noh, K.L., Chaudhari, Q.M., Serpedin, E., et al.: Novel clock phase offset and skew estimation using two-way timing message exchanges for wireless sensor networks. IEEE Trans. Commun. **54**(4), 766–777 (2007)
9. Levy, C., Pinchas, M.: Maximum likelihood estimation of clock skew in IEEE 1588 with fractional Gaussian noise. Math. Probl. Eng. **1**, 1–24 (2015)
10. Leng, M., Wu, Y.C.: On clock synchronization algorithms for wireless sensor networks under unknown delay. IEEE Trans. Veh. Technol. **9**(1), 182–190 (2010)
11. Ahmad, A., Serpedin, E., Nounou, H., et al.: Joint node localization and time-varying clock synchronization in wireless sensor networks. IEEE Trans. Wirel. Commun. **12**(10), 5322–5333 (2013)
12. Leng, M., Wu, Y.C.: Low-complexity maximum-likelihood estimator for clock synchronization of wireless sensor nodes under exponential delays. IEEE Trans. Sig. Process. **9**(10), 4860–4870 (2011)
13. Sun, W., Brännström, F., Ström, E.G.: On clock offset and skew estimation with exponentially distributed delays. In: Proceedings of IEEE International Conference on Communications (ICC), Budapest, pp. 1872–1877 (2013)
14. Jeske, D.R.: Jackknife bias correction of a clock offset estimator. In: Minguez, R., Sarabia, J.-M., Balakrishnan, N., Arnold, B.C. (eds.) Advances in Mathematical and Statistical Modeling. Statistics for Industry and Technology, pp. 245–254. Springer, New York (2008). https://doi.org/10.1007/978-0-8176-4626-4_18
15. Hajikhani, M., Kunz, T., Schwart, H.: A recursive method for bias estimation in asymmetric packet-based networks. In: Proceedings of IEEE International Symposium on Precis on Clock Synchronization for Measurement, Control, and Communication (ISPCS), Austin, TX, pp. 47–52 (2014)
16. Hajikhani, M., Kunz, T., Schwart, H.: A recursive method for clock synchronization in asymmetric packet-based networks. IEEE ACM Trans. Netw. **24**, 2332–2342 (2015)
17. Wu, M., Sun, Z., Zhou, Z.: A time synchronization method for WSN based on time bias correction. In: Proceedings of the 32nd Chinese Control Conference (CCC), Xi'an, China, pp. 7456–7461 (2013)

18. Pawar, P.M., Nielsen, R.H., Prasad, N.R., et al.: A hybrid algorithm for efficient wireless sensor network time synchronization. In: Proceedings of the 4th International Conference on Wireless Communications, Vehicular Technology, Information Theory and Aerospace and Electronics Systems (VITAE), Aalborg, pp. 1–5 (2014)

19. Zhang, J., Lin, S., Liu, D.: Cluster-based time synchronization protocol for wireless sensor networks. In: Sun, X., et al. (eds.) ICA3PP 2014. LNCS, vol. 8630, pp. 700–711. Springer, Cham (2014). https://doi.org/10.1007/978-3-319-11197-1_55

20. Noh, K.L., Serpedin, E.: Pairwise broadcast clock synchronization for wireless sensor networks. In: Proceedings of the IEEE International Symposium on a World of Wireless, Mobile and Multimedia Networks, Espoo, Finland, pp. 1–6 (2007)

21. Cheng, K.Y., Liu, K.S., Wu, Y.C., et al.: A distributed multihop time synchronization protocol for wireless sensor networks using pairwise broadcast synchronization. IEEE Trans. Wirel. Commun. 4, 1764–1772 (2009)

22. Albu, R., Labit, Y., Gayraud, T., et al.: An energy efficient clock synchronization protocol for wireless sensor networks. In: Proceedings of the Wireless Days (2010 IFIP), Venice, pp. 1–5 (2010)

23. Qianfeng, C., Menglei, Z., Mingwu, Y.: A new energy-efficient time synchronization protocol in wireless sensor networks. In: Proceedings of the IEEE International Conference on Computer and Information Technology (CIT), Xi'an, China, pp. 684–688 (2014)

24. Chaudhari, Q.M., Serpedin, E., Kim, J.S.: Energy-efficient estimation of clock offset for inactive nodes in wireless sensor networks. IEEE Trans. Inf. Theory 6(1), 582–596 (2010)

25. Cao, X.: Joint estimation of clock skew and offset in pairwise broadcast synchronization mechanism. IEEE Trans. Commun. 61(6), 2508–2521 (2013)

26. Benzaïd, C., Bagaa, M., Younis, M.: An efficient clock synchronization protocol for wireless sensor networks. In: Proceedings of International Wireless Communications and Mobile Computing Conference (IWCMC), Nicosia, pp. 718–723 (2014)

27. Chen, Z., Huang, Y., Li, D.: One feedback-based time synchronization method for wireless sensor networks. In: Proceedings of 27th Chinese Control and Decision Conference (CCDC 2015), Qingdao, China, pp. 1459–1464 (2015)

28. Solis, R., Borkar, V.S., Kumar, P.R.: A new distributed time synchronization protocol for multihop wireless networks. In: Proceedings of the 45th IEEE Conference on Decision and Control, San Diego, CA, pp. 2734–2739 (2006)

29. Du, J., Wu, Y.C.: Distributed clock skew and offset estimation in wireless sensor networks: asynchronous algorithm and convergence analysis. IEEE Trans. Wirel. Commun. 2(11), 5908–5917 (2013)

30. Etzlinger, B., Wymeersch, H., Springer, A.: Cooperative synchronization in wireless networks. IEEE Trans. Sig. Process. 2(11), 2837–2849 (2013)

31. Du, J., Wu, Y.C.: Fully distributed clock skew and offset estimation in wireless sensor networks. In: Proceedings of IEEE International Conference on Acoustics, Speech and Signal Processing, Vancouver, BC, pp. 4499–4503 (2013)

32. Zennaro, D., Ahmad, A., Vangelista, L., et al.: Network-wide clock synchronization via message passing with exponentially distributed link delays. IEEE Trans. Commun. 61, 2012–2024 (2013)

33. Wu, J., Bai, Y., Zhang, L.: Distributed time synchronization in wireless sensor networks via second-order consensus algorithms. J. Tianjin Univ. 21, 113–121 (2015)

34. Panigrahi, N., Khilar, P.M.: An evolutionary based topological optimization strategy for consensus based clock synchronization protocols in wireless sensor network. Swarm Evol. Comput. **22**, 66–85 (2015)
35. Terraneo, F., Leva, A., Seva, S., et al.: Reverse flooding: exploiting radio interference for efficient propagation delay compensation in WSN clock synchronization. Real-Time Syst. Symp. **1**, 175–184 (2015)

Collaborative Filtering Recommendation Considering Seed Life Cycle for BT Download Websites

Yue Liu[✉], Fei Cai, Qi Sun, and Yiming Zhu

School of Computer Engineering and Science, Shanghai University, 99 Shangda Road,
Shanghai, BaoShan District, China
yliu@staff.shu.edu.cn

Abstract. This paper researches the effect of seed life circle imposed on the results of recommendation system, and proposes a seed life circle based collaborative filtering method, including LCKNNCF (Life Circle K-Nearest Neighbor Collaborative Filtering) and LCSVD (Life Circle Singular Value Decomposition). This method gives some definitions like unit of time, seeds download index, download index matrix firstly, and then classifies the seeds according to the matrix. Secondly, LSSVM (Least Squares Support Vector Machines) is used to construct the life cycle function based on the download index matrix of the seeds in same category. Thirdly, the classifier ensemble is employed to classify the new arrival seeds into the corresponding category and calculates their life values which can be used to rank the candidate recommendation sets obtained by traditional collaborative filtering methods in order to improve the recommendation quality. Finally, experiments conducted on download website data set obtained from LehuBT demonstrate the effectiveness of our method outperforms the existing techniques significantly.

Keywords: Recommender system · Seed life cycle · Collaborative filtering

1 Introduction

There are many BT download websites based on IPv6 in colleges (e.g., LehuBT[1], neu6[2], byrBT[3], CGBT[4]). This kind of websites which has a large number of stable users and resources uses Private Tracker (PT) strategy and encourages users to upload their resources due to the user's download permissions determined by the number of resources they shared. The existing websites adopt the traditional recommendation technology such as top recommended and so on. However, the resources recommended by these methods on the page are universal resources which can be seen by each user and the page can only display limited resources, resulting in users cannot find the resources they

---

[1] http://bt.shu6.edu.cn/.
[2] http://bt.neu6.edu.cn/.
[3] http://bt.byr.cn/.
[4] http://cgbt.cnd.

© Springer Nature Singapore Pte Ltd. 2018
K. Li et al. (Eds.): ISICA 2017, CCIS 873, pp. 562–575, 2018.
https://doi.org/10.1007/978-981-13-1648-7_49

are interested in. Personalized recommendation technologies are proposed to satisfy the user's individual needs [1]. Through analyzing the past users' behaviors, the recommendation systems can set up appropriate item sifting and ranking models, and then items are recommended to the users by the models obtained. Personal recommendation systems can help users get the information more quickly and help the enterprise improve customer loyalty by making better user experiences.

The research on personal recommendation systems has been the hot field of data mining and machine learning [2]. Nowadays there exists lots of personalized recommendation systems. According to the recommendation technologies they use, recommendation systems can be classified into rule-based recommendation systems [3], content-based recommendation systems [4], collaborative filtering (CF) recommendation systems [5], and hybrid recommendation systems [6]. All these systems have successful applications such as music recommendation systems, movie recommendation systems, e-commerce, etc. CF is most widely used in recommender systems due to its high accuracy, good scalability, and ability to execute without content analysis for feature extraction. The core idea behind CF is that users who shared common interests in the past would still prefer similar items in future [2]. There are two primary approaches to build the CF-based recommender systems, namely user-based CF [7] and item-based CF [2]. User-based CF algorithms rely on the fact that each user belongs to a certain group whose members share the same interests. Item-based CF algorithms recommend users the items that are similar to what they have already interested in by building an item-item similarity matrix. However, comparing to the other websites which have used personalized recommendation systems, BT download websites have some special characteristics, i.e. the frequency and quantity of seed uploading are greater, and the new seeds were downloaded/viewed by a high proportion. More specifically, the number of download of a new video/torrent is 0, and then the number of download will increase over time when more and more users find and download it. The download possibility of the seed is reduced because most people will not download the seed again. Finally, the download amount will become to 0. This process is similar to the process of product life cycle [14]. The traditional recommendation algorithms considered that the different items were same important to users, and took no consideration for the items' life cycle characteristic.

To address this problem, this paper proposes two novel personal recommendation methods named Life-Circle K-Nearest Neighbor Collaborative Filtering (LCKNNCF) and Life Circle Singular Value Decomposition (LCSVD) considering the seed life cycle for BT download websites, which gives some definitions like unit of time, seeds download index, download index matrix firstly, and then classifies the seeds according to the matrix. Secondly, LSSVM (Least Squares Support Vector Machines) is used to construct the life cycle function based on the download index matrix of the seeds in same category. Thirdly, the classifier ensemble is employed to classify the new seeds into the corresponding category and calculates their life values. Fourthly, concerning LCKNNCF and LCSVD, candidate recommendation sets are obtained via traditional CF methods. And then it classifies the seeds of the set, calculates the life-circle values, and ranks them according to their life value. The Top-N items would be recommended to users, so that it recommends the items with high life-circle values to users which improves the recommendation accuracy. Finally, experiments conducted on download website data set obtained from LehuBT demonstrate the effectiveness of our method outperforms the existing techniques significantly.

The rest of this article is organized as follows. Section 2 presents the related literatures. Our proposed approach for recommendation system is illustrated in Sect. 3. In Sect. 4, experiment evaluations are conducted to compare our approach with other methods. Finally, conclusions and future research directions are presented in Sect. 5.

2 Related Works

Considering the influence of time factor in recommendation system, there are more and more researchers concerned about the time factors and obtained achievements. Xia et al. [8] defined the time decay function $\theta(x)$ based on item-based CF, the difference is the new similarity computing method based on the three different time decay functions. The method shows that the closer the time of two items purchased by a user, the higher similarity they share. But they did not consider that the closer behaviors are more important. Koren [9] suggested a model based on formally optimizing a global cost function and introduced a SVD-based latent factor model. Xiang and Yang [10] considered four main time effects, such as the time bias, the user bias shifting, the item bias shifting, and the user preference shifting. His work added these effects to SVD, got the final factor model timeSVD. Koren [11] took both overall time and segmentation time into account, and proposed a timeSVD++ model. Gaillard and Renders [12] proposed an incremental matrix completion method, which allows the factors related to both users and items to adapt "on-line" to drift in users' preferences and shift in items' perception or use. Salah et al. [13] proposed an incremental CF approach which is based on a weighted version of the online spherical K-means algorithm, and is able to handle the frequent changes within CF data in a very short time. Although all of the above works considered the user preference shifting of a single user, they did take the public interest into account. In other words, those methods ignored the fact that the items' vitality is becoming weaker as the time goes by. That's why we introduce the concept of product life cycle.

The product life-cycle theory is an economic theory that was developed by Vernon [14], and the inspiration of it comes from biology which thinks all living things go through a cycle of birth, growth, maturity, and death. As a kind of tactical and empirical model, product life cycle is often used to direct product planning, and market analysis. The product life cycle theory has been applied to many industries and has proved useful in identifying future strategies for products and services [15], so as to recommendation system. Nguyen [16] found that the life cycles of users affect the prediction accuracies and the performance of recommendation algorithms. A demand forecasting model of short life cycle products was proposed by Basallotriana et al. [17], whose life cycle function is constructed by weighted linear regression. Kim et al. [18] adopted some installed base concepts to depict the life cycle of consumer electronics, and combined it with the demand forecasting method. But, there is no related work in collaborative filtering based on item life cycle.

In the real world, the public's interest in a specific item changes regularly over time. Therefore, the capability to adapt to these changes precisely is an open issue. This paper is committed to study and handle the issue.

3   A Life-Circle Based Collaborative Filtering Method

3.1   Problem Statement and Definition

The product life cycle is defined as the period that starts with the initial product design (research and development) and ends with the withdrawal of the product from the marketplace. It is characterized by specific stages, including research, development, introduction, maturity, decline, and finally obsolescence as the product is removed from the market (discontinued). The typical process of life cycle is shown in Fig. 1(a). Seeds' in LehuBT share the same pattern of products in the market (see Fig. 1(b)). The traditional recommendation algorithms considered that the degree of attraction of different seeds is same important to users, and took no consideration for the seeds' life cycle characteristic. So, we apply the conception of product life cycle to the BT download websites, compare seed to product, and generate the seed life cycle. For the sake of brevity, we will refer to all the BT download websites. Therefore, the seed life cycle is defined as Definition 1.

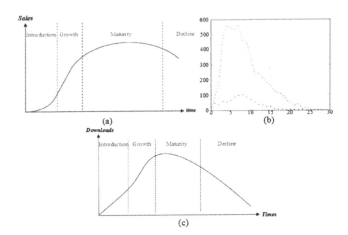

Fig. 1. (a) A typical process of product life cycle; (b) site downloads over time in LehuBt; (c) the seed life cycle.

Definition 1 (Seed Life Cycle). In the seed life cycle, the development stages of a seed from birth to death can be divided into four stages: introduction, growth, maturity, and decline. These develop with a gradual rise, peak, plateau, and then a fall. The downloads history of a new seed will follow an S-shaped curve over time and that the seed will pass through the four stages of sales growth depicted in Fig. 1(c).

During the introduction stage, the seed's novelty dictates that it will initially achieve only low downloads volumes and slow downloads growth, during which the slope of the downloads curve will remain relatively flat. If the seed is attractive, this initial stage gives way to the growth stage, during which the upward slope of the sales curve increases. When the seed gains popularity, it has been downloaded by most of the people who are interested in it. This is the maturity stage, during which the downloads curve again flattens. Finally, the decline stage arrives when the seed is no longer attractive, as the downloads curve slopes downward.

For each seed, its current vitality value is denoted by the probability of being downloaded at this moment. So it is imperative to construct a life cycle function to calculate this vitality value.

Definition 2 (Unit Time). The time span of a seed from being uploaded to nobody download it is defined as t_{train}. Integer multiples of 7 days is taken as the unit time span to conduct the experiment respectively. Segment the time span t_{train} averagely, each segment is a unit time.

Definition 3 (Download Index). Let d_{total} denote the total downloads within the time span t_{total}. Let t_i denote the i_{th} time period, d_i denotes the corresponding downloads for the time unit t_i, d_i and d_{total} can be obtained by using statistical analysis on database. Download index $p_{i,j}$ is the proportion of downloads $d_{i,j}$ of seed T_i in unit time t_j to total downloads in t_{train}:

$$p_{i,j} = d_{i,j}/d_{i,total} \tag{1}$$

where $d_{i,total}$ is the total downloads of T_i in T_{train}.

Definition 4 (Download Index Matrix). The download index vector of seed T_i is denoted by vector $T_i(p_{i0}, p_{i1}, \ldots, p_{in})$, then the download index matrix is denoted by the following matrix:

$$\begin{pmatrix} p_{11} & \cdots & p_{1n} \\ \vdots & \ddots & \vdots \\ p_{m1} & \cdots & p_{mn} \end{pmatrix} \tag{2}$$

We take a natural day as the unit time. The processed seed download index matrix is shown in Table 1.

Table 1. Download index matrix.

T_{id}	T_1	T_2	T_3	...	T_n
1	0.43	0.30	0.13	...	0.02
2	0.58	0.03	0.03	...	0.03
3	0.28	0.28	0.04	...	0.08
...	...	...	...	...	...
n	0.27	0.22	0.16	...	0.04

3.2 Method Process

In this section, we first propose a seed classifier construction method to classify the seed into different clusters by using ensemble learning, and construct the life cycle function of each cluster by a life cycle construction method. Then we develop the collaborative filtering method based on KNN and SVD respectively. The flowchart of our method is shown in Fig. 2.

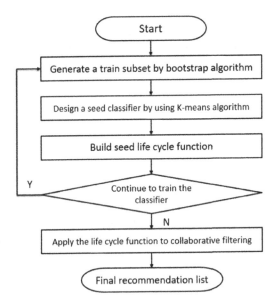

Fig. 2. Flowchart of life based collaborative filtering.

(1) Seed Clustering

According to the current classification of download websites, the life cycles of different categories have both similarities and differences, even it shows different features in the same category. While constructing life cycle function according to the current classification is unacceptable, our method clusters seeds by the proportion of seed's downloads over a period of time in total downloads instead of considering seed's specific categories and contents. Then constructing life cycle function by the result of the clustering.

In this paper, K-Means clustering is used to cluster the seeds into different categories. K-means algorithm needs to manually set the K values, in order to get the best clustering results, Indicator D and CH [19] are used to evaluate clustering effects.

$$D(NC) = \frac{\min_{x \in C_i, y \in C_j} d(x, y)}{\max_{x, y \in}} \tag{3}$$

$$CH(NC) = \frac{\frac{1}{NC} \sum_{i=1}^{n} n_i d^2 (c_i, c)}{\frac{1}{n - NC} \sum_{i=1}^{NC} \sum_{x \in C_i} d^2 (x, c_i)} \qquad (4)$$

where n denotes the total number of seed in the dataset, NC denotes the total number of cluster, C_i denotes the i_{th} cluster, c_i denotes the cluster center of i_{th} cluster.

(2) Construction of Seed Life Cycle Function

Different categories present different characteristics of life cycle, it's necessary to learn each life cycle curve respectively then get the corresponding life cycle function.

Least squares support vector machine (Least Squares Support Vector Machines, LSSVM) is used in this paper to construct the life cycle function. LSSVM is introduced into the idea of least squares, and adopts the error sum of squares loss function in the objective function instead insensitive loss function of the traditional SVM, it replaces the inequality constraints in SVM by equality constraints, which convert the quadratic programming (QP) optimization problem in SVM into a set of linear equation. The optimal function of LSSVM is only used for solving the linear equation, requires low computational complexity. And the most important thing is that it avoids the penalty factor C selection problems in SVM, which greatly simplifies the problem, improve the learning efficiency [20]. The Radial Based Function (RBF) neural network is adopted in this paper. Because the RBF kernel function has the unique best approximation property, no local minimum problem, and the advantage of fast learning and fast convergence. Using SVM to learn life cycle function, and construct life cycle functions respectively.

(3) Ensemble Learning Based Seed Classifier Construction Method

By clustering the existing data, each class gets a different life cycle function. However, there are a lot of new seeds that are lately uploaded, and needed to be classified.

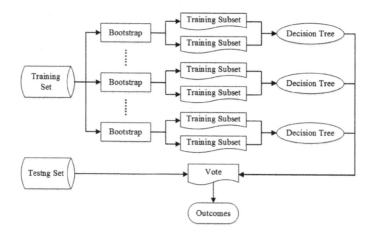

Fig. 3. The process of classifier construction based on ensemble learning.

There are many different ways to the structure of the classifier, such as decision tree, Bayesian classification, SVM. But the learning ability and generalization ability of a single classifier are limited, this article adopts the way of ensemble learning to structure classifier. Shown in Fig. 3.

The detail steps are as follow.

Step 1: Divides the source data set into the training set and the test set, the training set is used for single classifier modeling, and test set method is used for the final evaluation.

Step 2: Using Bootstrap algorithm [20] to repeatable sampling the original training set to produce a full-scale Bootstrap sample set. Then the collection is further divided into two disjoint subsets: training subsets and validation subsets. This is the famous Cross Validation method, which often used to guide model selection.

Step 3: Builds decision tree respectively by using the training subsets

Step 4: Integrates the individual decision tree by a simple vote. In the end, the integrated classifier prediction results of input sample x show as follows:

$$\bar{f}(x_i) = \text{Most}(f_j(x_j)(j \in \Omega)) \tag{5}$$

where Ω denotes the set of these classifiers, $f_j(x_i)$ denotes the forecast output of j_{th} classifier to the input sample x_i. $\bar{f}(x_i)$ is the conclusion of x_i under the integration program Ω.

(4) Life-Cycle Based Collaborative Filtering Method

The traditional collaborative filtering hasn't considering the change of the item vitality as time goes by and treats all the items in the transaction equally. In this paper, we analyze seed life cycle and propose a collaborative filtering algorithm based on seed life cycle, which will respectively be discussed in two different ways: KNN (UserCF and ItemCF) [7] and SVD [21].

KNN algorithm contains user-based collaborative filtering and item-based collaborative filtering, this section takes a user-based algorithm as an example. Traditional user-based KNN model based on the assumption of all items in the set have the same possibility of been downloaded, then get the Top-N recommendation from the candidate set. Item-based KNN is similar to the user-based KNN.

This paper presents a new collaborative filtering method: Life-Circle K-Nearest Neighbor Collaborative Filtering (LCKNNCF). Firstly, cluster the seeds into different categories, and get different seed life cycle function respectively. Secondly, the classifier is trained by the seed, and each seed belongs to a category. Finally, calculated the vitality of each seed in the candidate set which is calculated by the traditional KNN algorithm, and the Top-N recommendation is sorted by the vitality value.

Define unit time and download index, from the time that seed being uploaded, calculation seeds' per unit time downloads, and get the download index, all the seeds were calculated composition of seed download index matrix. Then transform the download index matrix to user-seed importance degree matrix, and calculate the similarity of seeds based on time factor. Finally, the new Top-N is recommended by KNN algorithm. The detail steps are as follow.

Seed-based LCKNNCF is similarly as TimeUserCF, only the seed exchange with the users. LCSVD is similarly as LCKNNCF, only change the step of UserCF to SVD.

4 Experiments

4.1 Experimental Dataset

LehuBT is one of the best resource sharing system in Chinese universities and has nearly 100000 users and millions of seeds. Users can upload seeds or videos to this website. The experiments conducted on the dataset obtained from LehuBT are used to demonstrate the effectiveness of our method.

We sort LehuBT dataset into four copies of the data (A, B, C and D) according to seed download records, uploaded time, categories, definition, subtitles and other information. Hereinto, Dataset A is seed download records, the time span of data is one month, is the same as dataset in second chapter, each user downloaded at least 10 seeds, which each seed has downloaded by at least 10 users, including 1046 users, 1849 seeds, 15376 records, the sparsity is 99.22%; Dataset B is the proportion of each time period downloads of seeds which are uploaded after January 1, 2012 to the total amount of downloads in each time span, a total of 8906 records; Dataset C is the attribute data of seeds in dataset B, 8906 records; Dataset D is user's actual download list after the one mouth time span, where each user downloaded at least 5 seeds.

Using B dataset first, clustering the seeds into different categories and constructing the life cycle function of each class by the result of clustering.

Using C dataset to analysis the classification of seed, by analyzing seeds' natural attributes to get the class it belongs, and then getting the life cycle state of the seed according to the life cycle function. Using A dataset for collaborative filtering recommendation, calculating the life cycle values of the candidates, and getting the final weight of each seed for a top-N recommendation.

4.2 Experimental Results and Discussion

(1) Seed Clustering and Structure Life Cycle Function

When clustering the seeds, choosing different parameters to get different clustering results. In order to analyze different life cycle characteristics of different categories, we chose many different sets of parameters. Through statistical analysis, when the selection of d was 7, 14, 21, 28, the average downloads of the seed was no longer change after the average 2 weeks, choosing 14 be the time span d. Considering that if the category number k is too large, the difference between categories is not clear, Take the K value of 2, 3, 4 for the experiment, and contrast indicator D and indicator CH, showing in the Table 2.

Table 2. Contrast indicator D and CH in different cluster numbers.

k	D	CH
2	0.185	0.049
3	0.750	0.810
4	0.570	0.390

By the above-mentioned, when k = 3 the clustering has the best effect, which minimizes differences within the class and maximizes differences between classes.

After the clustering, the LSSVM curves are used to fit the clustering centers of each class, and the life cycle curves of each class are obtained, and the fitting curves are obtained as shown in Fig. 4. It can be seen that the curves of LS-SVM with different parameters all obtain high degrees of fitting.

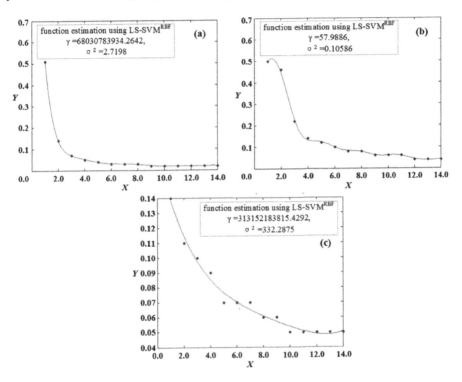

Fig. 4. The curve fitting results of the three kinds of the life cycle.

(2) Seed Classification Based on Ensemble Learning

By using the classifier based on ensemble learning to classify the seeds. Three input attributes are used in this experiment: uploading time (U), seed type (c), definition (P). In the process of program execution, the original data set is dynamically divided into training subset, verification subset, verification set and testing set, and the number of samples is 2:1:1:1. We chose SVM, BP, Bayes, decision tree as the baseline to compare the performance with our results. Ensemble learning classifier classification accuracy rate is the highest, in which the average accuracy is 46.67% higher than C4.5 which achieved the highest accuracy of the single classifier. This means that using ensemble learning can effectively improve the classification accuracy of the seeds.

Table 3 and Fig. 5 show the classification accuracy of different methods. We can see that BP has the same classification accuracy on different clusters, while other methods have different ones. Hereinto, SVM obtains the lowest classification accuracy (0.21).

BP, Bayes and C4.5 have the average classification accuracy on three clusters of 0.24, 0.33 and 0.45, respectively. Especially, the classifier based on ensemble learning is much higher than a single classifier in accuracy. Its average classification accuracy is twice as large as that of SVM and is 0.15 more than that of C4.5. Ensemble learning through the integration of multiple weak learners into a strong learner, avoiding the complex, tedious and time-consuming process of building strong learner in machine learning process. We use the method of Bootstrap to train the individual classifiers, and then produce the final classification results by using a relative majority vote to obtain the good effect.

Table 3. Different classification accuracy.

	C4.5	BP	Bayes	SVM	EnClassify
Cluster 1	0.42	0.24	0.31	0.27	0.61
Cluster 2	0.43	0.24	0.37	0.21	0.69
Cluster 3	0.50	0.24	0.31	0.15	0.67
Average	0.45	0.24	0.33	0.21	0.66

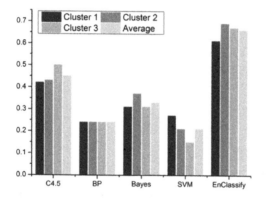

Fig. 5. Compare of all kinds of classifier.

(3) Seed Classification Based on Ensemble Learning
Respectively using KNN algorithm based on memory and SVD algorithm based on the model as the initial recommendation algorithms. The algorithms which considering the life cycle will be named respectively for LCUserCF, LCSeedCF and LCSVD. Top-5, Top-10, Top-20 is taken from the final results, and the parameters of the KNN is the best value we've got in chapter three. We chose KNN algorithm and SVD algorithm as the baseline to evaluate the experimental results in three aspects: accuracy, recall rate and coverage rate, respectively shown in Tables 4, 5 and 6.

Table 4. Accuracy of all kinds of algorithm.

	UserCF	ItemCF	SVD	LC UserCF	LC SeedCF	LCSVD
Top-5	4.46%	4.13%	6.02%	5.12%	5.06%	6.71%
Top-10	4.77%	4.55%	6.13%	5.43%	5.37%	6.92%
Top-20	5.49%	5.23%	6.48%	6.24%	6.11%	7.08%

Table 5. Recall rate of all kinds of algorithm.

	UserCF	ItemCF	SVD	LC UserCF	LC SeedCF	LCSVD
Top-5	13.52%	12.26%	16.02%	14.17%	13.21%	17.52%
Top-10	14.33%	13.06%	16.23%	15.51%	14.02%	17.39%
Top-20	14.39%	13.35%	15.93%	14.52%	14.52%	17.86%

Table 6. Coverage of all kinds of algorithm.

	UserCF	ItemCF	SVD	LC UserCF	LC SeedCF	LCSVD
Top-5	27.30%	23.26%	30.02%	25.13%	25.01%	32.33%
Top-10	29.01%	25.62%	33.13%	30.59%	26.98%	35.89%
Top-20	30.30%	25.86%	35.48%	33.75%	27.57%	38.11%

Experimental results show that accuracy, recall rate, coverage of each algorithm has been improved after we introduce the concept of seed life cycle. The accuracy of UserCF, ItemCF and SVD increased by 14.8%, 22.5%, 11.5%, the recall rate increased by 4.81%, 7.75%, 8.34%, and the coverage rate increased by 11.4%, 6.61%, 7.41%, respectively. From the perspective of the precision rate and recall rate, the recommendation algorithm which takes seed life cycle into consideration can select a more reasonable recommendation set. Some of the seeds which are interested by the user have upgraded their ranking in the recommendation set by means of the calculation of life cycle. The enhancement of the coverage indicates the consideration of life cycle can make the long tail better, which means more seeds are exposed.

5   Conclusions

Seeds of BT download websites downloaded at different times are not of equal importance and have their own life-circle, which attract fewer users as time goes by. It is why the traditional collaborative filtering methods cannot get good recommendation results in these BT download websites. So, this paper proposes a seed life circle based collaborative filtering method, including LCKNNCF and LCSVD for BT download websites. Firstly, to overcome the problem that different types of seeds have different life cycles, this method defines parameters like unit time and seed downloads index to calculate downloads matrix, then classifies the seeds according to the matrix, and finally builds a life-circle function for seeds of different categories. Secondly, in order to allocate the seeds to the corresponding categories more accurately, we construct a classifier based on ensemble learning to classify the seeds into different categories, and then calculate the vitality of seeds by using their corresponding life-circle functions. Finally, the Top-N seeds obtained by LCKNNCF and LCSVD would be recommended to users. We conduct several experiments on the data collected from the BT download website, LehuBT. Performances on accuracy, coverage and recall rate demonstrate that our method gets more precise recommendations after considering the life cycle.

Acknowledgements. This research is supported by the Natural Science Foundation of Shanghai (Grant No. 16ZR1411200), and Shanghai Leading Academic Discipline Project (Grant No. J50103).

References

1. Rich, E.: User modeling via stereotypes*. Cogn. Sci. **3**(4), 329–354 (1979)
2. Liu, J.-G., et al.: Overview of the evaluated algorithms for the personal recommendation systems. Complex Syst. Complex. Sci. **6**(3), 1–10 (2009)
3. Agrawal, R., Imielinski, T., Swami, A.N.: Mining association rules between sets of items in large databases. In: SIGMOD Conference Proceedings of the 1993 ACM SIGMOD International Conference on Management of Data, pp. 207–216 (1993)
4. Mock, K.: Dynamic email organization via relevance categories. In: Proceedings IEEE International Conference on TOOLS with Artificial Intelligence 1999, pp. 399–405 (1999)
5. Goldberg, D., Nichols, D., Oki, B.M., Terry, D.: Using collaborative filtering to weave an information tapestry. Commun. ACM **35**, 61–70 (1992)
6. Burke, R.: Hybrid web recommender systems. In: Brusilovsky, P., Kobsa, A., Nejdl, W. (eds.) The Adaptive Web. LNCS, vol. 4321, pp. 377–408. Springer, Heidelberg (2007). https://doi.org/10.1007/978-3-540-72079-9_12
7. Resnick, P., Iacovou, N., Suchak, M., Bergstrom, P., Riedl, J.: GroupLens: an open architecture for collaborative filtering of netnews. In: Proceedings of the 1994 ACM Conference on Computer Supported Cooperative Work, pp. 175–186 (1994)
8. Xia, C., Jiang, X., Liu, S., Luo, Z., Yu, Z.: Dynamic item-based recommendation algorithm with time decay. In: 2010 Sixth International Conference on Natural Computation (ICNC), pp. 242–247 (2010)
9. Koren, Y.: Factorization meets the neighborhood: a multifaceted collaborative filtering model. In: Proceedings of the 14th ACM SIGKDD International Conference on Knowledge Discovery and Data Mining, 24–27 August 2008, Las Vegas, Nevada, USA, pp. 426–434 (2008)
10. Xiang, L., Yang, Q.: Time-dependent models in collaborative filtering based recommender system. In: Main Conference Proceedings 2009 IEEE/WIC/ACM International Conference on Web Intelligence WI 2009, 15–18 September 2009, Milan, Italy, pp. 450–457 (2009)
11. Koren, Y.: Collaborative filtering with temporal dynamics. Commun. ACM **53**, 89–97 (2010)
12. Gaillard, J., Renders, J.-M.: Time-sensitive collaborative filtering through adaptive matrix completion. In: Hanbury, A., Kazai, G., Rauber, A., Fuhr, N. (eds.) ECIR 2015. LNCS, vol. 9022, pp. 327–332. Springer, Cham (2015). https://doi.org/10.1007/978-3-319-16354-3_35
13. Salah, A., Rogovschi, N., Nadif, M.: A dynamic collaborative filtering system via a weighted clustering approach. Neurocomputing **175**, 206–215 (2015)
14. Vernon, R.: International investment and international trade in product cycle. Int. Exec. **8**, 16 (1966)
15. Lancaster, G., Wesenlund, I.: A product life cycle theory for international trade: an empirical investigation. Eur. J. Mark. **18**, 72–89 (2013)
16. Nguyen, T.T.: Improving recommender systems: user roles and lifecycles. In: Proceedings of the 8th ACM Conference on Recommender systems, pp. 417–420 (2014)
17. Basallotriana, M.J., Rodrlguezsarasty, J.A., Benitezrestrepo, H.D.: Analogue-based demand forecasting of short life-cycle products: a regression approach and a comprehensive assessment. Int. J. Prod. Res. **55**, 2336–2350 (2017)

18. Kim, T.Y., Dekker, R., Heij, C.: Spare part demand forecasting for consumer goods using installed base information. Comput. Ind. Eng. **103**, 201–215 (2017)
19. Caliński, T., Harabasz, J.: A dendrite method for cluster analysis. Commun. Stat. **3**, 1–27 (1974)
20. Suykens, J.A.K., Vandewalle, J.: Least squares support vector machine classifiers. Neural Process. Lett. **9**, 293–300 (1999)
21. Bell, R.M., Koren, Y.: Scalable collaborative filtering with jointly derived neighborhood interpolation weights. In: Seventh IEEE International Conference on Data Mining ICDM 2007, pp. 43–52 (2007)

Centralized Access Control Scheme Based on OAuth for Social Networks

Yue Liu[1(✉)], Wei Gao[2], and Jingyun Liao[1]

[1] College of Information Engineering, Jiangxi College of Applied Technology,
Ganzhou 341000, Jiangxi, China
liuyue1229@qq.com
[2] Institute of Information Technology,
Guilin University of Electronic Technology, Guilin 541000, Guangxi, China

Abstract. Focusing on the issue of owning multiple social network accounts but failing to achieve centralized access control, a centralized access control scheme based on OAuth protocol for multiple social network services was proposed. Firstly, it is to build a communication channel between one main social network website and third-party social network services through OAuth protocol; Secondly, it is to use cryptographic algorithms to achieve privacy protection. Compared with currently adopted access control system, the proposed scheme realized centralized access control within multiple social network services to be able to effectively reduce the complexity of managing access control in social networks.

Keywords: Social network · Access control · Centralized management

1 Introduction

With the rapid growth of people's need to be virtually social connected, social networks are thriving at a fascinating speed. Most SNS (Social Network Services) allow users to set up and maintain their friend relationships which can then be utilized to achieve resource access control, i.e., friends within a certain relationship can obtain privileges to access protected resources. For example, Facebook users can share their new posts to a certain categorization of people (e.g. Friends, Specific friends, Friends expect). However, it should be noted that the establishment and maintenance of relationships is a continuous and very time-consuming process, besides, repeating this process on various SNS is tedious. Another fact not to be ignored is that some major SNS (e.g. Facebook, Twitter) have dominated the Social Network [1] while other sites focus on some specific social needs, all sites connect together and form the present social network graph. Therefore, it would tremendously improve users' experience of SNS if users can utilize their friend relationships and categorizations maintained on one prevalent and trusted social network to achieve centralized access control on other sites. Here we use an example to illustrate the above conception: Alice who is a Facebook user and has several well-maintained friend lists with numbers of friends in each list. When Alice registers on another social network site, say Flickr, to deposit some photos from the last travel with her Facebook friends and finds out that if she wants to share these photos only with her friends, all of them have to join Flickr and make a friend

© Springer Nature Singapore Pte Ltd. 2018
K. Li et al. (Eds.): ISICA 2017, CCIS 873, pp. 576–587, 2018.
https://doi.org/10.1007/978-981-13-1648-7_50

request to Alice. Afterwards, Alice will need to accept all her friend requests and add them to a group (e.g. Facebook friends) on Flickr, then she will be able to share these photos only with this group which resides her Facebook friends. Basically, it is precisely the same thing (i.e. setting up friend relationships and categorizations) as what she has been doing on Facebook. The current solution is quite time-consuming and inconvenient, so if Alice could simply utilize her friend relationships and categorizations which are carefully tended on Facebook to control access to her protected photos on Flickr, it would be very convenient.

Before proposing any solution, we should note that, as Carminati and Ferrari emphasized that "relationships are in general sensitive resources whose privacy should be properly guaranteed even if they are instrumental to perform access control", we ought to preserve user's information, shared resources and even the privacy of SNS when reaching a centralized access control solution. Shehab et al. proposed an access control framework to manage third-party applications [3], but their framework only applies to access control of user's profile information but not to the control of user's relationships. Barbara et al. presented a semi-decentralized architecture where user's relationships are encoded into certificates and stored by a certificate server [4], but their architecture implies extra burden on user's end, which violates the principle of "smart network services and dumb clients". Camenisch et al. proposed an SNS-based access control mechanism and three security definitions [5], but their mechanism cannot be directly applied to current SNS.

In this paper, by following the three security definitions in Camenisch's proposal and utilizing OAuth 2.0 protocol with several cryptographic algorithms as the communication channel between SNS, we achieved a centralized access control scheme for social networks.

2 Problem Statement

For simplicity, we hereby call a social network site that provides a platform to share its user resource and to extend its service as **SN** for short, and call an external service provider who makes use of this platform as **SP**, whether it is an independent site or purely relying on SN.

Now we formally state the main problem in this paper: a major SN (e.g. Facebook) has a large number of users, each of who establishes and maintains their friend relationships and friend categorizations. Users control access to their protected resources deposited on the SN by utilizing the categorizations such that only friends within certain categorization can be granted access. A registered user of SN named U_i also registered with SP and now wants to share some resource deposited on SP with certain friends from SN. Thus, U_i sets up an access policy named plc, which defines which friends from SN can access U_i's resource on SP. Afterwards, user U_i wants to utilize policy plc to control access to his/her resource on SP.

As shown in Fig. 1, this paper proposed a scheme which allows SP to execute user's preset access control policy to control users from SN to access resource deposited on SP while preserving private information of the user (e.g. user information, access control policy).

Fig. 1. Problem model

For simplicity, we assume in this paper that user U_i and U_j are both registered users on SN but user U_j is not registered on SP, therefore SN evaluates whether U_j satisfies U_i's preset access control policy and SP eventually enforces SN's decision.

Based on the above problem statement, we make the following performance and security requirements:

1. SNS-based access control: it shall allow users to utilize their friend relationships and categorizations which are established and maintained on SN to control access to the protected resource on SP.
2. Privacy-preserving: it shall ensure that user information (including user identity, user's friend relationships, and preset access policy) on SN is hidden to SP in the entire process of access control.
3. Resource secrecy: it shall ensure that SN does not learn any information of the resource stored at SP.
4. Channel abuse-free: it shall ensure that neither scheme-executive parties could abuse the communication channel.

3   Preliminaries

Before we formally present the proposal, we review some basic knowledge that will be used in this paper.

3.1   OAuth 2.0 Protocol

Cited from the OAuth 2.0 specification: "The OAuth 2.0 authorization framework enables a third-party application to obtain limited access to an HTTP service, either on behalf of a resource owner by orchestrating an approval interaction between the resource owner and the HTTP service, or by allowing the third-party application to obtain access token its own behalf" [6]. OAuth 2.0 protocol has been widely adopted by major SN for service authorization, and the authorization flow of OAuth fully satisfies the communication requirements of SN, SP, and users. Therefore, we use OAuth 2.0 as the communication protocol for the centralized access control scheme.

There are four roles as defined in OAuth to denote interactive parties: *Resource Owner*, a party (in particular, an individual) that owns some protected resource and is capable of granting access to its resource; *Resource Server*, the server where Resource Owners deposit their resource, and is responsible for controlling access to these resources; *Authorization Server*, the server that issues access tokens under the permit from authenticated resource owners; *Client*, a third-party application that requests for access to protected resource on a resource server on behalf of the resource owner with the owner's authorization. In some cases, one party can play several roles in OAuth, for example, in our study, SP is a resource server when users deposit resources and meanwhile SP is a client when it asks resource owner to grant access to owner's friends from SN. Please refer to [6] for more information of OAuth 2.0.

3.2 Group Signatures

Group signatures [7] allows group members to sign anonymously while keeping the ability to verify whether a signature belongs to this group. Each group member is assigned a secret key with which they can sign a message, this secret key is obtained from the group master who sets up the group. According to whether the secret signing keys are pre-computed during the initiation process, or are dynamically signed to each new-joined member, the group is called static or dynamic. Group signatures support traceability where by possessing an opening token, one can reveal the member identity tagged to a signature. A group signature scheme can be described as the following five algorithms:

1. *GroupSetup*: a probabilistic algorithm that on taking a security parameter λ and an initial size of the group as input, outputs a public key *GroupPK* for signature verification, a master key *GroupMK* for new member enrollment, and an opening key *GroupOK* for signature reveal.

$$(\text{GroupPK}, \text{GroupMK}, \text{GroupOK}) \leftarrow \text{GroupSetup}(\lambda, \text{size}) \tag{1}$$

2. *GroupEnroll*: a probabilistic algorithm that on taking a master key *GroupMK* and a user identity U_i as input, outputs a private group signing key sk_i which is then given to the user for message signing.

$$sk_i \leftarrow \text{GroupEnroll}(\text{GroupMK}, U_i) \tag{2}$$

3. *GroupSign*: a signing algorithm that on taking a private signing key sk_i and a message *msg* as input, outputs a signature σ.

$$\sigma \leftarrow \text{GroupSign}(sk_i, \text{msg}) \tag{3}$$

4. *GroupSigVerify*: a deterministic algorithm that on taking a group signature σ, the message *msg* and the group public key *GroupPK* as input, outputs a Boolean result.

$$0/1 \leftarrow \text{GroupSigVerify}(\text{GroupPK}, \sigma, \text{msg}) \tag{4}$$

5. **GroupSigOpen**: a tracing algorithm that on taking a valid signature σ and the group opening key *GroupOK*, outputs the signer identity U_i or a failure symbol $\bot$.

$$U_i / \bot \leftarrow \text{GroupSigOpen}(\text{GroupOK}, \sigma) \tag{5}$$

We require that the group signature scheme to be adopted should possess the three security properties described in [8]: anonymity, traceability, and strong exculpability.

3.3   Aes

AES (Advanced Encryption Standard) [9] is a symmetric key encryption specification which makes use of one symmetric key for both encryption and decryption. AES allows several parties to communicate securely by pre-sharing a symmetric key of scalable size, thus without holding the correct secure key, no party could get the plaintext from the ciphertext. AES can be described as the following three algorithms:

1. **AESKeyGenerate**: a probabilistic algorithm that on taking a security parameter λ as input, outputs a symmetric key K for both message encryption and decryption.

$$K \leftarrow \text{AESKeyGenerate}(\lambda) \tag{6}$$

2. **Encrypt**: an encrypting algorithm that on taking a message block *msg* of 128 bits and a symmetric key K as input, outputs the ciphertext C.

$$C \leftarrow \text{Encrypt}(K, \text{msg}) \tag{7}$$

3. **Decrypt**: a decrypting algorithm that on taking a ciphertext block C of 128 bits and a valid symmetric key K which is used in the message encryption step as input, outputs the message *msg*, otherwise $\bot$ as symbol to indicate failure if the key K is invalid, i.e., not the same key as used in encryption process.

$$\text{msg} \leftarrow \text{Decrypt}(K, C) \tag{8}$$

We use conventional security notions for AES implementation: semantic security, and ciphertext indistinguishability. Security performance of AES has been proven in years of practice, so we will not elaborate here anymore.

3.4   One-Time Signature

One-time signature (or Lamport signature) [10] allows one to sign a message based on a key pair, i.e., a public key for verifying signatures and a private key for signing messages, that is used only once. One-time signature can be described as the following three algorithms:

1. **OTKeyGenerate**: a probabilistic algorithm that on taking a security parameter λ as input, outputs a one-time key pair consists of one public verifying key *OTPublicKey*

and one private signing key *OTPrivateKey* for both message encryption and decryption.

$$(\text{OTPublicKey}, \text{OTPrivateKey}) \leftarrow \text{OTKeyGenerate}(\lambda) \qquad (9)$$

2. *OTSign*: a signing algorithm that on taking a message *msg* of arbitrary length and a one-time private signing key *OTPrivateKey* as input, outputs the one-time signature *ots*.

$$\text{ots} \leftarrow \text{OTSign}(\text{OTPrivateKey}, \text{msg}) \qquad (10)$$

3. *OTVerify*: a deterministic algorithm that on taking a message of arbitrary length *msg*, a one-time signature *ots* and a one-time public verifying key *OTPublicKey* as input, output a Boolean result *0/1* with regard to the validity of the public key *OTPublicKey* and the one-time signature.

$$0/1 \leftarrow \text{OTVerify}(\text{OTPublicKey}, \text{ots}, \text{msg}) \qquad (11)$$

One-time signature should follow the security requirement of strongly one-time unforgeability. The token specification is left undefined in OAuth 2.0 framework where only a skeleton of the token structure is shown. Therefore, each party that implements OAuth can define its own token structure which normally consists of several information segments. In this context, each individual data segment is relatively independent, security concerns arise as a malicious party may change some of these segments in tokens without being found. Thus, we consider one-time signature for our access control scheme. By utilizing the one-time signature, all data segments in the token structure can then be bound together.

3.5 Commitment with NIZK Proof

Commitment Schemes. A commitment scheme [11] allows one party to commit to a chosen value while keeping the value confidential to other parties, the commitment can subsequently be revealed by the committing party. A commitment scheme can be described as the following three algorithms:

1. *CSetup*: a probabilistic algorithm that on taking a security parameter λ as input, output parameters *CPars* for both committing messages and revealing messages.

$$\text{Pars} \leftarrow \text{CSetup}(\lambda) \qquad (12)$$

2. *Commit*: a committing algorithm that on taking parameters *CPars* and a message *msg* as input, outputs a commitment *cmt* together with an opening key *cop*.

$$(\text{cmt}, \text{cop}) \leftarrow \text{Commit}(\text{CPars}, \text{msg}) \qquad (13)$$

3. **CReveal**: a deterministic algorithm that on taking parameters *CPars*, an opening key *cop*, a message *msg*, and a commitment *cmt* as input, outputs a Boolean result indicating whether the commitment *cmt* commits to the message *msg*.

$$0/1 \leftarrow \text{CReveal}(\text{CPars, cop, msg, cmt}) \tag{14}$$

A commitment scheme shall have two security properties: hiding, and binding. Through an exchange of a commitment containing the resource description, it helps us achieves the following: SP provides a commitment to resource, then the commitment forms part of owner token and requester token, by which keeping the resource secure to SN; (2) with the commitment to resource, SN is capable of binding either owner tokens or requester tokens to the resource, such that to prevent token forgeability and reusability.

Commitment schemes are frequently used for zero-knowledge proofs. *Zero-knowledge proofs* (or interactive zero-knowledge proofs) are defined as those proofs, that to prove a chosen statement is true to a verifier, and that convey no additional knowledge other than the correctness of the proposition in question [12]. By taking advantage of commitment schemes, a prover can specify all information in the fashion of commitments which are sent to a verifier, whilst a verifier can send a challenge as a commitment to the prover in advance, then the prover can only reveal what needs to be revealed to answer the challenge that is to be revealed by the verifier after he has received commitments from the prover.

None-Interactive Zero-Knowledge Proof. Although interactive zero-knowledge proofs of knowledge would suffice for our access control scheme in terms of functional requirements, it requires multiple information exchange flows (challenge-and-response) between SN and SP, which will definitely overwhelm SN considering the huge number of users making use of this service simultaneously. This is why we have to consider none-interactive *zero-knowledge proofs of knowledge* (NIZK PoK). Blum et al. showed that it is possible to convince the verifier without interactions between the prover and the verifier by sharing a common reference string [13]. NIZK PoK, specifically for our access control scheme, can be described as the following three algorithms:

1. **ProofGenerate**: a proof generating algorithm that on taking two commitments *cmt1* and *cmt2*, two opening key *cop1* and *cop1* corresponding to the two commitments respectively and a message *msg* that the two commitments open as input, outputs a NIZK proof π or indicating inputs are invalid with regard to the commitments, opening keys and the message committed.

$$\pi \leftarrow \text{ProofGenerate}\{(\text{msg, cop1, cop2}) : \text{CReveal}(\text{CPars, cop1, msg, cmt1}) \\ = \text{CReveal}(\text{CPars, cop2, msg, cmt2}) = 1\} \tag{15}$$

2. *ProofVerify*: a deterministic algorithm that on taking two commitment *cmt1* and *cmt2* and a NIZK proof π as input, output a Boolean result *0/1*.

$$0/1 \leftarrow \text{ProofVerify}(\pi, \text{cmt1}, \text{cmt2}) \tag{16}$$

4 Centralized Access Control Scheme

Here we elaborate the construction of the OAuth based centralized access control scheme.

4.1 Scheme Setup

Taking a security parameter λ and an initial size of a group as input, SN runs the algorithm (1) (6) (9) and (12), then SN publishes GroupPK, OTPublicKey and enrolls user U_i by running (2). At last, SN securely possesses GroupOK, the symmetric key K, and OTPrivateKey.

4.2 Generate Owner Token (OT)

The scheme generates an owner token (ot) which represents the ownership of a resource protected by an access control policy as shown in Fig. 2.

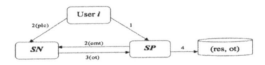

Fig. 2. OT generation steps

1. User i wants to deposit resource at SP;
2. SP (client) executes OAuth flow with i and SN (authorization server) and sends a commitment cmt1 to the resource by running (13) along with the authorization request, in this process, user i sets access policy plc with SN;
3. SN encrypts plc by (7) and gets ciphertext C, signs C and cmt1 on behalf of user i by (3) and gets signature σ_i. Then SN binds C, cmt1, and σi together by (10) and gets one-time signature ots1.

   Now SN possesses the owner token as ot = (σ_i, C, cmt1, ots1, OTPublicKey1), and sends to SP.

4.3 Verify Owner Token (OT)

Upon receiving of an owner token ot, SP first verifies it by checking whether the commitment cmt1 opens to the same resource res that the user i wants to deposit through (14), then verifying the group signature by (4), and verifying the one-time signature by (11). If all are valid, SP then stores ot in a database with the resource res.

4.4   Generate Requester Token (RT)

The scheme generates a requester token (rt) which represents a user's access request for a protected resource as shown in Fig. 3.

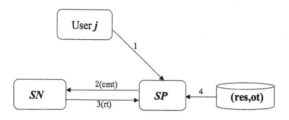

Fig. 3. RT generation steps

Step 1: User j (a registered user at SN) wants to access protected resource res owned by user i on SP; Step 2: SP(client) executes OAuth flow with j and SN (authorization server) and sends a commitment cmt2 to the resource by running (13) along with the authorization request; Step 3: SN signs cmt2 on behalf of user j by (3) to get a signature σ_j, and binds cmt2 and σ_j by (10) to get one-time signature ots2.

Now SN possesses the requester token rt = (σj, cmt2, ots2, OTPublicKey2), and sends to SP.

4.5   Generate NIZK Proof

Upon receiving of a requester token rt, SP validates the token by verifying the group signature (4), verifying the commitment (14) and verifying the one-time signature (11).

If rt is valid, SP produces a NIZK proof π by running (15) for convincing SN that the two commitments in ot and rt are pointing to the same resource;

4.6   Evaluate

Firstly, SP retrieves ot associated with the resource res and subsequently, SP acts as Resource Server who triggers the token validation process (can be done by a simple web service) with Authorization Server SN.

Secondly, SP queries for policy evaluation service with ot, rt and NIZK proof; Then, SN verifies the tokens and the NIZK proof by verifying the group signatures (4), verifying the one-time signatures (11) and verifying the NIZK proof (16). If all are valid, SN then retrieves the user identities i and j by opening the group signatures (5) and retrieves policy plc associated with the token by (8).

Finally, SN is able to evaluate the policy plc, and SN will return a Boolean result 0/1 to SP. SP grants access to the resource res to user j if 1 is returned, refuses otherwise.

5 Performance Analysis

5.1 Efficiency

We discuss the efficiency of our access control scheme by utilizing the following algorithm implementations, we are not limiting implementation methods of our scheme here but only for illustration purpose.

By adopting the realization of group signature scheme by Bichsel et al. [14], the one-time signature scheme by Lee et al. [15], the commitment scheme by Sandhya et al. [16], the NIZK proof scheme by Groth et al. [17] and AES with 128-bit block, we produced owner token as (σ_i, *C, cmt, ots, OTPublicKey*) and requester token as (σ_j, *cmt, ots, OTPublicKey*), and the efficiency of token generation and verification is shown in Table 1 and Fig. 4, the efficiency of communication between SN, SP and users is shown in Table 2.

Table 1. Efficiency of token generation and verification

Efficiency	Token size (bit)	Generation time (Avg. in ms)	Verification time (Avg. in ms)
Owner token	2274	≈650	≈760
Requester token	2146	≈580	≈740

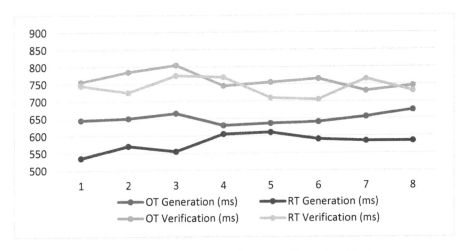

Fig. 4. Experiment results of token generation and verification

Table 2. Efficiency of communication channel

Efficiency	Resource deposit	Resource access
HTTP transaction (number of times)	5	6
Time consumption (Avg. in ms)	≈1535	≈1720

5.2 Security Analysis

Here we explain how our scheme achieves security requirements briefly.

Privacy-preserving is achieved by anonymity of group signature schemes, semantic security and ciphertext indistinguishability of AES. The anonymity of group signature guarantees that the identity of signer will not be disclosed to third-parties; Semantic security together with ciphertext indistinguishability of AES make sure that it is infeasible for SP or third-parties to reveal any information from the access policy encrypted at SN side, or to differentiate two encrypted policies. Thus, we ensure users are communicating with SP anonymously.

Resource secrecy is achieved by hiding of commitment schemes and NIZK proof schemes. Hiding of commitment schemes ensures that no malicious parties will be able to recover resource information. With NIZK PoK schemes, the zero-knowledge property ascertains the knowledge that two commitments open to the same message will not be conveyed to SN or third-parties. Thus, resource information is protected against SN or third-parties.

Channel abuse-free is achieved by traceability of group signature schemes, strongly one-time unforgeability of one-time signature schemes. Traceability of group signature makes sure that only SN has the ability to create valid signatures for tokens, thus there exist no malicious parties capable of faking tokens as SN. Strongly one-time unforgeability of one-time signature schemes further guarantees that no matter what party holds a valid token, it is infeasible for them to reuse any data fragment within the token because all data are bound by a one-time signature. Thus, our scheme imposes strong unforgeability of tokens. And whenever a party initiates communication with others, it must pass verification, therefore, no other parties can abuse the communication channel besides from benign SN and *SP*.

6 Conclusion

Access control is an imperative topic in social network, and centralized access control scheme is a popular research direction at all times. This paper proposed a centralized access control scheme based on OAuth 2.0 for controlling access to external social network service providers by utilizing friend relationships and categorizations from one main social network site in a privacy-preserving manner. We introduced several cryptographic algorithms to achieve privacy-preserving, resource secrecy, and channel abuse-free. However, OAuth protocol is not specially designed for access control, it causes that the proposed scheme requires multiple times of Internet interaction, better communication protocols are to be studied.

Acknowledgement. This work was jointly supported by Natural Science Foundation of China (61773296), the Education Department of Jiangxi Province of China Science and Technology research projects with the Grant No. GJJ151433, GJJ161687, GJJ161688 and GJJ161691.

References

1. Bright, L.F., Kleiser, S.B., Grau, S.L.: Too much Facebook? An exploratory examination of social media fatigue. Comput. Hum. Behav. **44**(C), 148–155 (2015)
2. Carminati, B., Ferrari, E.: Privacy-aware access control in social networks: issues and solutions. In: Nin, J., Herranz, J. (eds.) Privacy and Anonymity in Information Management Systems. AI&KP. Springer, London (2009). https://doi.org/10.1007/978-1-84996-238-4_9
3. Shehab, M., Squicciarini, A., Ahn, G.J., et al.: Access control for online social networks third party applications. Comput. Secur. **31**(8), 897–911 (2012)
4. Carminati, B., Ferrari, E., Perego, A.: Enforcing access control in web-based social networks. ACM Trans. Inf. Syst. Secur. (TISSEC) **13**(1), 6 (2009)
5. Camenisch, J., Karjoth, G., Neven, G., Preiss, F.S.: Anonymously sharing Flickr pictures with Facebook friends. In: Proceedings of the 12th ACM Workshop on Privacy in the Electronic Society (2013)
6. Hardt, D. (ed.): The OAuth 2.0 Authorization Framework. IETF, RCF 6749 (2012)
7. Chaum, D., van Heyst, E.: Group signatures. In: Davies, D.W. (ed.) EUROCRYPT 1991. LNCS, vol. 547, pp. 257–265. Springer, Heidelberg (1991). https://doi.org/10.1007/3-540-46416-6_22
8. Ateniese, G., Camenisch, J., Joye, M., Tsudik, G.: A practical and provably secure coalition-resistant group signature scheme. In: Bellare, M. (ed.) CRYPTO 2000. LNCS, vol. 1880, pp. 255–270. Springer, Heidelberg (2000). https://doi.org/10.1007/3-540-44598-6_16
9. Daemen, J., Rijmen, V.: The Design of Rijndael: AES-the Advanced Encryption Standard. Springer, Heidelberg (2002). https://doi.org/10.1007/978-3-662-04722-4
10. Lamport, L.: Constructing digital signatures from a one-way function, vol. 238. Technical report CSL-98, SRI International (1979)
11. Pedersen, T.P.: Non-interactive and information-theoretic secure verifiable secret sharing. In: Feigenbaum, J. (ed.) CRYPTO 1991. LNCS, vol. 576, pp. 129–140. Springer, Heidelberg (1992). https://doi.org/10.1007/3-540-46766-1_9
12. Goldwasser, S., Micali, S., Rackoff, C.: The knowledge complexity of interactive proof systems. SIAM J. Comput. **18**(1), 186–208 (1989)
13. Blum, M., Feldman, P., Micali, S.: Non-interactive zero-knowledge and its applications. In: Proceedings of the Twentieth Annual ACM Symposium on Theory of Computing. ACM (1988)
14. Bichsel, P., Camenisch, J., Neven, G., Smart, Nigel P., Warinschi, B.: Get shorty via group signatures without encryption. In: Garay, J.A., De Prisco, R. (eds.) SCN 2010. LNCS, vol. 6280, pp. 381–398. Springer, Heidelberg (2010). https://doi.org/10.1007/978-3-642-15317-4_24
15. Lee, J., Kim, S., Cho, Y., et al.: HORSIC: an efficient one-time signature scheme for wireless sensor networks. Inf. Process. Lett. **112**(20), 783–787 (2012)
16. Sandhya, M., Prasad, M.V.N.K.: A bio-cryptosystem for fingerprints using Delaunay Neighbor Structures(DNS) and fuzzy commitment scheme. In: Advances in Signal Processing and Intelligent Recognition Systems. AISC, vol. 425, pp. 159–171. Springer, Cham (2016). https://doi.org/10.1007/978-3-319-28658-7_14
17. Groth, J., Sahai, A.: Efficient non-interactive proof systems for bilinear groups. In: Smart, N. (ed.) EUROCRYPT 2008. LNCS, vol. 4965, pp. 415–432. Springer, Heidelberg (2008). https://doi.org/10.1007/978-3-540-78967-3_24

Research on Localization Scheme of Wireless Sensor Networks Based on TDOA

Xuefeng Yang$^{(\boxtimes)}$, Junqi Ma, and Yuting Lu

Jiangxi College of Applied Technology, Ganzhou 341000, Jiangxi, China
303568969@qq.com

Abstract. This paper presents a localization scheme of wireless sensor networks, which makes use of TDOA positioning scheme based on extended Calman filter. Firstly, the measured TDOA value is used to locate the target, and then the estimated value of the target node obtained by the algorithm is filtered by the extended Calman observation. Taking four anchor nodes as an example, the simulation analysis is carried out. The proposed scheme does not require global synchronization among nodes, and can effectively reduce the additional hardware overhead of node design and reduce the power consumption and cost of nodes.

Keywords: Wireless sensor networks · TDOA algorithm
Extended Calman filter TDOA algorithm

1 Introduction

Wireless sensor network (wireless sensor network, Wireless Sensor Network (MEMS) is the MEMS system on chip, Micro-Electro-Mechanism-System) (SOC, System-On-Chip) and wireless communication technology, a new type of highly integrated information and conceived the acquisition and processing mode [1]. In many applications of sensor networks, an important issue that concerns users is where specific events occur in which location or region. The problem of node localization is the prerequisite for many applications in wireless sensor network, realize the localization of sensor nodes and has an important role in a variety of applications, and it is also one of the basic research in sensor network issues. TDOA (Time, Difference, of, Arrival) positioning technology is the most promising target location technology in WSN positioning system at present [2]. The TDOA method does not require global synchronization between each node, and the conversion of TOA to TDOA can effectively offset the error signal after the same reflector is introduced, so as to improve the positioning accuracy. In order to improve the positioning accuracy, this paper proposes a positioning algorithm for time measurement based on wireless sensor network, the basic idea of the method: using the Taylor series expansion algorithm is improved to estimate the initial position of the target node, and the value of centralized filtering estimation algorithm with extended Calman filter in the background PC.

© Springer Nature Singapore Pte Ltd. 2018
K. Li et al. (Eds.): ISICA 2017, CCIS 873, pp. 588–600, 2018.
https://doi.org/10.1007/978-981-13-1648-7_51

2 Network Model and Parameter Acquisition

In this paper, a simple model of outdoor sensor networks suitable for localization algorithms is introduced. The sensor network consists of many unknown location and random distribution of SN (sensor node) anchor node of sensor nodes and several known locations (beacon node), as shown below, all nodes are in stationary state. TDOA estimates the acquisition method as follows: the anchor node periodically to the range of the target SN node and other anchor nodes transmit radio beacon signal, if the target node SN in the anchor node range, we can be the monitoring area is divided into several small sub regions and increasing the anchor to deal with node. The following Fig. 1 has three anchor nodes of the sensor network as an example, assuming that the anchor nodes A, B, C two-dimensional position coordinates are $(x_1, y_1), (x_2, y_2), (x_3, y_3)$, to be measured SN node S coordinates (x, y), v for the RF signal transmission rate. r_{ab}, r_{ac} are the known distances between anchor nodes A, B and A, and C, and then R_1, R_2, R_3 is the unknown distance from node S to A, B, C, respectively.

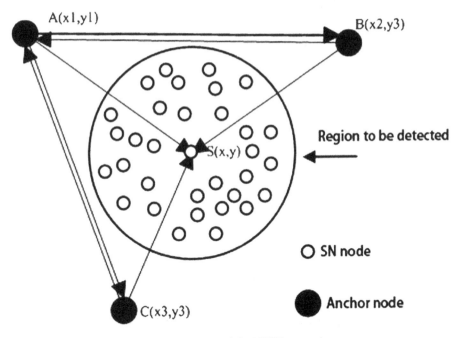

Fig. 1. Simple model of WSN network

Suppose A is the main anchor node, every T seconds, it will first send the signal to S, B and C nodes at the same time, and the S, B and C nodes will receive the beacon signal sent by A at t_1, t_b and t_c respectively. After receiving the beacon signal sent by B, the A will send a reply signal to A at $t_{bl}(t_{bl} \geq t_b)$. The signal will also be received by the S and C nodes, and the node S will receive the signal at t_2. The C node can only send a reply signal to the A at t_{cl} of receipt of the A and B signals. The same signal will also be received by the S and B nodes, and the node S will receive the signal at t_3. The arrival times of all the signals above are precisely measured based on the local clock of the node itself, so that we can obtain TDOA measurements without synchronization between nodes. Set $t_{b,a}, t_{c,a}$, respectively, node B, A arrival node S and node C, A arrive node S time difference, so that the TDOA measurements can be obtained by the following two formulas, respectively, based on the previous time measurement information:

$$t_{b,a} = t_2 - t_1 - (t_{bl} - t_b) - \frac{r_{ab}}{v} \tag{1}$$

$$t_{c,a} = t_3 - t_1 - (t_{cl} - t_c) - \frac{r_{ac}}{v} \tag{2}$$

In the effective TDOA measurement value, you can get the SN node to the anchor node between the two distance difference, multiple TDOA measurements can constitute a set location on the SN node hyperbolic equations, solving the equations can be obtained to estimate the location of target node.

3   TDOA Location Algorithm

3.1   Algorithm Mathematical Model

TDOA based localization technology for wireless sensor networks can establish a set of location equations after obtaining multiple TDOA measurements:

$$R_{i,1} = vt_{i,1} = R_i - R_1 = \sqrt{(x_i - x)^2 + (y_i - y)^2} \quad \sqrt{(x_1 - x)^2 + (y_1 - y)^2} \tag{3}$$
$$i = 2, \cdots, N$$

For the coordinate (x, y) of the target node to be measured, the anchor node coordinates (x_i, y_i), and the number of anchor nodes N, $t_{i,1}$ is the anchor signal arrival time difference between the main anchor node and the I anchor node. In geometry, each equation showed a hyperbola, if the TDOA parameter measurement is exact, which arrive at the anchor node for signal transmission (LOS) and no measurement error, then all the hyperbola will intersect at the same point, namely the target node (x, y) coordinates SN. However, in the actual wireless sensor networks spread multipath effect, usually exist in the environment of non line of sight (NLOS) and node processing delay, the TDOA parameter is bound to the measured error, then Eq. (3) is likely to be no solution. On the other hand, due to the number of participating general anchor points are more than the unknown equations in the least squares method, so it can make full use of the redundant information is an effective algorithm for solving such equations. The root operation Eq. (3) is nonlinear, nonlinear equations equivalent to nonlinear optimization problems, we must first carry on the linearized Taylor series expansion method proposed by Foy has the characteristics of high precision and strong robustness, etc. have been widely used in WSN positioning system. Although the

Taylor series expansion method can get accurate results, but need to be measured and the actual position of the initial node close to the estimated position to ensure the convergence of the algorithm, and the pre judgment on the case can not converge [3]. The positioning performance expansion method in comparison with the Fang algorithm and Taylor series, Taylor proposed an improved sequence can be used in outdoor WSN positioning system expansion method, and through in order to optimize the algorithm performance using the extended Calman filter method.

3.2 Improved Taylor Sequence Expansion Method

The Taylor sequence expansion method is a recursive algorithm that requires an initial estimate of position. In each iteration, the estimated position is improved by solving the local least squares (LS) solution of the TDOA measurement error. For a set of TDOA measurements, the algorithm starts with the Taylor series expansion of the formula (3) at the initial position (x_0, y_0) of the selected SN node, and consists of the two element Taylor formula:

$$
\begin{aligned}
f(x_0 + h, y_0 + k) = f(x_0, y_0) + \left(h\frac{\partial}{\partial x} + k\frac{\partial}{\partial y} \right) f(x_0, y_0) \\
+ \frac{1}{2!} \left(h\frac{\partial}{\partial x} + k\frac{\partial}{\partial y} \right)^2 f(x_0, y_0) + \cdots \\
+ \frac{1}{n!} \left(h\frac{\partial}{\partial x} + k\frac{\partial}{\partial y} \right)^n f(x_0, y_0) \\
+ \frac{1}{(n+1)!} \left(h\frac{\partial}{\partial x} + k\frac{\partial}{\partial y} \right)^{n+1} f(x_0 + \vartheta h, y_0 + \theta k)
\end{aligned}
\tag{4}
$$

The expansion of formula (3) and the removal of two orders or more components can be translated into:

$$
\psi = h_t - G_t \delta
\tag{5}
$$

For (5) the weighted least squares (WLS) algorithm is used to obtain the least squares estimate:

$$
\delta = \begin{bmatrix} \Delta x \\ \Delta y \end{bmatrix} = (G_t^T Q^{-1} G_t)^{-1} G_t^T Q^{-1} h_t
\tag{6}
$$

Q is the covariance matrix of TDOA measurements, In the initial iteration, $x = x_0$ and $y = y_0$ are available in season, and R_i and $G_t(i = 2, 3, \cdots, M)$ can be ordered in the next recursion

$$
x' = x_0 + \Delta x, \quad y' = y_0 + \Delta y
\tag{7}
$$

Repeat the above process until $\Delta x, \Delta y$ is small enough to meet a set threshold:

$$|\Delta x| + |\Delta y| < \delta \tag{8}$$

(x_0', y_0') is the estimated position of the target node (x, y) to be measured. The choice of the initial coordinates (x_0, y_0) of the target has a great influence on the convergence performance of the Taylor series expansion algorithm. In this paper, an initial coordinate selection algorithm based on least square estimation is proposed. The square of the Eq. (3) is simplified and can be obtained:

$$h_c = G_c z_c + \varepsilon_c \tag{9}$$

Of which:

$$z_c = [x, y, r]^T, r = R_1 = \sqrt{(X_1 - x)^2 + (Y_1 - y)^2} \tag{10}$$

According to the weighted least squares algorithm, the covariance matrix Q of the TDOA measurement error is approximated by the covariance matrix instead of the error e_c

$$z_c = (G_c^T Q^{-1} G_c)^{-1} G_c^T Q^{-1} h_c \tag{11}$$

According to the simulation results of the Chan algorithm, In most cases, the result of weighted least squares using Q instead of ψ_s is basically the same, so the least squares estimate based on the equation is believable. Therefore, the initial coordinate (x_0, y_0) estimate of the target can be expressed as:

$$\begin{cases} x_0 = z_c(1) \\ y_0 = z_c(2) \end{cases} \tag{12}$$

Cramer-Rao gives the theoretical lower bound of the error variance of unbiased parameter estimation. When the covariance matrix Q of TDOA measurement error is given, the lower bound of Cramer-Rao can be given by the following, in which the real coordinate (x_0, y_0) of the target node is given [4]:

$$CRLB = (G_0^T Q^{-1} G_0)^{-1} \begin{cases} G_0 = G|_{Ri=R_{0i}} & i = 1, 2, \cdots, N \\ \\ R_{0i} = \sqrt{(X_i - x_0)^2 + (Y_i - y_0)^2} \end{cases} \tag{13}$$

4   TDOA Positioning Technology Based on Filter Reconstruction

4.1   Calman Filtering Algorithm

Calman filter is a linear minimum mean square estimation algorithm, and it is the best filter in linear system. If the estimated coordinate location is (x_n, y_n) by the Fang algorithm and the Taylor sequence expansion method, the algorithm has a (x_n, y_n) coordinate estimate of $(\hat{x}_n, \hat{y}_n)$, an estimation value of $X_n = [x_n, y_n]^T$, and the equation of $X(k) = \varphi X(k-1) + \omega(k)$. φ for the transfer matrix, take $\varphi = \begin{bmatrix} 1 & 0 \\ 0 & 1 \end{bmatrix}$, the state noise $\omega(k)$ is white noise, the mean is zero; the observation equation $Z(k) = C(k)X(k) + V(k)$; measurement noise $V(k)$ is white noise; its mean is zero; $\omega(k)$ and $V(k)$ are independent of each other; its covariance matrix is $E[\omega(k)\omega^T(k)] = Q(k)\delta_{k,j}$ $E[V(k)V^T(k)] = R(k)\delta_{k,j}$; among them:

$$\delta_{k,j} = \begin{cases} 1, & \text{当} k = j \text{时} \\ 0, & \text{当} k \neq j \text{时} \end{cases} \qquad E[\omega(k) \cdot V(k)] = 0 \qquad (14)$$

The discrete Calman filter recursive operation equations are as follows: Initial value:

$$\hat{X}(0), P(0), k = 0 \qquad (15)$$

State one step prediction:

$$\hat{X}(k+1|k) = \varphi(k+1, k)\hat{X}(k) \qquad (16)$$

The prediction of the covariance matrix of the calculated state estimation error:

$$P(k+1|k) = \varphi(k+1, k)P(k)\varphi^T(k+1, k) + G(k+1, k)Q(k)G^T(k+1, k) \qquad (17)$$

The optimal filter gain is calculated:

$$K(k+1) = P(k+1|k)C^T(k+1)[C(k+1)P(k+1|k)C^T(k+1) + R(k+1)]^{-1} \qquad (18)$$

Calculate the optimal filter value:

$$\hat{X}(k+1) = \hat{X}(k+1|k) + K(k+1)[Z(k+1) - C(k+1)\hat{X}(k+1|k)]$$

Calculate filter error variance:

$$P(k+1) = [I - K(k+1)C(k+1)]P(k+1|k) \tag{19}$$

Return (16) for the next filter.
Selection of filter initial value:
As the upper model

$$E[\tilde{X}(k|k)] = E[X(k) - \hat{X}(k|k)] = [I - K(k)H(k)]\Phi(k, k-1)E[\tilde{X}(k-1|k-1)]$$

shows that in order to guarantee k the instantaneous filtering unbiased, And so on $E[\tilde{X}(k|k)] = 0$, the instantaneous filtering unbiased must be guaranteed $k-1$. And so on $E[\tilde{X}(k-1|k-1)] = 0$. And so on, obviously, as long as $k = 0$ instantaneous filter unbiased, can guarantee all instantaneous filter unbiased, want to ask the initial value unbiased, that is, $E[X(0) - \hat{X}(0|0)] = 0$ advisable $\hat{X}(0|0) = E[X(0)] = m_0$, at this time $P(0|0) = P_0 = Var(X(0))$.

4.2   Extended Calman Filtering Algorithm

The Kalman filter requires both the state equation and the observation equation of the system to be linear, but the problems in practical engineering can not be described by a simple linear system, so it is necessary to study the nonlinear filtering problem. Normally, we can linearize the nonlinear equation, and apply the Kalman filter to filter it. The extended Kalman filter studied here is a commonly used method. Set up the discrete nonlinear system model [5]:

$$X(k+1) = (X(h), k) + \Gamma[X(k), k]W(k) \tag{20}$$

$$Z(k+1) = h[X(k+1), k+1] + V(k+1) \tag{21}$$

The assumption of noise and initial value, as mentioned above, is based on the Calman filtering algorithm. The extended Kalman filter is to expand nonlinear function $\varphi(\bullet)$ around the filter value $\hat{X}(k|k)$ into Taylor series. After more than two items are omitted, the linearized model of the nonlinear system is obtained. Obtained by the system equation of state (20):

$$X(k+1) \approx \varphi[\hat{X}(k|k, k)] + \frac{\partial \varphi}{\partial X}\Big|_{X(k)=\hat{X}(k|k)}[X(k) - \hat{X}(k|k)] + \Gamma[X(k), k]W(k)$$

$$\frac{\partial \varphi}{\partial X}\Big|_{X(k)=\hat{X}(k|k)} = \Phi(k+1, k)$$

$$\varphi[\hat{X}(k|k, k)] - \frac{\partial \varphi}{\partial X}\Big|_{X(k)=\hat{X}(k|k)}\hat{X}(k|k) = u(k)$$

Then the equation of state becomes:

$$Z(k+1) = \Phi(k+1,k)X(k) + u(k) + \Gamma[\hat{X}(k|k),k]W(k) \tag{22}$$

The nonlinear function $h(\bullet)$ in the observation Eq. (21) is expanded around the prediction value $\hat{X}(k+1|k)$ into Taylor series, and after more than two entries are obtained:

Then the observation equation becomes:

$$Z(k+1) \approx h[\hat{X}(k+1|k,k+1)] + \frac{\partial h}{\partial X}\bigg|_{\hat{X}(k+1|k)} [X(k+1) - \hat{X}(k+1|k)] + V(k+1)$$

$$\Leftrightarrow \quad \frac{\partial \varphi}{\partial X}\bigg|_{\hat{X}(k+1|k)} = H(k+1)$$

$$h[\hat{X}(k+1|k,k+1)] - \frac{\partial \varphi}{\partial X}\bigg|_{\hat{X}(k+1|k)} \hat{X}(k+1|k) = y(k+1) \tag{23}$$

By the state linearized Eq. (22) and the observation Eq. (23) shows that the system is similar with control Kalman filter input form, application front with basic equations of input Kalman filter control available:

$$\begin{cases} \hat{X}(k+1,k) = \Phi(k+1,k)\hat{X}(k|k) + u(k) \\ P(k+1,k) = \Phi(k+1,k)P(k|k)\Phi^T(k+1,k) + \Gamma[\hat{X}(k|k),k]Q(k)\Gamma^T[\hat{X}(k|k),k] \\ K(k+1) = P(k+1|k)H^T(k+1)[H(k+1)P(k+1|k)H^T(k+1) + R(k+1)]^1 \\ \hat{X}(k+1|k+1) = \hat{X}(k+1|k) + K(k+1)[Z(k+1) \quad y(k+1) \quad H(k+1)\hat{X}(k+1|k)] \\ P(k+1|k+1) = [I \quad K(k+1)H(k+1)]P(k+1|k) \end{cases} \tag{24}$$

5 Simulation Results and Performance Analysis

This paper mainly based on the positioning error measurement of time difference scheme in theory from three aspects: processing delay node receiver, wireless multipath fading channel and the influence of non line of sight (NLOS Non, Line of Sight) communication. The processing delay of receiver from antenna receiving signal to the signal used by the receiver accurate decoding time, the delay is determined by the receiver circuit, it is usually considered to be constant or fluctuating within a small range, the error can be ignored. Therefore, we mainly consider multipath fading and the errors introduced by NLOS in TDOA simulation.

Simulation conditions: four anchor nodes (shown in Fig. 1) participating in TDOA measurements are set with coordinate positions (0, 0), (100, 0), (0100), (100100). In the Matlab6.5 simulation software, 200 groups of TDOA data are collected and simulated in the Gauss noise environment. The simulation results are shown below (Figs. 2 and 3):

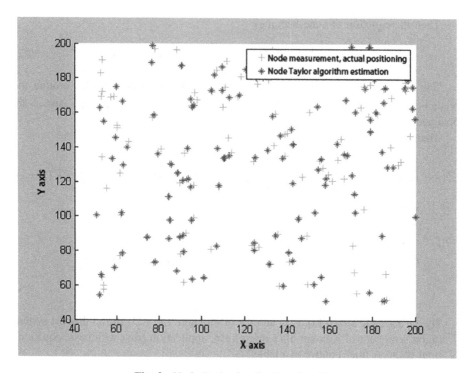

Fig. 2. Node Taylor localization algorithm

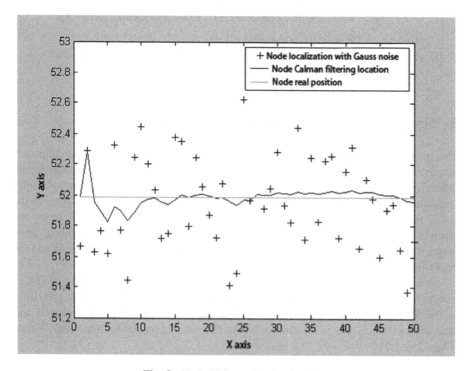

Fig. 3. Node Kalman filtering location

On the basis of the above simulation, 200 sets of TDOA data are collected and simulated in the Gauss noise environment first. There is no NLOS error at this time, and the influence of TDOA measurement error on the positioning results is considered only. The measurement error obeys the ideal Gauss distribution, the mean is 0, and the standard deviations are 1 m, 2 m, 3 m, 4 m and 5 m respectively. Then, the algorithm is simulated under the influence of NLOS. The simulation results are shown in the following diagram:

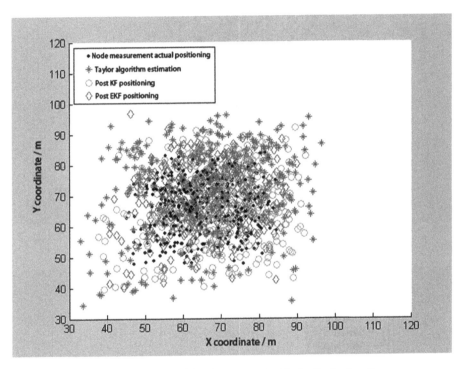

Fig. 4. The actual position is compared with the Taylor location.

The simulation results show that the improved Taylor sequence expansion method has better localization performance than the Fang algorithm in the case of three anchor nodes participating in the localization, both in the LOS and NLOS environments. As shown in Figs. 4 and 5, after EKF optimization, the localization result is closer to the target node. With the increase of error variance, the localization accuracy of the algorithm decreases, while the root mean square error of the localization in NLOS environment is larger than that of LOS. Such as the Taylor sequence is shown in Fig. 9 improved expansion measurement values obtained through extended Calman filter, the positioning accuracy is better than the traditional Calman filter, reduce the influence of measurement error on the positioning accuracy, close to TDOA CRLB (Cramer-Rao) lower limit value (Figs. 6 and 7).

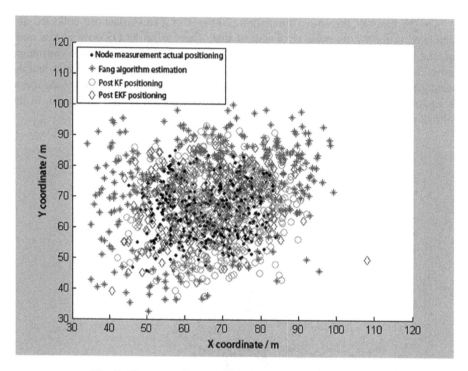

Fig. 5. Compares the actual location with the Fang location

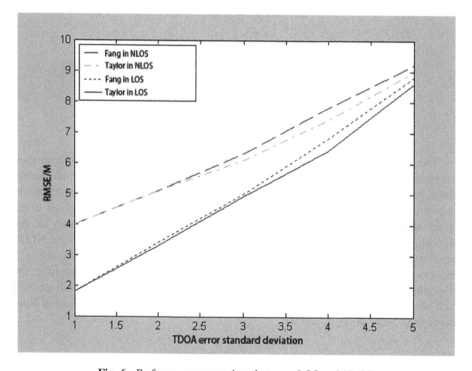

Fig. 6. Performance comparison between LOS and NLOS

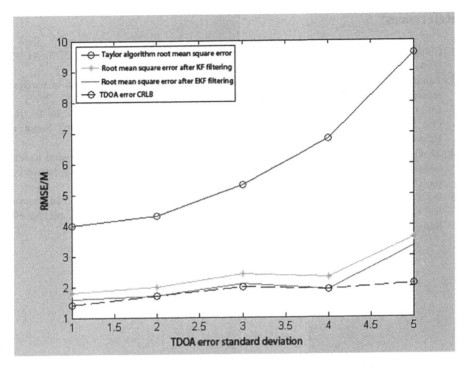

Fig. 7. Comparison of KF and EKF filtering under NLOS

6 Conclusion

This paper focuses on the application of TDOA localization algorithm in outdoor wireless sensor networks. The proposed scheme does not require global synchronization among nodes, reduces the additional hardware overhead of node design, and reduces the power consumption and cost of nodes. Because SN nodes to be monitored only passively listening to radio frequency signals of three anchor nodes, do not need to take the initiative to launch the RF signal, it will effectively reduce the communication overhead of SN nodes, reducing channel congestion, reduce node power consumption. The performance of the algorithm shows that the analysis and computer simulation, this paper first proposed by Taylor series expansion algorithm improved to estimate the initial position of the target node, then the estimated value as the observed extended Calman filter value positioning scheme optimization in the background PC concentration, while saving energy and bandwidth to a certain extent with high precision and has good application prospects in WSN positioning system.

Acknowledgement. This work was jointly supported by Natural Science Foundation of China (61773296), the Education Department of Jiangxi Province of China Science and Technology research projects with the Grant No. GJJ151433, GJJ161687, GJJ161688 and GJJ161691.

References

1. Cullar, D., Estrin, D., Strvastava, M.: Overview of sensor network. Computer **37**(8), 41–49 (2004)
2. Jianqi, L., Qinruo, W., Jiafu, W., et al.: Towards real-time indoor localization in wireless sensor networks. In: Xingang, L., Yang, L.T., Min, C., et al. (eds.), pp. 877–884 (2012)
3. Cui, L., Wang, F., Luo, H., Ju, H., Li, T.: A pervasive sensor node architecture. In: Jin, H., Gao, Guang R., Xu, Z., Chen, H. (eds.) NPC 2004. LNCS, vol. 3222, pp. 565–567. Springer, Heidelberg (2004). https://doi.org/10.1007/978-3-540-30141-7_84
4. Zhang, P., Martonosi, M.: LOCALE: Collaborative localization estimation for sparse mobile sensor networks. In: International Conference on Information Processing in Sensor Networks, pp. 195–206 (2008)
5. Gour, P., Sarje, A.: Localization in wireless sensor networks with ranging error. In: Buyya, R., Thampi, S.M. (eds.) Intelligent Distributed Computing. AISC, vol. 321, pp. 55–69. Springer, Cham (2015). https://doi.org/10.1007/978-3-319-11227-5_6

Author Index

Printed in the United States
By Bookmasters